Applied Mathematical Sciences

Volume 59

Applied Mathematical Sciences

1. *John:* Partial Differential Equations, 4th ed.
2. *Sirovich:* Techniques of Asymptotic Analysis.
3. *Hale:* Theory of Functional Differential Equations, 2nd ed.
4. *Percus:* Combinatorial Methods.
5. *von Mises/Friedrichs:* Fluid Dynamics.
6. *Freiberger/Grenander:* A Short Course in Computational Probability and Statistics.
7. *Pipkin:* Lectures on Viscoelasticity Theory.
8. *Giacaglia:* Perturbation Methods in Non-linear Systems.
9. *Friedrichs:* Spectral Theory of Operators in Hilbert Space.
10. *Stroud:* Numerical Quadrature and Solution of Ordinary Differential Equations.
11. *Wolovich:* Linear Multivariable Systems.
12. *Berkovitz:* Optimal Control Theory.
13. *Bluman/Cole:* Similarity Methods for Differential Equations.
14. *Yoshizawa:* Stability Theory and the Existence of Periodic Solution and Almost Periodic Solutions.
15. *Braun:* Differential Equations and Their Applications, Fourth Edition.
16. *Lefschetz:* Applications of Algebraic Topology.
17. *Collatz/Wetterling:* Optimization Problems.
18. *Grenander:* Pattern Synthesis: Lectures in Pattern Theory, Vol. I.
19. *Marsden/McCracken:* Hopf Bifurcation and Its Applications.
20. *Driver:* Ordinary and Delay Differential Equations.
21. *Courant/Friedrichs:* Supersonic Flow and Shock Waves.
22. *Rouche/Habets/Laloy:* Stability Theory by Liapunov's Direct Method.
23. *Lamperti:* Stochastic Processes: A Survey of the Mathematical Theory.
24. *Grenander:* Pattern Analysis: Lectures in Pattern Theory, Vol. II.
25. *Davies:* Integral Transforms and Their Applications, Third Edition.
26. *Kushner/Clark:* Stochastic Approximation Methods for Constrained and Unconstrained Systems.
27. *de Boor:* A Practical Guide to Splines, Revised Edition.
28. *Keilson:* Markov Chain Models-Rarity and Exponentiality.
29. *de Veubeke:* A Course in Elasticity.
30. *Sniatycki:* Geometric Quantization and Quantum Mechanics.
31. *Reid:* Sturmian Theory for Ordinary Differential Equations.
32. *Meis/Markowitz:* Numerical Solution of Partial Differential Equations.
33. *Grenander:* Regular Structures: Lectures in Pattern Theory, Vol. III.
34. *Kevorkian/Cole:* Perturbation Methods in Applied Mathematics.
35. *Carr:* Applications of Centre Manifold Theory.
36. *Bengtsson/Ghil/Källén:* Dynamic Meteorology: Data Assimilation Methods.
37. *Saperstone:* Semidynamical Systems in Infinite Dimensional Spaces.
38. *Lichtenberg/Lieberman:* Regular and Chaotic Dynamics, 2nd ed.
39. *Piccini/Stampacchia/Vidossich:* Ordinary Differential Equations in R^n.
40. *Naylor/Sell:* Linear Operator Theory in Engineering and Science.
41. *Sparrow:* The Lorenz Equations: Bifurcations, Chaos, and Strange Attractors.
42. *Guckenheimer/Holmes:* Nonlinear Oscillations, Dynamical Systems, and Bifurcations of Vector Fields.
43. *Ockendon/Taylor:* Inviscid Fluid Flows.
44. *Pazy:* Semigroups of Linear Operators and Applications to Partial Differential Equations.
45. *Glashoff/Gustafson:* Linear Operations and Approximation: An Introduction to the Theoretical Analysis and Numerical Treatment of Semi-Infinite Programs.
46. *Wilcox:* Scattering Theory for Diffraction Gratings.
47. *Hale et al.*: Dynamics in Infinite Dimensions.
48. *Murray:* Asymptotic Analysis.
49. *Ladyzhenskaya:* The Boundary-Value Problems of Mathematical Physics.
50. *Wilcox:* Sound Propagation in Stratified Fluids.
51. *GolubitskylSchaeffer:* Bifurcation and Groups in Bifurcation Theory, Vol. I.
52. *Chipot:* Variational Inequalities and Flow in Porous Media.
53. *Majda:* Compressible Fluid Flow and System of Conservation Laws in Several Space Variables.
54. *Wasow:* Linear Turning Point Theory.
55. *Yosida:* Operational Calculus: A Theory of Hyperfunctions.
56. *Chang/Howes:* Nonlinear Singular Perturbation Phenomena: Theory and Applications.
57. *Reinhardt:* Analysis of Approximation Methods for Differential and Integral Equations.
58. *Dwoyer/Hussaini/Voigt (eds):* Theoretical Approaches to Turbulence.

(continued after index)

J.A. Sanders F. Verhulst J. Murdock

Averaging Methods in Nonlinear Dynamical Systems

Second Edition

Jan A. Sanders
Faculteit Exacte
Wetenschappen Divisie
Wiskunde en Informatica
Free University of Amsterdam
The Netherlands
1081 De Boelelaan
Amsterdam 1081 HV

Ferdinand Verhulst
Mathematisch Instituut
State University of Utrecht
The Netherlands
6 Budapestlaan
Utrecht 3584 CD
f.verhulst@math.uu.nl

James Murdock
Dept. Mathematics
Iowa State University
Iowa, USA
478 Carver Hall
Ames 50011
jmurdock@iastate.edu

Mathematics Subject Classification (2000): 34C29, 58F30

ISBN-13: 978-1-4419-2376-9 eISBN-13: 978-0-387-48918-6

9 8 7 6 5 4 3 2 1

springer.com

Preface

Preface to the Revised 2nd Edition

Perturbation theory and in particular normal form theory has shown strong growth during the last decades. So it is not surprising that we are presenting a rather drastic revision of the first edition of the averaging book. Chapters 1 – 5, 7 – 10 and the Appendices A, B and D can be found, more or less, in the first edition. There are, however, many changes, corrections and updates.

Part of the changes arose from discussions between the two authors of the first edition and Jim Murdock. It was a natural step to enlist his help and to include him as an author.

One noticeable change is in the notation. Vectors are now in bold face, with components indicated by light face with subscripts. When several vectors have the same letter name, they are distinguished by superscripts. Two types of superscripts appear, plain integers and integers in square brackets. A plain superscript indicates the degree (in x), order (in ε), or more generally the "grade" of the vector (that is, where the vector belongs in some graded vector space). A superscript in square brackets indicates that the vector is a sum of terms beginning with the indicated grade (and going up). A precise definition is given first (for the case when the grade is order in ε) in Notation 1.5.2, and then generalized later as needed. We hope that the superscripts are not intimidating; the equations look cluttered at first, but soon the notation begins to feel familiar.

Proofs are ended by $\square$, examples by $\diamondsuit$, remarks by $\heartsuit$.

Chapters 6 and 11 – 13 are new and represent new insights in averaging, in particular its relation with dynamical systems and the theory of normal forms. Also new are surveys on invariant manifolds in Appendix C and averaging for PDEs in Appendix E.

We note that the physics literature abounds with averaging applications and methods. This literature is often useful as a source of interesting mathematical ideas and problems. We have chosen not to refer to these results as all of them appear to be formal, proofs of asymptotic validity are generally not included. Our goal is to establish the foundations and limitations of the methods in a rigorous manner. (Another point is that these physics results are usually missing out on the subtle aspects of resonance phenomena at higher order approximations and normalization that play an essential part in modern nonlinear analysis.)

When preparing the first and the revised edition, there were a number of private communications; these are not included in the references. We mention results and remarks by Ellison, Lebovitz, Noordzij and van Schagen.

We owe special thanks to Theo Tuwankotta who made nearly all the figures and to André Vanderbauwhede who was the perfect host for our meeting in Gent.

Ames
Amsterdam
Utrecht

James Murdock
Jan Sanders
Ferdinand Verhulst

Preface to the First Edition

In this book we have developed the asymptotic analysis of nonlinear dynamical systems. We have collected a large number of results, scattered throughout the literature and presented them in a way to illustrate both the underlying common theme, as well as the diversity of problems and solutions. While most of the results are known in the literature, we added new material which we hope will also be of interest to the specialists in this field.

The basic theory is discussed in chapters 2 and 3. Improved results are obtained in chapter 4 in the case of stable limit sets. In chapter 5 we treat averaging over several angles; here the theory is less standardized, and even in our simplified approach we encounter many open problems. Chapter 6 deals with the definition of normal form. After making the somewhat philosophical point as to what the right definition should look like, we derive the second order normal form in the Hamiltonian case, using the classical method of generating functions. In chapter 7 we treat Hamiltonian systems. The resonances in two degrees of freedom are almost completely analyzed, while we give a survey of results obtained for three degrees of freedom systems.
The appendices contain a mix of elementary results, expansions on the theory and research problems. In order to keep the text accessible to the reader we have not formulated the theorems and proofs in their most general form, since it is our own experience that it is usually easier to generalize a simple theorem, than to apply a general one. The exception to this rule is the general averaging theory in chapter 3.

Since the classic book on nonlinear oscillations by Bogoliubov and Mitropolsky appeared in the early sixties, no modern survey on averaging has been

published. We hope that this book will remedy this situation and also will connect the asymptotic theory with the geometric ideas which have been so important in modern dynamics. We hope to be able to extend the scope of this book in later versions; one might e.g. think of codimension two bifurcations of vectorfields, the theory of which seems to be nearly complete now, or resonances of vectorfields, a difficult subject that one has only very recently started to research in a systematic manner.
In its original design the text would have covered both the qualitative and the quantitative theory of dynamical systems. While we were writing this text, however, several books appeared which explained the qualitative aspects better than we could ever hope to do. To have a good understanding of the geometry behind the kind of systems we are interested in, the reader is referred to the monographs of V. Arnol′d [8], R. Abraham and J.E. Marsden [1], J. Guckenheimer and Ph. Holmes [116]. A more classical part of qualitative theory, existence of periodic solutions as it is tied in with asymptotic analysis, has also been omitted as it is covered extensively in the existing literature (see e.g. [121]).

A number of people have kindly suggested references, alterations and corrections. In particular we are indebted to R. Cushman, J.J. Duistermaat, W. Eckhaus, M.A. Fekken, J. Schuur (MSU), L. van den Broek, E. van der Aa, A.H.P. van der Burgh, and S.A. van Gils. Many students provided us with lists of mathematical or typographical errors, when we used preliminary versions of the book for courses at the 'University of Utrecht', the 'Free University, Amsterdam' and at 'Michigan State University'.

We also gratefully acknowledge the generous way in which we could use the facilities of the Department of Mathematics and Computer Science of the Free University in Amsterdam, the Department of Mathematics of the University of Utrecht, and the Center for Mathematics and Computer Science in Amsterdam.

Amsterdam, Utrecht,
Summer 1985

Jan Sanders
Ferdinand Verhulst

List of Figures

List of Tables

List of Algorithms

Contents

Map of the book

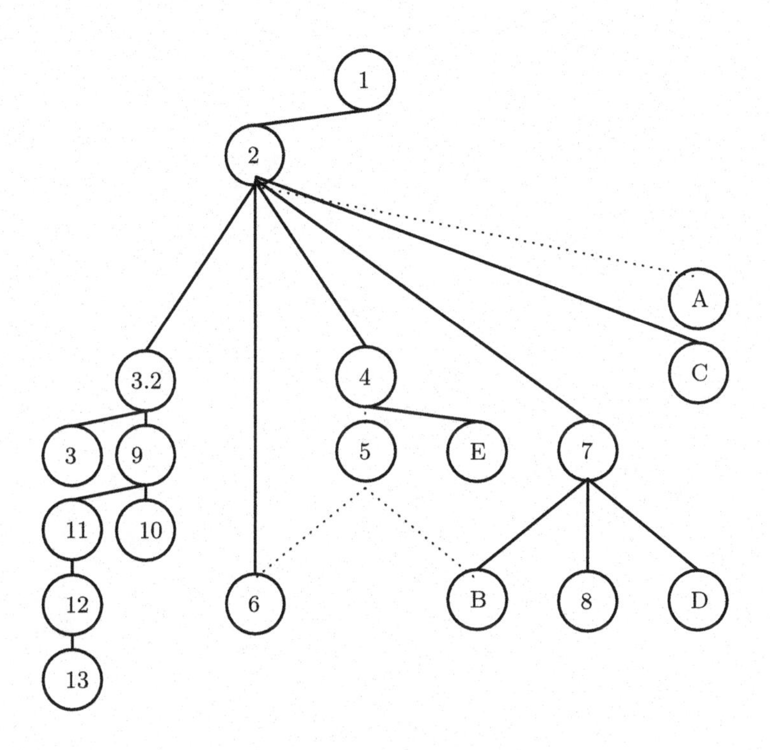

Fig. 0.1: The map of the book

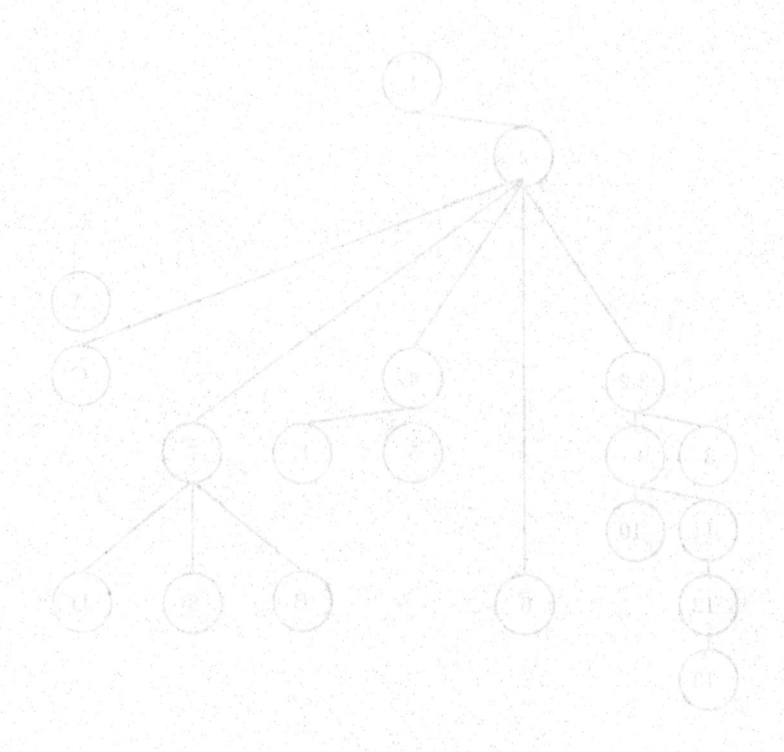

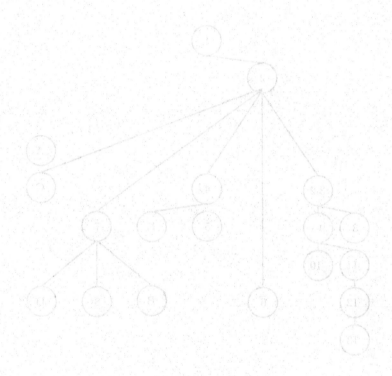

1

Basic Material and Asymptotics

1.1 Introduction

In this chapter we collect some material which will play a part in the theory to be developed in the subsequent chapters. This background material consists of the existence and uniqueness theorem for initial value problems based on contraction and, associated with this, continuation results and growth estimates.
The general form of the equations which we shall study is

$$\dot{\boldsymbol{x}} = \mathbf{f}(\boldsymbol{x}, t, \varepsilon),$$

where $\boldsymbol{x}$ and $\mathbf{f}(\boldsymbol{x}, t, \varepsilon)$ are vectors, elements of $\mathbb{R}^n$. All quantities used will be real except if explicitly stated otherwise.
Often we shall assume $\boldsymbol{x} \in D \subset \mathbb{R}^n$ with D an open, bounded set. The variable $t \in \mathbb{R}$ is usually identified with time; We assume $t \geq 0$ or $t \geq t_0$ with t_0 a constant. The parameter ε plays the part of a small parameter which characterizes the magnitude of certain perturbations. We usually take ε to satisfy either $0 \leq \varepsilon \leq \varepsilon_0$ or $|\varepsilon| \leq \varepsilon_0$, but even when $\varepsilon = 0$ is not in the domain, we may want to consider limits as $\varepsilon \downarrow 0$. We shall use $\mathsf{D}_{\boldsymbol{x}}\mathbf{f}(\boldsymbol{x}, t, \varepsilon)$ to indicate the derivative with respect to the spatial variable $\boldsymbol{x}$; so $\mathsf{D}_{\boldsymbol{x}}\mathbf{f}(\boldsymbol{x}, t, \varepsilon)$ is the matrix with components $\partial \mathrm{f}_i / \partial x_j(\boldsymbol{x}, t, \varepsilon)$. For a vector $\boldsymbol{u} \in \mathbb{R}^n$ with components $u_i, i = 1, \ldots, n$, we use the norm

$$\| \boldsymbol{u} \| = \textstyle\sum_{i=1}^{n} |u_i|. \tag{1.1.1}$$

For the $n \times n$-matrix A, with elements a_{ij} we have

$$\| A \| = \sum_{i,j=1}^{n} |a_{ij}|.$$

Any pair of vector and matrix norms satisfying $\|A\boldsymbol{x}\| \leq \|A\|\|\boldsymbol{x}\|$ may be used instead, such as the Euclidean norm for vectors and its associated operator norm for matrices, $\|A\| = \sup\{\|A\boldsymbol{x}\| : \|\boldsymbol{x}\| = 1\}$.

In the study of differential equations most vectors depend on variables. To estimate vector functions we shall nearly always use the sup norm. For instance for the vector functions arising in the differential equation formulated above we put

$$\| \mathbf{f} \|_{\sup} = \sup_{\boldsymbol{x} \in D, 0 \le t \le T, 0 < \varepsilon \le \varepsilon_0} \| \mathbf{f}(\boldsymbol{x}, t, \varepsilon) \| .$$

A system of differential equations on $\mathbb{R}^{2n}$ is called a **Hamiltonian system** with n **degrees of freedom** if it has the form

$$\begin{bmatrix} \dot{q}_i \\ \dot{p}_i \end{bmatrix} = \begin{bmatrix} \frac{\partial \mathrm{H}}{\partial p_i} \\ -\frac{\partial \mathrm{H}}{\partial q_i} \end{bmatrix}, \tag{1.1.2}$$

where $(q_1, \ldots, q_n, p_1, \ldots, p_n)$ are the coordinates on $\mathbb{R}^{2n}$ and $\mathrm{H} : \mathbb{R}^{2n} \to \mathbb{R}$ is a function called the **Hamiltonian** for the system[1]. Such systems appear occasionally throughout the book, and are studied intensively in Chapters 9 and 10, but we assume familiarity with the most basic facts about these systems. In particular, when dealing with Hamiltonian systems we often use special coordinate changes $(\boldsymbol{q}, \boldsymbol{p}) \leftrightarrow (\boldsymbol{Q}, \boldsymbol{P})$ that preserve the property of being Hamiltonian, and transform a system with Hamiltonian $\mathrm{H}(\boldsymbol{q}, \boldsymbol{p})$ into one with Hamiltonian $\mathrm{K}(\boldsymbol{Q}, \boldsymbol{P}) = \mathrm{H}(\boldsymbol{q}(\boldsymbol{Q}, \boldsymbol{P}), \boldsymbol{p}(\boldsymbol{Q}, \boldsymbol{P}))$. Such coordinate changes are associated with **symplectic mappings** but were known traditionally as **canonical transformations**.

1.2 The Initial Value Problem: Existence, Uniqueness and Continuation

The vector functions $\mathbf{f}(\boldsymbol{x}, t, \varepsilon)$ arising in our study of differential equations will have certain properties with respect to the variables $\boldsymbol{x}$ and t and the parameter ε. With respect to the 'spatial variable' $\boldsymbol{x}$, $\mathbf{f}$ will always satisfy a Lipschitz condition:

Notation 1.2.1 *Let* $G = D \times [t_0, t_0 + T] \times (0, \varepsilon_0]$.

Definition 1.2.2. *The vector function* $\mathbf{f} : G \to \mathbb{R}^n$ *satisfies a* **Lipschitz condition** *in* $\boldsymbol{x}$ *with* **Lipschitz constant** $\lambda_{\mathbf{f}}$ *if we have*

$$\| \mathbf{f}(\boldsymbol{x}_1, t, \varepsilon) - \mathbf{f}(\boldsymbol{x}_2, t, \varepsilon) \| \le \lambda_{\mathbf{f}} \| \boldsymbol{x}_1 - \boldsymbol{x}_2 \|,$$

where $\lambda_{\mathbf{f}}$ *is a constant. If* $\mathbf{f}$ *is periodic with period* T, *the Lipschitz condition will hold for all time.*

[1]The H is in honor of Christiaan Huygens.

It is well known that if $\mathbf{f}$ is of class C^1 on an open set U in $\mathbb{R}^n$, and D is a subset of U with compact and convex closure $\overline{D}$, $\mathbf{f}$ will satisfy a Lipschitz condition on $\overline{D}$ with $\lambda_{\mathbf{f}} = \max\{\|\mathsf{Df}(x)\| : x \in \overline{D}\}$. (The proof uses the mean value theorem for the scalar functions $\mathrm{g}_i(s) = \mathrm{f}_i(\boldsymbol{x}_1 + s(\boldsymbol{x}_2 - \boldsymbol{x}_1), t, \varepsilon)$ for $0 \leq s \leq 1$.) The following lemma (with proof contributed by J. Ellison) shows that convexity is not necessary. (This is a rather technical issue and the reader can skip the proof of this lemma on first reading.)

Lemma 1.2.3. *Suppose that* $\mathbf{f}$ *is* C^1 *on* U*, as above, and* $\overline{D}$ *is compact (but not necessarily convex). Then* $\mathbf{f}$ *is still Lipschitz on* $\overline{D}$*.*

Proof For convenience we suppress the dependence on t and ε. Since $\overline{D}$ is compact, there exists $M > 0$ such that $\|\mathbf{f}(\boldsymbol{x}_1) - \mathbf{f}(\boldsymbol{x}_2)\| \leq M$ for $\boldsymbol{x}_1, \boldsymbol{x}_2 \in \overline{D}$. Again by compactness, construct a finite set of open balls B_i with centers $\boldsymbol{p}_i$ and radii r_i (in the norm $\|\ \|$), such that each B_i is contained in U and such that the smaller balls B_i' with centers $\boldsymbol{p}_i$ and radii $r_i/3$ cover $\overline{D}$. Let $\lambda_{\mathbf{f}}^i$ be a Lipschitz constant for $\mathbf{f}$ in B_i, let $\lambda_{\mathbf{f}}^0 = \max_i \lambda_{\mathbf{f}}^i$, and let $\delta = \min_i r_i/3$. Observe that if $\boldsymbol{x}_1, \boldsymbol{x}_2 \in \overline{D}$ and $\|\boldsymbol{x}_1 - \boldsymbol{x}_2\| \leq \delta$, then $\boldsymbol{x}_1$ and $\boldsymbol{x}_2$ belong to the same ball B_i (in fact $\boldsymbol{x}_1$ belongs to some B_i' and then $\boldsymbol{x}_2 \in B_i$), and therefore $\|\mathbf{f}(\boldsymbol{x}_1) - \mathbf{f}(\boldsymbol{x}_2)\| \leq \lambda_{\mathbf{f}}^0 \|\boldsymbol{x}_1 - \boldsymbol{x}_2\|$. Now let $\lambda_{\mathbf{f}} = \max\{\lambda_{\mathbf{f}}^0, M/\delta\}$. We claim that $\|\mathbf{f}(\boldsymbol{x}_1) - \mathbf{f}(\boldsymbol{x}_2)\| \leq \lambda_{\mathbf{f}}^0 \|\boldsymbol{x}_1 - \boldsymbol{x}_2\|$ for all $\boldsymbol{x}_1, \boldsymbol{x}_2 \in \overline{D}$. If $\|\boldsymbol{x}_1 - \boldsymbol{x}_2\| \leq d$, this has already been proved (since $\lambda_{\mathbf{f}}^0 \leq \lambda_{\mathbf{f}}$). If $\|\boldsymbol{x}_1 - \boldsymbol{x}_2\| > \delta$, then

$$\|\mathbf{f}(\boldsymbol{x}_1) - \mathbf{f}(\boldsymbol{x}_2)\| \leq M = \frac{M}{\delta}\delta \leq \lambda_{\mathbf{f}}\delta < \lambda_{\mathbf{f}}\|\boldsymbol{x}_1 - \boldsymbol{x}_2\|.$$

This completes the proof of the lemma. □

We are now able to formulate a well-known existence and uniqueness theorem for initial value problems.

Theorem 1.2.4 (Existence and uniqueness). *Consider the differential equation*

$$\dot{\boldsymbol{x}} = \mathbf{f}(\boldsymbol{x}, t, \varepsilon).$$

We are interested in solutions $\mathbf{x}$ *of this equation with initial value* $\mathbf{x}(t_0) = \boldsymbol{a}$. *Let* $D = \{\boldsymbol{x} \in \mathbb{R}^n | \parallel \boldsymbol{x} - \boldsymbol{a} \parallel < d\}$*, inducing* G *by Notation 1.2.1, and* $\mathbf{f} : G \to \mathbb{R}^n$*. We assume that*

1. $\mathbf{f}$ *is continuous on* G*,*
2. $\mathbf{f}$ *satisfies a Lipschitz condition as in Definition 1.2.2.*

Then the initial value problem has a unique solution $\mathbf{x}$ *which exists for* $t_0 \leq t \leq t_0 + \inf(T, d/M)$ *where* $M = \sup_G \parallel \mathbf{f} \parallel = \parallel \mathbf{f} \parallel_{\text{sup}}$.

Proof The proof of the theorem can be found in any book seriously introducing differential equations, for instance Coddington and Levinson [59], Roseau [228] or Chicone [54]. □

Note that the theorem guarantees the existence of a solution on an interval of time which depends explicitly on the norm of $\mathbf{f}$. Additional assumptions enable us to prove continuation theorems, that is, with these assumptions one can obtain existence for larger intervals or even for all time. In the sequel we shall often meet equations in the so called standard form

$$\dot{\boldsymbol{x}} = \varepsilon \mathbf{g}^1(\boldsymbol{x}, t),$$

where the superscript reflects the ε-degree. (We often use integer superscripts in place of subscripts to avoid confusion with components of vectors. These superscripts are not to be taken as exponents.) Here, if the conditions of the existence and uniqueness theorem have been satisfied, we find that the solution exists for $t_0 \leq t \leq t_0 + \inf(T, d/M)$ with

$$M = \varepsilon \sup_{\boldsymbol{x} \in D} \sup_{t \in [t_0, t_0+T)} \| \mathbf{g}^1 \| .$$

This means that the size of the interval of existence of the solution is of the order C/ε with C a constant. This conclusion, in which ε is a small parameter, involves an asymptotic estimate of the size of an interval; such estimates will be made precise in Section 1.4.

1.3 The Gronwall Lemma

Closely related to contraction is the idea behind an inequality derived by Gronwall.

Lemma 1.3.1 (General Gronwall Lemma). *Suppose that for $t_0 \leq t \leq t_0 + T$ we have*

$$\varphi(t) \leq \alpha + \int_{t_0}^{t} \beta(s)\varphi(s)ds,$$

where φ and β are continuous and $\beta(t) > 0$. Then

$$\varphi(t) \leq \alpha \exp \int_{t_0}^{t} \beta(s)ds$$

for $t_0 \leq t \leq t_0 + T$.

Proof Let

$$\Phi(t) = \alpha + \int_{t_0}^{t} \beta(s)\varphi(s)ds.$$

Then $\varphi(t) \leq \Phi(t)$ and $\dot{\Phi}(t) = \beta(t)\varphi(t)$, so (since $\beta(t) > 0$) we have $\dot{\Phi}(t) - \beta(t)\Phi(t) \leq 0$. This differential inequality may be handled exactly as one would solve the corresponding differential equation (with $\leq$ replaced by $=$). That is, it may be rewritten as

$$\frac{d}{dt}\left(\Phi(t)e^{-\int_{t_0}^{t}\beta(s)ds}\right) \leq 0,$$

and then integrated from t_0 to t, using $\Phi(t_0) = \alpha$, to obtain

$$\Phi(t)e^{-\int_{t_0}^{t}\beta(s)ds} - \alpha \leq 0,$$

which may be rearranged into the desired result. □

Remark 1.3.2. The lemma may be generalized further to allow α to depend on t, provided we assume α is differentiable and $\alpha(t) \geq 0$, $\dot{\alpha}(t) > 0$. See [54]. ♡

Lemma 1.3.3 (Specific Gronwall lemma). *Suppose that for* $t_0 \leq t \leq t_0 + T$

$$\varphi(t) \leq \delta_2(t - t_0) + \delta_1 \int_{t_0}^{t} \varphi(s)\,\mathrm{d}s + \delta_3,$$

with $\varphi(t)$ *continuous for* $t_0 \leq t \leq t_0 + T$ *and constants* $\delta_1 > 0$, $\delta_2 \geq 0$, $\delta_3 \geq 0$ *then*

$$\varphi(t) \leq (\delta_2/\delta_1 + \delta_3)e^{\delta_1(t-t_0)} - \delta_2/\delta_1$$

for $t_0 \leq t \leq t_0 + T$.

Proof This has the form of Lemma 1.3.1 with $\alpha = \delta_1/\delta_2 + \delta_3$ and $\beta(t) = \delta_1$ for all t, and the result follows at once (changing back to $\varphi(t)$.) □

1.4 Concepts of Asymptotic Approximation

In the following sections we shall discuss those concepts and elementary methods in asymptotics which are necessary prerequisites for the study of slow-time processes in nonlinear oscillations.

In considering a function defined by an integral or defined as the solution of a differential equation with boundary or initial conditions, approximation techniques can be useful. In the applied mathematics literature no single theory dominates but many techniques can be found based on a great variety of concepts leading in general to different results. We mention here the methods of numerical analysis, approximation by orthonormal function series in a Hilbert space, approximation by convergent series and the theory of asymptotic approximations. Each of these methods can be suitable to understand an explicitly given problem. In this book we consider problems where the theory of asymptotic approximations is useful and we introduce the necessary concepts in detail.

One of the first examples of an **asymptotic approximation** was discussed by Euler [86], or [87, pp. 585-617], who studied the series

$$\sum_{n=0}^{\infty}(-1)^n n! x^n$$

with $x \in \mathbb{R}$. This series clearly diverges for all $x \neq 0$. We shall see in a moment why Euler would want to study such a series in the first place, but first we remark that if $x > 0$ is small, the individual terms decrease in absolute value rapidly as long as $nx < 1$. Euler used the truncated series to approximate the function given by the integral

$$\int_0^{\infty} \frac{e^{-s}}{1+sx}\, ds.$$

We return to Euler's example at the end of Section 1.4. Poincaré ([219, Chapter 8]) and Stieltjes [251] gave the mathematical foundation of using a divergent series in approximating a function. The theory of asymptotic approximations has expanded enormously ever since, but curiously enough only few authors concerned themselves with the foundations of the methods. Both the foundations and the applications of asymptotic analysis have been treated by Eckhaus [82]; see also Fraenkel [103].

We are interested in perturbation problems of the following kind: consider the differential equation

$$\dot{\boldsymbol{x}} = \mathbf{f}(\boldsymbol{x}, t, \varepsilon). \tag{1.4.1}$$

As usual, let $\boldsymbol{x}, \boldsymbol{a} \in \mathbb{R}^n$, $t \in [t_0, \infty)$ and $\varepsilon \in (0, \varepsilon_0]$ with ε_0 a small positive parameter. If the vector field $\mathbf{f}$ is sufficiently smooth in a neighborhood of $(\boldsymbol{a}, t_0) \in \mathbb{R}^n \times \mathbb{R}$, the initial value problem has a unique solution $\mathbf{x}_\varepsilon(t)$ for small values of ε on some interval $[t_0, \hat{t})$ (cf. Theorem 1.2.4);

Some of the problems arising in this approximation process can be illustrated by the following examples. Consider the first-order equation with initial value

$$\dot{x} = x + \varepsilon, \quad \mathrm{x}_\varepsilon(0) = 1.$$

The solution is $\mathrm{x}_\varepsilon(t) = (1+\varepsilon)e^t - \varepsilon$. We can rearrange this expression with respect to ε:

$$\mathrm{x}_\varepsilon(t) = e^t + \varepsilon(e^t - 1).$$

This result suggests that the function e^t is an approximation in some sense for $\mathrm{x}_\varepsilon(t)$ if t is not too large. In defining the concept of approximation one certainly needs a consideration of the domain of validity. A second simple example also shows that the solution does not always depend on the parameter ε in a smooth way:

$$\dot{x} = -\frac{\varepsilon x}{\varepsilon + t}, \quad \mathrm{x}_\varepsilon(0) = 1.$$

The solution reads

$$\mathrm{x}_\varepsilon(t) = \left(\frac{\varepsilon}{\varepsilon + t}\right)^\varepsilon.$$

To characterize the behavior of the solution with ε for $t \geq 0$ one has to divide $\mathbb{R}^+$ into different domains. For instance, it is sometimes possible to write

$$\mathrm{x}_\varepsilon(t) = 1 + \varepsilon \log \varepsilon - \varepsilon \log t + \mathcal{O}(\varepsilon/t),$$

where $\mathcal{O}(\varepsilon/t)$ is small compared to the other terms. ($\mathcal{O}$ will be defined more carefully below.) This expansion is possible when t is confined to an ε-dependent interval I_ε such that ε/t is small. (For instance, if $I_\varepsilon = (\sqrt{\varepsilon}, \infty)$ then $t \in I_\varepsilon$ implies $\varepsilon/t < \sqrt{\varepsilon}$.) Of course, this expansion does not satisfy the initial condition. Such problems about the domain of validity and the form of the expansions arise in classical mechanics; for some more realistic examples see [274]. To discuss these problems one has to introduce several concepts.

Definition 1.4.1. *A function $\delta(\varepsilon)$ will be called an* **order function** *if $\delta(\varepsilon)$ is continuous and positive in $(0, \varepsilon_0]$ and if $\lim_{\varepsilon \downarrow 0} \delta(\varepsilon)$ exists.*

Sometimes we use subscripts such as i in $\delta_i(\varepsilon), i = 1, 2, \ldots$. In many applications we shall use the set of order functions $\{\varepsilon^n\}_{n=1}^\infty$; however also order functions such as $\varepsilon^q, q \in \mathbb{Q}$ will play a part. To compare order functions we use Landau's symbols:

Definition 1.4.2. *Let $\varphi(t, \varepsilon)$ be a real- or vector valued function defined for $\varepsilon > 0$ (or $\varepsilon \geq 0$) and for $t \in I_\varepsilon$. The expression* **for $\varepsilon \downarrow 0$** *means that there exists an $\varepsilon_0 > 0$ such that the relevant statement holds for all $\varepsilon \in (0, \varepsilon_0]$). We define the symbols $\mathcal{O}(\cdot)$ and $o(\cdot)$ as follows.*

1. *We say that $\varphi(t, \varepsilon) = \mathcal{O}(\delta(\varepsilon))$ for $\varepsilon \downarrow 0$ if there exist constants $\varepsilon_0 > 0$ and $k > 0$ such that $\|\varphi(t, \varepsilon)\| \leq k|\delta(\varepsilon)|$ for all $t \in I_\varepsilon$, for $0 < \varepsilon < \varepsilon_0$.*
2. *We say that $\varphi(t, \varepsilon) = o(\delta(\varepsilon))$ for $\varepsilon \downarrow 0$ if*

$$\lim_{\varepsilon \downarrow 0} \frac{\|\varphi(t, \varepsilon)\|}{\delta(\varepsilon)} = 0,$$

uniformly for $t \in I_\varepsilon$. (That is, for every $\alpha > 0$ there exists $\beta > 0$ such that $\|\varphi(t, \varepsilon)\|/\delta(\varepsilon) < \alpha$ if $t \in I_\varepsilon$ and $0 < \varepsilon < \beta$.)

3. *We say that $\delta_1(\varepsilon) = o(\delta_2(\varepsilon))$ for $\varepsilon \downarrow 0$ if $\lim_{\varepsilon \downarrow 0} \delta_1(\varepsilon)/\delta_2(\varepsilon) = 0$.*

In all problems we shall consider ordering in a neighborhood of $\varepsilon = 0$ so in estimates we shall often omit 'for $\varepsilon \downarrow 0$'.

Examples 1.4.3 *The following show the usage of the symbols $\mathcal{O}(\cdot)$ and $o(\cdot)$.*

1. *$\varepsilon^n = o(\varepsilon^m)$ for $\varepsilon \downarrow 0$ if $n > m$;*
2. *$\varepsilon \sin(1/\varepsilon) = \mathcal{O}(\varepsilon)$ for $\varepsilon \downarrow 0$;*
3. *$\varepsilon^2 \log \varepsilon = o(\varepsilon^2 \log^2 \varepsilon)$ for $\varepsilon \downarrow 0$;*
4. *$e^{-1/\varepsilon} = o(\varepsilon^n)$ for $\varepsilon \downarrow 0$ and all $n \in \mathbb{N}$.* ♢

Now $\delta_1(\varepsilon) = o(\delta_2(\varepsilon))$ implies $\delta_1(\varepsilon) = \mathcal{O}(\delta_2(\varepsilon))$; for instance $\varepsilon^2 = o(\varepsilon)$ and $\varepsilon^2 = \mathcal{O}(\varepsilon)$ as $\varepsilon \downarrow 0$. It is useful to introduce the notion of a sharp estimate of order functions:

Definition 1.4.4 (Eckhaus [82]). *We say that $\delta_1(\varepsilon) = \mathcal{O}_\sharp(\delta_2(\varepsilon))$ for $\varepsilon \downarrow 0$ if $\delta_1(\varepsilon) = \mathcal{O}(\delta_2(\varepsilon))$ and $\delta_1(\varepsilon) \neq o(\delta_2(\varepsilon))$ for $\varepsilon \downarrow 0$.*

Example 1.4.5. One has $\varepsilon \sin(1/\varepsilon) = \mathcal{O}_\sharp(\varepsilon)$, $\varepsilon \log \varepsilon = \mathcal{O}_\sharp(2\varepsilon \log \varepsilon + \varepsilon^3)$. ♢

The real variable t used in the initial value problem (1.4.1) will be called **time**. Extensive use shall also be made of **time-like variables** of the form $\tau = \delta(\varepsilon)t$ with $\delta(\varepsilon) = \mathcal{O}(1)$.

We are now able to estimate the order of magnitude of functions $\varphi(t, \varepsilon)$, also written $\varphi_\varepsilon(t)$, defined in an interval I_ε, $\varepsilon \in (0, \varepsilon_0]$.

Definition 1.4.6. *Suppose that $\varphi_\varepsilon : I_\varepsilon \to \mathbb{R}^n$ for $0 < \varepsilon \leq \varepsilon_0$. Let $\|\cdot\|$ be the Euclidean metric on $\mathbb{R}^n$ and let $|\cdot|$ be defined by*

$$|\varphi_\varepsilon| = \sup\{\|\varphi_\varepsilon(t)\| : t \in I_\varepsilon\}.$$

(Notice that this norm depends on ε and could be written more precisely as $|\cdot|_\varepsilon$.) Let δ be an order function. Then:

1. *$\varphi_\varepsilon = \mathcal{O}(\delta(\varepsilon))$ in I_ε if $|\varphi_\varepsilon| = \mathcal{O}(\delta(\varepsilon))$ for $\varepsilon \downarrow 0$;*
2. *$\varphi_\varepsilon = o(\delta(\varepsilon))$ in I_ε if $\lim_{\varepsilon\downarrow 0} |\varphi_\varepsilon|/\delta(\varepsilon) = 0$;*
3. *$\varphi_\varepsilon = \mathcal{O}_\sharp(\delta(\varepsilon))$ in I_ε if $\varphi_\varepsilon = \mathcal{O}(\delta(\varepsilon))$ and $\varphi_\varepsilon \neq o(\delta(\varepsilon))$.*

It is customary to say that the estimates defined in this way are **uniform** *or* **uniformly valid** *on I_ε, because of the use of $|\cdot|$, which makes the estimates independent of t.*

Of course, one can give the same definitions for spatial variables.

Example 1.4.7. We wish to estimate the order of magnitude of the error we make in approximating $\sin(t + \varepsilon t)$ by $\sin(t)$ on the interval I_ε. If I_ε is $[0, 2\pi]$ we have for the difference of the two functions

$$\sup_{t\in[0,2\pi]} |\sin(t + \varepsilon t) - \sin(t)| = \mathcal{O}(\varepsilon).$$

Remark 1.4.8. An additional complication is that in many problems the boundaries of the interval I_ε depend on ε in such a way that the interval becomes unbounded as ε tends to 0. For instance in the example above we might wish to compare $\sin(t + \varepsilon t)$ with $\sin(t)$ on the interval $I_\varepsilon = [0, 2\pi/\varepsilon]$. We obtain in the sup norm

$$\sin(t + \varepsilon t) - \sin(t) = \mathcal{O}_\sharp(1)$$

(with $\mathcal{O}_\sharp$ as defined in Definition 1.4.4). ♡

Suppose $\delta(\varepsilon) = o(1)$ and we wish to estimate φ_ε on $I_\varepsilon = [0, L/\delta(\varepsilon)]$ with L a constant independent of ε. Such an estimate will be stated as $\varphi_\varepsilon = \mathcal{O}(\delta_0(\varepsilon))$ as $\varepsilon \downarrow 0$ on I_ε, or else as $\varphi_\varepsilon(t) = \mathcal{O}(\delta_0(\varepsilon))$ as $\varepsilon \downarrow 0$ on I_ε. The first form, without the t, is preferable, but is difficult to use in an example such as

$$\sin(t + \varepsilon t) - \sin(t) = \mathcal{O}(1)$$

as $\varepsilon \downarrow 0$ on I_ε. We express such estimates often as follows:

Definition 1.4.9. *We say that* $\varphi_\varepsilon(t) = \mathcal{O}(\delta(\varepsilon))$ *as* $\varepsilon \downarrow 0$ *on the* **time scale** $\delta(\varepsilon)^{-1}$ *if the estimate holds for* $0 \leq \delta(\varepsilon)t \leq L$ *with* L *a constant independent of* ε.

An analogous definition can be given for $o(\delta_0(\varepsilon))$-estimates. Once we are able to estimate functions in terms of order functions we are able to define asymptotic approximations.

Definition 1.4.10. *We define* **asymptotic approximations** *as follows.*

1. $\psi_\varepsilon(t)$ *is an asymptotic approximation of* $\varphi_\varepsilon(t)$ *on the interval* I_ε *if*

$$\varphi_\varepsilon(t) - \psi_\varepsilon(t) = o(1)$$

as $\varepsilon \downarrow 0$, *uniformly for* $t \in I_\varepsilon$. *Or rephrased for time scales:*

2. $\psi_\varepsilon(t)$ *is an asymptotic approximation of* $\varphi_\varepsilon(t)$ *on the time scale* $\delta(\varepsilon)^{-1}$ *if*

$$\varphi_\varepsilon - \psi_\varepsilon = o(1)$$

as $\varepsilon \downarrow 0$ *on the time scale* $\delta(\varepsilon)^{-1}$.

In general one obtains as approximations asymptotic series (or expansions) on some interval I_ε. An **asymptotic series** is an expression of the form

$$\varphi(t, \varepsilon) \sim \sum_{j=1}^{\infty} \delta_j(\varepsilon) \varphi^j(t, \varepsilon) \tag{1.4.2}$$

in which $\delta_j(\varepsilon)$ are order functions with $\delta_{j+1} = o(\delta_j)$. Such a series is not expected to converge, but instead one has

$$\varphi(t, \varepsilon) = \sum_{j=1}^{m} \delta_j(\varepsilon) \varphi^j(t, \varepsilon) + o(\delta_m(\varepsilon)) \text{ on } I_\varepsilon$$

for each m in $\mathbb{N}$, or, more commonly, the stronger condition

$$\varphi(t, \varepsilon) = \sum_{j=1}^{m} \delta_j(\varepsilon) \varphi^j(t, \varepsilon) + \mathcal{O}(\delta_{m+1}(\varepsilon)) \text{ on } I_\varepsilon,$$

often stated as "the error is of the order of the first omitted term."

Example 1.4.11. Consider, on $I = [0, 2\pi]$,

$$\varphi_\varepsilon(t) = \sin(t + \varepsilon t),$$
$$\tilde{\varphi}_\varepsilon(t) = \sin(t) + \varepsilon t \cos(t) - \frac{1}{2}\varepsilon^2 t^2 \sin(t).$$

The order functions are $\delta_n(\varepsilon) = \varepsilon^{n-1}$, $n = 1, 2, 3, \dots$ and clearly

$$\varphi_\varepsilon(t) - \tilde{\varphi}_\varepsilon(t) = o(\varepsilon^2) \text{ on } I,$$

so that $\tilde{\varphi}_\varepsilon(t)$ is a third-order asymptotic approximation of $\varphi_\varepsilon(t)$ on I. Asymptotic approximations are not unique. Another third-order asymptotic approximation of $\varphi_\varepsilon(t)$ on I is

$$\psi_\varepsilon(t) = \sin(t) + \varepsilon \varphi_{2_\varepsilon}(t) - \frac{1}{2}\varepsilon^2 t^2 \sin(t),$$

with $\varphi_{2_\varepsilon}(t) = \sin(\varepsilon t)\cos(t)/\varepsilon$. The functions $\varphi_{n_\varepsilon}(t)$ are not determined uniquely as is immediately clear from the definition. ♢

More serious is that for a given function different asymptotic approximations may be constructed with different sets of order functions. Consider an example given by Eckhaus ([82, Chapter 1]):

$$\varphi_\varepsilon(t) = (1 - \frac{\varepsilon}{1+\varepsilon} t)^{-1}, \quad I = [0, 1].$$

One easily shows that the following expansions are asymptotic approximations of φ_ε on I:

$$\psi_{1_\varepsilon}(t) = \sum_{n=0}^{m} (\frac{\varepsilon}{1+\varepsilon})^n t^n,$$
$$\psi_{2_\varepsilon}(t) = 1 + \sum_{n=1}^{m} \varepsilon^n t(t-1)^{n-1}.$$

Although asymptotic series in general are not unique, special forms of asymptotic series can be unique. A series of the form (1.4.2) in which each φ^n is independent of ε is called a **Poincaré asymptotic series**.

Theorem 1.4.12. *If $\varphi(t, \varepsilon)$ has a Poincaré asymptotic series with order functions $\delta_1, \delta_2, \dots$ then this series is unique.*

Proof First, $\varphi(t, \varepsilon) = \delta_1(\varepsilon)\varphi^1(t) + o(\delta_1(\varepsilon))$. Dividing by δ_1 we have $\varphi/\delta_1 = \varphi^1 + o(1)$, and letting $\varepsilon \to 0$ gives

$$\varphi^1(t) = \lim_{\varepsilon \to 0} \frac{\varphi(t, \varepsilon)}{\delta_1(\varepsilon)},$$

which determines $\varphi^1(t)$ uniquely. Next, dividing $\varphi = \delta_1\varphi^1 + \delta_2\varphi^2 + o(\delta_2)$ by δ_2 and letting $\varepsilon \to 0$ gives

$$\varphi^2(t) = \lim_{\varepsilon \to 0} \frac{\varphi(t, \varepsilon) - \delta_1(\varepsilon)\varphi^1(t)}{\delta_2(\varepsilon)},$$

which fixes φ^2. It is clear how to continue. Because of these formulas, Poincaré asymptotic series are often called **limit process expansions**. □

Another special type of asymptotic series is one in which the φ^j depend on ε only through a second time variable $\tau = \varepsilon t$. The next theorem, due to Perko [217], shows that certain series of this type are unique. This theorem will be used in Section 3.5.

Theorem 1.4.13 (Perko[217]). *Suppose that the function $\varphi(t,\varepsilon)$ has an asymptotic expansion of the form*

$$\varphi(t,\varepsilon) \sim \varphi^0(\tau,t) + \varepsilon\varphi^1(\tau,t) + \varepsilon^2\varphi^2(\tau,t) + \cdots, \tag{1.4.3}$$

valid on an interval $0 \le t \le L/\varepsilon$ for some $L > 0$. Suppose also that each $\varphi^j(\tau,t)$ is defined for $0 \le \tau \le L$ and $t \ge 0$, and is periodic in t with some period T (for all fixed τ). Then there is only one such expansion.

Proof By considering the difference of two such expansions, it is enough to prove that if

$$0 \sim \varphi^0(\tau,t) + \varepsilon\varphi^1(\tau,t) + \varepsilon^2\varphi^2(\tau,t) + \cdots$$

then each $\varphi^j = 0$. This asymptotic series implies that $\varphi^0(\tau,t) = o(1)$. We claim that $\varphi^0(\tau,t) = 0$ for any $t \ge 0$ and any τ with $0 \le \tau \le L$. Let $t_j = t+jT$ and $\varepsilon_j = \tau/t_j$, and note that $\varepsilon_j \to 0$ as $j \to \infty$ and that $0 \le t_j \le L/\varepsilon_j$. Now $\|\varphi^0(\tau,t)\| = \|\varphi^0(\varepsilon_j t_j, t_j)\| \to 0$ as $j \to \infty$ (in view of the definition of $|\cdot|$, so $\varphi^0(\tau,t) = 0$. We see that φ^0 drops out of the series, and we can divide by ε and repeat the argument for φ^1 and higher orders. □

For the sake of completeness we return to the example discussed by Euler which was mentioned at the beginning of this section. Instead of x we use the variable $\varepsilon \in (0,\varepsilon_0]$. Basic calculus can be used to show that we may define the function φ_ε by

$$\varphi_\varepsilon = \int_0^\infty \frac{e^{-s}}{1+\varepsilon s}\,\mathrm{d}s, \quad \varepsilon \in (0,\varepsilon_0].$$

Transform $\varepsilon s = \tau$ to obtain

$$\varphi_\varepsilon = \frac{1}{\varepsilon}\int_0^\infty \frac{e^{-\tau/\varepsilon}}{1+\tau}\,\mathrm{d}\tau,$$

and by partial integration

$$\varphi_\varepsilon = \frac{1}{\varepsilon}\left[-\varepsilon\frac{e^{-\tau/\varepsilon}}{1+\tau}\Big|_0^\infty - \varepsilon\int_0^\infty \frac{e^{-\tau/\varepsilon}}{(1+\tau)^2}\,\mathrm{d}\tau \right],$$

and after repeated partial integration

$$\varphi_\varepsilon = 1 - \varepsilon + 2\varepsilon\int_0^\infty \frac{e^{-\tau/\varepsilon}}{(1+\tau)^3}\,\mathrm{d}\tau.$$

We may continue the process and define

$$\tilde{\varphi}_\varepsilon = \sum_{n=0}^{m} (-1)^n n! \varepsilon^n.$$

It is easy to see that

$$\varphi_\varepsilon = \tilde{\varphi}_\varepsilon + R_{m_\varepsilon},$$

with

$$R_{m_\varepsilon} = (-1)^{m+1}(m+1)!\varepsilon^m \int_0^\infty e^{-\tau/\varepsilon}(1+\tau)^{-(m+2)}\,\mathrm{d}\tau.$$

Transforming back to t we can show that

$$R_{m_\varepsilon} = \mathcal{O}(\varepsilon^{m+1}).$$

Therefore $\tilde{\varphi}_\varepsilon$ is an asymptotic approximation of $\varphi(\varepsilon)$. The expansion is in the set of order functions $\{\varepsilon^n\}_{n=1}^\infty$ and the series is divergent.

A final remark concerns the case for which one is able to prove that an asymptotic series converges. This does not imply that the series converges to the function to be studied: consider the simple example

$$\varphi_\varepsilon = \sin(\varepsilon) + e^{-1/\varepsilon}.$$

Taylor expansion of $\sin(\varepsilon)$ produces the series

$$\tilde{\varphi}_\varepsilon = \sum_{n=0}^{m} \frac{(-1)^n \varepsilon^{2n+1}}{(2n+1)!}$$

which is convergent for $m \to \infty$; $\tilde{\varphi}_\varepsilon$ is an asymptotic approximation of φ_ε as

$$\varphi_\varepsilon - \tilde{\varphi}_\varepsilon = \mathcal{O}(\varepsilon^{2m+3}), \quad \forall m \in \mathbb{N}.$$

However, the series does not converge to φ_ε, but instead to $\sin(\varepsilon)$. The term $e^{-1/\varepsilon}$ is called **flat** or **transcendentally small**.

In the theory of nonlinear differential equations this matter of convergence is of some practical interest. Usually the calculation of one or a few more terms in the asymptotic expansion is all that one can do within a reasonable amount of (computer) time. But there are examples in bifurcation theory which show this flat behavior, see for instance [242].

1.5 Naive Formulation of Perturbation Problems

We are interested in studying initial value problems of the type

$$\dot{\boldsymbol{x}} = \mathbf{f}(\boldsymbol{x}, t, \varepsilon), \quad \mathbf{x}(t_0) = \boldsymbol{a}, \tag{1.5.1}$$

with $\boldsymbol{x}, \boldsymbol{a} \in D \subset \mathbb{R}^n$, $t, t_0 \in [0, \infty)$, $\varepsilon \in (0, \varepsilon_0]$. The vector field $\mathbf{f}$ meets the conditions of the basic existence and uniqueness Theorem 1.2.4. Suppose that

$$\lim_{\varepsilon \downarrow 0} \mathbf{f}(\boldsymbol{x}, t, \varepsilon) = \mathbf{f}(\boldsymbol{x}, t, 0)$$

exists uniformly on $D \times I$ with I a subinterval $[t_0, A]$ of $[0, \infty)$. Then we can associate with problem (1.5.1) an **unperturbed problem**

$$\dot{\boldsymbol{y}} = \mathbf{f}(\boldsymbol{y}, t, 0), \quad \mathbf{y}(t_0) = \boldsymbol{a}, \tag{1.5.2}$$

and we wish to establish the relation between the solution of (1.5.1) and (1.5.2). The relation will be expressed in terms of asymptotic approximations as introduced in Section 1.4. Note that this treatment is only useful if if we do not know the solution of (1.5.1) and if we can solve (1.5.2). The last assumption is not trivial as (1.5.2) is in general still nonautonomous and nonlinear.

Example 1.5.1. Let $\mathrm{x}_\varepsilon(t)$ be the solution of

$$\dot{x} = -\varepsilon x, \quad \mathrm{x}_\varepsilon(0) = 1; \quad x \in [0,1], \quad t \in [0, \infty), \quad \varepsilon \in (0, \varepsilon_0].$$

The associated unperturbed problem is

$$\dot{y} = 0, \quad \mathrm{y}(0) = 1.$$

It follows that $\mathrm{x}_\varepsilon(t) = e^{-\varepsilon t}$, $y(t) = 1$ and $\mathrm{x}_\varepsilon(t) - \mathrm{y}(t) = \mathcal{O}(\varepsilon)$ on the time scale 1. ◊

This is an example of **regular perturbation theory** for an autonomous system of the form

$$\dot{\boldsymbol{x}} = \mathbf{f}(\boldsymbol{x}, \varepsilon),$$

with $\boldsymbol{x} \in \mathbb{R}^n$. It is typical of regular perturbation theory that its results are valid only on time scale 1. We now turn to a general description of this theory.

Assuming that $\mathbf{f}$ is smooth, the solution $\mathbf{x}(\boldsymbol{a}, t, \varepsilon)$ with $\mathbf{x}(\boldsymbol{a}, 0, \varepsilon) = \boldsymbol{a}$ is smooth and can be approximated by its Taylor polynomial of degree k in ε as follows:

$$\mathbf{x}(\boldsymbol{a}, t, \varepsilon) = \mathbf{x}^0(\boldsymbol{a}, t) + \varepsilon \mathbf{x}^1(\boldsymbol{a}, t) + \cdots + \varepsilon^k \mathbf{x}^k(\boldsymbol{a}, t) + \mathcal{O}(\varepsilon^{k+1}),$$

uniformly on any finite interval $I = [0, L]$. In other words,

$$\mathbf{x}(\boldsymbol{a}, t, \varepsilon) \sim \sum_{j=0}^{\infty} \varepsilon^j \mathbf{x}^j(\boldsymbol{a}, t).$$

The coefficient functions $\mathbf{x}^j(\boldsymbol{a}, t)$ can be calculated recursively by substituting the series into the differential equation and equating like powers of ε.

Notation 1.5.2 *If* f *is a smooth vector valued function of* ε *for* ε *near zero, we write the* k*th Taylor polynomial, or* k*-jet, of* f *as*

$$\mathsf{J}^k_\varepsilon \mathrm{f} = \mathrm{f}^0 + \varepsilon \mathrm{f}^1 + \cdots + \varepsilon^k \mathrm{f}^k,$$

where $\mathrm{f}^j = \mathrm{f}^{(j)}(0)/j!$ *is the Taylor coefficient. The Taylor series of* f *through degree* k*, with remainder, will be written*

$$\mathrm{f}(\varepsilon) = \mathrm{f}^0 + \varepsilon \mathrm{f}^1 + \cdots + \varepsilon^k \mathrm{f}^k + \varepsilon^{k+1} \mathrm{f}^{[k+1]}(\varepsilon).$$

Thus a plain superscript denotes a Taylor coefficient, while a superscript in square brackets denotes a remainder. The notation is easily extended to functions of additional variables. For instance, a time-dependent vector field can be expanded as

$$\mathbf{f}(\boldsymbol{x}, t, \varepsilon) = \mathbf{f}^0(\boldsymbol{x}, t) + \varepsilon \mathbf{f}^1(\boldsymbol{x}, t) + \cdots + \varepsilon^k \mathbf{f}^k(\boldsymbol{x}, t) + \varepsilon^{k+1} \mathbf{f}^{[k+1]}(\boldsymbol{x}, t, \varepsilon).$$

In this notation it is always true that

$$\mathbf{f}(\boldsymbol{x}, t, \varepsilon) = \mathbf{f}^{[0]}(\boldsymbol{x}, t, \varepsilon),$$

and if $\mathbf{f}^0(\boldsymbol{x}, t)$ *is identically zero (as is often the case in averaging problems), then*

$$\mathbf{f}(\boldsymbol{x}, t, \varepsilon) = \varepsilon \mathbf{f}^{[1]}(\boldsymbol{x}, t, \varepsilon).$$

From an algebraic point of view, the vector space V *of formal power series in* ε *may be viewed as either a graded space* ($\mathcal{V} = \mathcal{V}^0 + \mathcal{V}^1 + \cdots$, *where* $\mathcal{V}^j$ *is the space of functions of exact degree* j *in* ε) *or as a filtered space* ($\mathcal{V} = \mathcal{V}^{[0]} \supset \mathcal{V}^{[1]} \supset \cdots$, *where* $\mathcal{V}^{[j]}$ *is the space of formal power series having terms of degree* $\geq j$). *Then*

$$\varepsilon^j \mathbf{f}^j \in \mathcal{V}^j \qquad \text{and} \qquad \varepsilon^j \mathbf{f}^{[j]} \in \mathcal{V}^{[j]}.$$

If we have more than one algebraically generating object, for instance ε *and* $\delta(\varepsilon)$*, with no algebraic relation between the two of them, then we use something like*

$$\varepsilon \mathbf{f}^1 + \delta(\varepsilon) \mathbf{f}^{0,1} + \varepsilon \delta(\varepsilon) \mathbf{f}^{[1,1]}.$$

The next theorem generalizes this idea to more general order functions.

Lemma 1.5.3. *Consider the initial value problems*

$$\dot{\boldsymbol{x}} = \mathbf{f}^0(\boldsymbol{x}, t) + \delta(\varepsilon) \mathbf{f}^{[1]}(\boldsymbol{x}, t, \varepsilon), \quad \mathbf{x}(t_0) = \boldsymbol{a} \tag{1.5.3}$$

and

$$\dot{\boldsymbol{y}} = \mathbf{f}^0(\boldsymbol{y}, t), \quad \mathbf{y}(t_0) = \boldsymbol{a}, \tag{1.5.4}$$

in which $\mathbf{f}^0$ *and* $\mathbf{f}^{[1]}$ *are Lipschitz continuous with respect to* $\boldsymbol{x}$ *in* $D \subset \mathbb{R}^n$ *and continuous with respect to* $(\boldsymbol{x}, t, \varepsilon) \in G$*. As usual,* $\delta(\varepsilon)$ *is an order function. If* $\mathbf{f}^{[1]}(\boldsymbol{x}, t, \varepsilon) = \mathcal{O}(1)$ *on the time scale* 1 *we have*

$$\mathbf{x}(t) - \mathbf{y}(t) = \mathcal{O}(\delta(\varepsilon))$$

on the time scale 1.

Proof We write the differential equations (1.5.3) and (1.5.4) as integral equations

$$\mathbf{x}(t) = \boldsymbol{a} + \int_{t_0}^{t} (\mathbf{f}^0(\mathbf{x}(s), s) + \delta(\varepsilon)\mathbf{f}^{[1]}(\mathbf{x}(s), s, \varepsilon))\, \mathrm{d}s,$$

$$\mathbf{y}(t) = \boldsymbol{a} + \int_{t_0}^{t} \mathbf{f}^0(\mathbf{y}(s), s)\, \mathrm{d}s.$$

Subtracting the equations and taking the norm of the difference we have

$$\begin{aligned} &\| \mathbf{x}(t) - \mathbf{y}(t) \| \\ &= \| \int_{t_0}^{t} (\mathbf{f}^0(\mathbf{x}(s), s) - \mathbf{f}^0(\mathbf{y}(s), s) + \delta(\varepsilon)\mathbf{f}^{[1]}(\mathbf{x}(s), s, \varepsilon))\, \mathrm{d}s \| \\ &\leq \int_{t_0}^{t} \| \mathbf{f}^0(\mathbf{x}(s), s) - \mathbf{f}^0(\mathbf{y}(s), s) \| \, \mathrm{d}s + \delta(\varepsilon) \int_{t_0}^{t} \| \mathbf{f}^{[1]}(\mathbf{x}(s), s, \varepsilon) \| \, \mathrm{d}s. \end{aligned}$$

There exists a constant M with $\| \mathbf{f}^{[1]}(\boldsymbol{x}, s, \varepsilon) \| \leq M$ on G. The Lipschitz continuity of $\mathbf{f}^0$ with respect to $\boldsymbol{x}$ implies moreover

$$\| \mathbf{x}(t) - \mathbf{y}(t) \| \leq \lambda_{\mathbf{f}^0} \int_{t_0}^{t} \| \mathbf{x}(s) - \mathbf{y}(s) \| \, \mathrm{d}s + \delta(\varepsilon) M (t - t_0).$$

We apply the Gronwall Lemma 1.3.3 with $\delta_1(\varepsilon) = \lambda_{\mathbf{f}^0}$, $\delta_2(\varepsilon) = M\delta(\varepsilon)$, $\delta_3 = 0$ to obtain

$$\| \mathbf{x}(t) - \mathbf{y}(t) \| \leq \delta(\varepsilon) \frac{M}{\lambda_{\mathbf{f}^0}} e^{\lambda_{\mathbf{f}^0}(t - t_0)} - \delta(\varepsilon) \frac{M}{\lambda_{\mathbf{f}^0}}. \tag{1.5.5}$$

We conclude from this inequality that $\mathbf{y}$ is an asymptotic approximation of $\mathbf{x}$ with error $\delta(\varepsilon)$ if $\lambda_{\mathbf{f}^0}(t - t_0)$ is bounded by a constant independent of ε; so the approximation is valid on the time scale 1. Note that we have a larger time scale, for instance $\log(\delta(\varepsilon))$, if we admit larger errors, e.g. $\sqrt{\delta}$. We note that if one tries to improve the accuracy by choosing an improved associated equation (by including higher-order terms in ε), the time scale of validity is not extended. More specifically, assume that we may write, using Notation 1.5.2,

$$\dot{\boldsymbol{x}} = \mathbf{f}^0(\boldsymbol{x}, t) + \delta(\varepsilon)\mathbf{f}^1(\boldsymbol{x}, t) + \overline{\delta}(\varepsilon)\mathbf{f}^{[1,1]}(\boldsymbol{x}, t, \varepsilon),$$

with $\mathbf{f}^{[1,1]} = \mathcal{O}(1)$ and $\overline{\delta}(\varepsilon) = o(\delta(\varepsilon))$. Applying the same estimation technique with $\delta_1(\varepsilon) = \lambda_{\mathbf{f}^0}$ and $\delta_2(\varepsilon) = \overline{\delta}(\varepsilon)M$ the estimate (1.5.5) produces for $\mathbf{y}$, the solution of

$$\dot{\boldsymbol{y}} = \mathbf{f}^0(t, \boldsymbol{y}) + \delta(\varepsilon)\mathbf{f}^{[1]}(t, \boldsymbol{y}), \quad \mathbf{y}(t_0) = \boldsymbol{a},$$

the following estimate for the error of the approximation:

$$\mathbf{x}(t) - \mathbf{y}(t) = \mathcal{O}(\overline{\delta}(\varepsilon))$$

on the time scale 1. To extend the time scale of validity we need more sophisticated methods. □

1.6 Reformulation in the Standard Form

We consider the perturbation problem of the form

$$\dot{\boldsymbol{x}} = \mathbf{f}^0(\boldsymbol{x}, t) + \varepsilon \mathbf{f}^{[1]}(\boldsymbol{x}, t, \varepsilon), \quad \mathbf{x}(t_0) = \boldsymbol{a}, \tag{1.6.1}$$

and the unperturbed problem

$$\dot{\boldsymbol{z}} = \mathbf{f}^0(\boldsymbol{z}, t), \ \mathbf{z}(t_0) = \boldsymbol{a}. \tag{1.6.2}$$

We assume that (1.6.2) can be solved explicitly. The solution will depend on the initial value $\boldsymbol{a}$ and we write it as $\mathbf{z}(\boldsymbol{a}, t)$. So we have

$$\mathbf{z} = \mathbf{z}(\boldsymbol{\zeta}, t) \ , \ \mathbf{z}(\boldsymbol{\zeta}, t_0) = \boldsymbol{\zeta} \ , \ \boldsymbol{\zeta} \in \mathbb{R}^n.$$

We now consider this as a transformation (method of variation of parameters or variation of constants) as follows:

$$\boldsymbol{x} = \mathbf{z}(\boldsymbol{\zeta}, t). \tag{1.6.3}$$

Using (1.6.1) and (1.6.2) we derive the differential equation for $\boldsymbol{\zeta}$

$$\frac{\partial \mathbf{z}(\boldsymbol{\zeta}, t)}{\partial t} + \mathsf{D}_{\boldsymbol{\zeta}} \mathbf{z}(\boldsymbol{\zeta}, t) \cdot \frac{d\boldsymbol{\zeta}}{dt} = \mathbf{f}^0(\mathbf{z}(\boldsymbol{\zeta}, t), t) + \varepsilon \mathbf{f}^{[1]}(\mathbf{z}(\boldsymbol{\zeta}, t), t, \varepsilon).$$

Since $\mathbf{z}$ satisfies the unperturbed equation, the first terms on the left and right cancel out. If we assume that $\mathsf{D}_{\boldsymbol{\zeta}} \mathbf{z}(\boldsymbol{\zeta}, t)$ is nonsingular we may write

$$\dot{\boldsymbol{\zeta}} = \varepsilon (\mathsf{D}_{\boldsymbol{\zeta}} \mathbf{z}(\boldsymbol{\zeta}, t))^{-1} \cdot \mathbf{f}^{[1]}(\mathbf{z}(\boldsymbol{\zeta}, t), t, \varepsilon). \tag{1.6.4}$$

Equation (1.6.4) supplemented by the initial value of $\boldsymbol{\zeta}$ will be called a **perturbation problem in the standard form**.

In general, however, equation (1.6.4) will be messy. Consider for example the perturbed mathematical pendulum equation

$$\ddot{\phi} + \sin(\phi) = \varepsilon g(\phi, t, \varepsilon).$$

Equation (1.6.4) will in this case necessarily involve elliptic functions. Another difficulty of a more technical nature might be that the transformation introduces nonuniformities in the time-dependent behavior, so there is no Lipschitz constant λ independent of t. Still the standard form (1.6.4) may be useful to draw several general conclusions. A simple case in mathematical biology involving elementary functions is the following example.

Example 1.6.1. Consider two species living in a region with a restricted supply of food and a slight interaction between the species affecting their population density x_1 and x_2. We describe the population growth by the model

$$\frac{dx_1}{dt} = \beta_1 x_1 - x_1^2 + \varepsilon f_1(x_1, x_2), \quad \mathrm{x}_1(0) = a_1,$$
$$\frac{dx_2}{dt} = \beta_2 x_2 - x_2^2 + \varepsilon f_2(x_1, x_2), \quad \mathrm{x}_2(0) = a_2,$$

where the constants $\beta_i, a_i > 0$ and $\mathrm{x}_i(t) \geq 0$ for $i = 1, 2$. The solution of the unperturbed problem is

$$\mathrm{x}_i(t) = \frac{\beta_i}{1 + \frac{\beta_i - a_i}{a_i} e^{-\beta_i t}} = \frac{\beta_i a_i e^t}{\beta_i + a_i(e^{\beta_i t} - 1)}.$$

Applying (1.6.4) we get

$$\frac{d\zeta_i}{dt} = \varepsilon e^{-\beta_i t}(1 + \frac{\zeta_i}{\beta_i}(e^{\beta_i t} - 1))^2 f_i(\cdot, \cdot), \quad \zeta_i(0) = a_i, \quad i = 1, 2,$$

in which we abbreviated the expression for f_i. ♢

As has been suggested earlier on, the transformation may often be not practical, and one can see in this example why, since even if we take f_i constant, the right-hand side of the equation grows exponentially. There is however an important class of problems where this technique works well and we shall treat this in Section 1.7.

1.7 The Standard Form in the Quasilinear Case

The perturbation problem (1.6.1) will be called **quasilinear** if the equation can be written as

$$\dot{\boldsymbol{x}} = A(t)\boldsymbol{x} + \varepsilon \mathbf{f}^{[1]}(\boldsymbol{x}, t, \varepsilon), \tag{1.7.1}$$

in which $A(t)$ is a continuous $n \times n$-matrix. The unperturbed problem

$$\dot{\boldsymbol{y}} = A(t)\boldsymbol{y}$$

possesses n linearly independent solutions from which we construct the fundamental matrix $\Phi(t)$. We choose Φ such that $\Phi(t_0) = I$. We apply the variation of constants procedure

$$\boldsymbol{x} = \Phi(t)\boldsymbol{z},$$

and we obtain, using (1.6.4),

$$\dot{\boldsymbol{z}} = \varepsilon \Phi^{-1}(t)\mathbf{f}^{[1]}(\Phi(t)\boldsymbol{z}, t, \varepsilon). \tag{1.7.2}$$

If A is a constant matrix we have for the fundamental matrix

$$\Phi(t) = e^{A(t-t_0)}.$$

The standard form becomes in this case

$$\dot{\boldsymbol{z}} = \varepsilon e^{-A(t-t_0)} \mathbf{f}^{[1]}(e^{A(t-t_0)}\boldsymbol{z}, t, \varepsilon). \tag{1.7.3}$$

Clearly if the eigenvalues of A are not all purely imaginary, the perturbation equation (1.7.3) may present some serious problems even if $\mathbf{f}^{[1]}$ is bounded.

Remark 1.7.1. In the theory of forced nonlinear oscillations the perturbation problem may be of the form

$$\dot{\boldsymbol{x}} = \mathbf{f}^0(\boldsymbol{x}, t) + \varepsilon \mathbf{f}^{[1]}(\boldsymbol{x}, t, \varepsilon), \tag{1.7.4}$$

where $\mathbf{f}^0(\boldsymbol{x}, t) = A\boldsymbol{x} + \mathbf{h}(t)$ and A a constant matrix. The variation of constants transformation then becomes

$$\boldsymbol{x} = e^{A(t-t_0)}\boldsymbol{z} + e^{A(t-t_0)} \int_{t_0}^{t} e^{-A(s-t_0)} \mathbf{h}(s)\, ds. \tag{1.7.5}$$

The perturbation problem in the standard form is

$$\dot{\boldsymbol{z}} = \varepsilon e^{-A(t-t_0)} \mathbf{f}^{[1]}(\boldsymbol{x}, t, \varepsilon),$$

in which $\boldsymbol{x}$ still has to be replaced by expression (1.7.5). ♡

Example 1.7.2. In studying nonlinear oscillations one often considers the perturbed initial value problem

$$\ddot{x} + \omega^2 x = \varepsilon g(x, \dot{x}, t, \varepsilon)\ ,\ \mathrm{x}(t_0) = a_1\ ,\ \dot{\mathrm{x}}(t_0) = a_2. \tag{1.7.6}$$

Two independent solutions of the unperturbed problem $\ddot{y} + \omega^2 y = 0$ are $\cos(\omega(t - t_0))$ and $\sin(\omega(t - t_0))$. The variation of constants transformation becomes

$$\begin{aligned} x &= z_1 \cos(\omega(t - t_0)) + \frac{z_2}{\omega} \sin(\omega(t - t_0)), \\ \dot{x} &= -z_1 \omega \sin(\omega(t - t_0)) + z_2 \cos(\omega(t - t_0)). \end{aligned} \tag{1.7.7}$$

Note that the fundamental matrix is such that $\Phi(t_0) = I$. Equation (1.7.3) becomes in this case

$$\begin{aligned} \dot{z}_1 &= -\frac{\varepsilon}{\omega} \sin(\omega(t - t_0)) g(\cdot, \cdot, t, \varepsilon), \quad \mathrm{z}_1(t_0) = a_1, \\ \dot{z}_2 &= \varepsilon \cos(\omega(t - t_0)) g(\cdot, \cdot, t, \varepsilon), \quad \mathrm{z}_2(t_0) = a_2. \end{aligned} \tag{1.7.8}$$

The expressions for x and $\dot{x}$ have to be substituted in g on the dots. ◇

It may be useful to adopt a transformation which immediately provides us with equations for the variation of the amplitude r and the phase ϕ of the solution. We put

$$\begin{bmatrix} x \\ \dot{x} \end{bmatrix} = \begin{bmatrix} r\sin(\omega t - \phi) \\ r\omega\cos(\omega t - \phi) \end{bmatrix}. \tag{1.7.9}$$

The perturbation equations become

$$\begin{bmatrix} \dot{r} \\ \dot{\psi} \end{bmatrix} = \varepsilon \begin{bmatrix} \frac{1}{\omega}\cos(\omega t - \phi)g(\cdot,\cdot,t,\varepsilon) \\ \frac{1}{r\omega}\sin(\omega t - \phi)g(\cdot,\cdot,t,\varepsilon) \end{bmatrix} \tag{1.7.10}$$

The initial values for r and ϕ can be calculated from (1.7.9). It is clear that the perturbation formulation (1.7.10) may get us into difficulties in problems where the amplitude r can become small. In Sections 2.2–2.7 we show the usefulness of both transformation (1.7.7) and (1.7.9).

2

Averaging: the Periodic Case

2.1 Introduction

The simplest form of averaging is **periodic averaging**, which is concerned with solving a perturbation problem in the standard form

$$\dot{\boldsymbol{x}} = \varepsilon \mathbf{f}^1(\boldsymbol{x}, t) + \varepsilon^2 \mathbf{f}^{[2]}(\boldsymbol{x}, t, \varepsilon), \quad \mathbf{x}(0) = \boldsymbol{a}, \tag{2.1.1}$$

where $\mathbf{f}^1$ and $\mathbf{f}^{[2]}$ are T-periodic in t; see Notation 1.5.2 for the superscripts.

It seems natural to simplify the equation by truncating (dropping the ε^2 term) and averaging over t (while holding $\boldsymbol{x}$ constant), so we consider the **averaged equation**

$$\dot{\boldsymbol{z}} = \varepsilon \overline{\mathbf{f}}^1(\boldsymbol{z}), \quad \mathbf{z}(0) = \boldsymbol{a}, \tag{2.1.2}$$

with

$$\overline{\mathbf{f}}^1(\boldsymbol{z}) = \frac{1}{T} \int_0^T \mathbf{f}^1(\boldsymbol{z}, s)\, \mathrm{d}s.$$

The basic result is that (under appropriate technical conditions to be specified later in Section 2.8), the solutions of these systems remain close (of order ε) for a time interval of order $1/\varepsilon$:

$$\|\mathbf{x}(t) - \mathbf{z}(t)\| \le c\varepsilon \quad \text{for} \quad 0 \le t \le L/\varepsilon$$

for positive constants c and L. Two proofs of this result will be given in Section 2.8 below, and another in Section 4.2 (as a consequence of a more general averaging theorem for nonperiodic systems).

The procedure of averaging can be found already in the works of Lagrange and Laplace who provided an intuitive justification and who used the procedure to study the problem of secular[1] perturbations in the solar system.

[1] secular: pertaining to an age, or the progress of ages, or to a long period of time.

To many physicists and astronomers averaging seems to be such a natural procedure that they do not even bother to justify the process. However it is important to have a rigorous approximation theory, since it is precisely the fact that averaging seems so natural that obscures the pitfalls and restrictions of the method. We find for instance misleading results based on averaging by Jeans, [138, Section268], who studies the two-body problem with slowly varying mass; cf. the results obtained by Verhulst [274].

Around 1930 we see the start of precise statements and proofs in averaging theory. A historical survey of the development of the theory from the 18th century until around 1960 can be found in Appendix A. After this time many new results in the theory of averaging have been obtained. The main trends of this research will be reflected in the subsequent chapters.

2.2 Van der Pol Equation

In this and the following sections we shall apply periodic averaging to some classical problems. For more examples see for instance Bogoliubov and Mitropolsky [35]. Also we present several counter examples to show the necessity of some of the assumptions and restrictions that will be needed when the validity of periodic averaging is proved in Section 2.8. Consider the **Van der Pol equation**

$$\ddot{x} + x = \varepsilon \mathrm{g}(x, \dot{x}), \tag{2.2.1}$$

with initial values x_0 and $\dot{x}_0$ given and g a sufficiently smooth function in $D \subset \mathbb{R}^2$. This is a quasilinear system (Section 1.7) and we use the amplitude-phase transformation (1.7.9) to put the system in the standard form. Put

$$\begin{aligned} x &= r \sin(t - \phi), \\ \dot{x} &= r \cos(t - \phi). \end{aligned}$$

The perturbation equations (1.7.10) become

$$\begin{bmatrix} \dot{r} \\ \dot{\phi} \end{bmatrix} = \varepsilon \begin{bmatrix} \cos(t-\phi)\mathrm{g}(r\sin(t-\phi), r\cos(t-\phi)) \\ \frac{1}{r}\sin(t-\phi)\mathrm{g}(r\sin(t-\phi), r\cos(t-\phi)) \end{bmatrix}. \tag{2.2.2}$$

This is of the form

$$\dot{\boldsymbol{x}} = \varepsilon \mathbf{f}^1(\boldsymbol{x}, t),$$

with $\boldsymbol{x} = (r, \phi)$. We note that the vector field is 2π-periodic in t and that according to Theorem 2.8.1 below, if $\mathrm{g} \in C^1(D)$ we may average the right-hand side as long as we exclude a neighborhood of the origin (where the polar coordinates fail). Since the original equation is autonomous, the averaged equation depends only on r and we define the two components of the averaged vector field as follows:

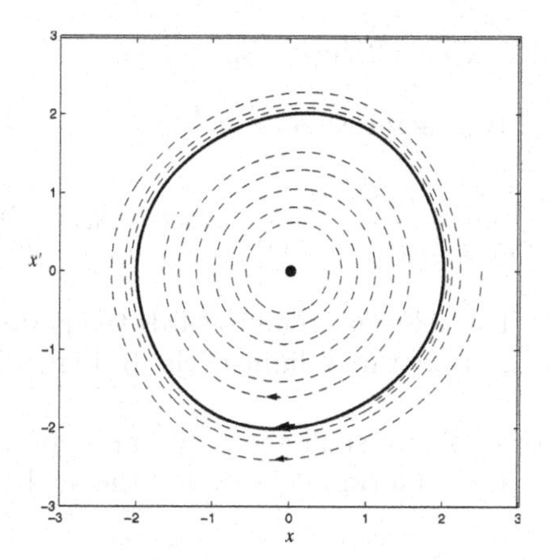

Fig. 2.1: Phase orbits of the Van der Pol equation $\ddot{x}+x=\varepsilon(1-x^2)\dot{x}$ where $\varepsilon=0.1$. The origin is a critical point of the flow, the limit-cycle (closed curve) corresponds to a stable periodic solution.

$$\begin{aligned}\overline{\mathrm{f}}_1^1(r) &= \frac{1}{2\pi}\int_0^{2\pi} \cos(s-\phi)\mathrm{g}(r\sin(t-\phi), r\cos(s-\phi))\,\mathrm{d}s\\ &= \frac{1}{2\pi}\int_0^{2\pi} \cos(s)\mathrm{g}(r\sin(s), r\cos(s))\,\mathrm{d}s,\end{aligned}$$

and

$$\overline{\mathrm{f}}_2^1(r) = \frac{1}{r}\frac{1}{2\pi}\int_0^{2\pi} \sin(s)\mathrm{g}(r\sin(s), r\cos(s))\,\mathrm{d}s.$$

An asymptotic approximation can be obtained by solving

$$\dot{\overline{r}} = \varepsilon\overline{\mathrm{f}}_1^1(\overline{r})\ , \quad \dot{\overline{\phi}} = \varepsilon\overline{\mathrm{f}}_2^1(\overline{r})$$

with appropriate initial values. Notice that this equation is of the form (2.1.2) with $\mathbf{z} = (\overline{r}, \overline{\phi})$. This is a reduction to the problem of solving a first-order autonomous system. We specify this for a famous example, the Van der Pol equation:

$$\ddot{x} + x = \varepsilon(1-x^2)\dot{x}.$$

We obtain

$$\dot{\overline{r}} = \frac{1}{2}\varepsilon\overline{r}(1-\frac{1}{4}\overline{r}^2), \quad \dot{\overline{\phi}} = 0.$$

If the initial value of the amplitude r_0 equals 0 or 2 the amplitude $\overline{r}$ is constant for all time. Here $r_0 = 0$ corresponds to an unstable critical point of the original equation, $r_0 = 2$ gives a periodic solution:

$$x(t) = 2\sin(t - \phi_0) + \mathcal{O}(\varepsilon) \tag{2.2.3}$$

on the time scale $1/\varepsilon$. In general we obtain

$$x(t) = \frac{r_0 e^{\frac{1}{2}\varepsilon t}}{(1 + \frac{1}{4}{r_0}^2(e^{\varepsilon t} - 1))^{\frac{1}{2}}} \sin(t - \phi_0) + \mathcal{O}(\varepsilon) \tag{2.2.4}$$

on the time scale $1/\varepsilon$. The solutions tend towards the periodic solution (2.2.3) and we call its phase orbit a (stable) limit-cycle. In Figure 2.1 we depict some of the orbits.

In the following example we shall show that an appropriate choice of the transformation into standard form may simplify the analysis of the perturbation problem.

2.3 A Linear Oscillator with Frequency Modulation

Consider an example of **Mathieu's equation**

$$\ddot{x} + (1 + 2\varepsilon\cos(2t))x = 0,$$

with initial values $x(0) = x_0$ and $\dot{x}(0) = \dot{x}_0$. We may proceed as in Section 2.2; however equation (2.2.1) now explicitly depends on t. The amplitude-phase transformation produces, with $g = -2\cos(2t)x$,

$$\begin{aligned}
\dot{r} &= -2\varepsilon r \sin(t - \phi)\cos(t - \phi)\cos(2t),\\
\dot{\phi} &= -2\varepsilon \sin^2(t - \phi)\cos(2t).
\end{aligned}$$

The right-hand side is 2π-periodic in t; averaging produces

$$\dot{\overline{r}} = \frac{1}{2}\varepsilon\overline{r}\sin(2\overline{\phi}), \quad \dot{\overline{\phi}} = \frac{1}{2}\varepsilon\cos(2\overline{\phi}).$$

To approximate the solutions of a time-dependent linear system we have to solve an autonomous nonlinear system. Here the integration can be carried out but it is more practical to choose a different transformation to obtain the standard form, staying inside the category of linear systems with linear transformations. We use transformation (1.7.7) with $\omega = 1$ and $t_0 = 0$:

$$x = z_1\cos(t) + z_2\sin(t), \quad \dot{x} = -z_1\sin(t) + z_2\cos(t).$$

The perturbation equations become (cf. formula (1.7.8))

$$\begin{aligned}
\dot{z}_1 &= 2\varepsilon\sin(t)\cos(2t)(z_1\cos(t) + z_2\sin(t)),\\
\dot{z}_2 &= -2\varepsilon\cos(t)\cos(2t)(z_1\cos(t) + z_2\sin(t)).
\end{aligned}$$

The right-hand side is 2π-periodic in t; averaging produces

$$\dot{\overline{z}}_1 = -\frac{1}{2}\varepsilon\overline{z}_2, \quad \overline{z}_1(0) = x_0,$$
$$\dot{\overline{z}}_2 = -\frac{1}{2}\varepsilon\overline{z}_1, \quad \overline{z}_2(0) = \dot{x}_0.$$

This is a linear system with solutions

$$\overline{z}_1(t) = \frac{1}{2}(x_0 + \dot{x}_0)e^{-\frac{1}{2}\varepsilon t} + \frac{1}{2}(x_0 - \dot{x}_0)e^{\frac{1}{2}\varepsilon t},$$
$$\overline{z}_2(t) = \frac{1}{2}(x_0 + \dot{x}_0))e^{-\frac{1}{2}\varepsilon t} - \frac{1}{2}(x_0 - \dot{x}_0)e^{\frac{1}{2}\varepsilon t}.$$

The asymptotic approximation for the solution $\mathrm{x}(t)$ of this Mathieu equation reads

$$\overline{\mathrm{x}}(t) = \frac{1}{2}(x_0 + \dot{x}_0)e^{-\frac{1}{2}\varepsilon t}(\cos(t) + \sin(t)) + \frac{1}{2}(x_0 - \dot{x}_0)e^{\frac{1}{2}\varepsilon t}(\cos(t) - \sin(t)).$$

We note that the equilibrium solution $\mathrm{x} = \dot{\mathrm{x}} = 0$ is unstable. In the following example an amplitude-phase representation is more appropriate.

2.4 One Degree of Freedom Hamiltonian System

Consider the equation of motion of a one degree of freedom Hamiltonian system

$$\ddot{x} + x = \varepsilon \mathrm{g}(x),$$

where g is sufficiently smooth. (This may be written in the form (1.1.2) with $n = 1$ by putting $q = x$, $p = \dot{x}$, and $H = (q^2 + p^2)/2 - \varepsilon F(q)$, where $F' = \mathrm{g}$.) Applying the formulae of Section 2.2 we obtain for the amplitude and phase the following equations

$$\dot{r} = \varepsilon \cos(t - \phi)\mathrm{g}(r \sin(t - \phi)),$$
$$\dot{\phi} = \varepsilon \frac{\sin(t - \phi)}{r}\mathrm{g}(r \sin(t - \phi)).$$

We have

$$\int_0^{2\pi} \cos(s - \phi)\mathrm{g}(r \sin(s - \phi))\, \mathrm{d}s = 0.$$

So the averaged equation for the amplitude is

$$\dot{\overline{r}} = 0,$$

i.e., in first approximation the amplitude is constant. This means that for a small Hamiltonian perturbation of the harmonic oscillator, the leading-order approximation has periodic solutions with a constant amplitude but in general a period depending on this amplitude, i.e. on the initial values. (In fact, the exact solution is also periodic, but our calculation does not prove this.) It is easy to verify that one can obtain the same result by using transformation (1.7.7) but the calculation is much more complicated.

Finally we remark that the transformation is not symplectic, since $dq \wedge dp = r\,dr \wedge d\psi$. We could have made it symplectic by taking $r = \sqrt{2\tau}$. Then we find that $dq \wedge dp = d\tau \wedge d\psi$. For more details, see Chapter 10.

2.5 The Necessity of Restricting the Interval of Time

Consider the equation

$$\ddot{x} + x = 8\varepsilon\dot{x}^2\cos(t),$$

with initial values $x(0) = 0$, $\dot{x}(0) = 1$. Reduction to the standard form using the amplitude-phase transformation (1.7.9) produces

$$\begin{aligned}\dot{r} &= 8\varepsilon r^2\cos^3(t-\psi)\cos(t), \quad r(0) = 1,\\ \dot{\psi} &= 8\varepsilon r\cos^2(t-\psi)\sin(t-\psi)\cos(t)\ , \ \psi(0) = 0.\end{aligned}$$

Averaging gives the associated system

$$\begin{aligned}\dot{\overline{r}} &= 3\varepsilon\overline{r}^2\cos(\overline{\psi}),\\ \dot{\overline{\psi}} &= -\varepsilon\overline{r}\sin(\overline{\psi}).\end{aligned}$$

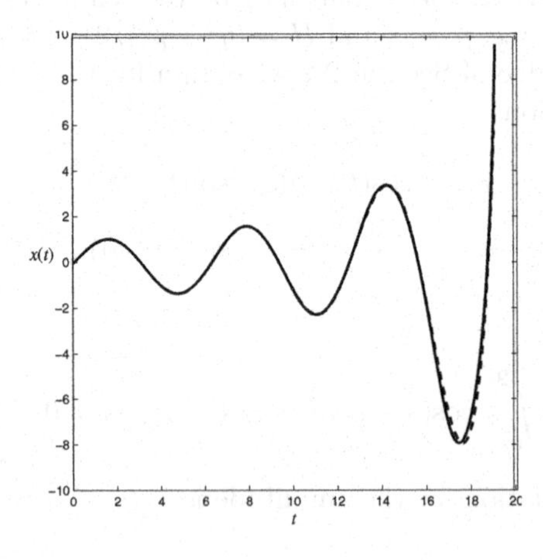

Fig. 2.2: Solution $x(t)$ of $\ddot{x} + x = \frac{2}{15}\dot{x}^2\cos(t)$, $x(0) = 0$, $\dot{x}(0) = 1$. The solution obtained by numerical integration has been drawn full line, the asymptotic approximation has been indicated by $- - -$.

Integration of the system and using the fact that $\dot{\overline{\psi}} = 0$ for the given initial conditions yields

$$\mathrm{x}(t) = \frac{\sin(t)}{1 - 3\varepsilon t} + \mathcal{O}(\varepsilon)$$

on the time scale $1/\varepsilon$. A similar estimate holds for the derivative $\dot{\mathrm{x}}$. The approximate solution is bounded if $0 \leq \varepsilon t \leq C < \frac{1}{3}$. In Figure 2.2 we depict the approximate solution and the solution obtained by numerical integration.

2.6 Bounded Solutions and a Restricted Time Scale of Validity

One might wonder whether the necessity to restrict the time scale is tied in with the characteristic of solutions becoming unbounded as in Section 2.5. A simple example suffices to contradict this.

Consider the equation

$$\ddot{x} + x = \varepsilon x, \quad \mathrm{x}(0) = 1, \quad \dot{\mathrm{x}}(0) = 0.$$

After amplitude-phase transformation and averaging as in Section 2.2 we obtain

$$\begin{aligned} \dot{\overline{r}} &= 0, \quad \overline{\mathrm{r}}(0) = 1, \\ \dot{\overline{\psi}} &= \frac{1}{2}\varepsilon, \quad \overline{\psi}(0) = \frac{1}{2}\pi. \end{aligned}$$

We have the approximations

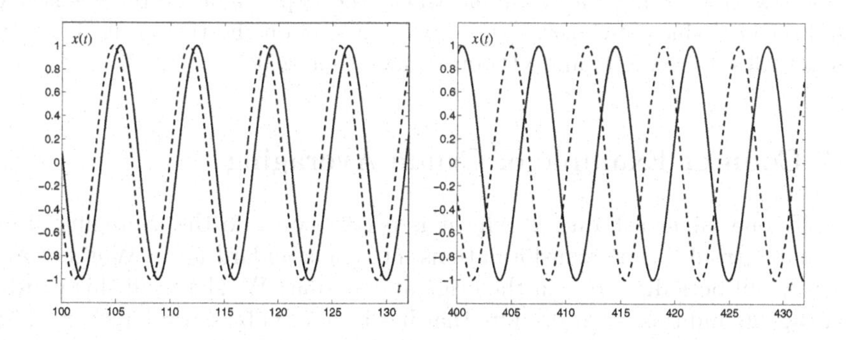

Fig. 2.3: Exact and approximate solutions of $\ddot{x}+x = \varepsilon x$, $\mathrm{x}(0) = 1$, $\dot{\mathrm{x}}(0) = 0$; $\varepsilon = 0.1$. The exact solution has been drawn full line, the asymptotic approximation has been indicated by $- \ - \ -$. Notice that the on the left the interval $100 - 130$ time units has been plotted, on the right the interval $400 - 430$.

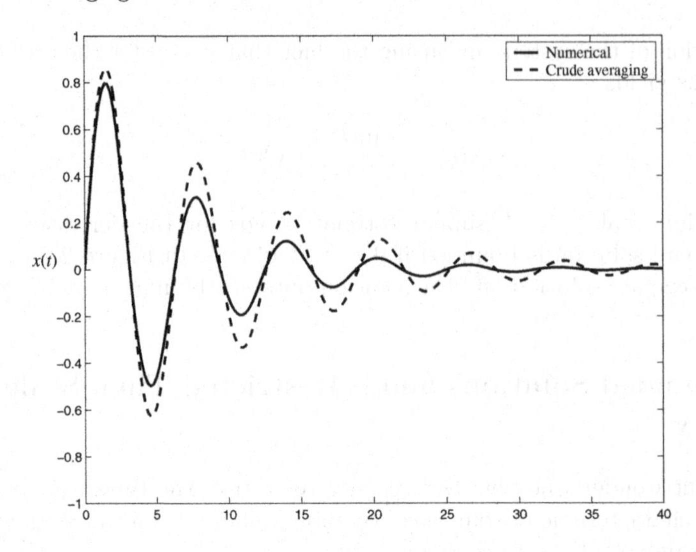

Fig. 2.4: Asymptotic approximation and solution obtained by "crude averaging" of $\ddot{x} + 4\varepsilon\cos^2(t)\dot{x} + x = 0$, $\mathrm{x}(0) = 0$, $\dot{\mathrm{x}}(0) = 1$; $\varepsilon = 0.1$. The numerical solution and the asymptotic approximation nearly coincide and they decay faster than the "crude approximation".

$$\overline{\mathrm{x}}(t) = \cos((1 - \frac{1}{2}\varepsilon)t), \quad \dot{\overline{\mathrm{x}}}(t) = -\sin((1 - \frac{1}{2}\varepsilon)t).$$

Since $\mathrm{x}(t) - \overline{\mathrm{x}}(t) = \mathcal{O}(\varepsilon)$ on the time scale $1/\varepsilon$ and

$$\mathrm{x}(t) = \cos((1 - \varepsilon)^{\frac{1}{2}} t),$$

it follows that we have an example where the approximation on $1/\varepsilon$ is not valid on $1/\varepsilon^2$ since obviously $\mathrm{x}(t) - \overline{\mathrm{x}}(t) = \mathcal{O}_\sharp(1)$ on the time scale $1/\varepsilon^2$. In Figure 2.3 we draw $\mathrm{x}(t)$ and $\overline{\mathrm{x}}(t)$ on various time scales.

2.7 Counter Example of Crude Averaging

Finally one might ask oneself why it is necessary to do the averaging after (perhaps) troublesome transformations into the standard form. Why not average small periodic terms in the original equation? We shall call this **crude averaging** and this is a procedure that has been used by several authors. The following counter example may serve to discourage this. Consider the equation

$$\ddot{x} + 4\varepsilon\cos^2(t)\dot{x} + x = 0,$$

with initial conditions $\mathrm{x}(0) = 0$, $\dot{\mathrm{x}}(0) = 1$. The equation corresponds to an oscillator with linear damping where the friction coefficient oscillates between

0 and 4ε. It seems perfectly natural to average the friction term to produce the equation

$$\ddot{z} + 2\varepsilon\dot{z} + z = 0\ ,\ \mathrm{z}(0) = 0\ ,\ \dot{\mathrm{z}}(0) = 1.$$

We expect $\mathrm{z}(t)$ to be an approximation of $\mathrm{x}(t)$ on some time scale. We have

$$\mathrm{z}(t) = \frac{1}{(1-\varepsilon^2)^{\frac{1}{2}}} e^{-\varepsilon t}\sin((1-\varepsilon^2)^{\frac{1}{2}}t).$$

It turns out that this is a poor result. To see this we do the averaging via the standard form as in Section 2.2. We obtain

$$\dot{\overline{r}} = -\frac{1}{2}\varepsilon\overline{r}(2+\cos(2\overline{\psi})), \quad \overline{\mathrm{r}}(0) = 1,$$
$$\dot{\overline{\psi}} = \frac{1}{2}\varepsilon\sin(2\overline{\psi}), \quad \overline{\psi}(0) = 0$$

and we have $\overline{\mathrm{r}}(t) = e^{-\frac{3}{2}\varepsilon t}$, $\overline{\psi}(t) = 0$. So

$$\mathrm{x}(t) = e^{-\frac{3}{2}\varepsilon t}\sin(t) + \mathcal{O}(\varepsilon)$$

on the time scale $1/\varepsilon$. Actually we shall prove in Chapter 5 that this estimate is valid on $[0,\infty)$. We have clearly

$$\mathrm{x}(t) - \mathrm{z}(t) = \mathcal{O}_\sharp(1)$$

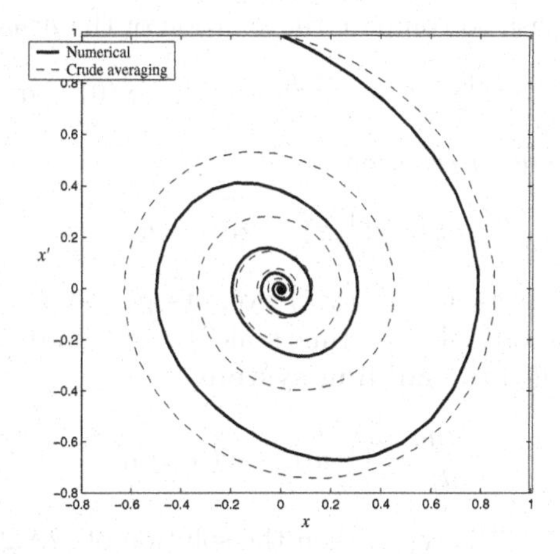

Fig. 2.5: Phase plane for $\ddot{x} + 4\varepsilon\cos^2(t)\dot{x} + x = 0$, $\mathrm{x}(0) = 0$, $\dot{\mathrm{x}}(0) = 1$; $\varepsilon = 0.1$. The phase-orbit of the numerical solution and the asymptotic approximation nearly coincide and have been represented by a full line; the crude approximation has been indicated by $- \ - \ -$.

on the time scale $1/\varepsilon$. In Figure 2.4 we depict $\mathrm{z}(t)$ and $\mathrm{x}(t)$ obtained by numerical integration. We could have plotted $e^{-\frac{3}{2}\varepsilon t}\sin(t)$ but this asymptotic approximation nearly coincides with the numerical solution. It turns out that if $\varepsilon = 0.1$

$$\sup_{t\geq 0} |\mathrm{x}(t) - e^{-\frac{3}{2}\varepsilon t}\sin(t)| \leq 0.015.$$

In Figure 2.5 we illustrate the behavior in the phase plane of the crude and the numerical solution.

2.8 Two Proofs of First-Order Periodic Averaging

In this section we will give two proofs of the basic theorem about first-order averaging, which was stated (somewhat vaguely) in Section 2.1. Both proofs are important, and ideas from both will play a role in the sequel. The proof that we give first is more recent, and is shorter, but relies on an inequality due to Besjes that is not obvious. The second proof is earlier (and so perhaps more natural), with roots in the work of Bogoliubov, but is longer, at least if all of the details are treated carefully. We assume that the differential equations are defined on all of $\mathbb{R}^n$ although it is easy to adapt the arguments to open subsets of $\mathbb{R}^n$. Because certain partial differential equations and functional differential equations can be viewed as ordinary differential equations in a Banach space, we include some remarks about this case, without attempting a complete treatment; see also Appendix E.

Recall that we wish to compare the solution of the *original equation*

$$\dot{\boldsymbol{x}} = \varepsilon \mathbf{f}^1(\boldsymbol{x}, t) + \varepsilon^2 \mathbf{f}^{[2]}(\boldsymbol{x}, t, \varepsilon), \quad \mathbf{x}(0) = \boldsymbol{a}, \tag{2.8.1}$$

with that of the *averaged equation*

$$\dot{\boldsymbol{z}} = \varepsilon \overline{\mathbf{f}}^1(\boldsymbol{z}), \quad \mathbf{z}(0) = \boldsymbol{a}, \tag{2.8.2}$$

where $\overline{\mathbf{f}}^1(\boldsymbol{z})$ is the average of $\mathbf{f}^1(\boldsymbol{z}, t)$ over its period T in t. Observe that introducing the new variable (or "time scale") $\tau = \varepsilon t$ into (2.8.2) removes the ε, giving what is called the **guiding system**

$$\frac{d\boldsymbol{w}}{d\tau} = \overline{\mathbf{f}}^1(\boldsymbol{w}), \quad \mathbf{w}(0) = \boldsymbol{a}. \tag{2.8.3}$$

If the solution of (2.8.3) is $\mathbf{w}(\tau)$, then the solution of (2.8.2) is

$$\mathbf{z}(t, \varepsilon) = \mathbf{w}(\varepsilon t). \tag{2.8.4}$$

That is, t enters into $\mathbf{z}(t, \varepsilon)$ only in the combination εt. For most of this section, we consider the initial point $\boldsymbol{a}$ to be fixed. This will be relaxed in Theorem 2.8.9.

All of the arguments in this section require, as a preliminary step, the choice of a connected, bounded open set D (with compact closure $\overline{D}$) containing $\boldsymbol{a}$, a constant $L > 0$, and a constant $\varepsilon_0 > 0$, such that the solutions $\mathbf{x}(t,\varepsilon)$ and $\mathbf{z}(t,\varepsilon)$ with $0 \le \varepsilon \le \varepsilon_0$ remain in D for $0 \le t \le L/\varepsilon$. (Further restrictions will be placed on ε_0 later.) There are two main ways to achieve this goal.

1. We may pick D and ε_0 arbitrarily (for instance, choosing an interesting region of phase space) and choose L in response to this. Since the right-hand sides of (2.8.1) and (2.8.2) are bounded by a constant times ε (for $0 \le \varepsilon \le \varepsilon_0$ and for $\boldsymbol{x}$ or $\boldsymbol{z}$ in $\overline{D}$), the existence of a suitable L is obvious.
2. Alternatively, L may be chosen arbitrarily and D and ε_0 chosen in response. For instance, if a solution of (2.8.3) exists for $0 \le \tau \le L$, and if D is a neighborhood of this solution segment, then there will exist ε_0 such that the solutions of (2.8.1) and (2.8.2) will remain in D for $0 \le t \le L/\varepsilon$ if $0 \le \varepsilon \le \varepsilon_0$.

All of this is usually abbreviated to a remark that "since $\boldsymbol{x}$ and $\boldsymbol{z}$ move at a rate $\mathcal{O}(\varepsilon)$, they remain bounded for time $\mathcal{O}(1/\varepsilon)$." In the infinite dimensional case, the closure of a bounded open set is not compact, and it is necessary to impose additional boundedness assumptions (for instance, on $\mathbf{f}^{[2]}(\boldsymbol{x},t,\varepsilon)$) at various places in the following arguments.

Recall from Section 1.2 that a periodic vector field of class C^1 satisfies a Lipschitz condition on compact sets for all time. (See Definition 1.2.2 and Lemma 1.2.3.) The Lipschitz property often fails in an infinite-dimensional setting, where even a linear operator can fail to be Lipschitz (for linear operators this is called being "unbounded," meaning unbounded on the unit sphere), but it can be imposed as an added assumption.

Theorem 2.8.1. *Suppose that $\mathbf{f}^1$ is Lipschitz continuous, $\mathbf{f}^{[2]}$ is continuous, and ε_0, D, and L are as above. Then there exists a constant $c > 0$ such that*

$$\|\mathbf{x}(t,\varepsilon) - \mathbf{z}(t,\varepsilon)\| < c\varepsilon$$

for $0 \le \varepsilon \le \varepsilon_0$ and $0 \le t \le L/\varepsilon$.

Proof Let $\mathbf{E}(t,\varepsilon) = \mathbf{x}(t,\varepsilon) - \mathbf{z}(t,\varepsilon) = \mathbf{x}(t,\varepsilon) - \mathbf{w}(\varepsilon t)$ denote the error. Calculating $\mathbf{E}$ from the differential equations for $\mathbf{x}$ and $\mathbf{z}$, and integrating, yields

$$\mathbf{E}(t,\varepsilon) = \varepsilon \int_0^t [\mathbf{f}^1(\mathbf{x}(s,\varepsilon),s) + \varepsilon^2 \mathbf{f}^{[2]}(\mathbf{x}(s,\varepsilon),s,\varepsilon) - \overline{\mathbf{f}}^1(\mathbf{w}(\varepsilon s))]\,\mathrm{d}s.$$

Omitting the arguments of $\mathbf{E}$, $\mathbf{x}$, and $\mathbf{w}$, the integrand may be written as

$$[\mathbf{f}^1(\mathbf{x},s) - \mathbf{f}^1(\mathbf{w},s)] + \varepsilon \mathbf{f}^{[2]}(\mathbf{x},s,\varepsilon) + [\mathbf{f}^1(\mathbf{w},s) - \overline{\mathbf{f}}^1(\mathbf{w})],$$

leading to

$$\|\mathbf{E}\| \leq \varepsilon \int_0^t \|\mathbf{f}^1(\mathbf{x}, s) - \mathbf{f}^1(\mathbf{w}, s)\| \, \mathrm{d}s + \varepsilon^2 \left\| \int_0^t \mathbf{f}^{[2]}(\mathbf{x}, s, e) \, \mathrm{d}s \right\|$$
$$+ \varepsilon \left\| \int_0^t [\mathbf{f}^1(\mathbf{w}, s) - \overline{\mathbf{f}}^1(\mathbf{w})] \, \mathrm{d}s \right\|.$$

In the first integral we use the Lipschitz constant $\lambda_{\mathbf{f}^1}$. Since $\mathbf{f}^{[2]}$ is continuous and periodic it is bounded on $\overline{D}$ for all time. The third integral is bounded by a constant times ε by Lemma 2.8.2 below. Thus we have

$$\|\mathbf{E}(t, \varepsilon)\| \leq \varepsilon \lambda_{\mathbf{f}^1} \int_0^t \|\mathbf{E}(s, \varepsilon)\| \, \mathrm{d}s + c_0 \varepsilon^2 t + c_1 \varepsilon$$

for suitable c_0 and c_1. It follows from the specific Gronwall Lemma 1.3.3 that

$$\|\mathbf{E}(t, \varepsilon)\| \leq \varepsilon (c_0 L + c_1) e^{\lambda_{\mathbf{f}^1} L}$$

for $0 \leq \varepsilon \leq \varepsilon_0$ and $0 \leq t \leq L/\varepsilon$. Taking $c = (c_0 L + c_1) e^{\lambda_{\mathbf{f}^1} L}$, the theorem is proved. □

The preceding proof depends on the following lemma of Besjes [31], applied to $\varphi = \mathbf{f}^1 - \overline{\mathbf{f}}^1$ and $\mathbf{x} = \mathbf{w}(\varepsilon t)$. (The assumption $\dot{\boldsymbol{x}} = \mathcal{O}(\varepsilon)$ is familiar from the beginning of this section. The lemma is stated in this generality for future use.)

Lemma 2.8.2. *Suppose that $\varphi(\boldsymbol{x}, s)$ is periodic in s with period T, has zero mean in s for fixed $\boldsymbol{x}$, is bounded for all s and for $\boldsymbol{x} \in D$, and has Lipschitz constant λ_φ in $\boldsymbol{x}$ for $\boldsymbol{x} \in D$. Suppose that $\mathbf{x}(t, \varepsilon)$ belongs to D for $0 \leq \varepsilon \leq \varepsilon_0$ and $0 \leq t \leq L/\varepsilon$ and satisfies $\dot{\boldsymbol{x}} = \mathcal{O}(\varepsilon)$. Then there is a constant $c_1 > 0$ such that*

$$\left\| \int_0^t \varphi(\mathbf{x}(s, \varepsilon), s) \, \mathrm{d}s \right\| \leq c_1$$

for $0 \leq \varepsilon \leq \varepsilon_0$ and $0 \leq t \leq L/\varepsilon$.

Proof First observe that if $\boldsymbol{x}$ were constant, the result would be trivial, not only for the specified range of t but for all t, because the integral would be periodic and c_1 could be taken to be its amplitude. In fact, $\boldsymbol{x}$ is not constant but varies slowly. We begin by dividing the interval $[0, t]$ into periods $[0, T], [T, 2T], \ldots, [(m-1)T, mT]$ and a leftover piece $[mT, t]$ that is shorter than a period. Then

$$\left\| \int_0^t \varphi(\mathbf{x}(s, \varepsilon), s) \, \mathrm{d}s \right\| \leq \sum_{i=1}^m \left\| \int_{(i-1)T}^{iT} \varphi(\mathbf{x}(s, \varepsilon), s) \, \mathrm{d}s \right\| + \left\| \int_{mT}^t \varphi(\mathbf{x}(s, \varepsilon), s) \, \mathrm{d}s \right\|.$$

Each of the integrals over a period can be estimated as follows (see discussion below):

$$
\begin{aligned}
\left\| \int_{(i-1)T}^{iT} \varphi(\mathbf{x}(s,\varepsilon),s)\,\mathrm{d}s \right\| &= \left\| \int_{(i-1)T}^{iT} [\varphi(\mathbf{x}(s,\varepsilon),s) - \varphi(\mathbf{x}((i-1)T,\varepsilon),s)]\,\mathrm{d}s \right\| \\
&\le \lambda_\varphi \int_{(i-1)T}^{iT} \|\mathbf{x}(s,\varepsilon) - \mathbf{x}((i-1)T,\varepsilon)\|\,\mathrm{d}s \\
&\le \lambda_\varphi \int_{(i-1)T}^{iT} c_2\varepsilon\,\mathrm{d}s \\
&\le \lambda_\varphi c_2 T\varepsilon.
\end{aligned}
$$

(The first equality holds because $\varphi(\mathbf{x}((i-1)T,s,\varepsilon)$ integrates to zero; the bound $\|\mathbf{x}(s,\varepsilon) - \mathbf{x}((i-1)T,\varepsilon)\| \le c_2\varepsilon$, for some c_2, follows from the slow movement of $\mathbf{x}$.) The final integral over a partial period is bounded by the maximum of $\|\varphi\|$ times T; call this c_3. Then

$$
\| \int_0^t \varphi(\mathbf{x}(s,\varepsilon),s)\,\mathrm{d}s\| \le m\lambda_\varphi c_2 T\varepsilon + c_3.
$$

But by the construction, $mT \le t \le L/\varepsilon$, so $m\lambda_\varphi c_2 T\varepsilon + c_3 \le \lambda_\varphi c_2 L + c_3$; take this for c_1. $\square$

The second (and more traditional) proof of first-order averaging introduces the notion of a **near-identity transformation**, which will be important for higher-order approximations in the next section. To avoid detailed hypotheses that must be modified when the order of approximation is changed, we assume that all functions are smooth, that is, infinitely differentiable. A near-identity transformation is actually a family of transformations depending on ε and reducing to the identity when $\varepsilon = 0$. In general, a near-identity transformation has the form

$$
\boldsymbol{x} = \mathbf{U}(\boldsymbol{y},t,\varepsilon) = \boldsymbol{y} + \varepsilon\mathbf{u}^{[1]}(\boldsymbol{y},t,\varepsilon), \tag{2.8.5}
$$

where $\mathbf{u}^{[1]}$ is periodic in t with period T; here $\boldsymbol{y}$ is the new vector variable that will replace $\boldsymbol{x}$. (For our immediate purposes, it is sufficient to take $\mathbf{u}^1(\boldsymbol{y},t)$ in (2.8.5), but the more general form will be used later.) The goal is to choose $\mathbf{u}^{[1]}$ so that (2.8.5) carries the original equation

$$
\dot{\boldsymbol{x}} = \varepsilon\mathbf{f}^1(\boldsymbol{x},t) + \varepsilon^2\mathbf{f}^{[2]}(\boldsymbol{x},t,\varepsilon) \tag{2.8.6}
$$

into the **full averaged equation**

$$
\dot{\boldsymbol{y}} = \varepsilon\overline{\mathbf{f}}^1(\boldsymbol{y}) + \varepsilon^2\mathbf{f}_\star^{[2]}(\boldsymbol{y},t,\varepsilon) \tag{2.8.7}
$$

for some $\mathbf{f}_\star^{[2]}$, induced by the transformation and also periodic in t. Now the *averaged equation* (or for extra clarity **truncated averaged equation**)

$$
\dot{\boldsymbol{z}} = \varepsilon\overline{\mathbf{f}}^1(\boldsymbol{z}) \tag{2.8.8}
$$

is obtained by deleting the last term and changing the variable name from $\boldsymbol{y}$ to $\boldsymbol{z}$. Here $\boldsymbol{z}$ is not a new variable related to $\boldsymbol{x}$ or $\boldsymbol{y}$ by any formula; instead, $\boldsymbol{z}$ is introduced just to distinguish the solutions of (2.8.7) from those of

(2.8.8). The proof of Theorem 2.8.1 using these equations will be broken into several lemmas, for easy reference in later arguments. The first establishes the validity of near-identity transformations, the second the existence of the particular near-identity transformation we need, the third estimates error due to truncation, and the fourth carries this estimate back to the original variables.

Lemma 2.8.3. *Consider (2.8.5) as a smooth mapping $\boldsymbol{y} \mapsto \mathbf{U}(\boldsymbol{y}, t, \varepsilon)$ depending on t and ε. For any bounded connected open set $D \subset \mathbb{R}^n$ there exists ε_0 such that for all $t \in \mathbb{R}$ and for all ε satisfying $0 \leq \varepsilon \leq \varepsilon_0$, this mapping carries D one-to-one onto its image $\mathbf{U}(D, t, \varepsilon)$. The inverse mapping has the form*

$$\boldsymbol{y} = \mathbf{V}(\boldsymbol{x}, t, \varepsilon) = \boldsymbol{x} + \varepsilon \mathbf{v}^{[1]}(\boldsymbol{x}, t, \varepsilon), \tag{2.8.9}$$

and is smooth in $(\boldsymbol{x}, t, \varepsilon)$.

The following proof uses the fact that the mapping $\mathbf{u}^{[1]}$, as in (2.8.5), is Lipschitz on $\overline{D}$ with some Lipschitz constant $\lambda_{\mathbf{u}^{[1]}}$. This will be true in the finite-dimensional case, by the same arguments discussed above in the case of $\mathbf{f}^1$ (Lemma 1.2.3, with an a priori bound on ε). Alternatively, Lemma 2.8.4 below shows that if $\mathbf{f}^1$ is Lipschitz with constant $\lambda_{\mathbf{f}^1}$ then the $\mathbf{u}^1$ that are actually used in averaging can be taken to be Lipschitz with constant $\lambda_{\mathbf{u}^1} = 2\lambda_{\mathbf{f}^1} T$. This second argument can be used in the infinite-dimensional case.

Proof First we show that $\mathbf{U}$ is one-to-one on D for small enough ε. Suppose $\mathbf{U}(\boldsymbol{y}_1, t, \varepsilon) = \mathbf{U}(\boldsymbol{y}_2, t, \varepsilon)$ with $0 \leq \varepsilon \leq 1/\lambda_{\mathbf{u}^1}$. Then $\boldsymbol{y}_1 + \varepsilon \mathbf{u}^{[1]}(\boldsymbol{y}_1, t, \varepsilon) = \boldsymbol{y}_2 + \varepsilon \mathbf{u}^{[1]}(\boldsymbol{y}_2, t, \varepsilon)$, so $\|\boldsymbol{y}_2 - \boldsymbol{y}_1\| = \varepsilon \|\mathbf{u}^{[1]}(\boldsymbol{y}_1, t, \varepsilon) - \mathbf{u}^{[1]}(\boldsymbol{y}_2, t, \varepsilon)\| \leq \varepsilon \lambda_{\mathbf{u}^1} \|\boldsymbol{y}_1 - \boldsymbol{y}_2\|$. If $\varepsilon \lambda_{\mathbf{u}^1} < 1$, we have shown that unless $\|\boldsymbol{y}_2 - \boldsymbol{y}_1\|$ vanishes, it is less than itself. Therefore $\boldsymbol{y}_1 = \boldsymbol{y}_2$, and $\mathbf{U}$ is one-to-one for $0 \leq \varepsilon \leq 1/\lambda_{\mathbf{u}^1}$. It follows that $\mathbf{U}$ maps D invertibly onto $\mathbf{U}(D, t, \varepsilon)$. It remains to check the smoothness and form of the inverse.

Since $\mathsf{D}_{\boldsymbol{y}} \mathbf{U}(\boldsymbol{y}, t, 0)$ is the identity matrix, the implicit function theorem implies that $\boldsymbol{x} = \mathbf{U}(\boldsymbol{y}, t, \varepsilon)$ is locally smoothly invertible in the form (2.8.9) for small enough ε. More precisely: each $\boldsymbol{y}_0 \in \mathbb{R}^n$ has a neighborhood on which $\mathbf{U}$ is invertible for ε in an interval that depends on $\boldsymbol{y}_0$. Since the closure of D is compact, it can be covered by a finite number k of these neighborhoods, with bounds $\varepsilon_1, \ldots, \varepsilon_k$ on ε. Let ε_0 be the minimum of $1/\lambda_{\mathbf{u}^1}, \varepsilon_1, \ldots, \varepsilon_k$. Then for $0 \leq \varepsilon \leq \varepsilon_0$ the local inverses (which are smooth and have the desired form) exist and must coincide with the global inverse obtained in the last paragraph.

□

Lemma 2.8.4. *There exist mappings $\mathbf{U}$ (not unique) such that (2.8.5) carries (2.8.6) to (2.8.7). In particular, $\mathbf{u}^{[1]}$ may be taken to have Lipschitz constant $2\lambda_{\mathbf{f}^1} T$ (where T is the period).*

Notation 2.8.5 *We use the notation $\mathsf{D}\mathbf{f} \cdot \mathbf{g}$ for the multiplication of $\mathbf{g}$ by the derivative of $\mathbf{f}$, not for the gradient of the inner product of $\mathbf{f}$ and $\mathbf{g}$!*

Proof If equations $\dot{\boldsymbol{x}} = \varepsilon \mathbf{f}^{[1]}(\boldsymbol{x}, t, \varepsilon)$ and $\dot{\boldsymbol{y}} = \varepsilon \mathbf{g}^{[1]}(\boldsymbol{y}, t, \varepsilon)$ are related by the coordinate change $\boldsymbol{x} = \mathbf{U}(\boldsymbol{y}, t, \varepsilon)$, then the chain rule implies that $\mathbf{f}^{[1]} = \mathbf{U}_t + \mathsf{D}\mathbf{U} \cdot \mathbf{g}^{[1]}$, or, stated more carefully, $\mathbf{f}^{[1]}(\mathbf{U}(\boldsymbol{y}, t, \varepsilon), t, \varepsilon) = \mathbf{U}_t(\boldsymbol{y}, t, \varepsilon) + \mathsf{D}\mathbf{U}(\boldsymbol{y}, t, \varepsilon) \cdot \mathbf{g}^{[1]}(\boldsymbol{y}, t, \varepsilon)$. Substituting the forms of $\mathbf{U}$, $\mathbf{f}^{[1]}$, and $\mathbf{g}^{[1]}$ given in (2.8.5), (2.8.6), and (2.8.7) and extracting the leading-order term in ε gives

$$\mathbf{u}_t^1(\boldsymbol{y}, t) = \mathbf{f}^1(\boldsymbol{y}, t) - \overline{\mathbf{f}}^1(\boldsymbol{y}), \tag{2.8.10}$$

often called the **homological equation** of averaging theory. In deriving (2.8.10) we have assumed what we want to prove, that is, that the desired $\mathbf{u}^1$ exists. The actual proof follows by reversing the steps. Consider (2.8.10) as an equation to be solved for $\mathbf{u}^1$. Since the right-hand side of (2.8.10) has zero mean, the function

$$\mathbf{u}^1(\boldsymbol{y}, t) = \int_0^t [\mathbf{f}^1(\boldsymbol{y}, s) - \overline{\mathbf{f}}^1(\boldsymbol{y})] \,\mathrm{d}s + \boldsymbol{\kappa}^1(\boldsymbol{y}) \tag{2.8.11}$$

will be periodic in t for any choice of the function $\boldsymbol{\kappa}^1$.

Now return to the chain rule calculation at the beginning of this proof, taking $\mathbf{f}^{[1]}$ as in (2.8.6), $\mathbf{u}^1$ as in (2.8.11), and considering $\mathbf{g}^{[1]}$ as to be determined. It is left to the reader to check that $\dot{\boldsymbol{y}} = \varepsilon \mathbf{g}^{[1]}$ must have the form (2.8.7) for some $\mathbf{f}_\star^{[2]}$. Finally, we check that if $\boldsymbol{\kappa}^1(\boldsymbol{y}) = 0$ then $\mathbf{u}^1$ has Lipschitz constant $2\lambda_{\mathbf{f}^{[1]}} T$. It is easy to check that $\overline{\mathbf{f}}^1$ has the same Lipschitz constant as $\mathbf{f}^1$. Since $\mathbf{u}^1$ is periodic in t, for each t there exists $t' \in [0, T]$ such that $\mathbf{u}^1(\boldsymbol{y}, t) = \mathbf{u}^1(\boldsymbol{y}, t')$. Then

$$\begin{aligned}
&\left\|\mathbf{u}^1(\boldsymbol{y}_1, t) - \mathbf{u}^1(\boldsymbol{y}_2, t)\right\| = \left\|\mathbf{u}^1(\boldsymbol{y}_1, t') - \mathbf{u}^1(\boldsymbol{y}_2, t')\right\| \\
&\le \int_0^{t'} \left[\left\|\mathbf{f}^1(\boldsymbol{y}_1, s) - \mathbf{f}^1(\boldsymbol{y}_2, s)\right\| + \left\|\overline{\mathbf{f}}^1(\boldsymbol{y}_1) - \overline{\mathbf{f}}^1(\boldsymbol{y}_2)\right\|\right] \mathrm{d}s \\
&\le \int_0^{t'} 2\lambda_{\mathbf{f}^{[1]}} \left\|\boldsymbol{y}_1 - \boldsymbol{y}_2\right\| \,\mathrm{d}s = 2\lambda_{\mathbf{f}^{[1]}} t' \left\|\boldsymbol{y}_1 - \boldsymbol{y}_2\right\| \\
&\le 2\lambda_{\mathbf{f}^{[1]}} T \left\|\boldsymbol{y}_1 - \boldsymbol{y}_2\right\|,
\end{aligned}$$

and this proves the lemma. □

The simplest way to resolve the ambiguity of (2.8.11) is to choose $\boldsymbol{\kappa}^1(\boldsymbol{y}) = 0$. (Warning: Taking $\boldsymbol{\kappa}^1(\boldsymbol{y}) = 0$ is not the same as taking $\mathbf{u}^1$ to have zero mean, which is another attractive choice, especially if the periodic functions are written as Fourier series.) Taking $\boldsymbol{\kappa}^1(\boldsymbol{y}) = 0$ has the great advantage that it makes $\mathbf{U}(\boldsymbol{y}, 0, \varepsilon) = \boldsymbol{y}$, so that initial conditions (at time $t = 0$) need not be transformed when changing coordinates from $\boldsymbol{x}$ to $\boldsymbol{y}$. In addition, $\mathbf{U}(\boldsymbol{y}, mT, \varepsilon) = \boldsymbol{y}$ at each **stroboscopic time** mT (for integers m). For this reason, choosing $\boldsymbol{\kappa}^1(\boldsymbol{y}) = 0$ is called **stroboscopic averaging**.

Remark 2.8.6. We mention for later use here that the composition of two such transformations again fixes the initial conditions. If we allow this type of transformation only it is the choice of a subgroup of all the formal near-identity transformations. ♡

We now introduce the following specific solutions:

1. $\mathbf{x}(t,\varepsilon)$ denotes the solution of (2.8.6) with initial condition $\mathbf{x}(0,\varepsilon)=\boldsymbol{a}$.
2. $\mathbf{y}(t,\varepsilon)$ denotes the solution of (2.8.7) with initial condition
$$\mathbf{y}(0,\varepsilon)=\mathbf{V}(\boldsymbol{a},0,\varepsilon)=\boldsymbol{a}+\varepsilon\mathbf{v}^{[1]}(\boldsymbol{a},0,\varepsilon)=\boldsymbol{a}+\varepsilon\mathbf{b}(\varepsilon). \tag{2.8.12}$$
If stroboscopic averaging is used, this reduces to $\mathbf{y}(0,\varepsilon)=\boldsymbol{a}$. Notice that
$$\mathbf{x}(t,\varepsilon)=\mathbf{U}(\mathbf{y}(t,\varepsilon),t,\varepsilon). \tag{2.8.13}$$
3. $\mathbf{z}(t,\varepsilon)$ denotes the solution of (2.8.8) with $\mathbf{z}(t,0)=\boldsymbol{a}$. Notice the double truncation involved here: both the differential equation and (in the nonstroboscopic case) the initial condition for $\mathbf{y}$ are truncated to obtain the differential equation and initial condition for $\mathbf{z}$. This solution $\mathbf{z}(t,\varepsilon)$ is traditionally called the **first approximation** to $\mathbf{x}(t,\varepsilon)$.
4. $\mathbf{U}(\mathbf{z}(t,\varepsilon),t,\varepsilon)$ is often called the **improved first approximation** to $\mathbf{x}(t,\varepsilon)$. It is natural to regard $\mathbf{z}(t,\varepsilon)$ as an approximation to $\mathbf{y}(t,\varepsilon)$, and in view of (2.8.13) it seems natural to use $\mathbf{U}(\mathbf{z}(t,\varepsilon),t,\varepsilon)$ as an approximation to $\mathbf{x}(t,\varepsilon)$. But $\mathbf{z}(t,\varepsilon)$ is already an $\mathcal{O}(\varepsilon)$-approximation to $\mathbf{x}(t,\varepsilon)$ for time $\mathcal{O}(1/\varepsilon)$, and applying $\mathbf{U}$ makes an $\mathcal{O}(\varepsilon)$ change, *so the order of approximation is not actually improved.* (This point will become clearer when we consider higher-order averaging in the next section.)

The next lemma and theorem estimate certain differences between these solutions. The reader is invited to replace the order symbols $\mathcal{O}$ by more precise statements as in Theorem 2.8.1.

Lemma 2.8.7. *The solutions* $\mathbf{y}(t,\varepsilon)$ *and* $\mathbf{z}(t,\varepsilon)$ *defined above satisfy*
$$\|\mathbf{y}(t,\varepsilon)-\mathbf{z}(t,\varepsilon)\|=\mathcal{O}(\varepsilon)$$
for time $\mathcal{O}(1/\varepsilon)$.

Proof We have
$$\mathbf{y}(t,\varepsilon)=\boldsymbol{a}+\varepsilon\mathbf{b}^{[1]}(\varepsilon)+\int_0^t[\varepsilon\overline{\mathbf{f}}^1(\mathbf{y}(s,\varepsilon),s)+\varepsilon^2\mathbf{f}_\star^{[2]}(\mathbf{y}(s,\varepsilon),s,\varepsilon)]\,\mathrm{d}s,$$
and
$$\mathbf{z}(t,\varepsilon)=\boldsymbol{a}+\int_0^t\varepsilon\overline{\mathbf{f}}^1(\mathbf{z}(s,\varepsilon))\,\mathrm{d}s.$$
Letting $\mathbf{E}(t,\varepsilon)=\mathbf{y}(t,\varepsilon)-\mathbf{z}(t,\varepsilon)$, it follows that
$$\|\mathbf{E}(t,\varepsilon)\|\le\varepsilon\|\mathbf{b}(\varepsilon)\|+\varepsilon\lambda_{\overline{\mathbf{f}}^1}\int_0^t\|\mathbf{E}(s,\varepsilon)\|\,\mathrm{d}s+\varepsilon^2Mt,$$
where M is the bound for $\mathbf{f}_\star^{[2]}$ on $\overline{D}$. The theorem follows from the specific Gronwall inequality Lemma 1.3.3. □

Theorem 2.8.8. *The solutions* $\mathbf{x}(t,\varepsilon)$ *and* $\mathbf{z}(t,\varepsilon)$ *defined above satisfy the estimate* $\|\mathbf{x}(t,\varepsilon)-\mathbf{z}(t,\varepsilon)\| = \mathcal{O}(\varepsilon)$ *for time* $\mathcal{O}(1/\varepsilon)$.

This reproves Theorem 2.8.1.

Proof By the triangle inequality, $\|\mathbf{x}(t,\varepsilon)-\mathbf{z}(t,\varepsilon)\| \le \|\mathbf{x}(t,\varepsilon)-\mathbf{y}(t,\varepsilon)\| + \|\mathbf{y}(t,\varepsilon)-\mathbf{z}(t,\varepsilon)\|$. The first term is $\mathcal{O}(\varepsilon)$ for all time by (2.8.13) and (2.8.5), and the second is $\mathcal{O}(\varepsilon)$ for time $\mathcal{O}(1/\varepsilon)$ by Lemma 2.8.7. □

An important variation of the basic averaging theorem deals with the uniformity of the estimate with respect to the initial conditions: if $\boldsymbol{a}$ is varied in D, can we use the same c and L? The best answer seems to be the following. The proof requires only small changes (or see [201, Theorem 6.2.3]).

Theorem 2.8.9. *Let* K *be a compact subset of* $\mathbb{R}^n$ *and let* D *be a bounded open connected subset containing* K *in its interior. Let* L *be given arbitrarily, and for each* $\boldsymbol{a} \in K$ *let* $L_{\boldsymbol{a}}$ *be the largest real number less than or equal to* L *such that* $w(\boldsymbol{a},\tau)$ *belongs to* K *for* $0 \le \tau \le L_{\boldsymbol{a}}$. *Then there exist* c *and* ε_0 *such that*

$$\|\mathbf{x}(\boldsymbol{a},t,\varepsilon)-\mathbf{z}(\boldsymbol{a},t,\varepsilon)\| < c\varepsilon$$

for $0 \le t \le L_{\boldsymbol{a}}/\varepsilon$ *and* $0 \le \varepsilon \le \varepsilon_0$.

2.9 Higher-Order Periodic Averaging and Trade-Off

Averaging of order k, described in this section, has two purposes: to obtain $\mathcal{O}(\varepsilon^k)$ error estimates valid for time $\mathcal{O}(1/\varepsilon)$, with $k > 1$, and (under more restrictive assumptions) to obtain (weaker) $\mathcal{O}(\varepsilon^{k-j})$ error estimates for (longer) time $\mathcal{O}(1/\varepsilon^{j+1})$. In the latter case we say that we have **traded off** j orders of accuracy for longer time of validity. (This idea of trade-off arises more naturally in the theory of Poincaré–Lindstedt expansions, as briefly explained in Section 3.5 below.)

2.9.1 Higher-Order Periodic Averaging

We continue with the simplifying assumption that all functions are smooth and defined on $\mathbb{R}^n$. There are once again two ways of proving the main theorem, one due to Ellison, Sáenz, and Dumas ([84]) using the Besjes inequality (Lemma 2.8.2) and a traditional one along the lines initiated by Bogoliubov. However, this time both proofs make use of near-identity transformations. The following lemma generalizes Lemma 2.8.4, and is formulated to include what is needed in both types of proofs.

Lemma 2.9.1. *Given the system*

$$\dot{\boldsymbol{x}} = \varepsilon\mathbf{f}^1(\boldsymbol{x},t) + \cdots + \varepsilon^k\mathbf{f}^k(\boldsymbol{x},t) + \varepsilon^{k+1}\mathbf{f}^{[k+1]}(\boldsymbol{x},t,\varepsilon), \qquad (2.9.1)$$

with period T *in* t, *there exists a transformation*

$$\boldsymbol{x} = \mathbf{U}(\boldsymbol{y}, t, \varepsilon) = \boldsymbol{y} + \varepsilon \mathbf{u}^1(\boldsymbol{y}, t) + \cdots + \varepsilon^k \mathbf{u}^k(\boldsymbol{y}, t), \tag{2.9.2}$$

also periodic, such that

$$\dot{\boldsymbol{y}} = \varepsilon \mathbf{g}^1(\boldsymbol{y}) + \cdots + \varepsilon^k \mathbf{g}^k(\boldsymbol{y}) + \varepsilon^{k+1} \mathbf{g}^{[k+1]}(\boldsymbol{y}, t, \varepsilon). \tag{2.9.3}$$

Here $\mathbf{g}^1$ *equals the average* $\overline{\mathbf{f}}^1$ *of* $\mathbf{f}^1$, *and* $\mathbf{g}^2, \ldots, \mathbf{g}^k$ *are independent of* t *but not unique (since they depend on choices made in obtaining the* $\mathbf{u}^i$*). There is an algorithm to compute these functions in the following order:* $\mathbf{g}^1, \mathbf{u}^1, \mathbf{g}^2, \mathbf{u}^2, \ldots, \mathbf{g}^k, \mathbf{u}^k$. *In particular, it is possible to compute the (autonomous) truncated averaged equation*

$$\dot{\boldsymbol{z}} = \varepsilon \mathbf{g}^1(\boldsymbol{z}) + \cdots + \varepsilon^k \mathbf{g}^k(\boldsymbol{z}) \tag{2.9.4}$$

without computing the last term of (2.9.2). If the shorter transformation

$$\boldsymbol{\xi} = \widehat{\mathbf{U}}(\boldsymbol{z}, t, \varepsilon) = \boldsymbol{z} + \varepsilon \mathbf{u}^1(\boldsymbol{z}, t) + \cdots + \varepsilon^{k-1} \mathbf{u}^{k-1}(\boldsymbol{z}, t) \tag{2.9.5}$$

is applied to (2.9.4), the result is the following modification of the original equation, in which $\mathbf{h}^k(\cdot, t)$ *has zero mean:*

$$\dot{\boldsymbol{\xi}} = \varepsilon \mathbf{f}^1(\boldsymbol{\xi}, t) + \cdots + \varepsilon^k [\mathbf{f}^k(\boldsymbol{\xi}, t) + \mathbf{h}^k(\boldsymbol{\xi}, t)] + \varepsilon^{k+1} \widetilde{\mathbf{f}}^{[k+1]}(\boldsymbol{\xi}, t, \varepsilon). \tag{2.9.6}$$

Proof As in Lemma 2.8.3, the transformation (2.9.2) is invertible and defines a legitimate coordinate change. When this coordinate change is applied to (2.9.1) the result has the form (2.9.3), except that in general the $\mathbf{g}^j$ will depend on t. The calculations are messy, and are best handled in a manner to be described in Section 3.2 below, but for each j the result has the form $\mathbf{g}^j = \mathbf{K}^j + \partial \mathbf{u}^j / \partial t$, where $\mathbf{K}^j$ is a function built from $\mathbf{f}^1, \ldots, \mathbf{f}^j$, the previously calculated $\mathbf{u}^1, \ldots, \mathbf{u}^{j-1}$, and their derivatives. The first two of the $\mathbf{K}^j$ are given by

$$\begin{pmatrix} \mathbf{K}^1(\boldsymbol{y}, t) \\ \mathbf{K}^2(\boldsymbol{y}, t) \end{pmatrix} = \begin{pmatrix} \mathbf{f}^1(\boldsymbol{y}, t) \\ \mathbf{f}^2(\boldsymbol{y}, t) + \mathsf{D}_{\boldsymbol{y}} \mathbf{f}^1(\boldsymbol{y}, t) \cdot \mathbf{u}^1(\boldsymbol{y}, t) - \mathsf{D}_{\boldsymbol{y}} \mathbf{u}^1(\boldsymbol{y}, t) \cdot \mathbf{g}^1(\boldsymbol{y}) \end{pmatrix}. \tag{2.9.7}$$

(This recursive expression assumes $\mathbf{g}^1$ is calculated before $\mathbf{K}^2$ is formed. That is, $\mathbf{g}^1$ may be replaced by $\mathbf{f}^1 - \mathbf{u}^1_t$.) Thus, if $\mathbf{g}^j$ is to be independent of t, we must have

$$\frac{\partial \mathbf{u}^j}{\partial t}(\boldsymbol{y}, t) = \mathbf{K}^j(\boldsymbol{y}, t) - \mathbf{g}^j(\boldsymbol{y}). \tag{2.9.8}$$

This *homological equation* has the same form as (2.8.10), and is solvable in the same way: take $\mathbf{g}^j = \overline{\mathbf{K}}^j$, so that the right-hand side has zero mean, and integrate with respect to t. This determines $\mathbf{u}^j$ up to an additive "constant" $\boldsymbol{\kappa}^2(\boldsymbol{y})$; after the first case $j = 1$ (where $\mathbf{K}^1 = \mathbf{f}^1$ and the homological equation is the same as in first-order averaging) the previously chosen constants enter into $\mathbf{K}^j$, making $\mathbf{g}^j$ nonunique. The remainder of the proof is to reverse the steps and check that, with the $\mathbf{g}^j$ and $\mathbf{u}^j$ constructed in this way, (2.9.2) actually carries (2.9.1) into (2.9.3) for some $\mathbf{g}^{[k+1]}$. In the case of (2.9.5) and (2.9.6) the last homological equation is replaced by

$$0 = \mathbf{K}^k + \mathbf{h}^k - \mathbf{g}^k.$$

(This depends upon the internal structure of $\mathbf{K}^k$. In fact $\mathbf{K}^k = \mathbf{f}^k+$ terms independent of $\mathbf{f}^k$, so that when $\mathbf{h}^k$ is added to $\mathbf{f}^k$ it is also added to $\mathbf{K}^k$.) Taking $\mathbf{g}^k = \overline{\mathbf{K}}^k$ as usual, it follows that $\mathbf{h}^k = \mathbf{K}^k - \overline{\mathbf{K}}^k$ has zero mean. □

Choosing $\boldsymbol{\kappa}^j(\boldsymbol{y}) = 0$ for all j once again leads to "stroboscopic" averaging, in which both the "short" and "long" transformations ($\widehat{\mathbf{U}}$ and $\mathbf{U}$) reduce to the identity at stroboscopic times. If stroboscopic averaging is used, the natural way to construct an approximation to the solution $\mathbf{x}(\boldsymbol{a}, t, \varepsilon)$ of (2.9.1) with initial value $\mathbf{x}(\boldsymbol{a}, 0, \varepsilon) = \boldsymbol{a}$ is to solve the truncated averaged equation (2.9.4) with $\mathbf{z}(\boldsymbol{a}, 0, \varepsilon) = \boldsymbol{a}$, and pass this solution $\mathbf{z}(\boldsymbol{a}, t, \varepsilon)$ back through the "short" transformation $\widehat{\mathbf{U}}$ to obtain

$$\boldsymbol{\xi}(\boldsymbol{a}, t, \varepsilon) = \widehat{\mathbf{U}}(\mathbf{z}(\boldsymbol{a}, t, \varepsilon), t, \varepsilon). \tag{2.9.9}$$

This is an important difference between first- and higher-order averaging: it is not possible to use $\mathbf{z}(\boldsymbol{a}, t, \varepsilon)$ directly as an approximation to $\mathbf{x}(\boldsymbol{a}, t, \varepsilon)$. (In the first-order case, we did not need to define $\widehat{\mathbf{U}}$ because it reduces to the identity.) If stroboscopic averaging is not used, the $\mathbf{z}$ equation must be solved not with initial condition $\boldsymbol{a}$, but with an ε-dependent initial condition $\mathbf{V}(\boldsymbol{a}, 0, \varepsilon)$, where $\boldsymbol{y} = \mathbf{V}(\boldsymbol{x}, t, \varepsilon)$ is the inverse of the coordinate change $\mathbf{U}$. (It is not necessary to calculate $\mathbf{V}$ exactly, only its power series in ε to sufficient order, which can be done recursively.) In the following proofs we assume stroboscopic averaging for convenience, but the theorems remain true in the general case.

Theorem 2.9.2. *The exact solution* $\mathbf{x}(\boldsymbol{a}, t, \varepsilon)$ *and its approximation* $\boldsymbol{\xi}(\boldsymbol{a}, t, \varepsilon)$*, defined above, are related by*

$$\|\mathbf{x}(\boldsymbol{a}, t, \varepsilon) - \boldsymbol{\xi}(\boldsymbol{a}, t, \varepsilon)\| = \mathcal{O}(\varepsilon^k)$$

for time $\mathcal{O}(1/\varepsilon)$ *and* ε *small.*

Proof For a proof using the Besjes inequality, write

$$\mathsf{J}^k_\varepsilon \mathbf{f}^{[1]} = \varepsilon \mathbf{f}^1(\boldsymbol{x}, t) + \cdots + \varepsilon^k \mathbf{f}^k(\boldsymbol{x}, t),$$

so that (2.9.6) reads

$$\dot{\boldsymbol{\xi}} = \mathsf{J}^k_\varepsilon \mathbf{f}^{[1]}(t, \boldsymbol{\xi}, \varepsilon) + \varepsilon^k \mathbf{h}^k(t, \boldsymbol{\xi}) + \varepsilon^{k+1} \widetilde{\mathbf{f}}^{[k+1]}(t, \boldsymbol{\xi}, \varepsilon).$$

Let $\mathbf{E}(t, \varepsilon) = \mathbf{x}(\boldsymbol{a}, t, \varepsilon) - \boldsymbol{\xi}(\boldsymbol{a}, t, \varepsilon)$. Then

$$\|\mathbf{E}\| \leq \varepsilon \lambda_{\mathsf{J}^k_\varepsilon \mathbf{f}^{[1]}} \int_0^t \|\mathbf{E}(s, \varepsilon)\| \, \mathrm{d}s + \varepsilon^k \left\| \int_0^t \mathbf{h}^k(\boldsymbol{\xi}, s) \, \mathrm{d}s \right\|$$
$$+ \varepsilon^{k+1} \int_0^t \|\mathbf{f}^{[k+1]}(\mathbf{x}, s, \varepsilon) - \widetilde{\mathbf{f}}^{[k+1]}(\boldsymbol{\xi}, s, \varepsilon)\| \, \mathrm{d}s,$$

and the remainder of the proof is like that of Theorem 2.8.1. Complete details, including minimal hypotheses on the $\mathbf{f}^j$, are given in [84]. See Appendix E, [28] and [27] for an application of this argument to an infinite-dimensional case.

For a traditional proof along the lines of Lemma 2.8.7 and Theorem 2.8.8, one first shows by Gronwall that $\|\mathbf{y}-\mathbf{z}\| = \mathcal{O}(\varepsilon^k)$ for time $\mathcal{O}(1/\varepsilon)$. Next, writing $\mathbf{U}(\mathbf{y})$ for $\mathbf{U}(\mathbf{y}(\boldsymbol{a},t,\varepsilon),t,\varepsilon)$ and similarly for other expressions, we have $\boldsymbol{x} = \mathbf{U}(\boldsymbol{y})$, $\boldsymbol{\xi} = \widehat{\mathbf{U}}(\mathbf{z})$, and therefore

$$\|\mathbf{x}-\boldsymbol{\xi}\| \le \|\mathbf{U}(\mathbf{y}) - \widehat{\mathbf{U}}(\mathbf{y})\| + \|\widehat{\mathbf{U}}(\mathbf{y}) - \widehat{\mathbf{U}}(\mathbf{z})\|.$$

By the definitions of $\mathbf{U}$ and $\widehat{\mathbf{U}}$, the first term is $\mathcal{O}(\varepsilon^k)$ for as long as $\mathbf{y}$ remains in D, which is at least for time $\mathcal{O}(1/\varepsilon)$. The second term is $\mathcal{O}(\varepsilon^k)$ for time $\mathcal{O}(1/\varepsilon)$, by applying the Lipschitz constant of $\widehat{\mathbf{U}}$ to the previous estimate for $\|\mathbf{y}-\mathbf{z}\|$. □

For the reader's convenience, we restate this theorem in the second order case with complete formulas. Beginning with the system

$$\dot{\boldsymbol{x}} = \varepsilon \mathbf{f}^1(\boldsymbol{x},t) + \varepsilon^2 \mathbf{f}^2(\boldsymbol{x},t) + \varepsilon^3 \mathbf{f}^{[3]}(\boldsymbol{x},t,\varepsilon),$$

put $\mathbf{g}^1(\boldsymbol{y}) = \overline{\mathbf{f}}^1(\boldsymbol{y})$. Then put

$$\mathbf{u}^1(\boldsymbol{y},t) = \int_0^t [\mathbf{f}^1(\boldsymbol{y},s) - \mathbf{g}^1(\boldsymbol{y})]\,\mathrm{d}s,$$

define $\mathbf{K}^2$ as in (2.9.7), and set $\mathbf{g}^2(\boldsymbol{y}) = \overline{\mathbf{K}}^2(\boldsymbol{y})$. Let $\mathbf{z}(t,\varepsilon)$ be the solution of

$$\dot{\boldsymbol{z}} = \varepsilon \mathbf{g}^1(\boldsymbol{z}) + \varepsilon^2 \mathbf{g}^2(\boldsymbol{z}), \quad \mathbf{z}(0) = \mathbf{a}.$$

Then the solution of the original problem is

$$\mathbf{x}(t,\varepsilon) = \mathbf{z}(t,\varepsilon) + \varepsilon \mathbf{u}^1(\mathbf{z}(t,\varepsilon),t) + \mathcal{O}(\varepsilon^2)$$

for time $\mathcal{O}(1/\varepsilon)$.

Summary 2.9.3 *We have presented two ways of justifying first-order and higher-order averaging, a more recent approach due to Besjes (in first-order averaging) and Ellison, Sáenz, and Dumas (in higher-order averaging), and a traditional one going back to Bogoliubov. What are the advantages and disadvantages of the two proof methods? In our judgment, the recent proof is best as far as the error estimate itself is concerned, but the traditional proof is better for qualitative considerations:*

1. *The recent proof uses only* $\widehat{\mathbf{U}}$ *(which must appear in any proof since it is needed to define the approximate solution) and not* $\mathbf{U}$*. Furthermore the error estimate does not use the Lipschitz constant for* $\widehat{\mathbf{U}}$*, as does the traditional proof.*

2. *The recent proof is applicable in certain infinite-dimensional settings where the traditional proof is not. To see this, notice that for averaging of order* $k > 1$ *we have not proved the existence of a Lipschitz constant for* $\widehat{\mathbf{U}}$ *in the infinite-dimensional case. (The question does not arise for first-order averaging, since there* $\widehat{\mathbf{U}}$ *is the identity.) The only apparent way to do a proof would be to imitate the proof of the Lipschitz constant* $2LT\varepsilon$ *for* $\mathbf{U}$ *in Lemma 2.8.4, but the calculation of* $\mathbf{U}$ *for* $k > 1$ *is too complicated and a Lipschitz condition on* $\mathbf{f}^{[1]}$ *does not appear to be passed on to* $\mathbf{U}$ *or* $\widehat{\mathbf{U}}$*. It is true that a Lipschitz constant for* $\mathbf{U}$ *or* $\widehat{\mathbf{U}}$ *still seems to be necessary to prove the invertibility of* $\mathbf{U}$ *or* $\widehat{\mathbf{U}}$ *(as in Lemma 2.8.3), but as Ellison, Sáenz, and Dumas point out, although this invertibility is nice, it is not strictly necessary for the validity of the error estimate. Without it, what we have shown is that (2.9.4), viewed as a vector field on* $\mathbb{R}^n \times S^1$*, is* $\widehat{\mathbf{U}}$*-related to (2.9.6). (In differential geometry, the pushforward of a vector field on a manifold* M *by a noninvertible map* $F : M \to N$ *is not a vector field, but if a vector field on* N *exists such that the pushforward of each individual vector in the field on* M *belongs to the field on* N*, the two fields are called* F*-related.) If we use stroboscopic averaging, there is no need to use an inverse map to match up the initial conditions, and the error estimate is still valid. A noninvertible map cannot be used for the conjugacy and shadowing results in Chapter 6.*
3. *The traditional proof involves solutions of three differential equations, for* $\mathbf{x}$, $\mathbf{y}$, *and* $\mathbf{z}$, *and provides a smooth conjugacy* $\mathbf{U}$ *sending* $\mathbf{y}$ *to* $\mathbf{x}$*. In Chapter 6 we will see that qualitative properties of* $\mathbf{z}$ *can be passed along to* $\mathbf{y}$ *by implicit function arguments based on truncation, and then passed to* $\mathbf{x}$ *via the conjugacy. The recent proof has a conjugacy* $\widehat{\mathbf{U}}$ *carrying* $\mathbf{z}$ *to* $\boldsymbol{\xi}$*, but the passage from* $\boldsymbol{\xi}$ *to* $\mathbf{x}$ *has not been studied with regard to its effect on qualitative properties. For this reason the traditional setup will be used in Chapter 6.*

2.9.2 Estimates on Longer Time Intervals

To obtain trade-off estimates for longer time intervals, it is necessary to assume that part of the averaged equation vanishes.

Theorem 2.9.4. *With the notation of Lemma 2.9.1, suppose that* $\mathbf{g}^1 = \mathbf{g}^2 = \cdots = \mathbf{g}^{\ell-1} = 0$, *(where* $\ell \leq k$*) . Then solutions of (2.9.1), (2.9.3), and (2.9.4) exist for time* $\mathcal{O}(1/\varepsilon^\ell)$*, and for each integer* $j = 0, 1, \ldots, \ell-1$ *the exact solution* $x(a, t, \varepsilon)$ *and the approximate solution* $\xi(a, t, \varepsilon)$ *defined by (2.9.9) satisfy the estimate*

$$\|\mathbf{x}(\boldsymbol{a}, t, \varepsilon) - \boldsymbol{\xi}(\boldsymbol{a}, t, \varepsilon)\| = \mathcal{O}(\varepsilon^{k-j})$$

for time $\mathcal{O}(1/\varepsilon^{j+1})$.

Proof Since the solutions x, y, and z of the three equations mentioned all move at a rate $\mathcal{O}(\varepsilon^\ell)$ on any compact set, the solutions exist and remain bounded for time $\mathcal{O}(1/\varepsilon^\ell)$. (The details may be handled in either of the two ways discussed in Section 2.8 in the paragraph after equation (2.8.4).) For the error estimate, we follow the "traditional" proof style, and let $\delta = \varepsilon^{\ell+1}\lambda_{\mathsf{J}_\varepsilon^{k-\ell+1}\mathbf{g}^\ell}$. A Gronwall argument shows that $\|\mathbf{y}(\boldsymbol{a},t,\varepsilon) - \mathbf{z}(\boldsymbol{a},t,\varepsilon)\| \leq \varepsilon^{k+1}B\delta^{-1}(e^{\delta t-1})$, where B is a bound for $\mathbf{f}^{[k+1]}$. Choose s_0 and c so that $e^s - 1 \leq cs$ for $0 \leq s \leq s_0$; then $e^{\delta t} - 1 \leq c\delta t$ for the time intervals occurring in the theorem, and $\|\mathbf{y}(\boldsymbol{a},t,\varepsilon) - \mathbf{z}(\boldsymbol{a},t,\varepsilon)\| = \mathcal{O}(\varepsilon^{k-j})$ for time $\mathcal{O}(1/\varepsilon^{j+1})$. Next apply $\widehat{\mathbf{U}}$. In fact, it suffices to omit more terms from $\mathbf{U}$ for the weaker estimates, as long as the omitted terms are not greater asymptotically than the desired error. See [197] and [264]. □

The most important case of Theorem 2.9.4 is when $\ell = k$ and $j = \ell - 1$. In this case (2.9.4) reduces to $\dot{z} = \varepsilon^\ell g^\ell(z)$ and we are trading as much accuracy as possible for increased length of validity, so the error is $\mathcal{O}(\varepsilon)$ for time $\mathcal{O}(1/\varepsilon^\ell)$. The next example illustrates this with $\ell = 2$ and $j = 1$.

2.9.3 Modified Van der Pol Equation

Consider the **Modified Van der Pol equation**

$$\ddot{x} + x - \varepsilon x^2 = \varepsilon^2(1 - x^2)\dot{x}.$$

We choose an amplitude-phase representation to obtain perturbation equations in the standard form: $(x,\dot{x}) \mapsto (r,\phi)$ by $x = r\sin(t-\phi)$, $\dot{x} = r\cos(t-\phi)$. We obtain

$$\begin{aligned}
\dot{r} &= \varepsilon r^2\cos(t-\phi)\sin(t-\phi)^2 + \varepsilon^2 r\cos(t-\phi)^2[1 - r^2\sin^2(t-\phi)^2],\\
\dot{\phi} &= \varepsilon r\sin(t-\phi)^3 + \varepsilon^2\sin(t-\phi)\cos(t-\phi)[1 - r^2\sin(t-\phi)^2].
\end{aligned}$$

The conditions of Theorem 2.9.4 have been satisfied with $\ell = k = 2$ and $j = 1$. The averaged equations describe the flow on the time scale $1/\varepsilon^2$ with error

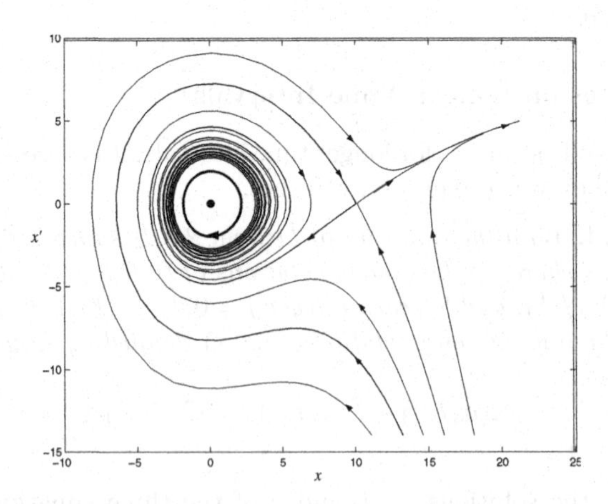

Fig. 2.6: The $(x,\dot{x})$-plane of the equation $\ddot{x} + x - \varepsilon x^2 = \varepsilon^2(1 - x^2)\dot{x}$; $\varepsilon = 0.1$.

$O(\varepsilon)$. The saddle point behavior has not been described by this perturbation approach as the saddle point coordinates are $(1/\varepsilon, 0)$. We put

$$\mathbf{u}^1(r,\phi,t) = \begin{bmatrix} \frac{1}{3}r^2\sin(t-\phi)^3 \\ -\frac{1}{3}r\cos(t-\phi)(2+\sin(t-\phi)^2) \end{bmatrix}.$$

After the calculation of $\mathsf{D}\mathbf{f}^1 \cdot \mathbf{u}^1$ and averaging we obtain

$$\dot{\overline{r}} = \frac{1}{2}\varepsilon^2\overline{r}(1-\frac{1}{4}\overline{r}^2), \quad \dot{\overline{\phi}} = \frac{5}{12}\varepsilon^2\overline{r}^2.$$

We conclude that as in the Van der Pol equation we have a stable periodic solution with amplitude $r = 2 + \mathcal{O}(\varepsilon)$ (cf. Section 2.2). The $O(\varepsilon)$-term in the original equation induces only a shifting of the phase-angle ϕ. For the periodic solution we have

$$x(t) = 2\cos(t - \frac{5}{3}\varepsilon^2 t) + \mathcal{O}(\varepsilon)$$

on the time scale $1/\varepsilon^2$. See Figure 2.6 for the phase-portrait. Important examples of Theorem 2.9.4 in the theory of Hamiltonian systems will be treated later on in Chapter 10.

2.9.4 Periodic Orbit of the Van der Pol Equation

In Section 2.2 we calculated the first-order approximation of the Van der Pol equation

$$\ddot{x} + x = \varepsilon(1-x^2)\dot{x}.$$

For the amplitude r and the phase ϕ the equations are

$$\dot{r} = \varepsilon r\cos(t-\phi)^2[1-r^2\sin(t-\phi)^2],$$
$$\dot{\phi} = \varepsilon\sin(t-\phi)\cos(t-\phi)[1-r^2\sin(t-\phi)^2].$$

Averaging over t (period 2π) yields

$$\dot{\overline{r}} = \frac{\varepsilon}{2}\overline{r}(1-\frac{\overline{r}^2}{4}), \quad \dot{\overline{\phi}} = 0,$$

producing a periodic solution of the original equation in $\mathrm{r}(0) = 2$. In the notation of Theorem 2.9.1 we have

$$\mathbf{u}^1(r,\phi,t) = \begin{bmatrix} \frac{1}{4}r\sin(2(t-\phi)) + \frac{1}{32}r^3\sin(4(t-\phi)) \\ -\frac{1}{4}\cos(2(t-\phi)) + \frac{1}{8}r^2\cos(2(t-\phi)) - \frac{1}{32}r^2\cos(4(t-\phi)) \end{bmatrix}.$$

For the equation, averaged to second-order, we obtain

$$\dot{\overline{r}} = \frac{1}{2}\varepsilon\overline{r}(1-\frac{1}{4}\overline{r}^2), \quad \dot{\overline{\phi}} = \frac{1}{8}\varepsilon^2(1-\frac{3}{2}\overline{r}^2+\frac{11}{32}\overline{r}^4),$$

where in the notation of Lemma 2.9.1 $\boldsymbol{z} = (\overline{r}, \overline{\phi})$. For the periodic solution we obtain (with $r(0) = 2$, $\phi(0) = 0$):

$$\overline{r} = 2\ , \ \overline{\phi} = \frac{1}{16}\varepsilon^2 t$$

and we have

$$\begin{bmatrix} r \\ \phi \end{bmatrix}(t) = \begin{bmatrix} 2 \\ \frac{1}{16}\varepsilon^2 t \end{bmatrix} + \varepsilon \begin{bmatrix} \frac{1}{2}\sin(2(t - \frac{1}{16}\varepsilon^2)) + \frac{1}{4}\sin(4(t - \frac{1}{16}\varepsilon^2)) \\ \frac{1}{4}\cos(2(t - \frac{1}{16}\varepsilon^2)) - \frac{1}{8}\cos(4(t - \frac{1}{16}\varepsilon^2)) \end{bmatrix} + \mathcal{O}(\varepsilon^2)$$

on the time scale $1/\varepsilon$. Note that $\mathbf{u}^1$, used in this example, has the property that its average over t is zero.

3

Methodology of Averaging

3.1 Introduction

This chapter provides additional details and variations on the method of averaging for periodic systems. Topics include methods of handling the "bookkeeping" of averaging calculations, averaging systems containing "slow time", ways to remove the nonuniqueness of the averaging transformation, and the relationship between averaging and the method of multiple scales.

3.2 Handling the Averaging Process

In the previous chapter the averaging procedure has been described in sufficient detail to prove the basic error estimates, but there remain questions about how to handle the details of the calculations efficiently. In this section we address two of those questions:

1. What is the best way to work with near-identity transformations and to compute their effect on a given differential equation? The answer we suggest is a particular version of Lie transforms (which is not the most popular version in the applied literature).
2. How difficult is it to solve the truncated averaged equations after they have been obtained? The answer is that it is equivalent to solving one autonomous nonlinear system of differential equations and a sequence of inhomogeneous linear systems. Furthermore, if the autonomous nonlinear system is explicitly solvable then the sequence of inhomogeneous linear systems is solvable by quadrature.

To develop Lie theory from scratch would be too lengthy for this book. Instead, we begin with a short discussion of Lie theory for linear systems, to motivate the definitions in the nonlinear case. Next we state the main definitions and results of Lie theory for nonlinear autonomous systems, with references but without proof. Then we will derive the Lie theory for periodically time-dependent systems (the case needed for averaging) from the autonomous case.

3.2.1 Lie Theory for Matrices

If W is any matrix, the family of matrices

$$T(s) = e^{sW}$$

for $s \in \mathbb{R}$ is called the **one-parameter group** generated by W. (The "parameter" is s, and it is a group because $T(s)T(t) = T(s+t)$.) The solution to the system of differential equations

$$\frac{d\boldsymbol{x}}{ds} = W\boldsymbol{x}$$

with initial condition $\boldsymbol{x}(0) = \boldsymbol{y}$ is given by

$$\mathbf{x}(s) = T(s)\mathbf{y}.$$

For our applications we are not concerned with the group property, but with the fact that when s is small, $T(s)$ is a near-identity transformation. To emphasize that s is small we set $s = \varepsilon$ and obtain

$$T(\varepsilon) = e^{\varepsilon W}.$$

At this point we also make a change in the traditional definition of generator: Instead of regarding the single matrix W as the generator of the family of matrices $T(\varepsilon)$, we call εW the generator. In this way the word "generator" becomes roughly the equivalent of "logarithm."

With these conventions in place, it is not difficult to generalize by allowing W to depend on ε, so that

$$T(\varepsilon) = e^{\varepsilon W(\varepsilon)} \tag{3.2.1}$$

is the family of near-identity transformations with generator $\varepsilon W(\varepsilon)$. (In the literature, one will find both $W(\varepsilon)$ and $\varepsilon W(\varepsilon)$ referred to as the generator. Our choice of $\varepsilon W(\varepsilon)$ is motivated by Baider's work in normal form theory, and agrees with the terminology in Chapter 13 below.) Notice that $\mathbf{x} = T(\varepsilon)\mathbf{y}$ is the result of solving

$$\frac{d\boldsymbol{x}}{ds} = W(\varepsilon)\boldsymbol{x} \tag{3.2.2}$$

with $\mathbf{x}(0) = \mathbf{y}$ to obtain $\mathbf{x}(s) = e^{sW(\varepsilon)}\mathbf{y}$, and then setting $s = \varepsilon$. This is not the same as solving the system

$$\frac{d\boldsymbol{x}}{d\varepsilon} = W(\varepsilon)\boldsymbol{x}, \tag{3.2.3}$$

where s does not appear; this is a nonautonomous linear system whose solution cannot be expressed using exponentials. The use of equations (3.2.1) and

(3.2.2) is often called **Hori's method**, while (3.2.3) is called the **method of Deprit**. In [203] these are referred to as **format 2b** and **format 2c** respectively (out of five formats for handling near-identity transformations, classified as formats 1a, 1b, 2a, 2b, and 2c). It is our opinion that of all approaches, format 2b (or Hori's method) is the best. The simplest reason for this is the fundamental importance of exponentials throughout mathematics. Also, in Hori's method the generator of the inverse transformation $T(\varepsilon)^{-1}$ is $-\varepsilon W(\varepsilon)$; there is no simple formula for the generator of the inverse transformation in Deprit's method. But the most important advantage of Hori's method is that it generalizes easily to an abstract setting of graded Lie algebras, and coincides in that context with the formulation used by Baider (Chapter 13). While we are on this subject, we mention that in Deprit's method it is customary to include factorials in the denominator when $W(\varepsilon)$ is expanded in a power series, while in Hori's method this is not done; this is just a convention and has no significance. For those who like the "triangle algorithm" of Deprit, it is possible to formulate Hori's method in triangular form as well; see [188] and [203, Appendix C].

3.2.2 Lie Theory for Autonomous Vector Fields

To pass from the linear case to the autonomous nonlinear case, the linear vector field $W(\varepsilon)\boldsymbol{x}$ appearing on the right-hand side of (3.2.2) is replaced by a nonlinear vector field

$$\mathbf{w}^{[1]}(\boldsymbol{x},\varepsilon) = \mathbf{w}^1(\boldsymbol{x}) + \varepsilon\mathbf{w}^2(\boldsymbol{x}) + \cdots .$$

The indexing reflects the fact that this will often appear in the form $\varepsilon\mathbf{w}^{[1]}(\boldsymbol{x},\varepsilon) = \varepsilon\mathbf{w}^1(\boldsymbol{x}) + \varepsilon^2\mathbf{w}^2(\boldsymbol{x}) + \cdots$. We say that $\varepsilon\mathbf{w}^{[1]}(\boldsymbol{x},\varepsilon)$ generates the family of near-identity transformations $\boldsymbol{x} = \mathbf{U}(\boldsymbol{y},\varepsilon)$ if $\mathbf{w}^{[1]}$ and $\mathbf{U}$ are related as follows. Let $\mathbf{x} = \boldsymbol{\Phi}(\mathbf{y},\varepsilon,s)$ be the solution of

$$\frac{d\boldsymbol{x}}{ds} = \mathbf{w}^{[1]}(\boldsymbol{x},\varepsilon) = \mathbf{w}^1(\mathbf{x}) + \varepsilon\mathbf{w}^2(\boldsymbol{x}) + \cdots \tag{3.2.4}$$

with initial conditions $\mathbf{x}(\varepsilon, s=0) = \mathbf{y}$; then

$$\mathbf{U}(\boldsymbol{y},\varepsilon) = \boldsymbol{\Phi}(\boldsymbol{y},\varepsilon,s=\varepsilon) = \boldsymbol{y} + \varepsilon\mathbf{u}^1(\boldsymbol{y}) + \varepsilon^2\mathbf{u}^2(\boldsymbol{y}) + \cdots . \tag{3.2.5}$$

The near-identity transformation $\mathbf{U}$ will be used as a coordinate change in differential equations involving time (as in Section 2.9). Notice that s in (3.2.4) is not to be interpreted as time; when s receives a value (as in the initial condition or in (3.2.5) we write "$s =$" inside the function to emphasize that this is a value of s, not of t. We assume that $\mathbf{w}^{[1]}$ is smooth (infinitely differentiable), so that its power series in ε exists but need not converge. The series for $\mathbf{U}$ can be generated recursively from the one for $\mathbf{w}^{[1]}$ by an algorithm given below, and in practice one carries the calculation only to some finite order. The version of Lie theory associated with (3.2.4) ("Hori's method") is set forth in [132], [188] (for Hamiltonian systems), [149], and [203, Appendix C], where it is called "format 2b" and is treated in detail. The practical use of the generator $\varepsilon\mathbf{w}^{[1]}(\cdot,\varepsilon)$ is based on two recursive algorithms. One generates the transformation $\mathbf{U}$ from $\mathbf{w}^{[1]}$, and the other computes the effect of $\mathbf{U}$ on a differential equation. Given $\varepsilon\mathbf{w}^{[1]}(\boldsymbol{x},\varepsilon)$, we define two differential operators $\mathcal{D}_{\mathbf{w}^{[1]}}$ and $\mathcal{L}_{\mathbf{w}^{[1]}}$ acting on mappings and vector fields $\mathbf{a}^{[0]}(\boldsymbol{y},\varepsilon)$ as follows (where we write D for $\mathsf{D}_{\boldsymbol{y}}$, since the latter notation can be confusing in situations where we change the name of the variable):

$$\mathcal{D}_{\mathbf{w}^{[1]}}\mathbf{a}^{[0]}(\boldsymbol{y},\varepsilon) = \mathsf{D}\mathbf{a}^{[0]}(\boldsymbol{y},\varepsilon)\cdot\mathbf{w}^{[1]}(\boldsymbol{y},\varepsilon) \tag{3.2.6}$$

and

$$\mathcal{L}_{\mathbf{w}^{[1]}}\mathbf{f}^{[0]}(\boldsymbol{y},\varepsilon) = \mathsf{D}\mathbf{f}^{[0]}(\boldsymbol{y},\varepsilon)\cdot\mathbf{w}^{[1]}(\boldsymbol{y},\varepsilon) - \mathsf{D}\mathbf{w}^{[1]}(\boldsymbol{y},\varepsilon)\cdot\mathbf{f}^{[0]}(\boldsymbol{y},\varepsilon). \tag{3.2.7}$$

This $\mathcal{D}_{\mathbf{w}^{[1]}}$ is the familiar operator of "differentiation of scalar fields along the flow of the vector field $\mathbf{w}^{[1]}(\boldsymbol{y},\varepsilon)$," applied componentwise to the components of the vector $\mathbf{a}^{[0]}$, while $\mathcal{L}_{\mathbf{w}^{[1]}}$ is the **Lie derivative** of a vector field along the flow of $\mathbf{w}^{[1]}(\boldsymbol{y},\varepsilon)$. These operators, and their exponentials occurring in the following theorem, should be considered as expanded in power series in ε and then applied to the power series of $\mathbf{a}^{[0]}(\boldsymbol{y},\varepsilon)$, $\mathbf{f}^{[0]}(\boldsymbol{y},\varepsilon)$, respectively. Thus, for instance, $\mathcal{L}_{\varepsilon\mathbf{w}^{[1]}} = \varepsilon\mathcal{L}_{\mathbf{w}^1} + \varepsilon^2\mathcal{L}_{\mathbf{w}^2} + \cdots$; since $\mathcal{L}_{\varepsilon\mathbf{w}^{[1]}}$ is based on $\varepsilon\mathbf{w}^{[1]}(\boldsymbol{y},\varepsilon)$, the degree in ε matches the superscript. This is to be applied to $\mathbf{f}^{[0]}(\boldsymbol{y},\varepsilon) = \mathbf{f}^0(\boldsymbol{y}) + \varepsilon\mathbf{f}^1(\boldsymbol{y}) + \cdots$, with $\mathcal{L}_{\mathbf{w}^i}\mathbf{f}^j = \mathsf{D}\mathbf{f}^j\cdot\mathbf{w}^i - \mathsf{D}\mathbf{w}^i\cdot\mathbf{f}^j$. Of course, $e^{\mathcal{L}_{\varepsilon\mathbf{W}}} = 1 + \mathcal{L}_{\varepsilon\mathbf{W}} + (1/2!)\mathcal{L}^2_{\varepsilon\mathbf{W}} + \cdots$.

Theorem 3.2.1. *The Taylor series of the transformation* $\mathbf{U}$ *generated by* $\mathbf{w}^{[1]}$ *is*

$$\boldsymbol{x} = e^{\varepsilon\mathcal{D}_{\mathbf{w}^{[1]}}}\boldsymbol{y}.$$

This transformation carries the differential equation $\dot{\boldsymbol{x}} = \mathbf{f}^{[0]}(\boldsymbol{x},\varepsilon)$ *into* $\dot{\boldsymbol{y}} = \mathbf{g}^{[0]}(\boldsymbol{y},\varepsilon)$, *where the Taylor series of* $\mathbf{g}^{[0]}$ *is*

$$\mathbf{g}^{[0]}(\boldsymbol{y},\varepsilon) = e^{\varepsilon\mathcal{L}_{\mathbf{w}^{[1]}}}\mathbf{f}^{[0]}(\boldsymbol{y},\varepsilon).$$

The second part of Theorem 3.2.1 has the following natural interpretation. The set of vector fields of the form $\mathbf{f}^{[0]}(\boldsymbol{y},\varepsilon) = \mathbf{f}^0(\boldsymbol{y}) + \varepsilon\mathbf{f}^1(\boldsymbol{y}) + \cdots$ forms a graded Lie algebra (graded by degree in ε) with Lie bracket $[\mathbf{f},\mathbf{g}] = \mathcal{L}_{\mathbf{f}}\mathbf{g}$. (The negative of this bracket is sometimes used, as in [203].) Then $\mathcal{L}_{\varepsilon\mathbf{w}^{[1]}}\mathbf{f} = [\varepsilon\mathbf{w}^{[1]},\mathbf{f}]$, and we may consider $\mathcal{L}_{\varepsilon\mathbf{w}^{[1]}}$ to be a natural action of a part of the Lie algebra (the part having no ε^0 term) on the whole Lie algebra. This leads naturally to Baider's generalization (mentioned above and in Chapter 13).

3.2.3 Lie Theory for Periodic Vector Fields

In order to apply these algorithms to averaging, it is necessary to incorporate periodic time dependence. This may be done by the common procedure of converting a nonautonomous system to an autonomous one of one higher dimension. Specifically, with the typical initial system (in which $\mathbf{f}^0 = 0$)

$$\dot{\boldsymbol{x}} = \varepsilon \mathbf{f}^{[1]}(\boldsymbol{x}, t, \varepsilon) = \varepsilon \mathbf{f}^1(\boldsymbol{x}, t) + \cdots + \varepsilon^k \mathbf{f}^k(\boldsymbol{x}, t) + \cdots \tag{3.2.8}$$

we associate the autonomous system

$$\begin{aligned} \begin{bmatrix} \dot{\boldsymbol{x}} \\ \dot{x}_{n+1} \end{bmatrix} &= \begin{bmatrix} 0 \\ 1 \end{bmatrix} + \varepsilon \begin{bmatrix} \mathbf{f}^{[1]}(\boldsymbol{x}, x_{n+1}, \varepsilon) \\ 0 \end{bmatrix} \\ &= \begin{bmatrix} 0 \\ 1 \end{bmatrix} + \varepsilon \begin{bmatrix} \mathbf{f}^1(\boldsymbol{x}, x_{n+1}) \\ 0 \end{bmatrix} + \cdots + \varepsilon^k \begin{bmatrix} \mathbf{f}^k(\boldsymbol{x}, x_{n+1}) \\ 0 \end{bmatrix} + \cdots, \end{aligned} \tag{3.2.9}$$

and consider the enlarged vector $(\boldsymbol{x}, x_{n+1}) = (x_1, \ldots, x_n, x_{n+1})$ as the $\boldsymbol{x}$ in Theorem 3.2.1. In order to force $x_{n+1} = t$ (without an additive constant) we always impose the initial condition $x_{n+1}(0, \varepsilon) = 0$ together with the initial condition on the rest of $\boldsymbol{x}$. Thus we deal with a restricted class of $(n+1)$-dimensional vector fields having $(n+1)$st component 1 and all other components periodic in x_{n+1}. The class of generators is subject to a different restriction: The $(n+1)$st component must be zero, because we want $y_{n+1} = x_{n+1}$ since both should equal t. That is, (3.2.4) becomes

$$\begin{bmatrix} \frac{d\boldsymbol{x}}{ds} \\ \frac{dx_{n+1}}{ds} \end{bmatrix} = \begin{bmatrix} \mathbf{w}^{[1]}(\boldsymbol{x}, x_{n+1}, \varepsilon) \\ 0 \end{bmatrix}, \tag{3.2.10}$$

with initial conditions $\mathbf{x}(\varepsilon, s = 0) = \mathbf{y}$ and $x_{n+1}(\varepsilon, s = 0) = y_{n+1}$. It is to be noted that in this application the set of generators (3.2.10) is no longer a subset of the set of vector fields, although both are subsets of the larger graded Lie algebra of all ε-dependent vector fields (3.2.9) on $\mathbb{R}^{n+1}$.

There is no need to introduce x_{n+1} in practice. Instead, just observe that the differential operator associated with (3.2.9) is

$$\mathbf{f}^{[0]} = \frac{\partial}{\partial x_{n+1}} + \sum_{i=1}^{n} \varepsilon \mathrm{f}_i^{[1]}(\boldsymbol{x}, x_{n+1}, \varepsilon) \frac{\partial}{\partial x_i},$$

and since $x_{n+1} = t$ this may be written

$$\frac{\partial}{\partial t} + \varepsilon \sum_{i=1}^{n} \mathrm{f}_i^{[1]}(\boldsymbol{x}, t, \varepsilon) \frac{\partial}{\partial x_i}. \tag{3.2.11}$$

Similarly, the differential operator associated with (3.2.10) is just

$$\varepsilon \sum_{i=1}^{n} \mathrm{w}_i^{[1]}(\boldsymbol{x}, t, \varepsilon) \frac{\partial}{\partial x_i}, \tag{3.2.12}$$

with no $\partial/\partial t$ term. (There is also no $\partial/\partial\varepsilon$ term, as would occur in the Deprit version of Lie theory.) The required Lie derivative may now be computed from the commutator bracket of (3.2.12) and (3.2.11), and it is

$$\mathcal{L}_{\varepsilon \mathbf{w}^{[1]}} \mathbf{f}^{[0]}(\boldsymbol{x}, t, \varepsilon) = -\varepsilon \mathbf{w}_t^{[1]} + \varepsilon^2 (\mathsf{D}\mathbf{f}^{[1]} \cdot \mathbf{w}^{[1]} - \mathsf{D}\mathbf{w}^{[1]} \cdot \mathbf{f}^{[1]}). \tag{3.2.13}$$

In the Deprit approach one would have to add a term $\varepsilon \frac{\partial \mathbf{f}^{[1]}}{\partial \varepsilon}$, or, at lowest order, $\varepsilon \mathbf{f}^1$. According to Theorem 3.2.1, the system (3.2.8) is transformed by generator $\mathbf{w}^{[1]}$ into

$$\dot{\boldsymbol{y}} = \varepsilon \mathbf{g}^{[1]}(\boldsymbol{y}, t, \varepsilon) = \varepsilon \mathbf{g}^1(\boldsymbol{y}, t) + \cdots, \tag{3.2.14}$$

where (as differential operators)

$$\mathbf{g}^{[0]}(\boldsymbol{y}, t, \varepsilon) = e^{\varepsilon \mathcal{L}_{\mathbf{w}^{[1]}}} \mathbf{f}^{[0]}(\boldsymbol{y}, t, \varepsilon). \tag{3.2.15}$$

To apply this to averaging, one wishes to make $\mathbf{g}^{[1]}$ independent of t. (Formally, this is done to all orders, but in practice, of course, we stop somewhere.) For this purpose we take $\mathbf{w}^{[1]} = \mathbf{w}^1 + \varepsilon \mathbf{w}^2 + \cdots$ to be unknown, and derive homological equations for the $\mathbf{w}^j$. Upon expanding the right-hand side of (3.2.15), it is seen that each degree in ε has the form $\varepsilon^j(\mathbf{K}^j - \mathbf{w}_t^j)$, where $\mathbf{K}^j$ is constructed from $\mathbf{f}^1, \ldots, \mathbf{f}^j$ and from $\mathbf{w}^1, \ldots, \mathbf{w}^{j-1}$ (and their derivatives). Therefore, equating the ε^j terms on both sides of (3.2.15) leads to a homological equation

$$\mathbf{w}_t^j = \mathbf{K}^j(t, \boldsymbol{y}) - \mathbf{g}^j(\boldsymbol{y}), \tag{3.2.16}$$

having the same form as (2.9.8) except that $\mathbf{w}^j$ appears instead of $\mathbf{u}^j$, and the $\mathbf{K}^j$ are formed differently. The first two $\mathbf{K}^j$ are given by

$$\begin{pmatrix} \mathbf{K}^1 \\ \mathbf{K}^2 \end{pmatrix} = \begin{pmatrix} \mathbf{f}^1 \\ \mathbf{f}^2 + \mathsf{D}\mathbf{f}^1 \cdot \mathbf{w}^1 - \mathsf{D}\mathbf{w}^1 \cdot \mathbf{f}^1 + \frac{1}{2}(\mathsf{D}\mathbf{w}^1 \cdot \mathbf{w}_t^1 - \mathsf{D}\mathbf{w}_t^1 \cdot \mathbf{w}^1) \end{pmatrix}. \tag{3.2.17}$$

As before, the system of homological equations can be solved recursively if each $\mathbf{g}^j$ is taken to be the mean of $\mathbf{K}^j$. One advantage of (3.2.16) over (2.9.8) is that we now have an algorithmic procedure, equation (3.2.15), to derive the $\mathbf{K}^j$.

3.2.4 Solving the Averaged Equations

Next we turn to the second question at the beginning of this section: How hard is it to solve the truncated averaged equations (2.9.4)? These equations are

$$\dot{\boldsymbol{z}} = \varepsilon \mathbf{g}^1(\boldsymbol{z}) + \cdots + \varepsilon^k \mathbf{g}^k(\boldsymbol{z}). \tag{3.2.18}$$

Remember that even the exact solution of this system has an error $\mathcal{O}(\varepsilon^k)$ over time $\mathcal{O}(1/\varepsilon)$ when compared to the solution of the full averaged equations (2.9.3), so it is sufficient to obtain an approximate solution of (3.2.18) with an error of the same order for the same time interval. (This remark must be modified in the case where a solution is sought that is valid for a longer time, such as $\mathcal{O}(1/\varepsilon^2)$, using the trade-off principle. The method that we describe here loses validity after time $\mathcal{O}(1/\varepsilon)$, so it discards some of the validity that the method of averaging has under the conditions when trade-off holds.) Introducing slow time $\tau = \varepsilon t$, the system becomes

$$\frac{d\boldsymbol{z}}{d\tau} = \mathbf{g}^1(\boldsymbol{z}) + \varepsilon \mathbf{g}^2(\boldsymbol{z}) + \cdots + \varepsilon^{k-1}\mathbf{g}^k(\boldsymbol{z}). \tag{3.2.19}$$

This system can be solved by regular perturbation theory with error $\mathcal{O}(\varepsilon^k)$ over any finite τ-interval, for instance $0 \le \tau \le 1$, which is the same as $0 \le t \le 1/\varepsilon$. (For more information on the regular perturbation method, see [201].) The procedure is to substitute

$$\boldsymbol{z} = \mathbf{z}^{[0]}(\tau, \varepsilon) = \mathbf{z}^0(\tau) + \varepsilon \mathbf{z}^1(\tau) + \cdots + \varepsilon^{k-1}\mathbf{z}^{k-1}(\tau) \tag{3.2.20}$$

into (3.2.19), expand, and equate like powers of ε, obtaining a sequence of equations

$$\begin{pmatrix} \frac{d\mathbf{z}^0}{d\tau} \\ \frac{d\mathbf{z}^1}{d\tau} \\ \vdots \end{pmatrix} = \begin{pmatrix} \mathbf{g}^1(\mathbf{z}^0) \\ \mathsf{D}\mathbf{g}^1(\mathbf{z}^0(\tau)) \cdot \mathbf{z}^1 + \mathbf{g}^2(\mathbf{z}^0(\tau)) \\ \vdots \end{pmatrix}, \tag{3.2.21}$$

in which each equation after the first has the form

$$\frac{d\mathbf{z}^j}{d\tau} = A(\tau) \cdot \mathbf{z}^j + \mathbf{q}^{j+1}(\tau), \tag{3.2.22}$$

with $A(\tau) = \mathsf{D}\mathbf{g}^1(\mathbf{z}^0(\tau))$. Therefore the determination of $\mathbf{z}^{[0]}$ (to sufficient accuracy) is reduced to the solution of a single autonomous nonlinear system for $\mathbf{z}^0$ and a sequence of inhomogeneous linear systems. The equation for $\mathbf{z}^0$ is exactly the *guiding system* defined in (2.8.3) for first-order averaging.

Furthermore, suppose that this guiding system is explicitly solvable as a function of τ and its initial conditions $\boldsymbol{a}$, and let the solution be $\boldsymbol{\varphi}(\boldsymbol{z}_0, \tau)$, so that

$$\frac{d}{d\tau}\boldsymbol{\varphi}(\boldsymbol{z}_0, \tau) = \mathbf{g}^1(\boldsymbol{\varphi}(\boldsymbol{z}_0, \tau)). \tag{3.2.23}$$

Then we claim that the inhomogeneous linear equations (3.2.22) are solvable by quadrature. For if we set $X(\boldsymbol{z}_0, \tau) = \mathsf{D}\boldsymbol{\varphi}(\boldsymbol{z}_0, \tau)$, then (for any fixed $\boldsymbol{z}_0$) $X(\boldsymbol{z}_0, \tau)$ is a fundamental solution matrix of the homogeneous linear system $d\boldsymbol{x}/d\tau = A(\boldsymbol{z}_0, \tau)\boldsymbol{x}$, and (as is the case for any linear system) its inverse $Y(\boldsymbol{z}_0, \tau) = X(\boldsymbol{z}_0, \tau)^{-1}$ is a fundamental solution matrix of the adjoint system $d\boldsymbol{y}/d\tau = -\boldsymbol{y}A(\boldsymbol{z}_0, \tau)$ (where $\boldsymbol{y}$ is a row vector). These matrices X and Y provide the general solution of (3.2.22) by quadrature in the form

$$\mathbf{z}^j(\tau) = X(\boldsymbol{z}_0, \tau)\left\{\mathbf{z}^j(0) + \int_0^\tau Y(\boldsymbol{z}_0, \sigma)\mathbf{q}^{j+1}(\sigma)\, \mathrm{d}\sigma\right\}. \tag{3.2.24}$$

Notice that if $\boldsymbol{\varphi}(\boldsymbol{z}_0, \tau)$ has been computed, no additional work is needed to obtain X and Y. Indeed, X can be obtained (as stated above) by differentiating $\boldsymbol{\varphi}$, and $Y(\boldsymbol{z}_0, \tau) = X(\boldsymbol{\varphi}(\boldsymbol{z}_0, \tau), -\tau)$ since $\boldsymbol{\varphi}(\cdot, -\tau)$ is the inverse mapping of $\boldsymbol{\varphi}(\cdot, \tau)$.

The solutions obtained by this method are two-time scale solutions, that is, they involve time only through the expressions t and εt. By Theorem 1.4.13, they must coincide with the two-time scale solutions discussed below in Section 3.5. It was pointed out above that we have discarded any validity beyond time $\mathcal{O}(1/\varepsilon)$ that the averaging solution may have. Thus the method of averaging is seen to be superior to the two scale method in this respect. It is sometimes possible to gain validity for longer time by a different choice of scales, but only when the method of averaging already possesses the desired validity. For further discussion of this point see Section 3.5.

In conclusion it should be noted that even if (3.2.19) cannot be solved in closed form by this method, it can be solved numerically much more efficiently than the original unaveraged equations. This is because solving it for time 1 is equivalent to solving the original equations for time $1/\varepsilon$.

3.3 Averaging Periodic Systems with Slow Time Dependence

Many technical and physical applications of the theory of asymptotic approximations concern problems in which certain quantities exhibit slow variation with time. Consider for instance a pendulum with variable length, a spring with varying stiffness or a mechanical system from which mass is lost. From the point of view of first-order averaging, the application of the preceding theory is simple (unless there are passage through resonance problems, cf. Chapter 7), the technical obstructions however can be considerable. We illustrate this as follows. Suppose the system has been put in the form

$$\dot{\boldsymbol{x}} = \varepsilon \mathbf{f}^1(\boldsymbol{x}, \varepsilon t, t)\ , \quad \mathbf{x}(0) = \boldsymbol{a},$$

with $\boldsymbol{x} \in \mathbb{R}^n$. Introduce the new independent variable

$$\tau = \varepsilon t.$$

Then we have the $(n+1)$-dimensional system in the standard form

$$\begin{bmatrix} \dot{\boldsymbol{x}} \\ \dot{\tau} \end{bmatrix} = \varepsilon \begin{bmatrix} \mathbf{f}^1(\boldsymbol{x}, \tau, t) \\ 1 \end{bmatrix}, \quad \begin{bmatrix} \mathbf{x} \\ \tau \end{bmatrix}(0) = \begin{bmatrix} \boldsymbol{a} \\ 0 \end{bmatrix}, \tag{3.3.1}$$

Suppose we may average the vector field over t, then an approximation can be obtained by solving the initial value problem

$$\begin{bmatrix} \dot{\boldsymbol{y}} \\ \dot{\tau} \end{bmatrix} = \varepsilon \begin{bmatrix} \overline{\mathbf{f}}^1(\boldsymbol{y}, \tau) \\ 1 \end{bmatrix}, \quad \begin{bmatrix} \mathbf{y} \\ \tau \end{bmatrix}(0) = \begin{bmatrix} \boldsymbol{a} \\ 0 \end{bmatrix},$$

or

$$\dot{\boldsymbol{y}} = \varepsilon \overline{\mathbf{f}}^1(\boldsymbol{y}, \varepsilon t), \quad \mathbf{y}(0) = \boldsymbol{a}.$$

So the recipe is simply: average over t, keeping εt and $\boldsymbol{x}$ fixed, and solve the resulting equation. In practice this is not always so easy; we consider a simple example below in Section 3.3.1. Mitropolsky devoted a book [190] to the subject with many more details and examples. Some problems with slowly varying time in celestial mechanics are considered in Appendix D.

For higher-order averaging, the setup is a little more complicated. If the usual near-identity transformation is applied to (3.3.1), the variable τ (which is treated as a new component of $\boldsymbol{x}$, say x_{n+1}) is transformed along with the rest of $\boldsymbol{x}$, so that the new variable replacing τ may no longer equal εt. It can be shown that with a correct choice of some of the arbitrary constants (that arise in solving the homological equations), this can be avoided. While this makes it clear that the previous theorems justifying higher-order averaging remain valid in this setting, and do not need to be reproved, it is better in practice to do the calculations in the following (equivalent) way, which contains no risk of transforming τ. We present the details to second-order, but there is no difficulty in continuing to higher orders.

Take the initial system to be

$$\dot{\boldsymbol{x}} = \varepsilon \mathbf{f}^1(\boldsymbol{x}, \tau, t) + \varepsilon^2 \mathbf{f}^2(\boldsymbol{x}, \tau, t) + \cdots , \tag{3.3.2}$$

with $\tau = \varepsilon t$, where $\mathbf{f}^i$ is 2π-periodic in t. Introduce a coordinate change

$$\boldsymbol{x} = \boldsymbol{y} + \varepsilon \mathbf{u}^1(\boldsymbol{y}, \tau, t) + \varepsilon^2 \mathbf{u}^2(\boldsymbol{y}, \tau, t), \tag{3.3.3}$$

with $\mathbf{u}^i$ periodic in t, and seek $\mathbf{u}^i$ so that

$$\dot{\boldsymbol{y}} = \varepsilon \mathbf{g}^1(\boldsymbol{y}, \tau) + \varepsilon^2 \mathbf{g}^2(\boldsymbol{y}, \tau) + \cdots . \tag{3.3.4}$$

That is, we seek to eliminate t, but not τ, so that (3.3.4) is not autonomous, although it has been simplified. The usual calculations (along the lines of Lemma 2.8.4) lead to the following homological equations:

$$\begin{pmatrix} \mathbf{u}_t^1(\boldsymbol{y}, \tau, t) \\ \mathbf{u}_t^2(\boldsymbol{y}, \tau, t) \end{pmatrix} = \begin{pmatrix} \mathbf{f}^1(\boldsymbol{y}, \tau, t) - \mathbf{g}^1(\boldsymbol{y}, \tau) \\ \mathbf{f}^2 + \mathsf{D}\mathbf{f}^1 \cdot \mathbf{u}^1 - \mathsf{D}\mathbf{u}^1 \cdot \mathbf{g}^1 - \mathbf{u}_\tau^1 - \mathbf{g}^2 \end{pmatrix}. \tag{3.3.5}$$

(Each function in the last equation is evaluated at $(t, \tau, \boldsymbol{y})$ except $\mathbf{g}^i$, which are evaluated at $(\tau, \boldsymbol{y})$.) The first equation is solvable (with periodic $\mathbf{u}^1$) only if we take

$$\mathbf{g}^1(\boldsymbol{y}, \tau) = \frac{1}{2\pi} \int_0^{2\pi} \mathbf{f}^1(\boldsymbol{y}, \tau, s)\, \mathrm{d}s, \tag{3.3.6}$$

and the solution is

$$\mathbf{u}^1(\boldsymbol{y}, \tau, t) = \int_0^t \widetilde{\mathbf{f}}^1(\boldsymbol{y}, \tau, s)\, \mathrm{d}s + \boldsymbol{\kappa}^1(\boldsymbol{y}, \tau), \tag{3.3.7}$$

where $\widetilde{\mathbf{f}}^1 = \mathbf{f}^1 - \mathbf{g}^1$ and $\boldsymbol{\kappa}^1$ is arbitrary. All functions appearing in the second homological equation are now fixed except $\mathbf{g}^2$ and $\mathbf{u}^2$, and these may be obtained by repeating the same procedure (and introducing a second arbitrary vector $\boldsymbol{\kappa}^2$). The simplest choice of $\boldsymbol{\kappa}^1$ and $\boldsymbol{\kappa}^2$ is zero, but there is a much more interesting choice which we will investigate in Section 3.4 below.

3.3.1 Pendulum with Slowly Varying Length

Consider a pendulum with slowly varying length and some other perturbations to be specified later on. If we put the mass and the gravitational constant equal to one, and if we put $l = l(\varepsilon t)$ for the length of the pendulum, we have according to [190] the equation

$$\frac{d}{dt}(l^2(\varepsilon t)\dot{x}) + l(\varepsilon t)x = \varepsilon g(x, \dot{x}, \varepsilon t),$$

with initial values given. The first problem is to put this equation in standard form. If $\varepsilon = 0$ we have a harmonic oscillator with frequency $\omega_0 = l(0)^{-\frac{1}{2}}$ and solutions of the form $\cos(\omega_0 t + \phi)$. This inspires us to introduce another time-like variable

$$s = \int_0^t l(\varepsilon\sigma)^{-\frac{1}{2}}\, d\sigma.$$

If $\varepsilon = 0$, s reduces to the natural time-like variable $\omega_0 t$. For s to be time-like we require $l(\varepsilon t)$ to be such that $s(t)$ increases monotonically and that $t \to \infty \Rightarrow s \to \infty$. We abbreviate $\varepsilon t = \tau$; note that since

$$\frac{d}{dt} = l^{-\frac{1}{2}}(\tau)\frac{d}{ds},$$

the equation becomes

$$\frac{d^2x}{ds^2} + x = \varepsilon l^{-1}(\tau) g(x, l^{-\frac{1}{2}}(\tau)\frac{dx}{ds}, \tau) - \frac{3}{2}\varepsilon l^{-\frac{1}{2}}(\tau)\frac{dl}{d\tau}\frac{dx}{ds}.$$

Introducing amplitude-phase coordinates by

$$x = r\sin(s - \phi), \quad x' = r\cos(s - \phi),$$

(with $x' = \frac{dx}{ds}$) produces the standard form

$$r' = \varepsilon\cos(s - \phi)\left(l^{-1}(\tau)g(x, l(\tau)^{-\frac{1}{2}}x', \tau) - \frac{3}{2}l(\tau)^{-\frac{1}{2}}\frac{dl}{d\tau}r\cos(s - \phi)\right),$$

$$\phi' = \frac{\varepsilon}{r}\sin(s - \phi)\left(l^{-1}(\tau)g(x, l(\tau)^{-\frac{1}{2}}x', \tau) - \frac{3}{2}l(\tau)^{-\frac{1}{2}}\frac{dl}{d\tau}r\cos(s - \phi)\right),$$

$$\tau' = \varepsilon l^{\frac{1}{2}}(t).$$

Initial values $\mathrm{r}(0) = r_0$, $\phi(0) = \phi_0$, and $\tau(0) = 0$. Averaging over s does not touch the last equation, so we still have $\tau = \varepsilon t$, as it should be. We consider two cases.

The Linear Case, $g = 0$

Averaging produces

$$\overline{r}' = -\frac{3}{4}\varepsilon\overline{r}l^{-\frac{1}{2}}\frac{dl}{d\tau}, \quad \overline{\phi}' = 0.$$

The first equation can be written as

$$\frac{d\overline{r}}{d\tau} = -\frac{3}{4}\overline{r}l^{-1}\frac{dl}{d\tau},$$

so we have on the time scale $1/\varepsilon$ in s

$$r(\tau) = r_0(\mathrm{l}(0)/\mathrm{l}(t))^{\frac{3}{4}} + \mathcal{O}(\varepsilon)\ ,\ \phi(\tau) = \phi_0 + \mathcal{O}(\varepsilon). \tag{3.3.8}$$

A Nonlinear Perturbation with Damping

Suppose that the oscillator has been derived from the mathematical pendulum so that we have a Duffing type of perturbation (coefficient μ); moreover we have small linear damping (coefficient σ). We put

$$g = \mu l(\varepsilon t)x^3 - \sigma l(\varepsilon t)\dot{x}$$

or

$$g = \mu l(\varepsilon t)r^3\sin(t-\phi)^3 - \sigma l^{\frac{1}{2}}(\varepsilon t)r\cos(t-\phi).$$

The standard form for r and ϕ becomes

$$r' = \varepsilon\cos(s-\phi)\left(\mu r^3\sin(t-\phi)^3 - \sigma l^{-\frac{1}{2}}r\cos(t-\phi) - \frac{3}{2}l(\tau)^{-\frac{1}{2}}\frac{dl}{d\tau}r\cos(s-\phi)\right),$$
$$\phi' = \frac{\varepsilon}{r}\sin(s-\phi)\left(\mu r^3\sin(t-\phi)^3 - \sigma l^{-\frac{1}{2}}r\cos(t-\phi) - \frac{3}{2}l(\tau)^{-\frac{1}{2}}\frac{dl}{d\tau}r\cos(s-\phi)\right),$$
$$\tau' = \varepsilon l^{\frac{1}{2}}(t).$$

Averaging produces for r and ϕ

$$\overline{r}' = -\frac{1}{2}\varepsilon\overline{r}l(\tau)^{-\frac{1}{2}}\left(\sigma + \frac{3}{2}\frac{dl}{d\tau}\right),$$
$$\overline{\phi}' = \frac{3}{8}\varepsilon\mu\overline{r}^2,$$
$$\tau' = \varepsilon l^{\frac{1}{2}}(t).$$

If $\overline{r}$ is known, $\overline{\phi}$ can be obtained by direct integration. The equation for $\overline{r}$ can be written as

$$\frac{d\overline{r}}{d\tau} = -\frac{1}{2}l^{-1}\overline{r}(\sigma + \frac{3}{2}\frac{dl}{d\tau}),$$

so we have with Theorem 2.8.1 on the time scale $1/\varepsilon$ in s

$$r(\tau) = r_0(\mathrm{l}(0)/\mathrm{l}(t))^{\frac{3}{4}} \exp(-\frac{1}{2}\sigma \int_0^\tau l^{-1}(u)\,\mathrm{d}u) + \mathcal{O}(\varepsilon).$$

Remark 3.3.1. Some interesting studies have been devoted to the equation

$$\ddot{y} + \omega^2(\varepsilon t)y = 0.$$

The relation with our example becomes clear when we put $g = 0$ and transform $x = y/l$. If l can be differentiated twice, we obtain

$$\ddot{y} + l^{-1}(1 - \ddot{l})y = 0.$$

3.4 Unique Averaging

The higher-order averaging methods described in Section 2.9 are not unique, because at each stage arbitrary "constants" of integration $\boldsymbol{\kappa}^j(\boldsymbol{y})$ or $\boldsymbol{\kappa}^j(\boldsymbol{y},\tau)$ are introduced. (These are constants in the sense that they are independent of t, except possibly in the form of slow time $\tau = \varepsilon t$.) These arbitrary functions appear first in the transformation, but then reappear at higher order in the averaged equations themselves. In this section we discuss three ways of determining the $\boldsymbol{\kappa}^j$ so as to force a unique result. Each of the methods has advantages. The first has already been discussed in Section 2.9, and leads to stroboscopic averaging. The second is well adapted to the use of Fourier series for periodic functions. The third, which we call *reduced averaging*, is rather remarkable. When reduced averaging is possible (which is not always the case), it produces an averaged equation (to any order) which coincides with the usual first-order averaged equation; in other words, by choosing the arbitrary quantities correctly, all of the higher-order terms in the averaged equation are made to vanish. Furthermore, the solution of an initial value problem given by reduced averaging coincides exactly with the solution obtained by a popular two-time scale method (presented in Section 3.5 below).

The starting point for our discussion is the homological equation (2.9.8), that is,

$$\frac{\partial \mathbf{u}^j}{\partial t}(\boldsymbol{y},t) = \mathbf{K}^j(\boldsymbol{y},t) - \mathbf{g}^j(\boldsymbol{y}). \tag{3.4.1}$$

Any 2π-periodic vector $\mathbf{h}$ of t (and perhaps other variables) can be decomposed into its **mean** $\overline{\mathbf{h}}$ and its **zero-mean part** $\widetilde{\mathbf{h}} = \mathbf{h} - \overline{\mathbf{h}}$. In this notation the solution of (3.4.1) is given by

$$\mathbf{g}^j = \overline{\mathbf{K}}^j \tag{3.4.2}$$

and

$$\mathbf{u}^j(\boldsymbol{y},t) = \int_0^t \widetilde{\mathbf{K}}^j(\boldsymbol{y},s)\,\mathrm{d}s + \boldsymbol{\kappa}^j(\boldsymbol{y}). \tag{3.4.3}$$

A uniqueness rule is simply a rule for choosing $\boldsymbol{\kappa}^j(\boldsymbol{y})$; it must be a rule that can be applied to all problems at all orders. Such a rule, used consistently for each j, leads to unique sequences $\boldsymbol{\kappa}^j$, $\mathbf{u}^j$, and $\mathbf{K}^j$ (for $j = 1, 2, \dots$), given an initial sequence $\mathbf{f}^j$. A uniqueness rule may reduce the class of admissible transformations (for instance, from all periodic near-identity transformations to just those that have mean zero). In this case systems that were equivalent (transformable into one another) before may become inequivalent after the uniqueness rule has been imposed. The most desirable form of uniqueness rule is one that avoids this difficulty and instead selects a unique representative of each equivalence class (under the original equivalence relation) and, to each system, assigns the (necessarily unique) transformation that brings it into the desired form. In this case we speak of a **hypernormalization rule**.

The simplest uniqueness rule is $\boldsymbol{\kappa}^j(\boldsymbol{y}) = 0$. As discussed in Section 2.9, this leads to an averaging transformation $\mathbf{U}$ that equals the identity at all stroboscopic times, which is very convenient for handling initial values. It may seem that this is all there is to say about the subject; why would anyone want another choice? But let us see.

If $\mathbf{h}$ is any 2π-periodic vector of t (and perhaps other variables), its Fourier series (in complex form) will be written

$$\mathbf{h}(t) = \sum_{\nu=-\infty}^{\infty} \mathbf{h}_\nu e^{i\nu t},$$

where $\mathbf{h}_\nu$ is a function of the other variables. Then the mean is $\overline{\mathbf{h}} = \mathbf{h}_0$ and the zero-mean part is

$$\widetilde{\mathbf{h}}(t) = \sum_{\nu \neq 0} \mathbf{h}_\nu e^{i\nu t}.$$

The **zero-mean antiderivative** of $\widetilde{\mathbf{h}}$ is

$$\sum_{\nu \neq 0} \frac{1}{i\nu} \mathbf{h}_\nu e^{i\nu t}.$$

In this notation, the following solution of (3.4.2) is just as natural as (3.4.3):

$$\mathbf{u}^j(\boldsymbol{y}, t) = \sum_{\nu \neq 0} \frac{1}{i\nu} \mathbf{K}^j_\nu(\boldsymbol{y}) e^{i\nu t} + \boldsymbol{\lambda}^j(\boldsymbol{y}), \tag{3.4.4}$$

where $\boldsymbol{\lambda}^j$ is an arbitrary function of $\boldsymbol{y}$ (playing the same role as $\boldsymbol{\kappa}^j$); notice that $\boldsymbol{\lambda}^j = \overline{\mathbf{u}}^j$. From this point of view, the simplest choice of the arbitrary "constant" is $\boldsymbol{\lambda}^j(\boldsymbol{y}) = 0$. (It is not hard to compute the expression for $\boldsymbol{\kappa}^j$ that makes (3.4.3) equal to (3.4.4) with $\boldsymbol{\lambda}^j = 0$, but it is not a formula one would want to use.)

In periodic averaging, which we have been studying, the advantage of convenience with Fourier series probably does not outweigh the advantage that stroboscopic averaging gives to the solution of initial value problems. But in quasiperiodic (or multi-frequency) averaging, studied in Chapter 7, stroboscopic averaging is impossible (because there are no stroboscopic times). The study of multi-frequency averaging hinges strongly on the denominators that arise in the analog of (3.4.4) and which can lead to the famous "small divisor problem." Therefore in multi-frequency averaging the preferred uniqueness rule is $\boldsymbol{\lambda}^j = 0$.

The third uniqueness rule of averaging, which we call **reduced averaging**, requires that the homological equations be written in a more detailed form than (3.4.1). Instead of presenting the general version, we carry the calculations to second order only, although there is no difficulty (other than the length of the equations) to handling higher orders. The idea is that at each stage we choose the arbitrary vector $\boldsymbol{\lambda}^j$ occurring in $\mathbf{u}^j$ in such a way that $\mathbf{K}^{j+1}$ has zero mean, so that $\mathbf{g}^{j+1} = \overline{\mathbf{K}}^{j+1} = 0$. This procedure, when it is possible, meets the requirements for a hypernormalization rule as defined above, since the full freedom of determining the $\boldsymbol{\lambda}^j$ is used to achieve a unique final form in each equivalence class.

We begin with (3.4.4), in the form

$$\mathbf{u}^1 = \widetilde{\mathbf{u}}^1 + \boldsymbol{\lambda}^1. \tag{3.4.5}$$

In view of (2.9.7), and using $\overline{\mathbf{f}}^1 = \mathbf{g}^1$, we may write

$$\mathbf{K}^2 = \mathbf{f}^2 + (\mathsf{D}\mathbf{g}^1 + \mathsf{D}\widetilde{\mathbf{f}}^1) \cdot (\widetilde{\mathbf{u}}^1 + \boldsymbol{\lambda}^1) - (\mathsf{D}\widetilde{\mathbf{u}}^1 + \mathsf{D}\boldsymbol{\lambda}^1) \cdot \mathbf{g}^1.$$

Since the terms $\mathsf{D}\mathbf{g}^1 \cdot \widetilde{\mathbf{u}}^1 + \mathsf{D}\widetilde{\mathbf{f}}^1 \cdot \boldsymbol{\lambda}^1 - \mathsf{D}\widetilde{\mathbf{u}}^1 \cdot \mathbf{g}^1$ have zero mean, we have

$$\overline{\mathbf{K}}^2 = \overline{(\mathbf{f}^2 + \mathsf{D}\widetilde{\mathbf{f}}^1 \cdot \widetilde{\mathbf{u}}^1)} + \mathsf{D}\mathbf{g}^1 \cdot \boldsymbol{\lambda}^1 - \mathsf{D}\boldsymbol{\lambda}^1 \cdot \mathbf{g}^1.$$

Setting $\mathbf{h}^2 = \mathbf{f}^2 + \mathsf{D}\widetilde{\mathbf{f}}^1 \cdot \widetilde{\mathbf{u}}^1$, our goal is to choose $\boldsymbol{\lambda}^1$ so that

$$\mathsf{D}\boldsymbol{\lambda}^1 \cdot \mathbf{g}^1 - \mathsf{D}\mathbf{g}^1 \cdot \boldsymbol{\lambda}^1 = \overline{\mathbf{h}}^2. \tag{3.4.6}$$

(At higher orders, the equations will have the form

$$\mathsf{D}\boldsymbol{\lambda}^j \cdot \mathbf{g}^1 - \mathsf{D}\mathbf{g}^1 \cdot \boldsymbol{\lambda}^j = \overline{\mathbf{h}}^{j+1},$$

where $\mathbf{h}^{j+1}$ is known, based on earlier calculations, at the time that it is needed.) If equation (3.4.6) is solvable for $\boldsymbol{\lambda}^1$, and this $\boldsymbol{\lambda}^1$ is used in forming $\mathbf{u}^1$, then we will have $\mathbf{g}^2 = \overline{\mathbf{K}}^2 = 0$. In a moment we will turn our attention to the solution of (3.4.6). But first let us assume that $\boldsymbol{\lambda}^1$ and the higher-order $\boldsymbol{\lambda}^j$ can be obtained, and see what form the solution of an initial value problem by reduced averaging takes.

The second-order reduced averaged system will be simply

$$\dot{\boldsymbol{z}} = \varepsilon \mathbf{g}^1(\boldsymbol{z}). \tag{3.4.7}$$

The "short" transformation (2.9.5) with $k = 2$ will be

$$\boldsymbol{\xi} = \widehat{\mathbf{U}}(\boldsymbol{z}, t, \varepsilon) = \boldsymbol{z} + \varepsilon \mathbf{u}^1(\boldsymbol{z}, t). \tag{3.4.8}$$

The solution of (3.4.7) with initial condition $\boldsymbol{z}_0$ will be $\boldsymbol{\varphi}(\boldsymbol{z}_0, \varepsilon t)$, where $\boldsymbol{\varphi}(\boldsymbol{z}_0, \tau)$ is the solution of the guiding system $d\boldsymbol{z}/d\tau = \mathbf{g}^1(\boldsymbol{z})$ as in (3.2.23). Then the second-order approximation for the solution of the original system with initial condition $\widehat{\mathbf{U}}(\boldsymbol{z}_0, \varepsilon)$ is the function

$$\boldsymbol{\xi}(\tau, t, \varepsilon) = \boldsymbol{\xi}^0(\tau) + \varepsilon \boldsymbol{\xi}^1(\tau, t) = \boldsymbol{\varphi}(\boldsymbol{z}_0, \tau) + \varepsilon \mathbf{u}^1(\boldsymbol{\varphi}(\boldsymbol{z}_0, \tau), t), \tag{3.4.9}$$

obtained by substituting $\boldsymbol{z} = \boldsymbol{\varphi}(\boldsymbol{z}_0, \tau)$ into (3.4.8). (Since we are not using stroboscopic averaging, this is not the solution with initial condition $\boldsymbol{z}_0$.) There are two things to notice about this solution $\boldsymbol{\xi}(\tau, t, \varepsilon)$: First, it is a two-time scale expression in time scales t and τ. (This will be discussed more fully in the next Section 3.5). Second, the calculation of $\boldsymbol{\xi}(\tau, t, \varepsilon)$ does not require the complete determination of the function $\boldsymbol{\lambda}^1$. Indeed, since $\mathbf{u}^1 = \widetilde{\mathbf{u}}^1 + \boldsymbol{\lambda}^1$, only the function

$$\mathbf{v}^1(\boldsymbol{z}_0, \tau) = \boldsymbol{\lambda}^1(\boldsymbol{\varphi}(\boldsymbol{z}_0, \tau)) \tag{3.4.10}$$

needs to be calculated to determine (3.4.9). We will now show that $\mathbf{v}^1$ is quite easy to calculate. (It is still necessary to discuss the *existence* of $\boldsymbol{\lambda}^1$ for the sake of fully justifying the method. But it is never necessary to calculate it. This remark holds true for higher orders as well.)

Suppose that $\boldsymbol{\lambda}^1$ satisfies (3.4.6) and that $\mathbf{v}^1$ is defined by (3.4.10). Then $\partial \mathbf{v}^1/\partial \tau = \mathsf{D}\boldsymbol{\lambda}^1 \cdot \mathbf{g}^1$, and (3.4.6) implies that

$$\frac{\partial \mathbf{v}^1}{\partial \tau}(\boldsymbol{z}_0, \tau) = A(\boldsymbol{z}_0, \tau)\mathbf{v}^1(\boldsymbol{z}_0, \tau) + \mathbf{H}^2(\boldsymbol{z}_0, \tau), \tag{3.4.11}$$

where $A(\boldsymbol{z}_0, \tau) = \mathsf{D}\mathbf{g}^1(\boldsymbol{\varphi}(\boldsymbol{z}_0, \tau))$ and $\mathbf{H}^2(\boldsymbol{z}_0, \tau) = \overline{\mathbf{h}}^2(\boldsymbol{\varphi}(\boldsymbol{z}_0, \tau))$. This is an inhomogeneous linear ordinary differential equation (in τ, with parameters $\boldsymbol{z}_0$) having the same linear term as in (3.2.22), so it is solvable by quadrature in the same manner as discussed there. (That is, the method of reduced averaging does not change the amount of work that must be done. It is still necessary to solve the guiding system and a sequence of inhomogeneous linear equations, only now these equations arise in connection with obtaining the $\mathbf{v}^j$.)

Finally, we discuss the existence of $\boldsymbol{\lambda}^1$ by considering how it can be constructed from a knowledge of the general solution for $\mathbf{v}^1$. Suppose that Σ is a hypersurface in $\mathbb{R}^n$ transverse to the flow $\boldsymbol{\varphi}$ of the guiding system; that is, $\mathbf{g}^1(\boldsymbol{y})$ is not tangent to Σ for any $\boldsymbol{y} \in \Sigma$ (and in particular, $\mathbf{g}^1(\boldsymbol{y}) \neq 0$). Suppose also that the region D in which the averaging is to be done is swept out (foliated) by arcs of orbits of $\boldsymbol{\varphi}$ crossing Σ. (Thus D is topologically the product of Σ with an interval. In particular, D cannot contain any complete periodic orbits, although it may contain arcs from periodic orbits.) Then there is no difficulty in creating $\boldsymbol{\lambda}^1$ from $\mathbf{v}^1$. We just take $\boldsymbol{\lambda}^1$ to be arbitrary (for instance, zero) on Σ, and use this as the initial condition for solving (3.4.11) for $z_0 \in \Sigma$. When z_0 is confined to Σ it has only $n-1$ independent variables, and together with τ this forms a coordinate system in D; $\mathbf{v}^1$ is simply $\boldsymbol{\lambda}^1$ expressed in this coordinate system.

On the other hand, suppose that we are interested in a neighborhood D of a rest point of $\mathbf{g}^1$. This is the usual case in nonlinear oscillations. For instance, in $n = 2$ we may have a rest point surrounded by a nested family of periodic orbits of $\mathbf{g}^1$. Then a curve drawn from the rest point provides a transversal Σ, but the solutions of (3.4.11) will in general not return to their original values after a periodic orbit is completed. In this case it is still possible to carry out the solution, but properly speaking, the reduced averaging is being done in a covering space of the actual system. (It follows that the transformation $\mathbf{U}$ in reduced averaging does not provide a conjugacy of the original and full averaged equations near the rest point. For this reason reduced averaging cannot be used in Chapter 6.

Finally, consider the extreme case in which $\mathbf{g}^1$ is identically zero. (This is the case in which trade-off is possible according to Theorem 2.9.4, giving validity for time $1/\varepsilon^2$.) Then there is no such thing as a transversal to the flow of $\mathbf{g}^1$, and the construction of $\boldsymbol{\lambda}^1$ from $\mathbf{v}^1$ is impossible. It turns out that in this case the proper two-time scale solution uses the time scales t and $\varepsilon^2 t$ and is valid for time $1/\varepsilon^2$. (See Section 3.5 below.)

The results in this section are due to Perko [217, Section4] and Kevorkian [147, p. 417], based on earlier remarks by Morrison, Kuzmak, and Luke. As shown in [147], the method of reduced averaging remains valid in those cases in which the slow time τ is present in the original system (as in Section 3.3 above).

3.5 Averaging and Multiple Time Scale Methods

A popular alternative to the method of averaging for periodic systems (with or without additional slow time dependence) is the method of multiple time scales, which is actually a collection of several methods. We will discuss a few of these briefly, and point out that one particular two-time scale method gives identical results with the method of reduced averaging described in Section 3.4. This result is due to Perko [217].

The **Poincaré–Lindstedt method** is probably the first multiple time scale method to have been introduced, and is useful only for periodic solutions. We will not treat it in any detail (see [201] or [281]), but it is worthwhile to discuss a few points as motivation for things that we will do. Suppose that some smooth differential equation depending on a small parameter ε has a family of periodic solutions $\boldsymbol{x}(\omega(\varepsilon)t, \varepsilon)$ of period $2\pi/\omega(\varepsilon)$ (where the vector $\boldsymbol{x}(\theta, \varepsilon)$ has period 2π in θ). The function $\omega(\varepsilon)$ and the vector $\boldsymbol{x}(\theta, \varepsilon)$ can be expanded as

$$\omega(\varepsilon) = \omega^0 + \varepsilon\omega^1 + \varepsilon^2\omega^2 + \cdots,$$
$$\mathbf{x}(\theta, \varepsilon) = \mathbf{x}^0(\theta) + \varepsilon\mathbf{x}^1(\theta) + \varepsilon^2\mathbf{x}^2(\theta) + \cdots;$$

notice that each $\mathbf{x}^i(\theta)$ has period 2π. The Poincaré–Lindstedt method is a way of calculating these expansions recursively from the differential equation. It is then natural to approximate the periodic solution by truncating both of these series at some point, obtaining $\widetilde{\omega}(\varepsilon)$ and $\widetilde{\boldsymbol{x}}(\theta, \varepsilon)$, and then putting

$$\boldsymbol{x}(\omega(\varepsilon)t, \varepsilon) \approx \widetilde{\mathbf{x}}(\widetilde{\omega}(\varepsilon)t, \varepsilon).$$

There are two sources of error in this approximation, a small error in amplitude due to the truncation of $\boldsymbol{x}$ and a growing error in phase due to the truncation of ω. It is clear that after a while all accuracy will be lost. In fact, if the truncations are at order k, the approximation has error $\mathcal{O}(\varepsilon^{k+1})$ for time $\mathcal{O}(1/\varepsilon)$, but can also be considered as having error $\mathcal{O}(\varepsilon^k)$ for time $\mathcal{O}(1/\varepsilon^2)$, $\mathcal{O}(\varepsilon^{k-1})$ for time $\mathcal{O}(1/\varepsilon^3)$, and so forth. We refer to this phenomenon as **trade-off of accuracy for length of validity**; it has already been discussed (for averaging) in Section 2.9.

The Poincaré–Lindstedt method is a **two-time scale method**, using the natural time variable t and a "strained" time variable $s = \widetilde{\omega}(\varepsilon)t$. Alternatively, s may be considered as containing several time variables, $t, \tau = \varepsilon t, \sigma = \varepsilon^2 t, \ldots$. Our further discussion will be directed to methods using these variables, and in particular, to two- and three-time scale methods using (t, τ) or (t, τ, σ).

Incidentally, the Poincaré–Lindstedt method is an excellent counter example to a common mistake. Many introductory textbooks suggest that an approximation that is *asymptotically ordered* is automatically *asymptotically valid* (or they do not even make a distinction between these concepts). Asymptotic ordering means that successive terms have the order in ε indicated by their coefficient, on a certain domain (which may depend on ε). In the case of the Poincaré–Lindstedt series

$$\widehat{\mathbf{x}}(\widetilde{\omega}(\varepsilon)t, \varepsilon) = \mathbf{x}^0(\widetilde{\omega}(\varepsilon)t) + \varepsilon\mathbf{x}^1(\widetilde{\omega}(\varepsilon)t) + \varepsilon^2\mathbf{x}^2(\widetilde{\omega}(\varepsilon)t) + \cdots + \varepsilon^k\mathbf{x}^k(\widetilde{\omega}(\varepsilon)t),$$

the ε^j term is $\mathcal{O}(\varepsilon^j)$ on the entire t axis, because $\mathbf{x}^j$ is periodic and therefore bounded. So the series is asymptotically ordered for all time. But it is not asymptotically valid for all t, that is, the error is not $\mathcal{O}(\varepsilon^{k+1})$ for all time. Instead, the error estimates are those given above, with trade-off, on ε-dependent intervals of various lengths.

Motivated by the Poincaré–Lindstedt method, and perhaps by the observation that averaging frequently gives solutions involving several time scales, various authors have constructed multiple time scale methods that successfully approximate solutions that are not necessarily periodic. For an overview of these methods, see [209, Chapter 6]. These methods are often formulated for second-order differential equations, or systems of these representing coupled oscillators, but it is also possible to apply them to systems in the standard form for the method of averaging, and this is the best way to compare them with the method of averaging. The method we present here is one that uses the time scales t and $\tau = \varepsilon t$; this method always gives the same result as the method of Section 3.2.4 above (and, when it is possible, the method of reduced averaging) and requires the solution of exactly the same equations. The results are asymptotically valid for time $\mathcal{O}(1/\varepsilon)$, exactly as for averaging.

It is often hinted in the applied literature that one can get validity for time $\mathcal{O}(1/\varepsilon^2)$ by adding a third time scale $\sigma = \varepsilon^2 t$, but this is not generally correct. The possibility of a three-time scale solution is investigated in [208]. It is shown that even formally, a three-time scale solution cannot always be constructed. The scalar equation $\dot{x} = \varepsilon(-x^3 + x^p \cos t + \sin t)$, for $0 < p < 3$, with initial condition $x(0) > 0$, has a bounded solution for all time, but this solution cannot be approximated for time $\mathcal{O}(1/\varepsilon^2)$ by the three-time scale method. A three-time scale solution is possible when the first-order average $\mathbf{g}^1$ vanishes identically, but in that case the inclusion of τ is unnecessary, and there is a solution using t and σ alone that is valid for time $1/\varepsilon^2$, duplicating the result of averaging with trade-off given in Section 3.3 above.

We now present the formal details of the two-time scale method. Consider the initial value problem

$$\dot{\boldsymbol{x}} = \varepsilon \mathbf{f}^1(\boldsymbol{x}, t), \quad \mathbf{x}(0) = \boldsymbol{a},$$

with $\boldsymbol{x} \in D \subset \mathbb{R}^n$; $\mathbf{f}^1(\boldsymbol{x}, t)$ is T-periodic in t and meets other requirements which will be formulated in the course of our calculations. Suppose that two time scales suffice for our treatment, t and τ, a fast and a slow time. We expand

$$\mathbf{x}(\tau, t, \varepsilon) = \sum_{j=0}^{k-1} \varepsilon^j \mathbf{x}^j(\tau, t), \tag{3.5.1}$$

in which t and τ are used as independent variables. The differential operator becomes

$$\frac{d}{dt} = \frac{\partial}{\partial t} + \varepsilon \frac{\partial}{\partial \tau}.$$

The initial values become

$$\mathbf{x}^0(0,0) = \boldsymbol{a},$$
$$\mathbf{x}^i(0,0) = 0, \quad i > 0.$$

Substitution of the expansion in the equation yields

$$\frac{\partial \mathbf{x}^0}{\partial t} + \varepsilon \frac{\partial \mathbf{x}^0}{\partial \tau} + \varepsilon \frac{\partial \mathbf{x}^1}{\partial t} + \cdots = \varepsilon \mathbf{f}^1(\mathbf{x}^0 + \varepsilon \mathbf{x}^1 + \cdots, t).$$

To expand $\mathbf{f}^1$ we assume $\mathbf{f}^1$ to be sufficiently differentiable:

$$\mathbf{f}^1(\mathbf{x}^0 + \varepsilon \mathbf{x}^1 + \cdots, t) = \mathbf{f}^1(\mathbf{x}^0, t) + \varepsilon \mathrm{D}\mathbf{f}^1(\mathbf{x}^0, t) \cdot \mathbf{x}^1 + \cdots.$$

Collecting terms of the same order in ε we have the system

$$\begin{aligned}\frac{\partial \mathbf{x}^0}{\partial t} &= 0,\\ \frac{\partial \mathbf{x}^1}{\partial t} &= -\frac{\partial \mathbf{x}^0}{\partial \tau} + \mathbf{f}^1(\mathbf{x}^0, t),\\ \frac{\partial \mathbf{x}^2}{\partial t} &= -\frac{\partial \mathbf{x}^1}{\partial \tau} + \mathsf{D}\mathbf{f}^1(\mathbf{x}^0, t) \cdot \mathbf{x}^1,\\ &\vdots\end{aligned}$$

which can be solved successively. Integrating the first equation produces

$$\mathbf{x}^0 = \mathbf{A}^0(\tau), \quad \mathbf{A}^0(0) = \mathbf{x}^0.$$

At this stage $\mathbf{A}^0(\tau)$ is still undetermined. Integrating the second equation produces

$$\mathbf{x}^1 = \int_0^t [-\frac{d\mathbf{A}^0}{d\tau} + \mathbf{f}^1(\mathbf{A}^0(\sigma), s)]\, ds + \mathbf{A}^1(\tau), \quad \mathbf{A}^1(0) = 0, \quad \sigma = \varepsilon s.$$

Note that the integration involves mainly $\mathbf{f}^1(\mathbf{A}^0, t)$ as a function of t, since the τ-dependent contribution will be small. We wish to avoid terms in the expansion that become unbounded with time t. We achieve this by the **non-secularity condition**

$$\int_0^T [-\frac{d\mathbf{A}^0}{d\tau} + \mathbf{f}^1(\mathbf{A}^0(\sigma), s)]\, ds = 0, \quad \mathbf{A}^1(\tau) \text{ bounded.}$$

(This condition prevents the emergence of "false secular terms" that grow on a time scale of $1/\varepsilon$ and destroy the asymptotic validity of the solution.) From the integral we obtain

$$\frac{d\mathbf{A}^0}{d\tau} = \frac{1}{T}\int_0^T \mathbf{f}^1(\mathbf{A}^0, s)\, ds, \quad \mathbf{A}^0(0) = \boldsymbol{a},$$

i.e., $\mathbf{A}^0(\tau)$ is determined by the same equation as in the averaging method.

Then we also know from Chapters 1 and 2 that

$$\mathbf{x}(t) = \mathbf{A}^0(\tau) + \mathcal{O}(\varepsilon) \text{ on the time scale } 1/\varepsilon.$$

Note that to determine the first term $\mathbf{x}^0$ we have to consider the expression for the second term $\mathbf{x}^1$. This process repeats itself in the construction of higher-order approximations. We abbreviate

$$\mathbf{x}^1 = \mathbf{u}^1(\mathbf{A}^0(\tau), t) + \mathbf{A}^1(\tau),$$

with

$$\mathbf{u}^1(\mathbf{A}^0(\tau), t) = \int_0^t [-\frac{d\mathbf{A}^0}{d\tau} + \mathbf{f}^1(\mathbf{A}^0(\tau), s)]\, \mathrm{d}s.$$

We obtain

$$\begin{aligned}
\mathbf{x}^2 &= \int_0^t [-\frac{\partial \mathbf{x}^1}{\partial \tau} + \mathsf{D}\mathbf{f}^1(\mathbf{x}^0, s) \cdot \mathbf{x}^1]\, \mathrm{d}s + \mathbf{A}^2(\tau) \\
&= \int_0^t [-\frac{\partial \mathbf{u}^1}{\partial \tau} - \frac{d\mathbf{A}^1}{d\tau} + \mathsf{D}\mathbf{f}^1(\mathbf{A}^0, s) \cdot \mathbf{u}^1(\mathbf{A}^0, s) + \mathsf{D}\mathbf{f}^1(\mathbf{A}^0, s) \cdot \mathbf{A}^1]\, \mathrm{d}s \\
&\quad + \mathbf{A}^2(\tau),
\end{aligned}$$

where $\mathbf{A}^2(0) = 0$. To obtain an expansion with terms bounded in time, we apply again a *nonsecularity condition*

$$\int_0^T [-\frac{\partial \mathbf{u}^1}{\partial \tau} - \frac{d\mathbf{A}^1}{d\tau} + \mathsf{D}\mathbf{f}^1(\mathbf{A}^0, s) \cdot \mathbf{u}^1(\mathbf{A}^0, s) + \mathsf{D}\mathbf{f}^1(\mathbf{A}^0, s) \cdot \mathbf{A}^1]\, \mathrm{d}s = 0.$$

By interchanging D and $\int$ we obtain

$$\frac{d\mathbf{A}^1}{d\tau} = \mathsf{D}\overline{\mathbf{f}}^1(\mathbf{A}^0) \cdot \mathbf{A}^1 + \frac{1}{T} \int_0^T [-\frac{\partial \mathbf{u}^1}{\partial \tau} + \mathsf{D}\mathbf{f}^1(\mathbf{A}^0, s) \cdot \mathbf{u}^1(\mathbf{A}^0, s)]\, \mathrm{d}s,$$

where $\mathbf{A}^1(0) = 0$ and $\overline{\mathbf{f}}^1(\mathbf{A}^0)$ is the average of $\mathbf{f}^1(\mathbf{A}^0, t)$. This is a linear inhomogeneous equation for $\mathbf{A}^1(\tau)$ with variable coefficients.

Theorem 3.5.1. *The solution (3.5.1) constructed in this way is valid with error $\mathcal{O}(\varepsilon^k)$ for time $\mathcal{O}(1/\varepsilon)$. It coincides with the solution obtained by averaging together with regular perturbation theory using (3.2.24), and also with the solution obtained by reduced averaging in cases in which this is possible.*

Proof The error estimate is proved in [208], and also in [217], without any use of averaging. The rest of the theorem follows from Theorem 1.4.13. In the early literature of the subject, this theorem was proved in many special cases by showing computationally that the two-time scale and (appropriately constructed) averaging solutions coincide, and then deduce validity of the two-time scale method from validity of averaging. □

We illustrate the problem of the time-interval of validity in relation to the choice of time scales.

Example 3.5.2. Consider the initial value problem

$$\begin{aligned}
\dot{x} &= \varepsilon^2 y, \quad x(0) = 0, \\
\dot{y} &= -\varepsilon x, \quad y(0) = 1.
\end{aligned}$$

Transforming $\tau = \varepsilon t$ and expanding the solutions, we obtain, after substitution into the equations and applying initial values, the expansions

$$x = \varepsilon\tau - \frac{1}{6}\varepsilon^2\tau^3 + \varepsilon^3 \cdots, \quad y = 1 - \frac{1}{2}\varepsilon\tau^2 + \varepsilon^3 \cdots .$$

It is easy to see that the truncated expansions provide us with asymptotic approximations on the time scale $1/\varepsilon$. The solutions of this initial value problem are easy to obtain:

$$x(t) = \varepsilon^{\frac{1}{2}} \sin(\varepsilon^{\frac{3}{2}} t), \quad y(t) = \cos(\varepsilon^{\frac{3}{2}} t).$$

For the behavior of the solutions on time-intervals longer than $1/\varepsilon$, the natural time variable is clearly $\varepsilon^{\frac{3}{2}} t$.

4

Averaging: the General Case

4.1 Introduction

This chapter will be concerned with the theory of averaging for equations in the standard form

$$\dot{\boldsymbol{x}} = \varepsilon \mathbf{f}^1(\boldsymbol{x}, t) + \varepsilon^2 \mathbf{f}^{[2]}(\boldsymbol{x}, t, \varepsilon).$$

In Chapter 1 we discussed how to obtain perturbation problems in the standard form and in Chapter 2 we studied averaging in the periodic case.

Many results in the theory of asymptotic approximations have been obtained in the Soviet Union from 1930 onwards. Earlier work of this school has been presented in the famous book by Bogoliubov and Mitropolsky [35] and in the survey paper by Volosov [283]. A brief glance at the main Soviet mathematical journals shows that many results on integral manifolds, equations with retarded argument, quasi- or almost- periodic equations etc. have been produced. See also the survey by Mitropolsky [191] and the book by Bogoliubov, Mitropolsky and Samoilenko [33]. In 1966 Roseau [228, Chapter 12] presented a transparent proof of the validity of averaging in the periodic case. Different proofs for both the periodic and the general case have been provided by Besjes [31] and Perko [217]. In the last paper moreover the relation between averaging and the multiple time scales method has been established.

Most of the work mentioned above is concerned with approximations on the time scale $1/\varepsilon$. Extension of the time scale of validity is possible if for instance one studies equations leading to approximations starting inside the domain of attraction of an asymptotically stable critical point. Extensions like this were studied by Banfi [20] and Banfi and Graffi [21]. Eckhaus [81] gives a detailed proof and new results for systems with attraction; later the proof could be simplified considerably by Eckhaus using a lemma due to Sanchez–Palencia, see [275]. Results on related problems have been obtained by Kirchgraber and Stiefel [149] who study periodic solutions and the part played by invariant manifolds. Systems with attraction will be studied in Chapter 5. Another type of problem where extension of the time scale is possible is provided by systems in resonance with as an important example Hamiltonian systems.

In the theory of averaging of periodic systems one usually obtains $\mathcal{O}(\varepsilon)$-approximations on the time scale $1/\varepsilon$. In the case of general averaging an order function $\delta(\varepsilon)$ plays a part; see Section 4.3. The order function δ is determined by the behavior of $\mathbf{f}^1(\boldsymbol{x}, t)$ and its average on a long time scale. In the original theorem by Bogoliubov and Mitropolsky an $o(1)$ estimate has been given. Implicitly however, an $\mathcal{O}(\sqrt{\delta})$ estimate has been derived in the proof. Also in the proofs by Besjes one obtains an $\mathcal{O}(\sqrt{\varepsilon})$ estimate in the general case but, using a different proof, an $\mathcal{O}(\varepsilon)$ estimate in the periodic case.

A curiosity of the proofs is that in restricting the proof of the general case to the case of periodic systems one still obtains an $\mathcal{O}(\sqrt{\varepsilon})$ estimate. It takes a special effort to obtain an $\mathcal{O}(\varepsilon)$ estimate for periodic systems. Eckhaus [81] introduces the concept of *local average* of a vector field to give a new proof of the validity of periodic and general averaging. In the general case Eckhaus obtains an $\mathcal{O}(\sqrt{\delta})$ estimate; on specializing to the periodic case one can apply the averaging repeatedly to obtain an $\mathcal{O}(\varepsilon^r)$ estimate where r approaches 1 from below.

In the sequel we shall use Eckhaus' concept of local average to derive in a simple way an $\mathcal{O}(\varepsilon)$ estimate in the periodic case under rather weak assumptions (Section 4.2). In the general case one obtains under similar assumptions an $\mathcal{O}(\sqrt{\delta(\varepsilon)})$ estimate (Section 4.3).

In Section 4.5 we present the theory of second-order approximation in the general case; we find here that the first-order approximation is valid with $\mathcal{O}(\delta)$-error in the general case if we require the vector field to be differentiable instead of only Lipschitz continuous.

Extensions of this theory to functional differential equations (delay equations) were obtained in [168, 167, 169, 170] by methods which are very similar to the ones employed in this chapter. We are not going to state our results in this generality, and refer the interested reader to the cited literature.

4.2 Basic Lemmas; the Periodic Case

In this section we shall derive some basic results which are preliminary for our treatment of the general theory of averaging.

Definition 4.2.1 (Eckhaus[81]). *Consider the continuous vector field* $\mathbf{f} : \mathbb{R} \times \mathbb{R}^n \to \mathbb{R}^n$. *We define the* **local average** $\mathbf{f}_T$ *of* $\mathbf{f}$ *by*

$$\mathbf{f}_T(\boldsymbol{x}, t) = \frac{1}{T}\int_0^T \mathbf{f}(\boldsymbol{x}, t+s)\,\mathrm{d}s.$$

Remark 4.2.2. T is a parameter which can be chosen, and can be made ε-dependent if we wish. Whenever we estimate with respect to ε we shall require $\varepsilon T = o(1)$, where ε is a small parameter. So we may choose for instance

$T = 1/\sqrt{\varepsilon}$ or $T = 1/|\varepsilon \log(\varepsilon)|$. Included is also $T = \mathcal{O}_\sharp(1)$ which we shall use for periodic $\mathbf{f}$. Note that the local average of a continuous vector field always exists. ♡

Lemma 4.2.3. *Consider the continuous vector field* $\mathbf{f} : \mathbb{R}^n \times \mathbb{R} \to \mathbb{R}^n$, *T-periodic in t. Then*

$$\mathbf{f}_T(\boldsymbol{x}, t) = \overline{\mathbf{f}}(\boldsymbol{x}) = \frac{1}{T}\int_0^T \mathbf{f}(\boldsymbol{x}, s)\,\mathrm{d}s.$$

Proof We write $\mathbf{f}_T(\boldsymbol{x}, t) = \frac{1}{T}\int_t^{t+T} \mathbf{f}(\boldsymbol{x}, s)\,\mathrm{d}s$. Partial differentiation with respect to t produces zero because of the T-periodicity of $\mathbf{f}$; it follows that $\mathbf{f}_T$ does not depend on t *explicitly*, so we may put $t = 0$, that is, $\mathbf{f}_T(\boldsymbol{x}, t) = \mathbf{f}_T(\boldsymbol{x}, 0) = \frac{1}{T}\int_0^T \mathbf{f}(\boldsymbol{x}, s)\,\mathrm{d}s = \overline{\mathbf{f}}(\boldsymbol{x})$. □

We shall now introduce vector fields which can be averaged in a general sense. Since most applications are for differential equations we impose some additional regularity conditions on the vector field.

Definition 4.2.4. *Consider the vector field* $\mathbf{f}(\boldsymbol{x}, t)$ *with* $\mathbf{f} : \mathbb{R}^n \times \mathbb{R} \to \mathbb{R}^n$, *Lipschitz continuous in* $\boldsymbol{x}$ *on* $D \subset \mathbb{R}^n$, $t \geq 0$; $\mathbf{f}$ *continuous in* t *and* $\boldsymbol{x}$ *on* $\mathbb{R}^+ \times D$. *If the average*

$$\overline{\mathbf{f}}(\boldsymbol{x}) = \lim_{T\to\infty} \frac{1}{T}\int_0^T \mathbf{f}(\boldsymbol{x}, s)\,\mathrm{d}s$$

exists and the limit is uniform in $\boldsymbol{x}$ *on compact sets* $K \subset D$, *then* $\mathbf{f}$ *is called a* **KBM-vector field** *(KBM stands for Krylov, Bogoliubov and Mitropolsky). Note, that usually the vector field* $\mathbf{f}(\boldsymbol{x}, t)$ *contains parameters. We assume that the parameters and the initial conditions are independent of* ε, *and that the limit is uniform in the parameters.*

We now formulate a simple estimate.

Lemma 4.2.5. *Consider the Lipschitz continuous map* $\boldsymbol{x} : \mathbb{R} \to \mathbb{R}^n$ *with Lipschitz constant* $\lambda_{\boldsymbol{x}}$, *then*

$$\|\mathbf{x}(t) - \mathbf{x}_T(t)\| \leq \frac{1}{2}\lambda_{\mathbf{x}} T.$$

Proof One has

$$\|\mathbf{x}(t) - \mathbf{x}_T(t)\| = \left\|\frac{1}{T}\int_0^T (\mathbf{x}(t) - \mathbf{x}(t+s))\,\mathrm{d}s\right\| \leq \frac{1}{T}\int_0^T \lambda_{\mathbf{x}} s\,\mathrm{d}s = \frac{1}{2}\lambda_{\mathbf{x}} T,$$

and this gives the desired estimate. □

Corollary 4.2.6. *Let* $\mathbf{x}(t)$ *be a solution of the equation*

$$\dot{\boldsymbol{x}} = \varepsilon \mathbf{f}^1(\boldsymbol{x}, t), \quad t \geq 0, \quad \boldsymbol{x} \in D \subset \mathbb{R}^n.$$

Let

$$M = \sup_{\boldsymbol{x} \in D} \sup_{0 \leq \varepsilon t \leq L} \left\| \mathbf{f}^1(\boldsymbol{x}, t) \right\| < \infty.$$

Then $\|\mathbf{x}(t) - \mathbf{x}_T(t)\| \leq \frac{1}{2}\varepsilon MT$ *(since* $\lambda_{\mathbf{x}} = \varepsilon M$*).*

In the following lemma we introduce a perturbation problem in the standard form.

Lemma 4.2.7. *Consider the equation*

$$\dot{\boldsymbol{x}} = \varepsilon \mathbf{f}^1(\boldsymbol{x}, t), \quad t \geq 0, \quad \boldsymbol{x} \in D \subset \mathbb{R}^n.$$

Assume

$$\left\| \mathbf{f}^1(\boldsymbol{x}, t) - \mathbf{f}^1(\boldsymbol{y}, t) \right\| \leq \lambda_{\mathbf{f}^1} \left\| \boldsymbol{x} - \boldsymbol{y} \right\|$$

for all $\boldsymbol{x}, \boldsymbol{y} \in D$ *(Lipschitz continuity) and* $\mathbf{f}^1$ *is continuous in* t *and* $\boldsymbol{x}$*; Let*

$$M = \sup_{\boldsymbol{x} \in D} \sup_{0 \leq \varepsilon t \leq L} \left\| \mathbf{f}^1(\boldsymbol{x}, t) \right\| < \infty.$$

The constants $\lambda_{\mathbf{f}^1}$*,* L *and* M *are supposed to be* ε*-independent. Since*

$$\mathbf{x}(t) = \boldsymbol{a} + \varepsilon \int_0^t \mathbf{f}^1(\mathbf{x}(s), s)\, \mathrm{d}s,$$

where $\mathbf{x}$ *is a solution of the differential equation, we have, with* t *on the time scale* $1/\varepsilon$

$$\left\| \mathbf{x}_T(t) - \boldsymbol{a} - \varepsilon \int_0^t \mathbf{f}_T^1(\mathbf{x}(s), s)\, \mathrm{d}s \right\| \leq \frac{1}{2}\varepsilon(1 + \lambda_{\mathbf{f}^1} L)MT$$

or

$$\mathbf{x}_T(t) = \boldsymbol{a} + \varepsilon \int_0^t \mathbf{f}_T^1(\mathbf{x}(s), s)\, \mathrm{d}s + \mathcal{O}(\varepsilon T).$$

Proof By definition

$$\begin{aligned}
\mathbf{x}_T(t) &= \boldsymbol{a} + \frac{\varepsilon}{T} \int_0^T \int_0^{t+s} \mathbf{f}^1(\mathbf{x}(\sigma), \sigma)\, \mathrm{d}\sigma\, \mathrm{d}s \\
&= \boldsymbol{a} + \frac{\varepsilon}{T} \int_0^T \int_s^{t+s} \mathbf{f}^1(\mathbf{x}(\sigma), \sigma)\, \mathrm{d}\sigma\, \mathrm{d}s + \varepsilon \mathbf{R}_1 \\
&= \boldsymbol{a} + \frac{\varepsilon}{T} \int_0^T \int_0^t \mathbf{f}^1(\mathbf{x}(\sigma + s), \sigma + s)\, \mathrm{d}\sigma\, \mathrm{d}s + \varepsilon \mathbf{R}_1 \\
&= \boldsymbol{a} + \frac{\varepsilon}{T} \int_0^t \int_0^T \mathbf{f}^1(\mathbf{x}(\sigma), \sigma + s)\, \mathrm{d}s\, \mathrm{d}\sigma + \varepsilon \mathbf{R}_1 + \varepsilon \mathbf{R}_2 \\
&= \boldsymbol{a} + \varepsilon \int_0^t \mathbf{f}_T^1(\mathbf{x}(\sigma), \sigma)\, \mathrm{d}\sigma + \varepsilon \mathbf{R}_1 + \varepsilon \mathbf{R}_2.
\end{aligned}$$

$\mathbf{R}_1$ and $\mathbf{R}_2$ have been defined implicitly and we estimate these quantities as follows:

$$\|\mathbf{R}_1\| = \left\| \frac{1}{T} \int_0^T \int_0^s \mathbf{f}^1(\mathbf{x}(\sigma), \sigma)\, \mathrm{d}\sigma\, \mathrm{d}s \right\| \leq \frac{1}{T} \int_0^T \int_0^s M\, \mathrm{d}\sigma\, \mathrm{d}s \leq \frac{1}{2} MT,$$

and

$$\begin{aligned}
\|\mathbf{R}_2\| &= \left\| \frac{1}{T} \int_0^t \int_0^T [\mathbf{f}^1(\mathbf{x}(\sigma + s), \sigma + s) - \mathbf{f}^1(\mathbf{x}(\sigma), \sigma + s)]\, \mathrm{d}s\, \mathrm{d}\sigma \right\| \\
&\leq \frac{\lambda_{\mathbf{f}^1}}{T} \int_0^t \int_0^T \|\mathbf{x}(\sigma + s) - \mathbf{x}(\sigma)\|\, \mathrm{d}s\, \mathrm{d}\sigma \\
&\leq \varepsilon \frac{\lambda_{\mathbf{f}^1}}{T} \int_0^t \int_0^T \int_\sigma^{\sigma+s} \|\mathbf{f}^1(\mathbf{x}(\zeta), \zeta)\|\, \mathrm{d}\zeta\, \mathrm{d}s\, \mathrm{d}\sigma \\
&\leq \varepsilon \frac{\lambda_{\mathbf{f}^1}}{T} \int_0^t \int_0^T M s\, \mathrm{d}s\, \mathrm{d}\sigma = \frac{1}{2} \varepsilon t \lambda_{\mathbf{f}^1} MT \leq \frac{1}{2} \lambda_{\mathbf{f}^1} LMT,
\end{aligned}$$

which completes the proof. □

The preceding lemmas enable us to compare solutions of two differential equations:

Lemma 4.2.8. *Consider the initial value problem*

$$\dot{\boldsymbol{x}} = \varepsilon \mathbf{f}^1(\boldsymbol{x}, t), \quad \mathbf{x}(0) = \boldsymbol{a},$$

with $\mathbf{f}^1 : \mathbb{R}^n \times \mathbb{R}$ *Lipschitz continuous in* $\boldsymbol{x}$ *on* $D \subset \mathbb{R}^n$, t *on the time scale* $1/\varepsilon$; $\mathbf{f}^1$ *continuous in* t *and* $\boldsymbol{x}$. *If* $\mathbf{y}$ *is the solution of*

$$\dot{\boldsymbol{y}} = \varepsilon \mathbf{f}^1_T(\boldsymbol{y}, t), \quad \mathbf{y}(0) = \boldsymbol{a},$$

then $\mathbf{x}(t) = \mathbf{y}(t) + \mathcal{O}(\varepsilon T)$ *on the time scale* $1/\varepsilon$.

Proof Writing the differential equation as an integral equation, we see that

$$\mathbf{x}(t) = \boldsymbol{a} + \varepsilon \int_0^t \mathbf{f}^1(\mathbf{x}(s), s)\, \mathrm{d}s.$$

With Corollary 4.2.6 and Lemma 4.2.7 we obtain

$$\begin{aligned}
&\left\| \mathbf{x}(t) - \boldsymbol{a} - \varepsilon \int_0^t \mathbf{f}^1_T(\mathbf{x}(s), s)\, \mathrm{d}s \right\| \\
&\leq \|\mathbf{x}(t) - \mathbf{x}_T(t)\| + \left\| \mathbf{x}_T(t) - \boldsymbol{a} - \varepsilon \int_0^t \mathbf{f}^1_T(\mathbf{x}(s), s)\, \mathrm{d}s \right\| \\
&\leq \varepsilon MT (1 + \frac{1}{2} \lambda_{\mathbf{f}^1} L).
\end{aligned}$$

It follows that $\mathbf{x}(t) = \boldsymbol{a} + \varepsilon \int_0^t \mathbf{f}_T(\mathbf{x}(s), s)\, \mathrm{d}s + \mathcal{O}(\varepsilon T)$. Since

$$\mathbf{y}(t) = \boldsymbol{a} + \varepsilon \int_0^t \mathbf{f}_T^1(\mathbf{y}(s), s)\, \mathrm{d}s,$$

we have

$$\mathbf{x}(t) - \mathbf{y}(t) = \varepsilon \int_0^t [\mathbf{f}_T^1(\mathbf{x}(s), s) - \mathbf{f}_T^1(\mathbf{y}(s), s)]\, \mathrm{d}s + \mathcal{O}(\varepsilon T),$$

and because of the Lipschitz continuity of $\mathbf{f}_T^1$ (inherited from $\mathbf{f}^1$)

$$\|\mathbf{x}(t) - \mathbf{y}(t)\| \le \varepsilon \int_0^t \lambda_{\mathbf{f}^1} \|\mathbf{x}(s) - \mathbf{y}(t)\|\, \mathrm{d}s + \mathcal{O}(\varepsilon T).$$

The Gronwall Lemma 1.3.3 yields

$$\|\mathbf{x}(t) - \mathbf{y}(t)\| = \mathcal{O}(\varepsilon T e^{\varepsilon \lambda_{\mathbf{f}^1} t}),$$

from which the lemma follows. □

Corollary 4.2.9. *At this stage it is a trivial application of Lemmas 4.2.3 and 4.2.8 to prove the Averaging Theorem 2.8.1 in the periodic case.*

Serious progress however can be made in the general case where we shall obtain sharp estimates while keeping the same kind of simple proofs as in this section.

4.3 General Averaging

To prove the fundamental theorem of general averaging we need a few more results.

Lemma 4.3.1. *If $\mathbf{f}^1$ is a KBM-vector field and assuming $\varepsilon T = o(1)$ as $\varepsilon \downarrow 0$, then on the time scale $1/\varepsilon$ one has*

$$\mathbf{f}_T^1(\boldsymbol{x}, t) = \overline{\mathbf{f}}^1(\boldsymbol{x}) + \mathcal{O}(\delta_1(\varepsilon)/(\varepsilon T)),$$

where

$$\delta_1(\varepsilon) = \sup_{\boldsymbol{x} \in D} \sup_{t \in [0, L/\varepsilon)} \varepsilon \left\| \int_0^t [\mathbf{f}^1(\boldsymbol{x}, s) - \overline{\mathbf{f}}^1(\boldsymbol{x})]\, \mathrm{d}s \right\| .$$

Remark 4.3.2. We call $\delta_1(\varepsilon)$ the **order function of $\mathbf{f}^1$**. In the periodic case $\delta_1(\varepsilon) = \varepsilon$. ♡

Notation 4.3.3 *In the general setup, we have to define inductively a number of order functions $\delta(\varepsilon)$. Let κ be a counter, starting at 0. Let $\delta(\varepsilon) = \varepsilon$ and increase κ by one.*

The general induction step runs as follows. Let I_κ be a multi-index, written as $I_\kappa = \iota_0|\dots|\iota_m, m < \kappa$, where we do not write trailing zeros. Each ι_j stands for the multiplicity of the order function $\delta_{I_j}(\varepsilon)$ in the expression

$$\delta_{I_j}(\varepsilon) = \sup_{\boldsymbol{x}\in D} \sup_{t\in[0,L/\varepsilon)} \pi_{I_j}(\varepsilon) \left\| \int_0^t [\mathbf{f}^{I_j}(\boldsymbol{x},s) - \overline{\mathbf{f}}^{I_j}(\boldsymbol{x})]\, \mathrm{d}s \right\|,$$

with

$$\pi_{I_j}(\varepsilon) = \prod_{k=0}^{j-1} \delta_{I_k}^{\iota_k}(\varepsilon).$$

By putting $j = \kappa$ in these formulae, we obtain the definition of δ_{I_κ}. If the right hand side does not exist for $j = \kappa$, the theory stops here. If it does exist, we proceed to define

$$\delta_{I_\kappa}(\varepsilon)\mathbf{u}^{I_\kappa}(\boldsymbol{w},t) = \pi_{I_\kappa}(\varepsilon) \int_0^t [\mathbf{f}^{I_\kappa}(\boldsymbol{w},s) - \overline{\mathbf{f}}^{I_\kappa}(\boldsymbol{w})]\, \mathrm{d}s.$$

This definition implies that $\mathbf{u}^{I_\kappa}$ is bounded by a constant, independent of ε. We then increase κ by 1 and repeat our induction step, as far as necessary for the estimates we want to obtain.

Proof

$$\begin{aligned}\mathbf{f}_T(\boldsymbol{x},t) - \overline{\mathbf{f}}(\boldsymbol{x}) &= \frac{1}{T}\int_0^T [\mathbf{f}(\boldsymbol{x},t+s) - \overline{\mathbf{f}}(\boldsymbol{x})]\, \mathrm{d}s = \frac{1}{T}\int_t^{t+T} [\mathbf{f}(\boldsymbol{x},s) - \overline{\mathbf{f}}(\boldsymbol{x})]\, \mathrm{d}s \\ &= \frac{1}{T}\int_0^{t+T} [\mathbf{f}(\boldsymbol{x},s) - \overline{\mathbf{f}}(\boldsymbol{x})]\, \mathrm{d}s - \frac{1}{T}\int_0^t [\mathbf{f}(\boldsymbol{x},s) - \overline{\mathbf{f}}(\boldsymbol{x})]\, \mathrm{d}s.\end{aligned}$$

We assumed $\varepsilon T = o(1)$, so if $\alpha = 0$ or T we have $\varepsilon\alpha = o(1)$ (implying that we can still use the same L as in the definition of $\delta_1(\varepsilon)$ for ε small enough) and

$$\int_0^{t+\alpha} [\mathbf{f}(\boldsymbol{x},s) - \overline{\mathbf{f}}(\boldsymbol{x})]\, \mathrm{d}s = \mathcal{O}(\delta(\varepsilon)/\varepsilon),$$

from which the estimate follows. □

Lemma 4.3.4 (Lebovitz 1987, private communication). *For a KBM-vector field $\mathbf{f}^1(\boldsymbol{x},t)$ with $\mathbf{f}^1 : \mathbb{R}^n \times \mathbb{R} \to \mathbb{R}^n$ one has $\delta_1(\varepsilon) = o(1)$.*

Proof Choosing $\mu > 0$ we have, because of the uniform existence of the limit for $T \to \infty$:

$$\left\| \frac{1}{T}\int_0^T [\mathbf{f}(\boldsymbol{x},s) - \overline{\mathbf{f}}(\boldsymbol{x})]\, \mathrm{d}s \right\| < \mu$$

for $T > T_\mu$ (independent of ε) and $\boldsymbol{x} \in D \subset \mathbb{R}^n$ (uniformly) or

$$\left\| \frac{1}{t}\int_0^t [\mathbf{f}(\boldsymbol{x},s) - \overline{\mathbf{f}}(\boldsymbol{x})]\, \mathrm{d}s \right\| < \mu$$

for $T_\mu < t < L/\varepsilon$, $\boldsymbol{x} \in D$ and ε small enough. It follows that

$$\frac{\varepsilon}{L}\left\|\int_0^t [\mathbf{f}(\boldsymbol{x}, s) - \overline{\mathbf{f}}(\boldsymbol{x})]\,\mathrm{d}s\right\| < \mu\varepsilon t/L < \mu,$$

with μ arbitrarily small. □

Lemma 4.3.5. *Let* $\mathbf{y}$ *be the solution of the initial value problem*

$$\dot{\boldsymbol{y}} = \varepsilon \mathbf{f}^1_T(\boldsymbol{y}, t), \quad \mathbf{y}(0) = \boldsymbol{a}.$$

We suppose $\mathbf{f}^1$ *is a KBM-vector field with* order function $\delta_1(\varepsilon)$ *(cf. Notation 4.3.3); let* $\mathbf{z}$ *be the solution of the initial value problem*

$$\dot{\boldsymbol{z}} = \varepsilon \overline{\mathbf{f}}^1(\boldsymbol{z}), \quad \mathbf{z}(0) = \boldsymbol{a}.$$

Then

$$\mathbf{y}(t) = \mathbf{z}(t) + \mathcal{O}(\delta_1(\varepsilon)/(\varepsilon T)),$$

with t *on the time scale* $1/\varepsilon$.

Proof Using Lemma 4.3.1 we see that

$$\begin{aligned}\mathbf{y}(t) - \mathbf{z}(t) &= \varepsilon \int_0^t [\mathbf{f}^1_T(\mathbf{y}(s), s) - \overline{\mathbf{f}}^1(\mathbf{z}(s))]\,\mathrm{d}s \\ &= \varepsilon \int_0^t [\overline{\mathbf{f}}^1(\mathbf{y}(s)) - \overline{\mathbf{f}}^1(\mathbf{z}(s))]\,\mathrm{d}s + \mathcal{O}(\delta_1(\varepsilon)t/T).\end{aligned}$$

Since $\overline{\mathbf{f}}^1$ is Lipschitz continuous with constant $\lambda_{\mathbf{f}^1}$, we obtain

$$\mathbf{y}(t) - \mathbf{z}(t) = \mathcal{O}(\delta_1(\varepsilon)/(\varepsilon T)),$$

from which the lemma follows. □

We are now able to prove the general averaging theorem:

Theorem 4.3.6 (general averaging). *Consider the initial value problems*

$$\dot{\boldsymbol{x}} = \varepsilon \mathbf{f}^1(\boldsymbol{x}, t), \quad \mathbf{x}(0) = \boldsymbol{a},$$

with $\mathbf{f}^1 : \mathbb{R}^n \times \mathbb{R} \to \mathbb{R}^n$ *and*

$$\dot{\boldsymbol{z}} = \varepsilon \overline{\mathbf{f}}^1(\boldsymbol{z}), \quad \mathbf{z}(0) = \boldsymbol{a},$$

where

$$\overline{\mathbf{f}}^1(\boldsymbol{x}) = \lim_{T\to\infty} \frac{1}{T}\int_0^T \mathbf{f}^1(\boldsymbol{x}, t)\,\mathrm{d}t,$$

and $\boldsymbol{x}, \boldsymbol{z}, \boldsymbol{a} \in D \subset \mathbb{R}^n$, $t \in [0, \infty)$, $\varepsilon \in (0, \varepsilon_0]$. *Suppose*

1. $\mathbf{f}^1$ *is a KBM-vector field with average* $\overline{\mathbf{f}}^1$ *and order function* $\delta_1(\varepsilon)$*;*
2. $\mathbf{z}(t)$ *belongs to an interior subset of* D *on the time scale* $1/\varepsilon$*;*

then

$$\mathbf{x}(t) - \mathbf{z}(t) = \mathcal{O}(\sqrt{\delta_1(\varepsilon)})$$

as $\varepsilon \downarrow 0$ *on the time scale* $1/\varepsilon$*.*

Proof Applying Lemmas 4.2.8 and 4.3.5, using the triangle inequality, we have on the time scale $1/\varepsilon$:

$$\mathbf{x}(t) = \mathbf{z}(t) + \mathcal{O}(\varepsilon T) + \mathcal{O}(\delta_1(\varepsilon)/(\varepsilon T)).$$

The errors are of the same order of magnitude if

$$\varepsilon^2 T^2 = \delta_1(\varepsilon),$$

so that

$$\mathbf{x}(t) = \mathbf{z}(t) + \mathcal{O}(\sqrt{\delta_1(\varepsilon)}),$$

if we let $T = \sqrt{\delta_1(\varepsilon)}/\varepsilon$. □

Remark 4.3.7. As before, Condition 2 has been used implicitly in the estimates. Note that since $\delta_1(\varepsilon) = o(1)$ (Lemma 4.3.4) it follows that $\varepsilon T = o(1)$, which was an implicit assumption; ♡

Remark 4.3.8. The general theory has been revisited recently in [11]. ♡

To understand the theory of general averaging it is instructive to analyze a few simple examples.

4.4 Linear Oscillator with Increasing Damping

Consider the equation

$$\ddot{x} + \varepsilon(2 - F(t))\dot{x} + x = 0,$$

with initial values given at $t = 0 : \mathrm{x}(0) = r_0$, $\dot{\mathrm{x}}(0) = 0$. $F(t)$ is a continuous function, monotonically decreasing towards zero for $t \to \infty$ with $F(0) = 1$. So the problem is simple: we start with an oscillator with damping coefficient ε, we end up (in the limit for $t \to \infty$) with an oscillator with damping coefficient 2ε. We shall show that on the time scale $1/\varepsilon$ the system behaves approximately as if it has the limiting damping coefficient 2ε, which seems an interesting result. To obtain the standard form, transform $(x, \dot{x}) \mapsto (r, \phi)$ by

$$x = r\cos(t + \phi), \quad \dot{x} = -r\sin(t + \phi).$$

We obtain

$$\begin{aligned}\dot{r} &= \varepsilon r \sin^2(t+\phi)(-2+F(t)), \quad \mathrm{r}(0) = r_0,\\ \dot{\phi} &= \varepsilon \sin(t+\phi)\cos(t+\phi)(-2+F(t)), \quad \phi(0) = 0.\end{aligned}$$

Averaging produces

$$\dot{\tilde{r}} = -\varepsilon\tilde{r}, \quad \dot{\tilde{\phi}} = 0,$$

so that

$$\begin{aligned}x(t) &= r_0 e^{-\varepsilon t}\cos(t) + \mathcal{O}(\sqrt{\delta_1(\varepsilon)}),\\ \dot{x}(t) &= -r_0 e^{-\varepsilon t}\sin(t) + \mathcal{O}(\sqrt{\delta_1(\varepsilon)}).\end{aligned}$$

To estimate δ_1 we note that $x = \dot{x} = 0$ is a globally stable attractor (one can use the Lyapunov function $\frac{1}{2}(x^2+\dot{x}^2)$ to show this if the mechanics of the problem is not already convincing enough). So the order of magnitude of δ_1 is determined by

$$\sup_D \sup_{t\in[0,\frac{L}{\varepsilon})} \varepsilon \left| \int_0^t [\sin^2(s+\phi)(-2+F(s))+1]\,\mathrm{d}s \right|$$

and

$$\sup_D \sup_{t\in[0,\frac{L}{\varepsilon})} \varepsilon \left| \int_0^t [\sin(s+\phi)\cos(s+\phi)(-2+F(s))]\,\mathrm{d}s \right| .$$

The second integral is bounded for all t so this contributes $\mathcal{O}(\varepsilon)$. The same holds for the part

$$\int_0^t (-2\sin^2(s+\phi)+1)\,\mathrm{d}s.$$

To estimate

$$\int_0^t F(s)\sin^2(s+\phi)\,\mathrm{d}s$$

we have to make an assumption about F. For instance if F decreases exponentially with time we have $\delta_1(\varepsilon) = \mathcal{O}(\varepsilon)$ and an approximation with error $\mathcal{O}(\sqrt{\varepsilon})$. If $F \approx t^{-s}(0 < s < 1)$ we have $\delta_1(\varepsilon) = \mathcal{O}(\varepsilon^s)$ and an approximation with error $\mathcal{O}(\varepsilon^{\frac{s}{2}})$. If $F(t) = (1+t)^{-1}$ we have $\delta_1(\varepsilon) = \mathcal{O}(\varepsilon|\log(\varepsilon)|)$. If $\delta_1(\varepsilon)$ is not $o(1)$, then we need to adapt the average in order to apply the theory. We remark finally that to describe the dependence of the oscillator on the initial damping we clearly need a different order of approximation.

4.5 Second-Order Approximations in General Averaging; Improved First-Order Estimate Assuming Differentiability

Higher-order approximations in the periodic case are well known and form an established theory with many applications. It turns out there is an unexpected profit: the first-order approximation is better than we proved it to be in Section 4.3 under the differentiability condition.

Lemma 4.5.1. *Suppose* $\mathbf{f}^1$ *is a KBM-vector field which has a Lipschitz continuous first derivative in* $\boldsymbol{x}$*;* $\boldsymbol{x} \in D \subset \mathbb{R}^n$*,* t *on the time scale* $1/\varepsilon$*;* $\boldsymbol{x}$ *is the solution of*

$$\dot{\boldsymbol{x}} = \varepsilon \mathbf{f}^1(\boldsymbol{x}, t), \quad \mathbf{x}(0) = \boldsymbol{a}.$$

We define $\boldsymbol{y}$ *by*

$$\mathbf{x}(t) = \boldsymbol{y}(t) + \delta_1(\varepsilon)\mathbf{u}^1(\mathbf{y}(t), t).$$

Then

$$\mathbf{y}(t) = \boldsymbol{a} + \varepsilon \int_0^t \overline{\mathbf{f}}^1(\mathbf{y}(s))\,\mathrm{d}s$$
$$+\varepsilon\delta_1(\varepsilon)\int_0^t \left(\mathsf{D}\mathbf{f}^1(\mathbf{y}(s), s)\cdot \mathbf{u}^1(\mathbf{y}(s), s) - \mathsf{D}\mathbf{u}^1(\mathbf{y}(s), s)\cdot \overline{\mathbf{f}}^1(\mathbf{y}(s))\right)\,\mathrm{d}s + \mathcal{O}(\delta_1^2)$$

on the time scale $1/\varepsilon$*.*

Proof This is a standard computation:

$$\begin{aligned}
\mathbf{y}(t) &= \mathbf{x}(t) - \delta_1(\varepsilon)\mathbf{u}^1(\mathbf{y}(t), t) \\
&= \boldsymbol{a} + \varepsilon \int_0^t \mathbf{f}^1(\mathbf{x}(s), s)\,\mathrm{d}s - \varepsilon \int_0^t \left(\mathbf{f}^1(\mathbf{y}(s), s) - \overline{\mathbf{f}}^1(\mathbf{y}(s))\right)\,\mathrm{d}s \\
&\quad -\delta_1(\varepsilon)\int_0^t \mathsf{D}\mathbf{u}^1(\mathbf{y}(s), s)\cdot \frac{d\mathbf{y}}{ds}\,\mathrm{d}s \\
&= \boldsymbol{a} + \varepsilon \int_0^t \overline{\mathbf{f}}^1(\mathbf{y}(s))\,\mathrm{d}s \\
&\quad +\varepsilon\delta_1(\varepsilon)\int_0^t \left(\mathsf{D}\mathbf{f}^1(\mathbf{y}(s), s)\cdot \mathbf{u}^1(\mathbf{y}(s), s) - \mathsf{D}\mathbf{u}^1(\mathbf{y}(s), s)\cdot \overline{\mathbf{f}}^1(\mathbf{y}(s))\right)\,\mathrm{d}s \\
&\quad +\mathcal{O}(\delta_1^2),
\end{aligned}$$

and we have obtained the result we claimed. □

Lemma 4.5.2. *Let* $\mathbf{y}$ *be defined as in Lemma 4.5.1 and let* $\mathbf{v}$ *be the solution of*

$$\dot{\boldsymbol{v}} = \varepsilon \overline{\mathbf{f}}^1(\boldsymbol{v}) + \varepsilon\delta_1(\varepsilon)\mathbf{f}_T^{1|1}(\boldsymbol{v}, t) \ , \ \mathbf{v}(0) = \boldsymbol{a},$$

where

$$\mathbf{f}^{1|1}(\boldsymbol{v}, t) = \mathsf{D}\mathbf{f}^1(\boldsymbol{v}, t) \cdot \mathbf{u}^1(\boldsymbol{v}, t) - \mathsf{D}\mathbf{u}^1(\boldsymbol{v}, t) \cdot \overline{\mathbf{f}}^1(\boldsymbol{v}).$$

Assume that $\mathbf{f}^{1|1}$ *is a KBM-vector field, then*

$$\mathbf{y}(t) = \mathbf{v}(t) + \mathcal{O}(\delta_1(\varepsilon)\,(\varepsilon T + \delta_1(\varepsilon)))$$

on the time scale $1/\varepsilon$.

Proof

$$\mathbf{v}(t) = \boldsymbol{a} + \varepsilon \int_0^t \overline{\mathbf{f}}^1(\mathbf{v}(s))\,\mathrm{d}s + \varepsilon\delta_1(\varepsilon) \int_0^t \mathbf{f}_T^{1|1}(\mathbf{v}(s), s)\,\mathrm{d}s.$$

In the same way as in the proof of Lemma 4.2.7 we obtain

$$\int_0^t \mathbf{f}_T^{1|1}(\mathbf{v}(s), s)\,\mathrm{d}s = \int_0^t \mathbf{f}^{1|1}(\mathbf{v}(s), s)\,\mathrm{d}s + \mathcal{O}(T),$$

which implies

$$\mathbf{v}(t) = \boldsymbol{a} + \varepsilon \int_0^t \overline{\mathbf{f}}^1(\mathbf{v}(s))\,\mathrm{d}s + \varepsilon\delta_1(\varepsilon) \int_0^t \mathbf{f}^{1|1}(\mathbf{v}(s), s)\,\mathrm{d}s + \mathcal{O}(\delta_1(\varepsilon)\varepsilon T).$$

Subtracting this from the estimate for $\mathbf{y}$ in Lemma 4.5.1 produces

$$\begin{aligned}\mathbf{y}(t) - \mathbf{v}(t) &= \varepsilon \int_0^t \left(\overline{\mathbf{f}}^1(\mathbf{y}(s)) - \overline{\mathbf{f}}^1(\mathbf{v}(s))\right)\,\mathrm{d}s \\ &+ \varepsilon\delta_1(\varepsilon) \int_0^t \left(\mathbf{f}^{1|1}(\mathbf{y}(s), s) - \mathbf{f}^{1|1}(\mathbf{v}(s), s)\right)\,\mathrm{d}s + \mathcal{O}(\delta_1^2(\varepsilon) + \delta_1(\varepsilon)\varepsilon T).\end{aligned}$$

Using the Lipschitz continuity of $\overline{\mathbf{f}}^1$ and $\mathbf{f}^{1|1}$ and applying the Gronwall Lemma 1.3.3 yields the desired result. □

For the analysis of second-order approximations we need one more lemma

Lemma 4.5.3. *Let* $\mathbf{u}$ *(not to be mistaken for* $\mathbf{u}^1$*) be the solution of*

$$\dot{\boldsymbol{u}} = \varepsilon \overline{\mathbf{f}}^1(\boldsymbol{u}) + \varepsilon\delta_1(\varepsilon)\overline{\mathbf{f}}^{1|1}(\boldsymbol{u}) \ , \ \mathbf{u}(0) = \boldsymbol{a}$$

($\overline{\mathbf{f}}^{1|1}$ *is the general average of* $\mathbf{f}^{1|1}$*, which is assumed to be KBM). Let* $\mathbf{v}$ *be defined as in Lemma 4.5.2, then*

$$\mathbf{v}(t) = \mathbf{u}(t) + \mathcal{O}(\frac{\delta_1(\varepsilon)\delta_{1|1}(\varepsilon)}{\varepsilon T})$$

on the time scale $1/\varepsilon$.

Proof

$$\begin{aligned}\mathbf{v}(t)-\mathbf{u}(t)&=\varepsilon\int_0^t \left(\overline{\mathbf{f}}^1(\mathbf{v}(s))-\overline{\mathbf{f}}^1(\mathbf{u}(s))\right)\,\mathrm{d}s\\&+\varepsilon\delta_1(\varepsilon)\int_0^t \left(\mathbf{f}_T^{1|1}(\mathbf{v}(s),s)-\overline{\mathbf{f}}^{1|1}(\mathbf{u}(s))\right)\,\mathrm{d}s.\end{aligned}$$

It follows from Lemma 4.3.1 and Notation 4.3.3 that

$$\mathbf{f}_T^{1|1}(\mathbf{v}(t),t)=\overline{\mathbf{f}}^{1|1}(\mathbf{v}(t))+\mathcal{O}(\frac{\delta_{1|1}(\varepsilon)}{\varepsilon T}).$$

From this result and the Lipschitz continuity of $\overline{\mathbf{f}}^1$ and $\overline{\mathbf{f}}^{1|1}$, one obtains

$$\begin{aligned}|\mathbf{v}(t)-\mathbf{u}(t)|&\le\varepsilon\lambda_{\overline{\mathbf{f}}^1}\int_0^t|\mathbf{v}(s)-\mathbf{u}(s)|\,\mathrm{d}s+\varepsilon\delta_1(\varepsilon)\lambda_{\overline{\mathbf{f}}^{1|1}}\int_0^t|\mathbf{v}(s)-\mathbf{u}(s)|\,\mathrm{d}s\\&+\mathcal{O}(\frac{\delta_1(\varepsilon)\delta_{1|1}(\varepsilon)}{\varepsilon T}\varepsilon t).\end{aligned}$$

Application of the Gronwall Lemma produces the estimate of the lemma. □

Theorem 4.5.4 (Second-order approximation in general averaging). *Consider the initial value problems*

$$\dot{\boldsymbol{x}}=\varepsilon\mathbf{f}^1(\boldsymbol{x},t),\quad \mathbf{x}(0)=\boldsymbol{a}$$

and

$$\dot{\boldsymbol{u}}=\varepsilon\overline{\mathbf{f}}^1(\boldsymbol{u})+\varepsilon\delta_1(\varepsilon)\overline{\mathbf{f}}^{1|1}(\boldsymbol{u}),\quad \mathbf{u}(0)=\boldsymbol{a},$$

with $\mathbf{f}^1:\mathbb{R}^n\times\mathbb{R}\to\mathbb{R}^n$, $\boldsymbol{x},\boldsymbol{u},\boldsymbol{a}\in D\subset\mathbb{R}^n$, $t\in[0,\infty)$, $\varepsilon\in(0,\varepsilon_0]$, *and*

$$\mathbf{f}^{1|1}(\boldsymbol{x},t)=\mathsf{D}\mathbf{f}^1(\boldsymbol{x},t)\mathbf{u}^1(\boldsymbol{x},t)-\mathsf{D}\mathbf{u}^1(\boldsymbol{x},t)\overline{\mathbf{f}}^1(\boldsymbol{x}).$$

Suppose

1. $\mathbf{f}^1$ *and* $\mathbf{f}^{1|1}$ *are KBM-vector fields (with average* $\overline{\mathbf{f}}^1$ *and* $\overline{\mathbf{f}}^{1|1}$*),*
2. $\mathbf{u}(t)$ *belongs to an interior subset of* D *on the time scale* $1/\varepsilon$.

Then, on the time scale $1/\varepsilon$,

$$\mathbf{x}(t)=\mathbf{u}(t)+\delta_1(\varepsilon)\mathbf{u}^1(\mathbf{u}(t),t)+\mathcal{O}(\sqrt{\delta_{1|1}(\varepsilon)}\min(\delta_1(\varepsilon),\sqrt{\delta_{1|1}(\varepsilon)})+\delta_1^2(\varepsilon)).$$

Proof With $\mathbf{y}$ defined as in Lemma 4.5.1 we have

$$\begin{aligned}&|\mathbf{x}(t)-(\mathbf{u}(t)+\delta_1(\varepsilon)\mathbf{u}^1(\mathbf{u}(t),t))|\\&=|\mathbf{y}(t)+\delta_1(\varepsilon)\mathbf{u}^1(\mathbf{y}(t),t)-(\mathbf{u}(t)+\delta_1(\varepsilon)\mathbf{u}^1(\mathbf{u}(t),t))|\\&\le(1+\lambda_{\mathbf{u}^1}\delta_1(\varepsilon))|\mathbf{y}(t)-\mathbf{u}(t)|,\end{aligned}$$

where we used the triangle inequality and the Lipschitz continuity of $\mathbf{u}^1$. Again using the triangle inequality and Lemmas 4.5.2 and 4.5.3 we obtain

$$|\mathbf{y}(t) - \mathbf{u}(t)| = \mathcal{O}(\delta_1(\varepsilon)[\varepsilon T + \delta_1(\varepsilon)]) + \mathcal{O}(\delta_1(\varepsilon)\delta_{1|1}(\varepsilon)/(\varepsilon T)).$$

We choose T such that the errors are of the same order, so

$$\varepsilon^2 T^2 = \max(\delta_1^2(\varepsilon) + \delta_{1|1}(\varepsilon)).$$

This choice produces the estimate of the theorem. □

A remarkable consequence of this theorem is an improved estimate for the first-order result of Theorem 4.3.6. However, this is an improvement obtained after making additional assumptions.

Theorem 4.5.5. *Consider the initial value problems*

$$\dot{\boldsymbol{x}} = \varepsilon \mathbf{f}^1(\boldsymbol{x}, t), \quad \mathbf{x}(0) = \boldsymbol{a}$$

and

$$\dot{\boldsymbol{y}} = \varepsilon \overline{\mathbf{f}}^1(\boldsymbol{y}), \quad \mathbf{y}(0) = \boldsymbol{a}.$$

Then, with the assumptions of Theorem 4.5.4,

$$\mathbf{x}(t) = \mathbf{y}(t) + \mathcal{O}(\delta_1(\varepsilon)).$$

Proof With the Gronwall Lemma 1.3.3 we have in the usual way

$$\mathbf{u}(t) = \mathbf{y}(t) + \mathcal{O}(\delta_1(\varepsilon))$$

on the time scale $1/\varepsilon$; $\mathbf{u}$ has been defined in Theorem 4.5.4. Also from Theorem 4.5.4 we have

$$\mathbf{x}(t) = \mathbf{u}(t) + \mathcal{O}(\delta_1(\varepsilon)).$$

The triangle inequality produces the desired result. □

An extension of Theorem 4.5.4 which is nearly trivial but useful can be made as follows.

Theorem 4.5.6. *Assume that* $\mathsf{D}\mathbf{f}^1$ *exists and is continuous. Consider the initial value problems*

$$\dot{\boldsymbol{x}} = \varepsilon \mathbf{f}^1(\boldsymbol{x}, t) + \varepsilon^2 \mathbf{f}^2(\boldsymbol{x}, t) + \varepsilon^3 \mathbf{f}^{[3]}(\boldsymbol{x}, t, \varepsilon), \quad \mathbf{x}(0) = \boldsymbol{a}$$

and

$$\dot{\boldsymbol{u}} = \varepsilon \overline{\mathbf{f}}^1(\boldsymbol{u}) + \varepsilon \delta_1(\varepsilon) \overline{\mathbf{f}}^{1|1}(\boldsymbol{u}) + \varepsilon^2 \overline{\mathbf{f}}^2(\boldsymbol{u}), \quad \mathbf{u}(0) = \boldsymbol{a},$$

with $\mathbf{f}^i : \mathbb{R}^n \times \mathbb{R} \to \mathbb{R}^n$ *for* $i = 1, 2$, $\mathbf{f}^{[3]} : \mathbb{R}^n \times \mathbb{R} \times (0, \varepsilon_0] \to \mathbb{R}^n$, $\boldsymbol{x}, \boldsymbol{u}, \boldsymbol{a} \in D \subset \mathbb{R}^n$, $t \in [0, \infty)$ *and* $\varepsilon \in (0, \varepsilon_0]$ *with*

$$\mathbf{f}^{1|1}(\boldsymbol{x}, t) = \mathsf{D}\mathbf{f}^1(\boldsymbol{x}, t) \cdot \mathbf{u}^1(\boldsymbol{x}, t) - \mathsf{D}\mathbf{u}^1(\boldsymbol{x}, t) \cdot \overline{\mathbf{f}}^1(\boldsymbol{x}).$$

Suppose

1. *For $I = 1, 2, 1|1$, the $\mathbf{f}^I$ are KBM-vector fields, with averages $\overline{\mathbf{f}}^I$,*
2. *$|\mathbf{f}^{[3]}(\boldsymbol{x}, t, \varepsilon)|$ is bounded by a constant uniformly on $D \times [0, \frac{L}{\varepsilon}) \times (0, \varepsilon_0]$,*
3. *$\mathbf{u}(t)$ belongs to an interior subset of D on the time scale $1/\varepsilon$.*

Then on the time scale $1/\varepsilon$ one obtains

$$\mathbf{x}(t) = \mathbf{u}(t) + \delta_1(\varepsilon)\mathbf{u}^1(\mathbf{u}(t), t) + \mathcal{O}(\delta_1(\varepsilon)\left(\delta_{1|1}^{\frac{1}{2}}(\varepsilon) + \delta_1(\varepsilon)\right) + \delta_2^{\frac{1}{2}}(\varepsilon)).$$

Proof With some small additions, the proof of Theorem 4.5.4 can be repeated. □

4.5.1 Example of Second-Order Averaging

We return to the linear oscillator with increasing damping (Section 4.4):

$$\ddot{x} + \varepsilon(2 - F(t))\dot{x} + x = 0, \quad \mathrm{x}(0) = r_0\ , \ \dot{\mathrm{x}}(0) = 0,$$

where $F(0) = 1$ and $F(t)$ decreases monotonically towards zero. Transforming $(x, \dot{x}) \mapsto (r, \phi)$ gives

$$\begin{bmatrix} \dot{r} \\ \dot{\phi} \end{bmatrix} = \varepsilon \begin{bmatrix} r\sin^2(t+\phi)(-2+F(t)) \\ \sin(t+\phi)\cos(t+\phi)(-2+F(t)) \end{bmatrix}, \quad \begin{bmatrix} \mathrm{r} \\ \phi \end{bmatrix}(0) = \begin{bmatrix} r_0 \\ 0 \end{bmatrix}.$$

General averaging produced the vector field $\overline{\mathbf{f}}^1(r, \phi) = (-r, 0)$. First we suppose

$$F(t) = \frac{1}{(1+t)^\alpha}, \quad \alpha > 0.$$

We see, by splitting the integral as $\int_0^\infty = \int_0^{\sqrt{T}} + \int_{\sqrt{T}}^\infty$, that

$$\begin{aligned} \delta_1(\varepsilon) &= \mathcal{O}(\varepsilon), \quad \alpha > 1, \\ \delta_1(\varepsilon) &= \mathcal{O}(\varepsilon \log(\varepsilon)), \quad \alpha = 1, \\ \delta_1(\varepsilon) &= \mathcal{O}(\varepsilon^\alpha), \quad 0 < \alpha < 1. \end{aligned}$$

Now in the notation of Theorem 4.5.4

$$\begin{aligned} \mathsf{Df}^1 &= \begin{bmatrix} \frac{1}{2}[1 - \cos(2(t+\phi))][-2+F(t)] & r\sin(2(t+\phi))[-2+F(t)] \\ 0 & \cos(2(t+\phi))[-2+F(t)] \end{bmatrix}, \\ \delta_1(\varepsilon)\mathbf{u}^1 &= \varepsilon \begin{bmatrix} \frac{r}{2}\int_0^t [F(s) - F(s)\cos(2(s+\phi)) + 2\cos(2(s+\phi))]\,\mathrm{d}s \\ \int_0^t [-\sin(2(s+\phi)) + \frac{1}{2}F(s)\sin(2(s+\phi))]\,\mathrm{d}s \end{bmatrix} \\ &= \varepsilon \begin{bmatrix} \frac{r}{2} I_1(\phi, t) \\ I_2(\phi, t) \end{bmatrix} \quad \text{(which defines } I_1 \text{ and } I_2\text{)}, \\ \delta_1(\varepsilon)\mathsf{Du}^1 &= \varepsilon \begin{bmatrix} \frac{1}{2} I_1(\phi, t) & \frac{r}{2}\frac{\partial}{\partial\phi} I_1(\phi, t) \\ 0 & \frac{\partial}{\partial\phi} I_2(\phi, t) \end{bmatrix}. \end{aligned}$$

To compute $\overline{\mathbf{f}}^{1|1}$ we have to average $\mathsf{D}\mathbf{f}^1 \cdot \mathbf{u}^1$ and $\mathsf{D}\mathbf{u}^1 \cdot \overline{\mathbf{f}}^1$. It is easy to see that for $\overline{\mathbf{f}}^{1|1}$ to exist we have the condition $\alpha > 1$. So if $F(t)$ does not decrease fast enough $(0 < \alpha \leq 1)$ the second-order approximation in the sense of Theorem 4.5.4 does not exist. The calculation of the second-order approximation in the case $\alpha > 1$ involves long expressions which we omit. We finally discuss the case $F(t) = e^{-t}$; note that $\delta_1(\varepsilon) = \mathcal{O}(\varepsilon)$. Again in the notation of Theorem 4.5.4 we have the same expressions as before, except that now

$$\begin{aligned} I_1(\phi, t) &= \sin(2(t+\phi)) - e^{-t} + 1 + \frac{1}{5}e^{-t}\cos(2(t+\phi)) \\ &\quad -\frac{1}{5}\cos(2\phi) - \frac{2}{5}e^{-t}\sin(2(t+\phi)) - \frac{3}{5}\sin(2\phi) \end{aligned}$$

and

$$\begin{aligned} I_2(\phi, t) &= \frac{1}{2}\cos(2(t+\phi)) - \frac{1}{10}e^{-t}\sin(2(t+\phi)) \\ &\quad -\frac{1}{5}e^{-t}\cos(2(t+\phi)) + \frac{1}{10}\sin(2\phi) - \frac{3}{10}\cos(2\phi). \end{aligned}$$

After calculating $\mathbf{f}^{1|1}$ and averaging we obtain

$$\overline{\mathbf{f}}^{1|1} = (0, -\frac{1}{2}),$$

so if $\boldsymbol{u}$ in Theorem 4.5.4 is written as $\boldsymbol{u} = (\overline{r}, \overline{\phi})$, then

$$\begin{bmatrix} \dot{\overline{r}} \\ \dot{\overline{\phi}} \end{bmatrix} = \varepsilon \begin{bmatrix} -\overline{r} \\ 0 \end{bmatrix} + \varepsilon^2 \begin{bmatrix} 0 \\ -\frac{1}{2} \end{bmatrix}, \quad \begin{bmatrix} \overline{\mathrm{r}} \\ \overline{\phi} \end{bmatrix}(0) = \begin{bmatrix} r_0 \\ 0 \end{bmatrix},$$

and $\overline{\mathrm{r}}(t) = r_0 e^{-\varepsilon t}$, $\overline{\phi}(t) = -\frac{1}{2}\varepsilon^2 t$. For the solution of the original perturbation problem we have

$$\mathrm{x}(t) = r_0 e^{-\varepsilon t}[1 + \frac{\varepsilon}{2} I_1(-\frac{1}{2}\varepsilon^2 t, t)]\cos(t - \frac{1}{2}\varepsilon^2 t + \varepsilon I_2(-\frac{1}{2}\varepsilon^2 t, t)) + \mathcal{O}(\varepsilon^{\frac{3}{2}})$$

on the time scale $1/\varepsilon$.

4.6 Application of General Averaging to Almost-Periodic Vector Fields

In this section we discuss some questions that arise in studying initial value problems of the form

$$\dot{\boldsymbol{x}} = \varepsilon \mathbf{f}^1(\boldsymbol{x}, t), \quad \mathbf{x}(0) = \boldsymbol{a},$$

with $\mathbf{f}^1$ almost-periodic in t. For the basic theory of almost-periodic functions we refer to an introduction by Harald Bohr [36]. More recent introduction with the emphasis on its use in differential equations have been given by Fink [100] and Levitan and Zhikov [173]. In this book averaging has been discussed for proving existence of almost-periodic solutions. Both qualitative and quantitative aspects of almost-periodic solutions of periodic and almost-periodic differential equations has been given extensive treatment by Roseau [229]. A simple example of an almost-periodic function is found by taking the sum of two periodic functions as in

$$f(t) = \sin(t) + \sin(2\pi t).$$

Several equivalent definitions are in use; we take a three step definition by Bohr.

Definition 4.6.1. *A subset S of $\mathbb{R}$ is called* **relatively dense** *if there exists a positive number L such that $[a, a+L] \bigcap S \neq$ for all $a \in \mathbb{R}$. The number L is called the* **inclusion length**.

Definition 4.6.2. *Consider a vector field $\mathbf{f}(t)$, continuous on $\mathbb{R}$, and a positive number ε; $\tau(\varepsilon)$ is a* **translation-number** *of $\mathbf{f}$ if*

$$\| \mathbf{f}(t + \tau(\varepsilon)) - \mathbf{f}(t) \| \leq \varepsilon \textit{ for all } t \in \mathbb{R}.$$

Definition 4.6.3. *The vector field $\mathbf{f}(t)$, continuous on $\mathbb{R}$, is called* **almost-periodic** *if for each $\varepsilon > 0$ a relatively dense set of translation-numbers $\tau(\varepsilon)$ exists.*

In the context of averaging the following result is basic.

Lemma 4.6.4. *Consider the continuous vector field $\mathbf{f} : \mathbb{R}^n \times \mathbb{R} \to \mathbb{R}^n$. If $\mathbf{f}(\boldsymbol{x}, t)$ is almost-periodic in t and Lipschitz continuous in $\boldsymbol{x}$, $\mathbf{f}$ is a KBM-vector field.*

Proof This is a trivial generalization of [36, Section 50]. □

It follows immediately that with the appropriate assumptions of Theorem 4.3.6 or 4.5.5 we can apply general averaging to the almost-periodic differential equation. Suppose the conditions of Theorem 4.5.5 have been satisfied, then introducing again the averaged equation

$$\dot{\boldsymbol{y}} = \varepsilon \overline{\mathbf{f}}^1(\boldsymbol{y}), \quad \mathbf{y}(0) = \boldsymbol{a},$$

we have

$$\mathbf{x}(t) = \mathbf{y}(t) + \mathcal{O}(\delta_1(\varepsilon))$$

on the time scale $1/\varepsilon$. We shall discuss the magnitude of the error $\delta_1(\varepsilon)$.

Many cases in practice are covered by the following lemma:

Lemma 4.6.5. *If we can decompose the almost-periodic vector field* $\mathbf{f}^1(\boldsymbol{x},t)$ *as a finite sum of* N *periodic vector fields*

$$\mathbf{f}^1(\boldsymbol{x},t) = \sum_{n=1}^{N} \mathbf{f}^1_n(\boldsymbol{x},t),$$

we have $\delta_1(\varepsilon) = \varepsilon$ *and, moreover,*

$$\mathbf{x}(t) = \mathbf{y}(t) + \mathcal{O}(\varepsilon).$$

Proof Interchanging the finite summation and the integration gives the desired result. □

A fundamental obstruction against generalizing this lemma to infinite sums is that in general

$$\int_0^t [\mathbf{f}^1(\boldsymbol{x},s) - \overline{\mathbf{f}}^1(\boldsymbol{x})]\, \mathrm{d}s,$$

with $\mathbf{f}^1$ almost-periodic, need not be bounded (see the example below). One might be tempted to apply the approximation theorem for almost-periodic functions: For each $\varepsilon > 0$ we can find $N(\varepsilon) \in \mathbb{N}$ such that

$$\| \mathbf{f}^1(\boldsymbol{x},t) - \sum_{n=1}^{N(\varepsilon)} \mathbf{f}^1_n(x) e^{i\lambda_n t} \| \leq \varepsilon$$

for $t \in [0,\infty)$ and $\lambda_n \in \mathbb{R}$ (cf. [36, Section 84]). In general however, N depends on ε, which destroys the possibility of obtaining an $\mathcal{O}(\varepsilon)$ estimate for δ_1. The difficulties can be illustrated by the following initial value problem.

4.6.1 Example

Consider the equation

$$\dot{x} = \varepsilon a(x) f(t), \quad \mathrm{x}(0) = x_0.$$

The function $a(x)$ is sufficiently smooth in some domain D containing x_0. We define the almost-periodic function

$$f(t) = \sum_{n=1}^{\infty} \frac{1}{2^n} \cos(t/2^n).$$

Note that this is a uniformly convergent series consisting of continuous terms so we may integrate $f(t)$ and interchange summation and integration. For the average of the right-hand side we obtain

$$\lim_{T\to\infty} \frac{a(x)}{T} \int_0^T f(s)\, \mathrm{d}s = 0.$$

So we have

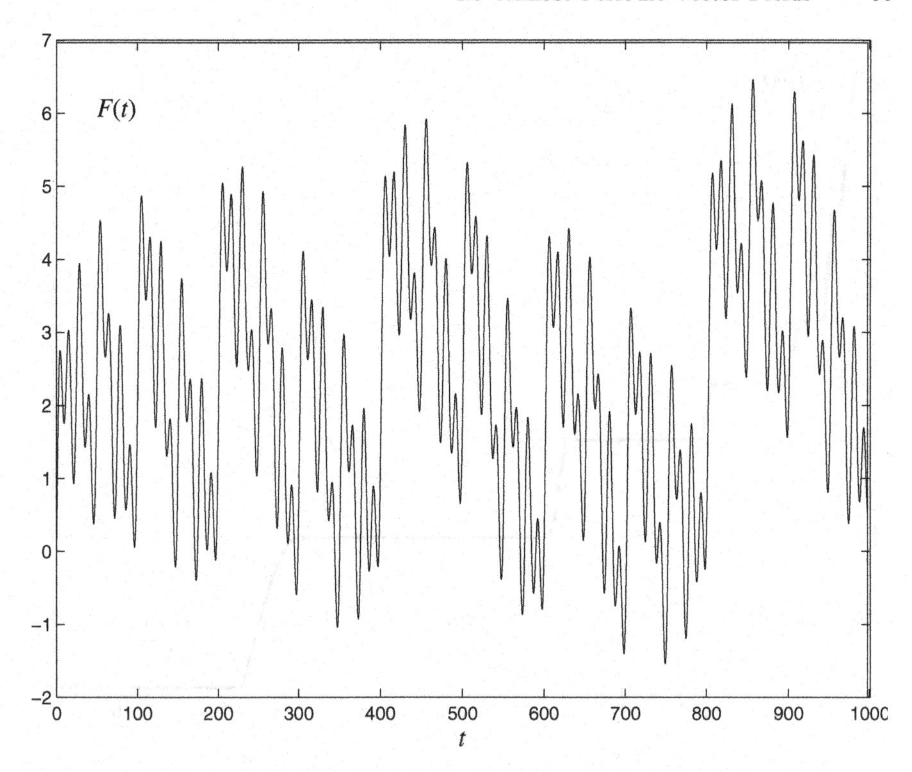

Fig. 4.1: $F(t) = \sum_{n=1}^{\infty} \sin(t/2^n)$ as a function of time on $[0, 10000]$. The function F is the integral of an almost-periodic function and is not uniformly bounded.

$$\mathrm{x}(t) = x_0 + \mathcal{O}(\delta_1(\varepsilon))$$

on the time scale $1/\varepsilon$. Suppose $\sup_{x \in D}\ a(x) = M$ then we have

$$\delta_1(\varepsilon) = \sup_{t \in [0, L/\varepsilon]} \varepsilon M \left|\sum\nolimits_{n=1}^{\infty} \sin(t/2^n)\right|.$$

It is easy to see that as $\varepsilon \downarrow 0$, $\delta_1(\varepsilon)/\varepsilon$ becomes unbounded, so the error in this almost-periodic case is larger than $\mathcal{O}(\varepsilon)$. In Figure 4.1 we illustrate this behavior of $\delta_1(\varepsilon)$ and $F(t) = \sum_{n=1}^{\infty} \sin(t/2^n)$. A simple example is in the case $a(x) = 1$; we have explicitly

$$\mathrm{x}(t) = x_0 + \varepsilon F(t).$$

The same type of error arises if the solutions are bounded. If we take for example $a(x) = x(1 - x)$, $0 < x_0 < 1$, we obtain

$$x(t) = \frac{x_0 e^{\varepsilon F(t)}}{1 - x_0 + x_0 e^{\varepsilon F(t)}}.$$

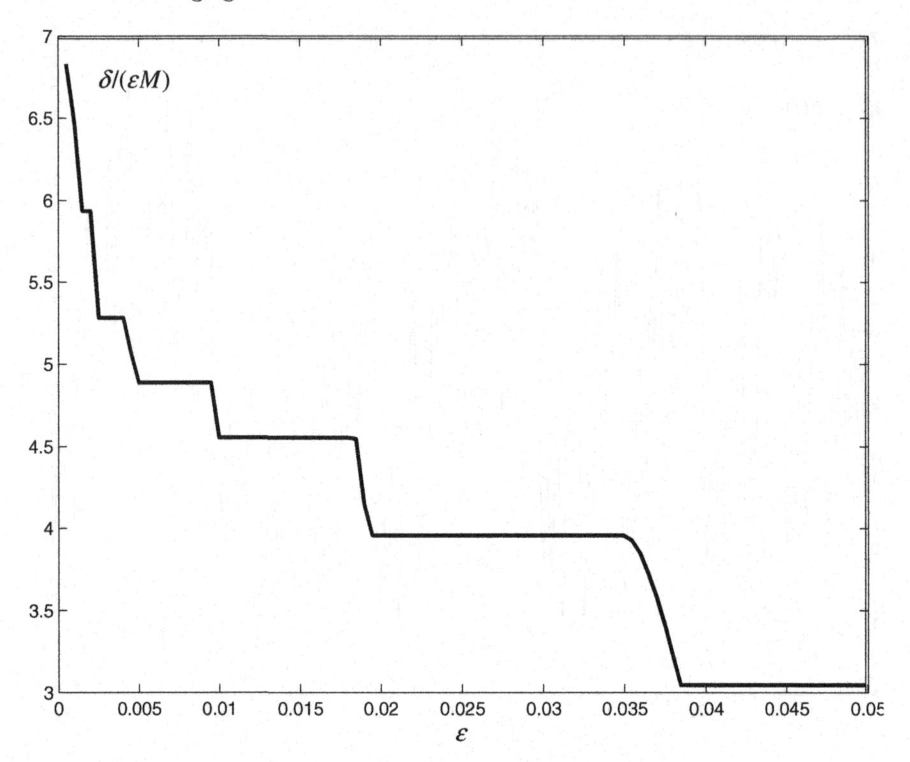

Fig. 4.2: The quantity $\delta_1/(\varepsilon M)$ as a function of ε obtained from the analysis of $F(t)$. Since ε decreases, $\delta_1/(\varepsilon M) = \sup_{0 \leq \varepsilon t \leq 1} F(t)$ increases.

Sometimes an $\mathcal{O}(\varepsilon)$-estimate can be obtained by studying the generalized Fourier expansion of an almost-periodic vector field

$$\mathbf{f}^1(\boldsymbol{x}, t) = \overline{\mathbf{f}}^1(\boldsymbol{x}) + \sum_{n=1}^{\infty} [\mathbf{a}_n^1(\boldsymbol{x}) \cos(\lambda_n t) + \mathbf{b}_n^1(\boldsymbol{x}) \sin(\lambda_n t)]$$

with $\lambda_n > 0$. We have:

Lemma 4.6.6. *Suppose the conditions of Theorem 4.5.5 have been satisfied for the initial value problems*

$$\dot{\boldsymbol{x}} = \varepsilon \mathbf{f}^1(\boldsymbol{x}, t), \quad \mathbf{x}(0) = \boldsymbol{a}$$

and

$$\dot{\boldsymbol{y}} = \varepsilon \overline{\mathbf{f}}^1(\boldsymbol{y}), \quad \mathbf{y}(0) = \boldsymbol{a}.$$

If $\mathbf{f}^1(\boldsymbol{x}, t)$ is an almost-periodic vector field with a generalized Fourier expansion such that $\lambda_n \geq \alpha > 0$ with α independent of n then

$$\mathbf{x}(t) = \mathbf{y}(t) + \mathcal{O}(\varepsilon)$$

on the time scale $1/\varepsilon$.

Proof If $\lambda_n \geq \alpha > 0$ we have that

$$\mathbf{I}(\boldsymbol{x}, t) = \int_0^t [\mathbf{f}^1(\boldsymbol{x}, s) - \overline{\mathbf{f}}^1(\boldsymbol{x})] \, ds$$

is an almost-periodic vector field; See [100, Chapter 4.8]. So $|\mathbf{I}(\boldsymbol{x}, t)|$ is bounded for $t \geq 0$ which implies $\delta_1(\varepsilon) = \mathcal{O}(\varepsilon)$. □

5

Attraction

5.1 Introduction

Averaging procedures for initial value problems, and their basic error estimates, have been established in the last two chapters under various circumstances. Usually the error estimates are valid for a time of order $1/\varepsilon$, although occasionally this can be extended to $1/\varepsilon^{j+1}$ for some integer $j > 1$ (see Theorem 2.9.4). In this chapter and the next, we investigate circumstances under which the validity of averaging can be extended still farther. Results can sometimes be obtained for all $t \geq 0$, or for all t such that the solution remains in a certain region. This chapter is concerned with solutions that approach a particular solution that is an attractor.

Chapter 6 generalizes this considerably by considering solutions influenced by one or more particular solutions that are hyperbolic, but not necessarily attracting. At the same time, Chapter 6 addresses questions of the qualitative behavior of the solutions, such as existence and stability of periodic solutions and existence of heteroclinic orbits connecting two hyperbolic orbits. The main result of this chapter, Theorem 5.5.1, proved here in a simple way, is reproved in Chapter 6 as a corollary of a more difficult result.

The idea of extending the error estimate for an approximate solution approaching an attractor is not limited to the method of averaging, but applies also to perturbation problems (called *regular perturbations*) that do not require averaging (or other special techniques such as matching). The next two sections are devoted to examples and theorem statements, first in the regular case and then for averaging. The proofs are in Sections 5.5 and 5.6.

The ideas presented in this chapter have been around for some time. Greenlee and Snow [113] proved the validity of approximations on the whole time-interval for harmonic oscillator equations with certain nonlinear perturbations under conditions which are compatible with the assumptions to be made in the next section (Theorem 5.5.1).

We mention the papers of Banfi and Graffi [20] and [21]. More detailed proofs were given by Balachandra and Sethna [18] and Eckhaus [81]. The proofs can be simplified a little by using a lemma due to Sanchez–Palencia [234] and this is the approach which we shall use in this chapter.

5.2 Equations with Linear Attraction

Consider again the initial value problem

$$\dot{\boldsymbol{x}} = \mathbf{f}^{[0]}(\boldsymbol{x}, t, \varepsilon), \quad \mathbf{x}(t_0) = \boldsymbol{a}$$

for $t \geq t_0; \boldsymbol{x}, \boldsymbol{a} \in D, 0 < \varepsilon \leq \varepsilon_0$. Suppose that $\mathbf{x} = 0$ is a solution of the equation (if we wish to study a particular solution $\mathbf{x} = \boldsymbol{\phi}(t)$ we can always shift to an equation for $\boldsymbol{y} = \boldsymbol{x} - \boldsymbol{\phi}(t)$, where the equation for $\boldsymbol{y}$ has the trivial solution).

Definition 5.2.1. *The solution* $\mathbf{x} = \mathbf{0}$ *of the equation is* **stable in the sense of Lyapunov** *if for every* $\varepsilon > 0$ *there exists a* $\delta > 0$ *such that*

$$\| \boldsymbol{a} \| \leq \delta \Rightarrow \| \mathbf{x}(t) \| < \varepsilon$$

for $t \geq t_0$.

The solution $\mathbf{x} = \mathbf{0}$ may have a different property which we call attraction:

Definition 5.2.2. *The solution* $\mathbf{x} = \mathbf{0}$ *of the equation is a* **attractor (positive)** *if there is a* $\delta > 0$ *such that*

$$\| \boldsymbol{a} \| < \delta \Rightarrow \lim_{t \to \infty} \mathbf{x}(t) = \mathbf{0}.$$

If the solution is stable and moreover an attractor we have a stronger type of stability:

Definition 5.2.3. *If the solution* $\mathbf{x} = \mathbf{0}$ *of the equation is* **stable in the sense of Lyapunov** *and* $\mathbf{x} = \mathbf{0}$ *is a (positive) attractor, the solution is* **asymptotically stable.**

It is natural to study the stability characteristics of a solution by linearizing the equation in a neighborhood of this solution. One may hope that the stability characteristics of the linear equation carry over to the full nonlinear equation. It turns out, however, that this is not always the case. Poincaré and Lyapunov considered some important cases in which the linear behavior with respect to stability is characteristic for the full equation. In the case which we discuss, the proof is obtained by estimating explicitly the behavior of the solutions in a neighborhood of $\boldsymbol{x} = \mathbf{0}$.

Theorem 5.2.4 (Poincaré–Lyapunov). *Consider the equation*

$$\dot{\boldsymbol{x}} = (A + B(t))\boldsymbol{x} + \mathbf{g}(\boldsymbol{x}, t), \quad \mathbf{x}(t_0) = \boldsymbol{a}, \quad t \geq t_0,$$

where $\boldsymbol{x}, \boldsymbol{a} \in \mathbb{R}^n$; A is a constant $n \times n$-matrix with all eigenvalues having strictly negative real part, $B(t)$ is a continuous $n \times n$-matrix with the property

$$\lim_{t\to\infty} \| B(t) \| = 0.$$

The vector field is continuous with respect to t and $\boldsymbol{x}$ and continuously differentiable with respect to $\boldsymbol{x}$ in a neighborhood of $\boldsymbol{x} = \mathbf{0}$; moreover

$$\mathbf{g}(\boldsymbol{x}, t) = o(\| \boldsymbol{x} \|) \text{ as } \| \boldsymbol{x} \| \to 0,$$

uniformly in t. Then there exist constants $C, t_0, \delta, \mu > 0$ such that if $\| \boldsymbol{a} \| < \delta/C$

$$\| \mathbf{x}(t) \| \leq C \| \boldsymbol{a} \| e^{-\mu(t-t_0)}, \quad t \geq t_0.$$

Remark 5.2.5. The domain $\| \boldsymbol{a} \| < \delta/C$ where the attraction is of exponential type will be called the Poincaré–Lyapunov domain of the equation at $\mathbf{0}$. ♡

Proof Note that in a neighborhood of $\boldsymbol{x} = \mathbf{0}$, the initial value problem satisfies the conditions of the existence and uniqueness theorem.

Since the matrix A has eigenvalues with all real parts negative, there exists a constant $\mu_0 > 0$ such that for the solution of the fundamental matrix equation

$$\dot{\Phi} = A\Phi, \quad \Phi(t_0) = I,$$

we have the estimate

$$\| \Phi(t) \| \leq Ce^{-\mu_0(t-t_0)}, \quad C > 0, \quad t \geq t_0.$$

The constant C depends on A only. From the assumptions on B and $\mathbf{g}$ we know that there exists $\eta(\delta) > 0$ such that for $\| \boldsymbol{x} \| \leq \delta$ one has

$$\| B(t) \| < \eta(\delta), \quad \| \mathbf{g}(\boldsymbol{x}, t) \| \leq \eta(\delta) \| \boldsymbol{x} \|,$$

for $t \geq t_0(\delta)$. Note that existence of the solution (in the Poincaré–Lyapunov domain) of the initial value problem is guaranteed on some interval $[t_0, \hat{t}]$. In the sequel we shall give estimates which show that the solution exists for all $t \geq t_0$. For the solution we may write the integral equation

$$\mathbf{x}(t) = \Phi(t)\boldsymbol{a} + \int_{t_0}^{t} \Phi(t - s + t_0)[B(s)\mathbf{x}(s) + \mathbf{g}(x(s), s)]\, \mathrm{d}s.$$

Using the estimates for Φ, B and $\mathbf{g}$ we have for $t \in [t_0, \hat{t}]$

$$\begin{aligned} \| \mathbf{x}(t) \| &\leq \| \Phi(t) \| \| \boldsymbol{a} \| \\ &\quad + \int_{t_0}^{t} \| \Phi(t - s + t_0) \| [\| B(s) \| \| \mathbf{x}(s) \| + \| \mathbf{g}(\mathbf{x}(s), s) \|]\, \mathrm{d}s \\ &\leq Ce^{-\mu_0(t-t_0)} \| \boldsymbol{a} \| + 2C\eta \int_{t_0}^{t} e^{-\mu_0(t-s)} \| \mathbf{x}(s) \| \, \mathrm{d}s \end{aligned}$$

or

$$e^{\mu_0(t-t_0)} \parallel \mathbf{x}(t) \parallel \leq C \parallel \boldsymbol{a} \parallel + 2C\eta \int_{t_0}^{t} e^{\mu_0(s-t_0)} \parallel \mathbf{x}(s) \parallel ds.$$

Using the Gronwall Lemma 1.3.3 (with $\delta_1 = 2C\eta$, $\delta_2 = 0, \delta_3 = C \parallel \boldsymbol{a} \parallel$) we obtain

$$e^{\mu_0(t-t_0)} \parallel \mathbf{x}(t) \parallel \leq C \parallel \boldsymbol{a} \parallel e^{2C\eta(t-t_0)}$$

or

$$\parallel \mathbf{x}(t) \parallel \leq C \parallel \boldsymbol{a} \parallel e^{(2C\eta - \mu_0)(t-t_0)}.$$

Put $\mu = \mu_0 - 2C\eta$; if δ (and therefore η) is small enough, μ is positive and we have

$$\parallel \mathbf{x}(t) \parallel \leq C \parallel \boldsymbol{a} \parallel e^{-\mu(t-t_0)}, \quad t \in [t_0, \hat{t}].$$

If δ small enough, we also have $\parallel \mathbf{x}(\hat{t}) \parallel \leq \delta$, so we can continue the estimation argument for $t \geq \hat{t}$; it follows that we may replace $\hat{t}$ by ∞ in our estimate. □

Corollary 5.2.6. *Under the conditions of the Poincaré–Lyapunov Theorem 5.2.4, $\boldsymbol{x} = \mathbf{0}$ is asymptotically stable.*

The exponential attraction of the solutions is even so strong that the difference between solutions starting in a Poincaré–Lyapunov domain will also decrease exponentially. This is the content of the following lemma.

Lemma 5.2.7. *Consider two solutions, $\mathbf{x}_1(t)$ and $\mathbf{x}_2(t)$ of the equation*

$$\dot{\boldsymbol{x}} = (A + B(t))\boldsymbol{x} + \mathbf{g}(\boldsymbol{x}, t)$$

for which the conditions of the Poincaré–Lyapunov Theorem 5.2.4 have been satisfied. Starting in the Poincaré–Lyapunov domain we have

$$\parallel \mathbf{x}_1(t) - \mathbf{x}_2(t) \parallel \leq C \parallel \mathbf{x}_1(t_0) - \mathbf{x}_2(t_0) \parallel e^{-\mu(t-t_0)}$$

for $t \geq t_0$ and constants $C, \mu > 0$.

Proof Consider the equation for $\mathbf{y}(t) = \mathbf{x}_1(t) - \mathbf{x}_2(t)$,

$$\dot{\boldsymbol{y}} = (A + B(t))\boldsymbol{y} + \mathbf{g}(\boldsymbol{y} + \mathbf{x}_2(t), t) - \mathbf{g}(\mathbf{x}_2(t), t),$$

with initial value $\mathbf{y}(t_0) = \mathbf{x}_1(t_0) - \mathbf{x}_2(t_0)$. We write the equation as

$$\dot{\boldsymbol{y}} = (A + B(t) + \mathsf{D}\mathbf{g}(\mathbf{x}_2(t), t))\, \boldsymbol{y} + \mathbf{G}(\mathbf{y}, t),$$

with

$$\mathbf{G}(\boldsymbol{y},t) = \mathbf{g}(\boldsymbol{y} + \mathbf{x}_2(t),t) - \mathbf{g}(\mathbf{x}_2(t),t) - \mathsf{D}\mathbf{g}(\mathbf{x}_2(t),t)\boldsymbol{y}.$$

Note that $\mathbf{G}(\mathbf{0},t) = \mathbf{0}$, $\lim_{t\to\infty} \mathbf{x}_2(t) = \mathbf{0}$ and as $\mathbf{g}$ is continuously differentiable with respect to $\boldsymbol{y}$,

$$\| \mathbf{G}(\boldsymbol{y},t) \| = o(\| \boldsymbol{y} \|),$$

uniformly for $t \geq t_0$. It is easy to see that the equation for $\boldsymbol{y}$ again satisfies the conditions of the Poincaré–Lyapunov Theorem 5.2.4; only the initial time may be shifted forward by a quantity which depends on $\mathsf{D}\mathbf{g}(\mathbf{x}_2(t),t)$. □

Remark 5.2.8. If t is large enough, we have

$$\| \mathbf{x}_1(t) - \mathbf{x}_2(t) \| \leq k \| \mathbf{x}_1(t_0) - \mathbf{x}_2(t_0) \|,$$

with $0 < k < 1$; we shall use this in Section 5.5. ♡

5.3 Examples of Regular Perturbations with Attraction

Solutions of differential equations can be attracted to a particular solution and this phenomenon may assume many different forms. Suppose for instance we consider an initial value problem in $\mathbb{R}^n$ of the form

$$\dot{\boldsymbol{x}} = A\boldsymbol{x} + \varepsilon\mathbf{f}^1(\boldsymbol{x},t), \quad \mathbf{x}(0) = \boldsymbol{a}.$$

The matrix A is constant and all the eigenvalues have negative real parts. If $\varepsilon = 0, \boldsymbol{x} = 0$ is an attractor; how do we approximate the solution if $\varepsilon \neq 0$ and what are the conditions to obtain an approximation of the solution on $[0,\infty)$? Also we should like to extend the problem to the case

$$\dot{\boldsymbol{x}} = A\boldsymbol{x} + \mathbf{g}^0(\boldsymbol{x}) + \varepsilon\mathbf{f}^1(\boldsymbol{x},t), \quad \mathbf{x}(0) = \boldsymbol{a},$$

where we suppose that the equation with $\varepsilon = 0$ has $\mathbf{x} = 0$ as an attracting solution with domain of attraction D. How do we obtain an approximation of the solution if $\boldsymbol{a} \in D$ and $\varepsilon \neq 0$?

5.3.1 Two Species

Consider the problem, encountered in Section 1.6.1, of two species with a restricted supply of food and a slight negative interaction between the species. The growth of the population densities x_1 and x_2 can be described by the system

$$\begin{aligned}\dot{x}_1 &= ax_1 - bx_1^2 - \varepsilon x_1x_2,\\ \dot{x}_2 &= cx_2 - dx_2^2 - \varepsilon e x_1x_2,\end{aligned}$$

where a,b,c,d,e are positive constants. Putting $\varepsilon = 0$ one notes that $(\frac{a}{b},\frac{c}{d})$ is a positive attractor; The domain of attraction D is given by $x_1 > 0, x_2 > 0$. In Figure 5.1 we give an example of the phase plane with $\varepsilon \neq 0$ and in Figure 5.2 with $\varepsilon = 0$.

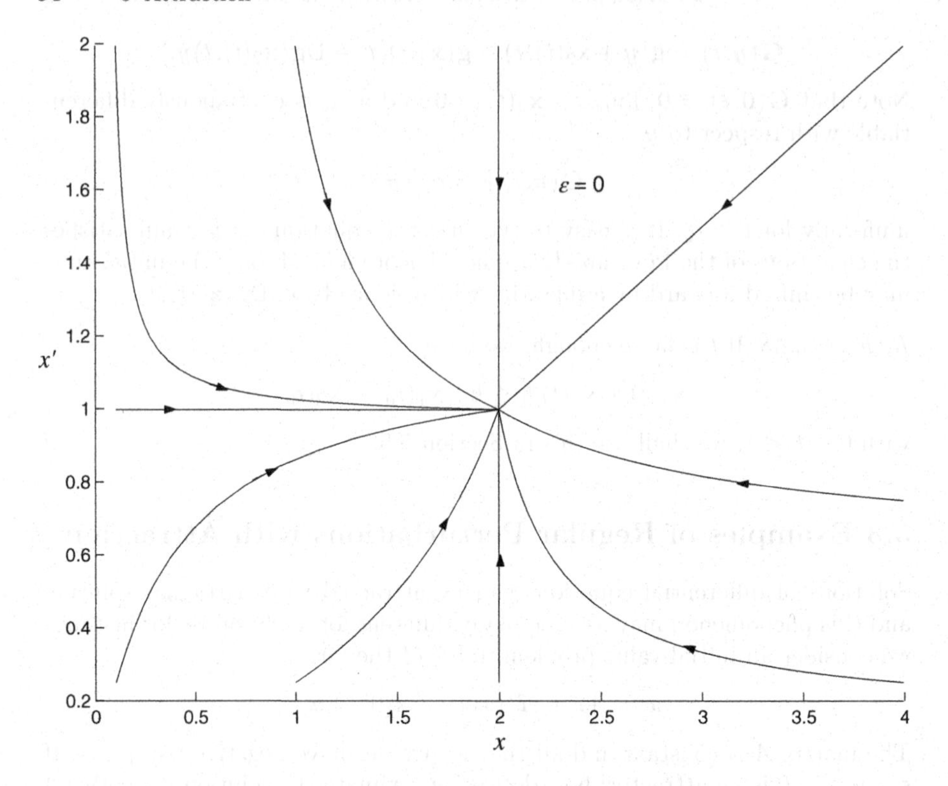

Fig. 5.1: Phase plane for the system $\dot{x}_1 = x_1 - \frac{1}{2}x_1^2 - \varepsilon x_1 x_2, \dot{x}_2 = x_2 - x_2^2 - \varepsilon x_1 x_2$ for $\varepsilon = 0$, i.e without interaction of the species.

5.3.2 A perturbation theorem

More generally, suppose that we started out with an equation of the form

$$\dot{\boldsymbol{x}} = \mathbf{f}^0(\boldsymbol{x}, t) + \varepsilon \mathbf{f}^1(\boldsymbol{x}, t)$$

and that the unperturbed equation ($\varepsilon = 0$) contains an attracting critical point while satisfying the conditions of the Poincaré–Lyapunov Theorem 5.2.4. In our formulation we shift the critical point to the origin.

Theorem 5.3.1. *Consider the equation*

$$\dot{\boldsymbol{x}} = \mathbf{f}^0(\boldsymbol{x}, t) + \varepsilon \mathbf{f}^1(\boldsymbol{x}, t), \quad \mathbf{x}(0) = \boldsymbol{a},$$

with $\boldsymbol{x}, \boldsymbol{a} \in \mathbb{R}^n$*;* $\mathbf{y} = 0$ *is an asymptotically stable solution in the linear approximation of the unperturbed equation*

$$\dot{\boldsymbol{y}} = \mathbf{f}^0(\boldsymbol{y}, t) = (A + B(t))\boldsymbol{y} + \mathbf{g}^0(\boldsymbol{y}, t),$$

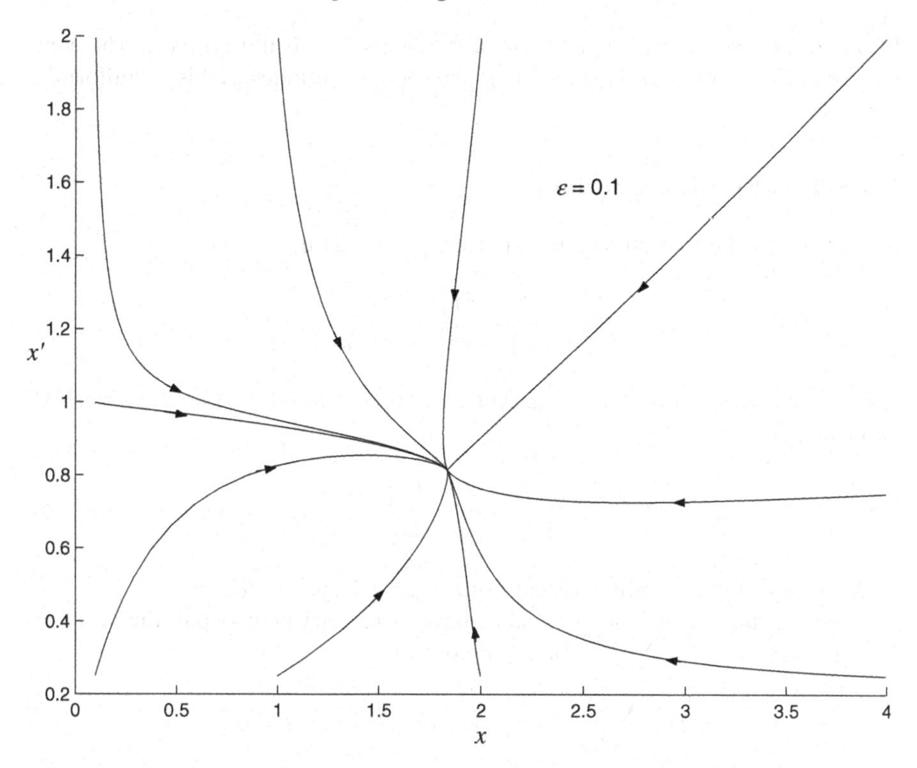

Fig. 5.2: Phase plane for the system $\dot{x}_1 = x_1 - \frac{1}{2}x_1^2 - \varepsilon x_1 x_2, \dot{x}_2 = x_2 - x_2^2 - \varepsilon x_1 x_2$ for $\varepsilon = 0.1$, i.e. with (small) interaction of the species.

with A a constant $n \times n$-matrix with all eigenvalues having negative real part, $B(t)$ is a continuous $n \times n$-matrix with the property

$$\lim_{t\to\infty} \parallel B(t) \parallel = 0.$$

D is the domain of attraction of $\boldsymbol{x} = 0$. The vector field $\mathbf{g}^0$ is continuous with respect to $\boldsymbol{x}$ and t and continuously differentiable with respect to $\boldsymbol{x}$ in $D \times \mathbb{R}^+$, while

$$\mathbf{g}^0(\boldsymbol{x}, t) = o(\parallel \boldsymbol{x} \parallel) \quad as \quad \parallel \boldsymbol{x} \parallel \to 0,$$

uniformly in t; here $\mathbf{f}^1(\boldsymbol{x}, t)$ is continuous in t and $\boldsymbol{x}$ and Lipschitz continuous with respect to $\boldsymbol{x}$ in $D \times \mathbb{R}^+$. Choosing the initial value $\boldsymbol{a}$ in the interior part of D and adding to the unperturbed equation $\mathbf{y}(0) = \boldsymbol{a}$ we have

$$\mathbf{x}(t) - \mathbf{y}(t) = \mathcal{O}(\varepsilon), \quad t \geq 0.$$

This theorem will be proved in Section 5.6. This is a zeroth-order result but of course, if the right-hand side of the equation is sufficiently smooth, we can improve the accuracy by straightforward expansions, as illustrated in the next example. So here a naive use of perturbation techniques yields a uniformly valid result.

5.3.3 Two Species, Continued

We return to the two interacting species, described by

$$\dot{x}_1 = ax_1 - bx_1^2 - \varepsilon x_1 x_2,$$
$$\dot{x}_2 = cx_2 - dx_2^2 - \varepsilon e x_1 x_2,$$

where a, b, c, d, e are positive constants, as treated in 5.3.1. The unperturbed equations are

$$\dot{x}_1 = ax_1 - bx_1^2,$$
$$\dot{x}_2 = cx_2 - dx_2^2,$$

with asymptotically stable critical point $\mathrm{x}_0 = a/b, \mathrm{y}_0 = c/d$.

The conditions of Theorem 5.3.1 have been satisfied; expanding $x(t) = \sum_{n=0}^{\infty} \varepsilon^n x_n(t)$, $y(t) = \sum_{n=0}^{\infty} \varepsilon^n y_n(t)$ we obtain

$$x(t) - \sum_{n=0}^{N} \varepsilon^n x_n(t) = \mathcal{O}(\varepsilon^{N+1}) \,, \quad t \geq 0,$$
$$y(t) - \sum_{n=0}^{N} \varepsilon^n y_n(t) = \mathcal{O}(\varepsilon^{N+1}) \,, \quad t \geq 0.$$

It is easy to compute $x_0(t), y_0(t)$; the higher-order terms are obtained as the solutions of linear equations.

5.4 Examples of Averaging with Attraction

Another attraction problem arises in the following way. Consider the problem

$$\dot{\boldsymbol{x}} = \varepsilon \mathbf{f}^1(\boldsymbol{x}, t), \quad \mathbf{x}(0) = \boldsymbol{a}.$$

Suppose that we may average and that the equation

$$\dot{\boldsymbol{z}} = \varepsilon \overline{\mathbf{f}}^1(\boldsymbol{z})$$

contains an attractor with domain of attraction D. Can we extend the time scale of validity of the approximation if we start the solution in D?

5.4.1 Anharmonic Oscillator with Linear Damping

Consider the anharmonic oscillator with linear damping:

$$\ddot{x} + x = -\varepsilon\dot{x} + \varepsilon x^3.$$

Putting $x = r\sin(t - \psi), \dot{x} = r\cos(t - \psi)$ we obtain (cf. Section 1.7)

$$\begin{aligned}\dot{r} &= \varepsilon r\cos(t-\psi)\left(-\cos(t-\psi) + r^2\sin(t-\psi)^3\right)\\ \dot{\psi} &= \varepsilon\sin(t+\psi)\left(-\cos(t-\psi) + r^2\sin(t-\psi)^3\right)\end{aligned}$$

or, upon averaging over t,

$$\begin{aligned}\dot{\overline{r}} &= -\frac{1}{2}\varepsilon\overline{r},\\ \dot{\overline{\psi}} &= \frac{3}{8}\varepsilon\overline{r}^2.\end{aligned}$$

We now ignore the ψ-dependence for the moment. Then this reduces to a scalar equation

$$\dot{\overline{r}} = -\frac{1}{2}\varepsilon\overline{r},$$

with attractor $r = 0$. Can we extend the time scale of validity of the approximation in ψ?

5.4.2 Duffing's Equation with Damping and Forcing

The second-order differential equation

$$\ddot{u} + a\dot{u} + u + bu^3 = c\cos\omega t$$

is called the **forced Duffing equation**, and is a central example in the theory of nonlinear oscillations. In order to create perturbation problems involving this equation, the parameters a, b, c, and ω are made into functions of a small parameter ε in such a way that the problem is solvable when $\varepsilon = 0$. There are several important perturbation problems that can be created in this way, but the one most often studied has a, b, c, and $\omega - 1$ proportional to ε; we may take $b = \varepsilon$ and write $a = \delta\varepsilon$, $c = \varepsilon A$, and $\omega = 1 + \varepsilon\beta$. The choices of a, b, and c are natural, since when $\varepsilon = 0$ the problem reduces to the solvable linear problem $\ddot{u} + u = 0$. The only thing that requires some explanation is the choice of ω.

Notice that when $\varepsilon = 0$ all solutions have period 2π, or (circular) frequency 1. Therefore if $\omega = 1$ there is exact **harmonic resonance** between the **forcing frequency** ω and the **free frequency** of the unperturbed solutions. It is natural to expect that the behavior of solutions will be different in the resonant and nonresonant cases, but in fact **near-resonance** also has an

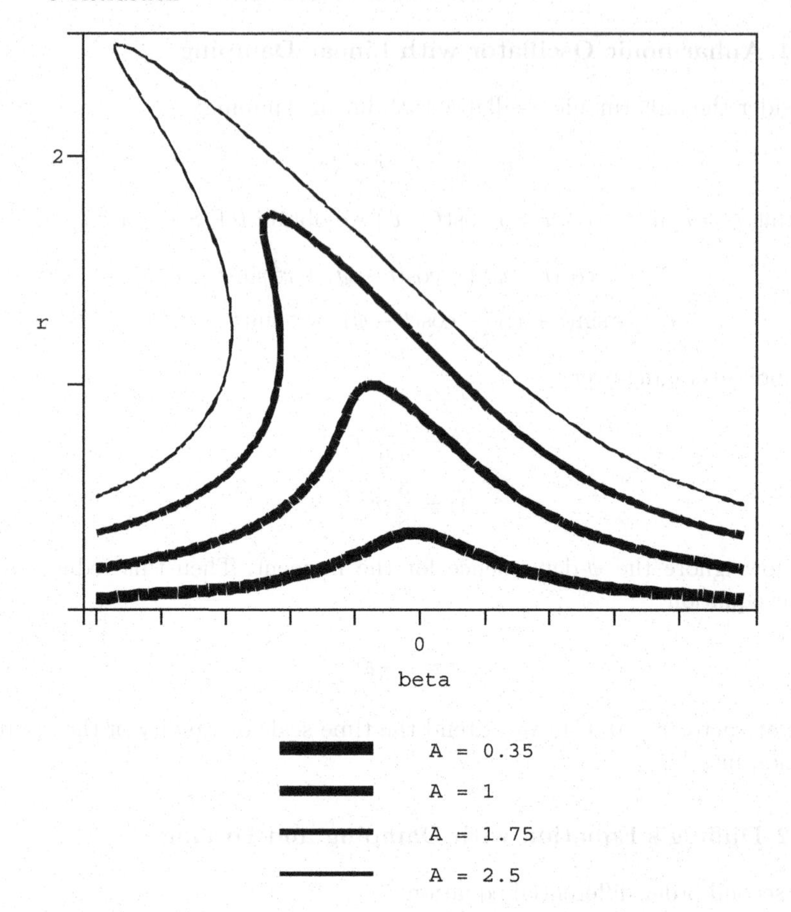

Fig. 5.3: Response curves for the harmonically forced Duffing equation.

effect. The assumption that $\omega = 1 + \varepsilon\beta$, with ε small, expresses the idea that ω is close to 1; β is called the **detuning parameter**. So the problem usually studied is

$$\ddot{u} + \varepsilon\delta\dot{u} + u + \varepsilon u^3 = \varepsilon A \cos(1 + \varepsilon\beta)t.$$

For a complete treatment, one also considers cases in which ω is close to various rational numbers p/q, referred to as subharmonic, superharmonic, and supersubharmonic resonances, but we do not consider these cases here. The phenomena associated with near-resonance are best explained in terms of resonance **horns** or **tongues**; see [201, Sections 4.5-7 and 6.4] for this and additional information about the Duffing equation.

To study this system by averaging, it must be converted into a first-order system in standard form. Since the standard form for (periodic) averaging assumes a period in t that is independent of ε, it is common to introduce a strained time variable $t^+ = (1+\varepsilon\beta)t$ and reexpress the equation using d/dt^+ in place of d/dt. But a more convenient approach is to place the detuning parameter β in the free frequency rather than in the forcing. Therefore the equation we consider here is the following:

$$\ddot{u} + \varepsilon\delta\dot{u} + (1+\varepsilon\beta)u + \varepsilon u^3 = \varepsilon A\cos t. \tag{5.4.1}$$

To prepare this equation for averaging, write it as a system

$$\begin{aligned}\dot{u} &= v\\ \dot{v} &= -u + \varepsilon(-\beta u - \delta v - u^3 + A\cos t).\end{aligned}$$

Then introduce rotating coordinates $\boldsymbol{x} = (x_1, x_2)$ in the u, v-plane by setting

$$u = x_1\cos t + x_2\sin t, \quad v = -x_1\sin t + x_2\cos t, \tag{5.4.2}$$

to obtain

$$\begin{aligned}\dot{x}_1 &= -\varepsilon(-\beta u - \delta v - u^3 + A\cos t)\sin t,\\ \dot{x}_2 &= \varepsilon(-\beta u - \delta v - u^3 + A\cos t)\cos t,\end{aligned}$$

where u and v stand for the expressions in (5.4.2). When these equations are written out in full and averaged, the result, using the notation of (2.8.8), is

$$\begin{aligned}\dot{z}_1 &= \varepsilon(-\frac{1}{2}\delta z_1 + \frac{1}{2}\beta z_2 + \frac{3}{8}z_1^2 z_2 + \frac{3}{8}z_2^3),\\ \dot{z}_2 &= \varepsilon(-\frac{1}{2}\beta z_1 - \frac{1}{2}\delta z_2 - \frac{3}{8}z_1^3 - \frac{3}{8}z_1 z_2^2 + \frac{1}{2}A).\end{aligned}$$

Changing time scale to $\tau = \varepsilon t$ and writing $' = d/d\tau$ yields the guiding system

$$\begin{aligned}w_1' &= -\frac{1}{2}\delta w_1 + \frac{1}{2}\beta w_2 + \frac{3}{8}w_1^2 w_2 + \frac{3}{8}w_2^3,\\ w_2' &= -\frac{1}{2}\beta w_1 - \frac{1}{2}\delta w_2 - \frac{3}{8}w_1^3 - \frac{3}{8}w_1 w_2^2 + \frac{1}{2}A.\end{aligned}$$

So far we have applied only the periodic averaging theory of Chapter 2. In order to apply the ideas of this chapter, we should look for rest points of this guiding system and see if they are attracting. The equation for rest points simplifies greatly if written in polar coordinates (r,θ), with $w_1 = r\cos\theta$, $w_2 = r\sin\theta$. (We avoided using polar coordinates in the differential equations because this coordinate system is singular at the origin.) The result is

$$0 = \beta r + \frac{3}{4}r^3 - A\cos\theta, \quad 0 = \delta r - A\sin\theta. \tag{5.4.3}$$

Eliminating θ yields the **frequency response curve**

$$\delta^2 r^2 + \left(\beta r + \frac{3}{4}r^3\right)^2 = A^2.$$

Graphs of r against β for various A are shown in Figure 5.3, in which points on solid curves represent attracting rest points (sinks) for the averaged equations, and points on dotted curves are saddles. Thus for certain values of the parameters there is one sink and for others there are two sinks and a saddle. For initial conditions in the basin of attraction of a sink, the theory of this chapter implies that the approximate solution given by averaging is valid for all future time. In Chapter 6 it will be shown that the rest points we have obtained for the averaged equations correspond to periodic solutions of the original equations.

5.5 Theory of Averaging with Attraction

Consider the following differential equation

$$\dot{\boldsymbol{x}} = \varepsilon \mathbf{f}^1(\boldsymbol{x}, t),$$

with $\boldsymbol{x} \in D \subset \mathbb{R}^n$. Suppose that $\mathbf{f}^1$ is a periodic or, more generally, a KBM-vector field (Definition 4.2.4). The averaged equation is

$$\dot{\boldsymbol{z}} = \varepsilon \overline{\mathbf{f}}^1(\boldsymbol{z}).$$

We know from the averaging theorems in Chapters 2 and 4 that if we supply these equations with an initial value $\boldsymbol{a} \in D^o \subset D$, the solutions stay $\delta(\varepsilon)$-close on the time scale $1/\varepsilon$; here $\delta(\varepsilon) = \varepsilon$ in the periodic case, $o(1)$ in the KBM case. Suppose now that

$$\overline{\mathbf{f}}^1(0) = 0$$

and that $\boldsymbol{z} = 0$ is an attractor for all the solutions $\mathbf{z}(t)$ starting in D^o (if this statement holds for $\boldsymbol{z} = \boldsymbol{x}_c$ with $\overline{\mathbf{f}}^1(\boldsymbol{x}_c) = 0$, we translate this critical point to the origin). In fact we suppose somewhat more: we can write

$$\overline{\mathbf{f}}^1(\boldsymbol{z}) = A\boldsymbol{z} + \mathbf{g}^1(\boldsymbol{z}),$$

with $\mathsf{D}\mathbf{g}^1(0) = 0$ and A a constant $n \times n$-matrix with the eigenvalues having negative real parts only. The matrix A *does not* depend on ε; in related problems where it does, some special problems may arise, see Robinson [225] and Section 5.9 of this chapter. The vector field $\mathbf{g}^1$ represents the nonlinear part of $\overline{\mathbf{f}}^1$ near $\boldsymbol{z} = 0$. We have seen that the Poincaré–Lyapunov Theorem 5.2.4 guarantees that the solutions attract exponentially towards the origin. Starting in the Poincaré–Lyapunov neighborhood of the origin we have

$$\| \mathbf{z}(t) \| \leq C \| \boldsymbol{z}_0 \| e^{-\mu(t-t_0)},$$

with C and μ positive constants. Moreover we have for two solutions $\mathbf{z}_1(t)$ and $\mathbf{z}_2(t)$ starting in a neighborhood of the origin that from a certain time $t = t_0$ onwards

$$\| \mathbf{z}_1(t) - \mathbf{z}_2(t) \| \leq C \| \mathbf{z}_1(t_0) - \mathbf{z}_2(t_0) \| e^{-\mu(t-t_0)}.$$

See Theorem 5.2.4 and Lemma 5.2.7. We shall now apply these results together with averaging to obtain asymptotic approximations on $[0, \infty)$. Starting outside the Poincaré–Lyapunov domain, averaging provides us with a time scale $1/\varepsilon$ which is long enough to reach the domain $\| \boldsymbol{x} \| \leq \delta$ where exponential contraction takes place. A summation trick, in this context proposed by Sanchez–Palencia [234] will take care of the growth of the error on $[0, \infty)$. A different proof has been given by Eckhaus [81]; see also Sanchez–Palencia [235] where the method is placed in the context of Banach spaces and where one can also find a discussion of the perturbation of orbits in phase space. Another proof is given in Theorem 6.5.1.

Theorem 5.5.1 (Eckhaus/Sanchez–Palencia). *Consider the initial value problem*

$$\dot{\boldsymbol{x}} = \varepsilon \mathbf{f}^1(\boldsymbol{x}, t), \quad \mathbf{x}(0) = \boldsymbol{a},$$

with $\boldsymbol{a}, \boldsymbol{x} \in D \subset \mathbb{R}^n$. *Suppose* $\mathbf{f}^1$ *is a KBM-vector field producing the averaged equation*

$$\dot{\boldsymbol{z}} = \varepsilon \overline{\mathbf{f}}^1(\boldsymbol{z}), \quad \mathbf{z}(0) = \boldsymbol{a},$$

where $\boldsymbol{z} = 0$ *is an asymptotically stable critical point in the linear approximation,* $\overline{\mathbf{f}}^1$ *is moreover continuously differentiable with respect to* $\boldsymbol{z}$ *in* D *and has a domain of attraction* $D^o \subset D$. *For any compact* $K \subset D^o$ *there exists a* $c > 0$ *such that for all* $\boldsymbol{a} \in K$

$$\mathbf{x}(t) - \mathbf{z}(t) = \mathcal{O}(\delta(\varepsilon)), \quad 0 \leq t < \infty,$$

with $\delta(\varepsilon) = o(1)$ *in the general case and* $\mathcal{O}(\varepsilon)$ *in the periodic case.*

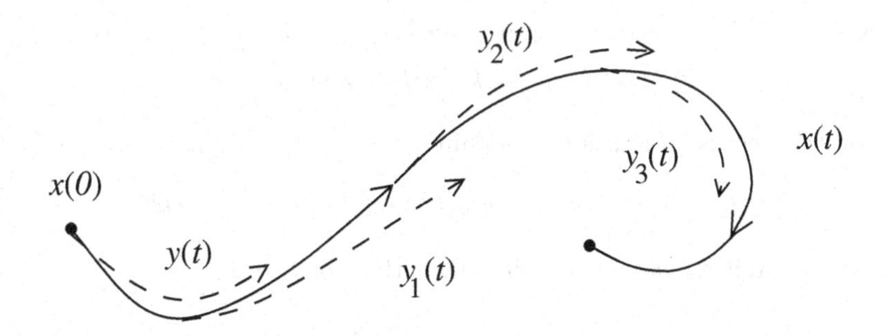

Fig. 5.4: Solution $\mathbf{x}$ starts in $\mathbf{x}(0)$ and attracts towards 0.

Proof Theorem 4.5.5 produces

$$\| \mathbf{x}(t) - \mathbf{z}(t) \| \leq \delta_1(\varepsilon) , \quad 0 \leq \varepsilon t \leq L,$$

with $\delta_1(\varepsilon) = o(1)$, the constant L is independent of ε. Putting $\tau = \varepsilon t$, $\frac{dz}{d\tau} = \overline{\mathbf{f}}^1(z)$ we know from Lemma 5.2.7 that from a certain time $\tau = T$ on, the flow is exponentially contracting and T does not depend on ε. Now we introduce the following partition of the time ($= t$)-axis

$$[0, \frac{T}{\varepsilon}] \bigcup [\frac{T}{\varepsilon}, \frac{2T}{\varepsilon}] \bigcup \cdots \bigcup [\frac{mT}{\varepsilon}, \frac{(m+1)T}{\varepsilon}] \bigcup \cdots , \quad m = 1, 2, \ldots .$$

On each segment $I_m = [\frac{mT}{\varepsilon}, \frac{(m+1)T}{\varepsilon}]$ we define $\mathbf{z}_{(m)}$ as the solution of

$$\dot{z} = \varepsilon \overline{\mathbf{f}}^1(z), \quad \mathbf{z}_{(m)}(\frac{mT}{\varepsilon}) = \mathbf{x}(\frac{mT}{\varepsilon}).$$

For all finite m we have from the averaging theorem

$$\| \mathbf{x}(t) - \mathbf{z}_{(m)}(t) \| \leq \delta_1(\varepsilon) , \quad t \in I_m. \tag{5.5.1}$$

If ε is small enough Lemma 5.2.7 produces on the other hand

$$\begin{aligned} \| \mathbf{z}(t) - \mathbf{z}_{(m)}(t) \|_{I_m} &\leq k \, \| \mathbf{z}(\frac{(m-1)T}{\varepsilon}) - \mathbf{z}_{(m)}(\frac{(m-1)T}{\varepsilon}) \| \\ &\leq k \, \| \mathbf{z}(t) - \mathbf{z}_{(m)}(t) \|_{I_{m-1}}, \end{aligned} \tag{5.5.2}$$

with $m = 1, 2, \ldots$ and $0 < k < 1$ and where $\mathbf{z}_{(m)}$ has been continued on I_{m-1} (the existence properties of the solutions permit this). The triangle inequality yields with (5.5.1) and (5.5.2)

$$\| \mathbf{x}(t) - \mathbf{z}(t) \|_{I_m} \leq \delta_1 + k \, \| \mathbf{z}(t) - \mathbf{z}_{(m)}(t) \|_{I_{m-1}}.$$

Using the triangle inequality again and (5.5.1), we obtain

$$\begin{aligned} \| \mathbf{x}(t) - \mathbf{z}(t) \|_{I_m} &\leq \delta_1(\varepsilon) + k \, \| \mathbf{x}(t) - \mathbf{z}(t) \|_{I_{m-1}} + k \, \| \mathbf{x}(t) - \mathbf{z}_{(m)}(t) \|_{I_{m-1}} \\ &\leq (1+k)\delta_1(\varepsilon) + k \, \| \mathbf{x}(t) - \mathbf{z}(t) \|_{I_{m-1}} . \end{aligned}$$

We use this recursion relation to obtain

$$\| \mathbf{x}(t) - \mathbf{z}(t) \|_{I_m} \leq (1+k)\delta_1(\varepsilon)(1 + k + k^2 + \cdots + k^m).$$

Taking the limit for $m \to \infty$ finally yields that for $t \to \infty$

$$\| \mathbf{x}(t) - \mathbf{z}(t) \| \leq \frac{1+k}{1-k}\delta_1(\varepsilon),$$

which completes the proof. □

Note that $\delta_1(\varepsilon)$ in the estimate is asymptotically the order function as it arises in the averaging theorem. So in the periodic case we have an $\mathcal{O}(\varepsilon)$ estimate for $t \in [0, \infty)$. This applies for instance to the second example of Section 5.3 if one transforms to polar coordinates. Somewhat more general: consider the autonomous system

$$\ddot{x} + x = \varepsilon g(x, \dot{x}).$$

The averaging process has been carried out in Section 2.2. Putting

$$x = r \sin(t - \psi), \quad \dot{x} = r \cos(t - \psi),$$

we obtained after averaging the equations

$$\frac{d\overline{r}}{dt} = \varepsilon \mathrm{f}_1^1(\overline{r}), \quad \frac{d\overline{\psi}}{dt} = \varepsilon \mathrm{f}_2^1(\overline{r}).$$

A critical point $\overline{r} = r_0$ of the first equation will never be asymptotically stable in the linear approximation as $\overline{\psi}$ has vanished from the equations. However, introducing polar coordinates $x = r \sin\theta$, $\dot{x} = r\cos\theta$ we obtain the scalar equation

$$\frac{d\overline{r}}{d\theta} = \varepsilon \mathrm{f}_1^1(\overline{r}).$$

If the critical point $\overline{r} = r_0$ is asymptotically stable in the linear approximation we can apply our theorem. E.g. for the Van der Pol equation (Section 2.2) we obtain

$$\frac{d\overline{r}}{d\theta} = \frac{1}{2}\varepsilon\overline{r}(1 - \frac{1}{4}\overline{r}^2).$$

There are two critical points: $\overline{r} = 0$ and $\overline{r} = 2$. The origin is unstable but $\overline{r} = 2$ (corresponding to the limit cycle) has eigenvalue $-\varepsilon$. Our theorem applies and we have

$$r(\theta) - \overline{r}(\theta) = \mathcal{O}(\varepsilon), \quad \theta \in [\theta_0, \infty)$$

for the orbits starting in D^o in the domain of attraction.

5.6 An Attractor in the Original Equation

Here we give the proof of Theorem 5.3.1 This case is even easier to handle than the case of averaging with attraction as the Poincaré–Lyapunov domain of the attractor is reached on a time scale of order 1.

Proof The solution $\mathbf{y}(t)$ will be contained in the Poincaré–Lyapunov domain around $\boldsymbol{y} = 0$ for $t \geq T$. Note that T does not depend on ε as the unperturbed equation does not depend on ε. We use the partition of the time axis

$$[0,T]\bigcup[T,2T]\bigcup\cdots\bigcup[mT,(m+1)T]\bigcup\cdots.$$

According to Lemma 1.5.3 we have

$$\mathbf{x}(t)-\mathbf{y}(t)=\mathcal{O}(\varepsilon),\quad 0\leq t\leq T.$$

From this point on we use exactly the same reasoning as formulated in Theorem 5.5.1. □

5.7 Contracting Maps

We shall now formulate the results of Section 5.5 in terms of mappings instead of vector fields. This framework enables us to recover Theorem 5.5.1; Moreover one can use this idea to obtain new results. Consider again a differential equation of the form

$$\dot{\boldsymbol{x}}=\varepsilon\mathbf{f}^1(\boldsymbol{x},t),\quad \boldsymbol{x}\in D\subset\mathbb{R}^n.$$

Supposing that $\mathbf{f}^1$ is a *KBM*-vector field we have the averaged equation

$$\dot{\boldsymbol{y}}=\varepsilon\overline{\mathbf{f}}^1(\boldsymbol{y}),\quad \boldsymbol{y}\in D^o\subset D.$$

Again $\overline{\mathbf{f}}^1(\boldsymbol{y})$ has an attracting critical point, say $\boldsymbol{y}=0$ and we know from Lemma 5.2.7 that under certain conditions there is a neighborhood Ω of $\boldsymbol{y}=0$ where the phase-flow is actually contracting exponentially. This provides us with a contracting map of Ω into itself. Indicating a solution $\mathbf{y}$ starting at $t=0$ in $\boldsymbol{y}_0$ by $\mathbf{y}(\boldsymbol{y}_0,t)$ we have the map

$$F_0(\boldsymbol{y}_0)=\mathbf{y}(\boldsymbol{y}_0,t_1),\quad \boldsymbol{y}_0\in\Omega\ ,\ t_1>0.$$

Here we have solutions $\mathbf{y}$ which approximate $\mathbf{x}(\boldsymbol{a},t)$ for $0\leq\varepsilon t\leq L$ if $\boldsymbol{a}$ and $\boldsymbol{y}_0$ are close enough. So we take $t_1=L/\varepsilon$ and we define

$$F_0(\boldsymbol{y}_0)=\mathbf{y}(\boldsymbol{y}_0,L/\varepsilon).$$

In the same way we define the map F_ε by

$$F_\varepsilon(\boldsymbol{a})=\mathbf{x}(\boldsymbol{a},L/\varepsilon).$$

If $\boldsymbol{a}-\boldsymbol{y}_0=o(1)$ as $\varepsilon\downarrow 0$ we have clearly

$$\parallel F_0(\boldsymbol{y}_0)-F_\varepsilon(\boldsymbol{a})\parallel\leq C\delta_1(\varepsilon)\text{ with }\delta_1(\varepsilon)=o(1).$$

We shall prove that for a contracting map F_0, repeated application of the maps F_0 and F_ε does not enlarge the distance between the iterates significantly. We define the iterates by the relations

$$F_\varepsilon{}^1(\boldsymbol{x}) = F_\varepsilon(\boldsymbol{x}),$$
$$F_\varepsilon{}^{m+1}(\boldsymbol{x}) = F_\varepsilon(F_\varepsilon{}^m(\boldsymbol{x})), \quad m = 1, 2, \ldots.$$

This will provide us with a theorem for contracting maps, analogous to Theorem 5.5.1. An application might be as follows. The equation for $\boldsymbol{y}$ written down above, is simpler than the equation for $\boldsymbol{x}$. Still it may be necessary to take recourse to numerical integration to solve the equation for $\boldsymbol{y}$. If the numerical integration scheme involves an estimate of the error on intervals of the time with length L/ε, we may envisage the numerical procedure as providing us with another map F_h which approximates the map F_0. Using the same technique as formulated in the proof of the theorem, one can actually show that the numerical approximations in this case are valid on $[0, \infty)$ and therefore also approximate the solutions of the original equation on the same interval. In the context of a study of successive substitutions for perturbed operator equations Van der Sluis [270] developed ideas which are related to the results discussed here. We shall split the proof into several lemmas to keep the various steps easy to follow.

Lemma 5.7.1. *Consider a family of maps $F_\varepsilon : D \to \mathbb{R}^n$, $\varepsilon \in [0, \varepsilon_0]$ with the following properties:*

1. *For all $\boldsymbol{x} \in D$ we have*

$$\| F_0(\boldsymbol{x}) - F_\varepsilon(\boldsymbol{x}) \| \leq \delta(\varepsilon),$$

   with $\delta(\varepsilon)$ an order function, $\delta = o(1)$ as $\varepsilon \downarrow 0$,
2. *There exist constants k and μ, $0 \leq k < 1$, $\mu \geq 0$ such that for all $\boldsymbol{x}, \boldsymbol{y} \in D$*

$$\| F_0(\boldsymbol{x}) - F_0(\boldsymbol{y}) \| \leq k \| \boldsymbol{x} - \boldsymbol{y} \| + \mu.$$

   This we call the **contraction-attraction property** *of the unperturbed flow,*
3. *There exists an interior domain $D^o \subset D$, invariant under F_0 such that the distance between the boundaries of D^o and D exceeds*

$$\frac{\mu + \delta(\varepsilon)}{1 - k},$$

Then, if $\| \boldsymbol{x} - \boldsymbol{y} \| \leq \frac{\mu+\delta(\varepsilon)}{1-k}$ and $\boldsymbol{x} \in D, \boldsymbol{y} \in D^o$, we have for $m \in \mathbb{N}$

$$\| F_\varepsilon^m(\boldsymbol{x}) - F_0^m(\boldsymbol{y}) \| \leq \frac{\mu + \delta(\varepsilon)}{1 - k}.$$

Proof We use induction. If $m = 0$ the statement is true; assuming that the statement is true for m we prove it for $m+1$, using the triangle inequality.

$$\begin{aligned} &\| F_\varepsilon^{m+1}(\boldsymbol{x}) - F_0^{m+1}(\boldsymbol{y}) \| \\ &\leq \| F_\varepsilon(F_\varepsilon^m(\boldsymbol{x})) - F_0(F_\varepsilon^m(\boldsymbol{x})) \| + \| F_0(F_\varepsilon^m(\boldsymbol{x})) - F_0(F_0^m(\boldsymbol{y})) \| . \end{aligned}$$

Since $\boldsymbol{y} \in D^o$ and D^o is invariant under F_0 we have $F_0^m(\boldsymbol{y}) \in D^o$. It follows from Assumption 3 and the induction hypothesis that $F_\varepsilon^m(\boldsymbol{x}) \in D$. So we can use Assumptions 1 and 2 to obtain the following estimate from the inequality above:

$$\begin{aligned} &\| F_\varepsilon^{m+1}(\boldsymbol{x}) - F_0^{m+1}(\boldsymbol{y}) \| \leq \delta(\varepsilon) + k \| F_\varepsilon^m(\boldsymbol{x}) - F_0^m(\boldsymbol{y}) \| + \mu \\ &\leq \delta(\varepsilon) + k\,\frac{\mu + \delta(\varepsilon)}{1-k} + \mu = \frac{\mu + \delta(\varepsilon)}{1-k}. \end{aligned}$$

This proves the lemma. □

5.8 Attracting Limit-Cycles

In this section we shall discuss problems where the averaged equation has a limit-cycle. It turns out that the theory for this case is like the case with the averaged system having a stable stationary point, except that it is not possible to approximate the angular variable (or the flow on the limit-cycle) uniformly on $[0, \infty)$; the approximation, however, is possible on intervals of length ε^{-N}, with N arbitrary (but, of course, ε-independent). We shall sketch only the results without giving proofs. For technical details the reader is referred to [239]. We consider systems of the form

$$\begin{bmatrix} \dot{\boldsymbol{x}} \\ \dot{\boldsymbol{\phi}} \end{bmatrix} = \begin{bmatrix} 0 \\ \boldsymbol{\Omega}^0(\boldsymbol{x}) \end{bmatrix} + \varepsilon \begin{bmatrix} \mathbf{X}^1(\boldsymbol{x}, \boldsymbol{\phi}) \\ \boldsymbol{\Omega}^1(\boldsymbol{x}, \boldsymbol{\phi}) \end{bmatrix}, \quad \begin{array}{l} \boldsymbol{x} \in D \subset \mathbb{R}^n, \\ \boldsymbol{\phi} \in \mathbb{T}^m. \end{array}$$

Example 5.8.1. As an example, illustrative for the theory, we shall take the Van der Pol equation

$$\ddot{x} + \varepsilon(x^2 - 1)\dot{x} + x = 0, \quad \mathrm{x}(0) = a_1, \quad \dot{\mathrm{x}}(0) = a_2, \quad \| \boldsymbol{a} \| \neq 0.$$

Introducing polar coordinates

$$x = r \sin\phi, \quad \dot{x} = r\cos\phi,$$

we obtain

$$\begin{bmatrix} \dot{r} \\ \dot{\phi} \end{bmatrix} = \begin{bmatrix} 0 \\ 1 \end{bmatrix} + \varepsilon \begin{bmatrix} \frac{1}{2} r[1 + \cos 2\phi - \frac{1}{4} r^2 (1 - \cos 4\phi)] \\ -\frac{1}{2}\sin 2\phi + \frac{1}{8} r^2 (2\sin 2\phi - \sin 4\phi) \end{bmatrix}$$

The second-order averaged equation of this vector field is (see Section 2.9.1)

$$\begin{bmatrix} \dot{r} \\ \dot{\phi} \end{bmatrix} = \begin{bmatrix} 0 \\ 1 \end{bmatrix} + \varepsilon \begin{bmatrix} \frac{1}{2}r(1-\frac{1}{4}r^2) \\ 0 \end{bmatrix} + \varepsilon^2 \begin{bmatrix} 0 \\ -\frac{1}{8}(\frac{11}{32}r^4 - \frac{3}{2}r^2 + 1) \end{bmatrix} + \mathcal{O}(\varepsilon^3).$$

Neglecting the $\mathcal{O}(\varepsilon^3)$ term, the equation for r represents a subsystem with attractor $r = 2$. The fact that the $\mathcal{O}(\varepsilon^3)$ term depends on another variable as well (i.e. on ϕ), is not going to bother us in our estimates since $\phi(t)$ is bounded (the circle is compact). This means that on solving the equation

$$\dot{\overline{r}} = \frac{1}{2}\varepsilon\overline{r}(1 - \frac{1}{4}\overline{r}^2), \quad \overline{\mathrm{r}}(0) = r_0 = \| \boldsymbol{a} \|$$

this is going to give us an $\mathcal{O}(\varepsilon)$-approximation to the r-component of the original solution, valid on $[0, \infty)$ (The fact that the ε^2-terms in the r-equation vanish, does in no way influence the results). Using this approximation, we can obtain an $\mathcal{O}(\varepsilon)$-approximation for the ϕ-component on $0 \le \varepsilon^2 t \le L$ by solving

$$\dot{\overline{\phi}} = 1 - \frac{1}{8}\varepsilon^2(\frac{11}{32}\overline{r}^4 - \frac{3}{2}\overline{r}^2 + 1), \quad \overline{\phi}(0) = \phi_0 = \arctan(a_1/a_2).$$

Although this equation is easy to solve the treatment can even be more simplified by noting that the attraction in the r-direction takes place on a time scale $1/\varepsilon$, while the slow fluctuation of ϕ occurs on a time scale $1/\varepsilon^2$. This has as a consequence that to obtain an $\mathcal{O}(\varepsilon)$-approximation $\overline{\phi}$ for ϕ on the time scale $1/\varepsilon^2$ we may take $\overline{r} = 2$ on computing $\overline{\phi}$. To prove this one uses an exponential estimate on $|\mathrm{r}(t) - 2|$ and the Gronwall inequality. Thus we are left with the following simple system

$$\begin{bmatrix} \dot{\overline{r}} \\ \dot{\hat{\phi}} \end{bmatrix} = \begin{bmatrix} 0 \\ 1 \end{bmatrix} + \varepsilon \begin{bmatrix} \frac{1}{2}\overline{r}(1-\frac{1}{4}\overline{r}^2) \\ 0 \end{bmatrix} + \varepsilon^2 \begin{bmatrix} 0 \\ -\frac{1}{16} \end{bmatrix}, \quad \begin{bmatrix} \overline{\mathrm{r}} \\ \hat{\phi} \end{bmatrix}(0) = \begin{bmatrix} r_0 \\ \phi_0 \end{bmatrix}.$$

For the general solution of the Van der Pol equation with $r_0 > 0$ we obtain

$$\mathrm{x}(t) = \frac{r_0 e^{\frac{1}{2}\varepsilon t}}{[1 + \frac{1}{4}{r_0}^2(e^{\varepsilon t} - 1)]^{\frac{1}{2}}} \sin(t - \frac{1}{16}\varepsilon^2 t + \phi_0) + \mathcal{O}(\varepsilon)$$

on $0 \le \varepsilon^2 t \le L$. There is no obstruction against carrying out the averaging process to any higher order to obtain approximations valid on longer time scales, to be expressed in inverse powers of ε. ◇

5.9 Additional Examples

To illustrate the theory in this chapter we shall discuss here examples which exhibit some of the difficulties.

5.9.1 Perturbation of the Linear Terms

We have excluded in our theory and examples the possibility of perturbing the linear part of the differential equation in such a way that the stability characteristics of the attractor change. This is an important point as is shown by the following adaptation of an example in Robinson [224]. Consider the linear system with constant coefficients

$$\dot{\boldsymbol{x}} = A(\varepsilon)\boldsymbol{x} + B(\varepsilon)\boldsymbol{x},$$

with

$$A(\varepsilon) = \begin{bmatrix} -\varepsilon^2 & \varepsilon \\ 0 & -\varepsilon^2 \end{bmatrix}, \quad B(\varepsilon) = \varepsilon^3 \begin{bmatrix} 0 & a^2 \\ 0 & 0 \end{bmatrix},$$

where a is a positive constant. Omitting the $\mathcal{O}(\varepsilon^3)$ term we find negative eigenvalues $(-\varepsilon^2)$ so that in this 'approximation' we have attraction towards the trivial solution. For the full equation we find eigenvalues $\lambda_\pm = -\varepsilon^2 \pm a\varepsilon^2$. So if $0 < a < 1$ we have attraction and $x = 0$ is asymptotically stable; if $a > 1$ the trivial solution is unstable. In both cases the flow is characterized by a time scale $1/\varepsilon^2$. This type of issue is discussed again in Section 6.8 under the topic of k-determined hyperbolicity.

5.9.2 Damping on Various Time Scales

Consider the equation of an oscillator with a linear and a nonlinear damping

$$\ddot{x} + \varepsilon^n a\dot{x} + \varepsilon\dot{x}^3 + x = 0, \quad 0 < a \in \mathbb{R}.$$

The importance of the linear damping is determined by the choice of n. We consider various cases.

The Case $n = 0$

Putting $\varepsilon = 0$ we have the equation $\ddot{y} + a\dot{y} + y = 0$. Applying Theorem 5.3.1 we have that if $\mathrm{y}(0) = \mathrm{x}(0), \dot{\mathrm{y}}(0) = \dot{\mathrm{x}}(0)$, then $\mathrm{y}(t)$ represents an $\mathcal{O}(\varepsilon)$-approximation of $\mathrm{x}(t)$ uniformly valid in time. A naive expansion

$$\mathrm{x}(t) = \mathrm{y}(t) + \varepsilon\mathrm{x}^1(t) + \varepsilon^2 \cdots$$

produces higher-order approximations with uniform validity.

The Case $n > 0$

If $n > 0$ we put $x = r\sin(t - \psi), \dot{x} = r\cos(t - \psi)$ to obtain

$$\begin{aligned} \dot{r} &= -\varepsilon^n ar\cos(t-\psi)^2 - \varepsilon r^3\cos(t-\psi)^4, \\ \dot{\psi} &= -\varepsilon^n a\sin(t-\psi)\cos(t-\psi) - \varepsilon r^2\sin(t-\psi)\cos(t-\psi)^3. \end{aligned}$$

Note that $\dot{\mathrm{r}}(t) \leq 0$.

The Case $n = 1$

The terms on the right-hand side are of the same order in ε; first-order averaging produces

$$\dot{\overline{r}} = -\frac{1}{2}\varepsilon a\overline{r} - \frac{3}{8}\varepsilon\overline{r}^3, \quad \dot{\overline{\psi}} = 0.$$

The solutions can be used as approximations valid on the time scale $1/\varepsilon$. However, we have attraction in the r-direction and we can proceed in the spirit of the results in the preceding section. Introducing polar coordinates by $\phi = t - \psi$ we find that we can apply Theorem 5.5.1 to the equation for the orbits ($\frac{dr}{d\phi} = \cdots$). So $\overline{r}$ represents a uniformly valid approximation of r; higher-order approximations can be used to obtain approximations for ψ or ϕ which are valid on a longer time scale than $1/\varepsilon$.

The Case $n = 2$

The difficulty is that we cannot apply the preceding theorems at this stage as we have no attraction in the linear approximation. The idea is to use the linear higher-order damping term nevertheless. Since the damping takes place on the time scale $1/\varepsilon^2$ the contraction constant κ looks like $k = e^{\mu\varepsilon}$ and therefore $\frac{1}{1-\kappa} = \mathcal{O}_\sharp(\frac{1}{\varepsilon})$. We lose an order of magnitude in ε in our estimate, but we can win an order of magnitude by looking at the higher-order approximation, which we are doing anyway, since we consider $\mathcal{O}(\varepsilon^2)$ terms. The amplitude component of the second-order averaged equation is

$$\dot{\overline{r}} = -\frac{3}{8}\varepsilon\overline{r}^3 - \frac{1}{2}\varepsilon^2 a\overline{r}, \quad \overline{\mathrm{r}}(0) = r_0$$

and we see that, with $\tilde{r} = \overline{r} + \varepsilon u^1(\overline{r}, \overline{\phi}, t)$,

$$\| \mathrm{r}(t) - \tilde{\mathrm{r}}(t) \| \leq C(L)\varepsilon^2 \text{ on } 0 \leq \varepsilon t \leq L,$$

where r is the solution of the nontruncated averaged equation. Using the contraction argument we find that

$$\| \mathrm{r}(t) - \tilde{\mathrm{r}}(t) \| \leq \frac{4C(L)}{aL}\varepsilon \text{ for all } t \in [0, \infty)$$

and therefore, since u^1 is uniformly bounded,

$$\| \mathrm{r}(t) - \overline{\mathrm{r}}(t) \| \leq \frac{4C(L)}{aL}\varepsilon \text{ for all } t \in [0, \infty).$$

This gives us the desired approximation of the amplitude for all time. The reader may want to apply the arguments in Appendix B to compute an approximation of the phase.

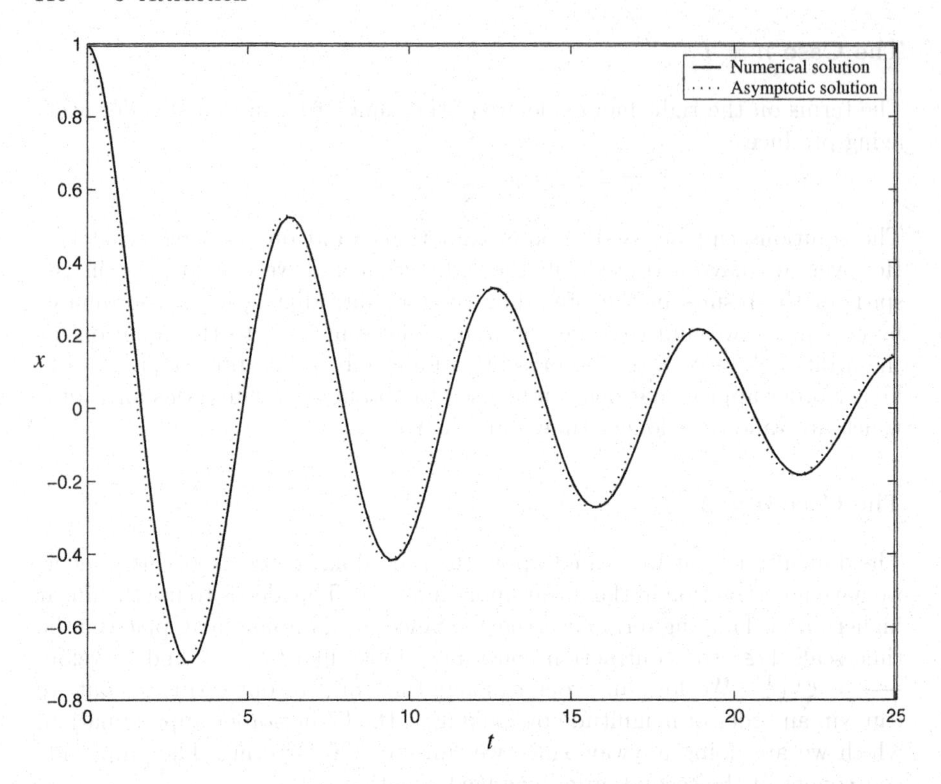

Fig. 5.5: Linear attraction on the time scale $1/\varepsilon^2$ for the equation $\ddot{x}+x+\varepsilon\dot{x}^3+3\varepsilon^2\dot{x} = 0$; $\varepsilon = 0.2$, $x(0) = 1$, $\dot{x}(0) = 0$. The solution obtained by numerical integration has been drawn full line. The asymptotic approximation is indicated by $- - --$ and has been obtained from the equation, averaged to second-order.

6

Periodic Averaging and Hyperbolicity

6.1 Introduction

The theory of averaging has both qualitative and quantitative aspects. From the earliest period, averaging was used not only to construct approximate solutions and estimate their error, but also to prove the existence of periodic orbits and determine their stability. With more recent developments in dynamical systems theory, it has become possible to study not only these local qualitative properties, but also global ones, such as the existence of connecting orbits from one periodic orbit to another. There are interactions both ways between the qualitative and quantitative sides of averaging: approximate solutions and their error estimates can be used in proving qualitative features of the corresponding exact solutions, and specific types of qualitative behavior allow the improvement of the error estimates to ones that hold for all time, rather than for time $\mathcal{O}(1/\varepsilon)$. These topics (for periodic averaging) are the subject of this chapter. For the most part this chapter depends only on Sections 2.8 and, occasionally, 2.9. The remainder of the book is independent of this chapter.

An example to motivate the chapter is presented in Section 6.2. Complete results are stated for this example, with references to the proofs in later sections. Some of the proofs become rather technical, but the example should make the meaning of the results clear.

Each of the theorems in Sections 6.3–6.7 about periodic averaging corresponds to an easier theorem in regular perturbation theory. The simplest way to present the proofs is to treat the regular case first, and then indicate the modifications necessary to handle the averaging case. For clarity, most of these sections are divided into two subsections. Reading only the "regular case" subsections will provide an introduction to Morse–Smale theory with an emphasis on shadowing. Reading both subsections will show how Morse–Smale theory interacts with averaging.

Remark 6.1.1. Morse–Smale theory deals with a particular class of flows on a smooth compact manifold satisfying hyperbolicity assumptions. These flows are **structurally stable**, meaning that when perturbed slightly, they remain *conjugate* to the unperturbed flow. In our treatment the manifolds are replaced by open sets Ω in $\mathbb{R}^n$ with compact closure, and the perturbations are smoothly dependent on a perturbation parameter (rather than just close in a suitable topology). For an introduction to Morse–Smale theory in its standard form, see [215]. The concept of shadowing is not usually considered part of Morse–Smale theory. It arises in the more general theory of Axiom A dynamical systems, in which the behavior called **chaos** occurs. The type of shadowing we study here is closely related, but occurs even in Morse–Smale systems, that do not exhibit chaos. ♡

For reference, we repeat the basic equations used in the classical proof of first-order averaging (Section 2.8): the *original system*

$$\dot{\boldsymbol{x}} = \varepsilon \mathbf{f}^1(\boldsymbol{x}, t) + \varepsilon^2 \mathbf{f}^{[2]}(\boldsymbol{x}, t, \varepsilon), \tag{6.1.1}$$

periodic in t with period 2π; the *(full) averaged equation*

$$\dot{\boldsymbol{y}} = \varepsilon \overline{\mathbf{f}}^1(\boldsymbol{y}) + \varepsilon^2 \mathbf{f}_\star^{[2]}(\boldsymbol{y}, t, \varepsilon); \tag{6.1.2}$$

the *truncated averaged equation*

$$\dot{\boldsymbol{z}} = \varepsilon \overline{\mathbf{f}}^1(\boldsymbol{z}); \tag{6.1.3}$$

and the *guiding system*

$$\boldsymbol{w}' = \overline{\mathbf{f}}^1(\boldsymbol{w}) \tag{6.1.4}$$

(where $' = d/d\tau$ with $\tau = \varepsilon t$). In addition, there is the coordinate transformation

$$\boldsymbol{x} = \mathbf{U}(\boldsymbol{y}, t, \varepsilon) = \boldsymbol{y} + \varepsilon \mathbf{u}^1(\boldsymbol{y}, t), \tag{6.1.5}$$

which is also periodic in t with period 2π and carries solutions of (6.1.2) to solutions of (6.1.1). The next paragraph outlines the required information about regular perturbation theory.

Regular perturbation theory, as defined in Section 1.5, deals with autonomous systems of the form

$$\dot{\boldsymbol{x}} = \mathbf{f}^{[0]}(\boldsymbol{x}, \varepsilon) = \mathbf{f}^0(\boldsymbol{x}) + \mathcal{O}(\varepsilon), \tag{6.1.6}$$

with $\boldsymbol{x} \in \mathbb{R}^n$, and consists in approximating the solution $\mathbf{x}(t, \boldsymbol{a}, \varepsilon)$ with $\mathbf{x}(0) = \boldsymbol{a}$ by its Taylor polynomial of degree k in ε as follows:

$$\mathbf{x}(\boldsymbol{a}, t, \varepsilon) = \mathbf{x}^0(\boldsymbol{a}, t) + \varepsilon \mathbf{x}^1(\boldsymbol{a}, t) + \cdots + \varepsilon^k \mathbf{x}^k(\boldsymbol{a}, t) + \mathcal{O}(\varepsilon^{k+1}) \tag{6.1.7}$$

uniformly for t in any finite interval $0 \le t \le T$. Notice that kth-order regular perturbation theory has error $\mathcal{O}(\varepsilon^{k+1})$ for time $\mathcal{O}(1)$, whereas kth-order averaging has error $\mathcal{O}(\varepsilon^k)$ for time $\mathcal{O}(1/\varepsilon)$. In averaging, the lowest-order approximation is first-order averaging, but in regular perturbation theory it is zeroth order. In this chapter we focus primarily on these lowest-order cases. There are recursive procedures in regular perturbation theory to calculate the coefficients $\mathbf{x}^i(\boldsymbol{a}, t)$ in (6.1.7), but we will not be concerned with these, except to say that the leading approximation $\mathbf{z}(\boldsymbol{a}, t) = \mathbf{x}^0(t)$ satisfies the unperturbed equation

$$\dot{\boldsymbol{z}} = \mathbf{f}^0(\boldsymbol{z}). \tag{6.1.8}$$

We call (6.1.8) the *guiding system*, because (6.1.8) plays the same role in the regular case that the guiding system plays in the averaging case. (There is no $\boldsymbol{w}$ variable, because there is no change of time scale.)

The outline for this chapter is as follows. Section 6.2 gives the motivating example. In Section 6.3 we present classical results showing that hyperbolic rest points of the guiding system correspond to rest points of the exact system in regular perturbation theory, and to hyperbolic periodic orbits in the averaging case. In Section 6.4 we investigate a fundamental reason why regular perturbation estimates break down after time $\mathcal{O}(1)$, and averaging estimates after time $\mathcal{O}(1/\varepsilon)$, and we begin to repair this situation. First we prove local topological conjugacy of the guiding system and the exact system near a hyperbolic rest point in the regular case (that is, we prove a version of the local structural stability theorem for Morse–Smale systems), and the appropriate extension to the averaging case. Then we show that the conjugacy provides *shadowing orbits* that satisfy error estimates on extended time intervals. Section 6.5 is an interlude, giving a different type of extended error estimate for solutions approaching an attractor. (This is the so-called Eckhaus/Sanchez–Palencia theorem, see Theorem 5.5.1 for a different proof.) In Sections 6.6 and 6.7 the conjugacy and shadowing results are extended still further, first to "dumbbell"-shaped neighborhoods containing two hyperbolic rest points of the guiding system, and then to larger interconnected networks of such points. These sections introduce the transversality condition that (together with hyperbolicity) is characteristic of Morse–Smale systems. Section 6.8 gives examples and discusses various degenerate cases in which the known results are incomplete.

6.2 Coupled Duffing Equations, An Example

Consider the following system of two coupled identical harmonically forced Duffing equations (see (5.4.1)):

$$\begin{aligned}
\ddot{u}_1 + \varepsilon\delta\dot{u}_1 + (1+\varepsilon\beta)u_1 + \varepsilon u_1^3 &= \varepsilon A\cos t + \varepsilon f(u_1, u_2),\\
\ddot{u}_2 + \varepsilon\delta\dot{u}_2 + (1+\varepsilon\beta)u_2 + \varepsilon u_2^3 &= \varepsilon A\cos t + \varepsilon g(u_1, u_2).
\end{aligned}$$

For simplicity in both the analysis and the geometry, we assume that f and g are polynomials containing only terms of even total degree, for instance,

$f(u_1, u_2) = (u_1 - u_2)^2$. The system of two second-order equations is converted to four first-order equations in the usual way, by setting $v_i = \dot{u}_i$. Next, rotating coordinates $\boldsymbol{x} = (x_1, x_2, x_3, x_4)$ are introduced in the (u_1, v_1)-plane and the (u_2, v_2)-plane by setting

$$u_1 = x_1 \cos t + x_2 \sin t, \quad v_1 = -x_1 \sin t + x_2 \cos t,$$
$$u_2 = x_3 \cos t + x_4 \sin t, \quad v_2 = -x_3 \sin t + x_4 \cos t.$$

The resulting system has the form

$$\dot{\boldsymbol{x}} = \varepsilon \begin{bmatrix} -F(\boldsymbol{x},t)\sin t \\ +F(\boldsymbol{x},t)\cos t \\ -G(\boldsymbol{x},t)\sin t \\ +G(\boldsymbol{x},t)\cos t \end{bmatrix}, \tag{6.2.1}$$

where

$$F(\boldsymbol{x},t) = -\beta u_1 - \delta v_1 - u_1^3 + f(u_1,u_2) + A\cos t,$$
$$G(\boldsymbol{x},t) = -\beta u_2 - \delta v_2 - u_2^3 + g(u_1,u_2) + A\cos t,$$

it being understood that u_1, u_2, v_1, v_2 are replaced by their expressions in $\boldsymbol{x}$ and t.

Upon averaging (6.2.1) over t and rescaling time by $\tau = \varepsilon t$, the following guiding system is obtained (with $' = d/d\tau$, and with variables renamed w as usual):

$$\boldsymbol{w}' = \begin{bmatrix} -\frac{1}{2}\delta w_1 + \frac{1}{2}\beta w_2 + \frac{3}{8}w_1^2 w_2 + \frac{3}{8}w_2^3 \\ -\frac{1}{2}\beta w_1 - \frac{1}{2}\delta w_2 - \frac{3}{8}w_1^3 - \frac{3}{8}w_1 w_2^2 + \frac{1}{2}A \\ -\frac{1}{2}\delta w_3 + \frac{1}{2}\beta w_4 + \frac{3}{8}w_3^2 w_4 + \frac{3}{8}w_4^3 \\ -\frac{1}{2}\beta w_3 - \frac{1}{2}\delta w_4 - \frac{3}{8}w_3^3 - \frac{3}{8}w_3 w_4^2 + \frac{1}{2}A \end{bmatrix}. \tag{6.2.2}$$

Since f and g contain only terms with even total degree, each term resulting from these in (6.2.1) is of odd total degree in $\cos t$ and $\sin t$, and therefore has zero average. Therefore the (w_1, w_2) subsystem in (6.2.2) is decoupled from the (w_3, w_4) subsystem. We introduce polar coordinates by $w_1 = r\cos\theta$, $w_2 = r\sin\theta$. The rest points for this subsystem then satisfy

$$0 = \beta r + \frac{3}{4}r^3 + A\cos\theta, \quad 0 = \delta r + A\sin\theta, \tag{6.2.3}$$

as in (5.4.3). For certain values of β, δ, and A there are three rest points, all hyperbolic (two sinks $\boldsymbol{p}_1$ and $\boldsymbol{p}_2$ and a saddle $\boldsymbol{q}$); from here on we assume that this is the case. The unstable manifold of q has two branches, one falling into $\boldsymbol{p}_1$ and the other into $\boldsymbol{p}_2$. The stable manifold of $\boldsymbol{q}$ forms a separatrix that forms the boundary of the basin of attraction of each sink. The unstable manifold of $\boldsymbol{q}$ intersects the stable manifolds of $\boldsymbol{p}_1$ and $\boldsymbol{p}_2$ transversely (in the (w_1, w_2)-plane), since the stable manifolds of the sinks are two-dimensional and their tangent vectors already span the plane. (Two submanifolds intersect transversely in an ambient manifold if their tangent spaces at each point of the intersection, taken together, span the tangent space to the ambient manifold.)

For the full guiding system (6.2.2) there are nine rest points: four "double sinks" $(\boldsymbol{p}_i, \boldsymbol{p}_j)$, four "saddle-sinks" $(\boldsymbol{p}_i, \boldsymbol{q})$ and $(\boldsymbol{q}, \boldsymbol{p}_i)$, and a "double saddle" $(\boldsymbol{q}, \boldsymbol{q})$. The stable (respectively, unstable) manifold of each rest point is the Cartesian product of the stable (respectively, unstable) manifolds of the (planar) rest points that make it up. The "diagram" of these rest points is a graph with nine vertices (corresponding to the rest points) with directed edges from one vertex to another if the unstable manifold of the first vertex intersects the stable manifold of the second. (The relation is transitive, so not all of the edges need to be drawn.) The diagram is shown in Figure 6.1.

All of these intersections are transverse. This is clear for intersections leading to double sinks, because the stable manifold of a double sink has dimension 4, so it needs to be checked only for intersections leading from the double saddle to a saddle-sink. Consider the case of $(\boldsymbol{q}, \boldsymbol{q})$ and $(\boldsymbol{p}_1, \boldsymbol{q})$. The stable manifold of $(\boldsymbol{p}_1, \boldsymbol{q})$ is three-dimensional and involves the full two dimensions of the first subsystem (in (w_1, w_2)) and the stable dimension of $\boldsymbol{q}$ (in (w_2, w_3)). The unstable manifold of $(\boldsymbol{q}, \boldsymbol{q})$ is two-dimensional and involves the unstable dimension of each copy of $\boldsymbol{q}$; the first copy of this unstable dimension is included in the stable manifold of $(\boldsymbol{p}_1, \boldsymbol{q})$, but the second copy is linearly independent of the stable manifold and completes the transversality.

All orbits approach one of the rest points as $t \to \infty$. Therefore it is possible to find a large open set Ω with compact closure containing the nine rest points with the property that all solutions entering Ω remain in Ω for all future time. It suffices to choose any large ball B containing the rest points, and let Ω be the union of all future half-orbits of points in B. Since all such orbits reenter B (if they leave it at all) in finite time, Ω constructed this way will be bounded. Another way is to find in each plane subsystem a future-invariant region bounded by arcs of solutions and segments transverse to the vector field, and let Ω be the Cartesian product of this region with itself.

From these remarks we can draw the following conclusions, on the basis of the theorems proved in this chapter:

1. The original system (6.2.1) has nine hyperbolic periodic orbits of period 2π, four of them stable. Approximations of these periodic solutions with error $\mathcal{O}(\varepsilon^k)$ for all time can be obtained by kth-order averaging. See Theorems 6.3.2 and 6.3.3.
2. The diagram of connecting orbits (intersections of stable and unstable manifolds) among these periodic orbits is the same as that in Figure 6.1. See Theorem 6.7.2.
3. Every approximate solution to an initial value problem obtained by first-order averaging that approaches one of the four stable periodic solutions is valid with error $\mathcal{O}(\varepsilon)$ for all future time. See Theorem 6.5.1.
4. Every approximate solution obtained by first-order averaging, having its guiding solution in Ω, is shadowed with error $\mathcal{O}(\varepsilon)$ by an exact solution for

all future time, although the approximate solution and its shadowing exact solution will not, in general, have the same initial values. See Theorem 6.7.1.

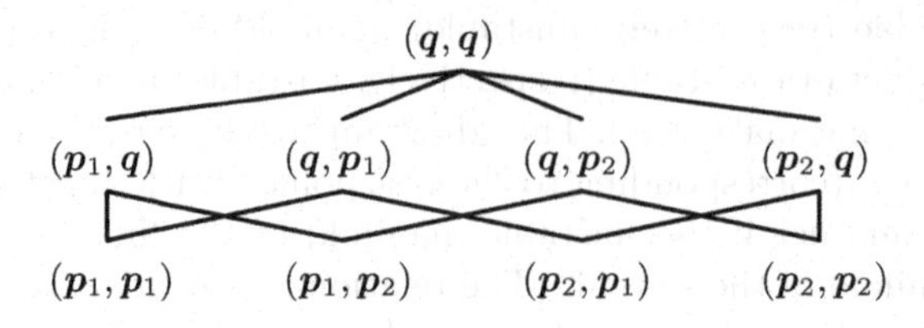

Fig. 6.1: Connection diagram for two coupled Duffing equations.

6.3 Rest Points and Periodic Solutions

A rest point $\boldsymbol{a}$ of an autonomous system $\dot{\boldsymbol{x}} = \mathbf{f}^0(\boldsymbol{x})$ is called **simple** if $\mathsf{D}\mathbf{f}^0(\boldsymbol{a})$ is nonsingular, and **hyperbolic** if $\mathsf{D}\mathbf{f}^0(\boldsymbol{a})$ has no eigenvalues on the imaginary axis (which includes having no zero eigenvalues, and is therefore stronger than simple). The **unstable dimension** of a hyperbolic rest point is the number of eigenvalues in the right half-plane. (Hyperbolic rest points have stable and unstable manifolds, and the unstable dimension equals the dimension of the unstable manifold). In this section we show that when the guiding system has a simple rest point, the full system has a rest point (in the regular case) or a periodic orbit (in the averaging case) for sufficiently small ε. Hyperbolicity and (with an appropriate definition for periodic orbits) unstable dimension are also preserved. Versions of these results are found in most differential equations texts; see [59, 121], and [54].

6.3.1 The Regular Case

For the regular case, governed by (6.1.6), the required theorem is very simple.

Theorem 6.3.1 (Continuation of rest points). *Let $\boldsymbol{a}_0$ be a rest point of (6.1.6) for $\varepsilon = 0$, so that $\mathbf{f}^0(\boldsymbol{a}_0) = 0$. If $\mathsf{D}\mathbf{f}^0(\boldsymbol{a}_0)$ is nonsingular, then there exists a unique smooth function $\boldsymbol{a}_\varepsilon$, defined for small ε, such that $\mathbf{f}^{[0]}(\boldsymbol{a}_\varepsilon, \varepsilon) = 0$; thus the rest point $\boldsymbol{a}_0$ "continues" (without bifurcation) to a rest point $\boldsymbol{a}_\varepsilon$ for ε near zero. If in addition $\mathsf{D}\mathbf{f}^0(\boldsymbol{a}_0, 0)$ is hyperbolic, then $\mathsf{D}\mathbf{f}^{[0]}(\boldsymbol{a}_\varepsilon, \varepsilon)$ is hyperbolic with the same unstable dimension.*

Proof The existence and smoothness of a unique $\boldsymbol{a}_\varepsilon$ for small ε follows from the implicit function theorem. In the hyperbolic case, let C_1 and C_2

be simple closed curves, disjoint from the imaginary axis and surrounding the eigenvalues for $\varepsilon = 0$ in the left and right half-planes, respectively. By Rouché's theorem the number of eigenvalues (counting multiplicity) inside C_1 and C_2 does not change as ε is varied (up to the point that an eigenvalue touches the curve). □

6.3.2 The Averaging Case

The analogue of Theorem 6.3.1 will be broken into two parts, one about existence and one about hyperbolicity. Let $\mathbf{x}(\boldsymbol{a}, t, \varepsilon)$ denote the solution of (6.1.1) with initial condition $\mathbf{x}(\boldsymbol{a}, 0, \varepsilon) = \boldsymbol{a}$.

Theorem 6.3.2 (Existence of periodic solutions). *Let $\boldsymbol{a}_0$ be a simple rest point of the guiding system (6.1.4), so that $\overline{\mathbf{f}}^1(\boldsymbol{a}_0) = 0$ and $\mathsf{D}\overline{\mathbf{f}}^1(\boldsymbol{a}_0)$ is nonsingular. Then there exists a unique smooth function $\boldsymbol{a}_\varepsilon$, defined for small ε, such that $\mathbf{x}(\boldsymbol{a}_\varepsilon, t, \varepsilon)$ is a periodic solution of the original system (6.1.1) with period 2π.*

Proof Let $\mathbf{y}(\boldsymbol{b}, t, \varepsilon)$ be the solution of (6.1.2) with $\mathbf{y}(\boldsymbol{b}, 0, \varepsilon) = \boldsymbol{b}$. This solution is periodic (for a fixed value of $\varepsilon \neq 0$) if

$$0 = \mathbf{y}(\boldsymbol{b}, 2\pi, \varepsilon) - \boldsymbol{b} = \varepsilon \int_0^{2\pi} [\overline{\mathbf{f}}^1(\mathbf{y}(\boldsymbol{b}, t, \varepsilon)) + \varepsilon \mathbf{f}_\star^{[2]}(\mathbf{y}(\boldsymbol{b}, t, \varepsilon), t, \varepsilon)]\, \mathrm{d}t.$$

Introduce the function

$$\mathbf{F}^{[0]}(\boldsymbol{b}, \varepsilon) = \int_0^{2\pi} [\overline{\mathbf{f}}^1(\mathbf{y}(\boldsymbol{b}, t, \varepsilon)) + \varepsilon \mathbf{f}_\star^{[2]}(\mathbf{y}(\boldsymbol{b}, t, \varepsilon), t, \varepsilon)]\, \mathrm{d}t.$$

(Notice the omission of the initial ε, which is crucial for the argument.) Then $\mathbf{y}(\boldsymbol{b}, t, \varepsilon)$ is periodic if $\mathbf{F}^{[0]}(\boldsymbol{b}, \varepsilon) = 0$. Now $\mathbf{F}^0(\boldsymbol{a}_0) = 2\pi \overline{\mathbf{f}}^1(\boldsymbol{a}_0) = 0$ by hypothesis, and $\mathsf{D}\mathbf{F}^0(\boldsymbol{a}_0) = 2\pi \mathsf{D}\overline{\mathbf{f}}^1(\boldsymbol{a}_0)$ is nonsingular. Therefore by the implicit function theorem there exists a unique $\boldsymbol{b}_\varepsilon$ with $\boldsymbol{b}_0 = \boldsymbol{a}_0$ such that $\mathbf{F}^{[0]}(\boldsymbol{b}_\varepsilon, \varepsilon) = 0$ for small ε. This implies that $\mathbf{y}(\boldsymbol{b}_\varepsilon, t, \varepsilon)$ is periodic for small ε. Let $\boldsymbol{a}_\varepsilon = \mathbf{U}(\boldsymbol{b}_\varepsilon, 0, \varepsilon)$. (If stroboscopic averaging is used, $\boldsymbol{a}_\varepsilon = \boldsymbol{b}_\varepsilon$.) Then $\boldsymbol{x}(\boldsymbol{a}_\varepsilon, t, \varepsilon) = \mathbf{U}(\mathbf{y}(\boldsymbol{b}_\varepsilon, t, \varepsilon), t, \varepsilon)$ is the required periodic solution of the original system. □

To study the hyperbolicity of the periodic solution it is helpful to introduce a new state space $\mathbb{R}^n \times S^1$ in which all of our systems are autonomous. Let θ be an angular variable tracing the circle S^1, and write the original system in the "suspended" form

$$\dot{\boldsymbol{x}} = \varepsilon \mathbf{f}^1(\boldsymbol{x}, \theta) + \varepsilon^2 \mathbf{f}^{[2]}(\boldsymbol{x}, \theta, \varepsilon), \quad \dot{\theta} = 1 \tag{6.3.1}$$

Technically, the periodic *solution* $\mathbf{x} = \mathbf{x}(\boldsymbol{a}_\varepsilon, t, \varepsilon)$ constructed in Theorem 6.3.2 gives rise to a periodic *orbit* of (6.3.1) supporting the family of solutions $(\mathbf{x}, \theta) = (\mathbf{x}(\boldsymbol{a}_\varepsilon, t, \varepsilon), t + \theta_0)$ with initial conditions $(\boldsymbol{a}_\varepsilon, \theta_0)$ for any θ_0; however, we will always take $\theta_0 = 0$. The same suspension to $\mathbb{R}^n \times S^1$ can be done for the full and truncated averaged systems as well. In the case of the truncated averaged system we get

$$\dot{\boldsymbol{z}} = \varepsilon \overline{\mathbf{f}}^1(\boldsymbol{z}), \quad \dot{\theta} = 1, \tag{6.3.2}$$

in which the two equations are uncoupled because the first equation is already autonomous. Notice that the rest point of the $\boldsymbol{z}$ equation is automatically a periodic solution in this context. The suspension will never be used for the guiding system, because the period in τ is not 2π but depends on ε.

Hyperbolicity for periodic solutions of autonomous equations is defined using their Floquet exponents. (See, for instance, [121].) Every periodic solution has one Floquet exponent equal to zero, namely the exponent along the direction of the orbit. Hyperbolicity is decided by the remaining exponents, which are required to lie off the imaginary axis. The *unstable dimension* is one plus the number of exponents in the right half-plane. (Hyperbolic periodic orbits have stable and unstable manifolds, which intersect along the orbit. The unstable dimension equals the dimension of the unstable manifold.)

Theorem 6.3.3 (Preservation of hyperbolicity). *If the hypotheses of Theorem 6.3.2 are satisfied, and in addition the rest point $\boldsymbol{a}_0$ of the guiding system is hyperbolic, then the periodic solution $(\mathbf{x}(\boldsymbol{a}_\varepsilon, t, \varepsilon), t)$ of (6.3.1) is hyperbolic, and its unstable dimension is one greater than that of the rest point.*

Proof The linear variational equation of (6.1.2) along its periodic solution $\mathbf{y}(\boldsymbol{b}_\varepsilon, t, \varepsilon)$ (constructed in Theorem 6.3.2) is obtained by putting $\boldsymbol{y} = \boldsymbol{b}_\varepsilon + \boldsymbol{\eta}$ into (6.1.2) and extracting the linear part in $\boldsymbol{\eta}$; since $\overline{\mathbf{f}}^1(\boldsymbol{b}_\varepsilon + \boldsymbol{\eta}) = \mathsf{D}\overline{\mathbf{f}}^1(\boldsymbol{b}_\varepsilon)\boldsymbol{\eta} + \mathcal{O}(\|\boldsymbol{\eta}\|^2)$ and $\mathsf{D}\overline{\mathbf{f}}^1(\boldsymbol{b}_\varepsilon)\boldsymbol{\eta} = A\boldsymbol{\eta} + \mathcal{O}(\varepsilon)$, where $A = \mathsf{D}\overline{\mathbf{f}}^1(\boldsymbol{a}_0)$, this variational equation has the form

$$\dot{\boldsymbol{\eta}} = \varepsilon A \boldsymbol{\eta} + \varepsilon^2 B(t, \varepsilon)\boldsymbol{\eta} \tag{6.3.3}$$

for some periodic matrix $B(t, \varepsilon)$ depending on both $\boldsymbol{b}_\varepsilon$ and $\mathsf{Df}_\star^{[2]}(\boldsymbol{b}_\varepsilon, t, \varepsilon)$. The Floquet exponents of this equation will equal the Floquet exponents of the periodic solution of the suspended system, omitting the exponent that is equal to zero. The principal matrix solution of (6.3.3) has the form

$$Q(t, \varepsilon) = e^{\varepsilon t A}[I + \varepsilon^2 V(t, \varepsilon)].$$

A logarithm of this principal matrix solution is given by

$$2\pi \Gamma(\varepsilon) = 2\pi\varepsilon A + \log[I + \varepsilon^2 V(2\pi, \varepsilon)],$$

where the logarithm is evaluated by the power series for $\log(1 + x)$. (Notice that it is not necessary to introduce a complex logarithm, as is sometimes needed in Floquet theory, because the principal matrix solution is close to the identity for ε small.) It follows that

$$\Gamma(\varepsilon) = \varepsilon A + \varepsilon^2 D(\varepsilon)$$

for some $D(\varepsilon)$. The Floquet exponents are the eigenvalues of $\Gamma(\varepsilon)$. The position of these eigenvalues with respect to the imaginary axis is the same as for the eigenvalues of $A + \varepsilon D(\varepsilon)$, and by Rouché's theorem (as in the proof of Theorem 6.3.1), the same as for the eigenvalues of A. For further details and generalizations, see [198, Lemmas 5.3, 5.4, and 5.5]. □

6.4 Local Conjugacy and Shadowing

In Section 2.8 it was proved that averaging approximations are, in general, valid for time $\mathcal{O}(1/\varepsilon)$, but the proofs did not show that the accuracy actually breaks down after that time; it is only the proofs that break down. In this section we will see a common geometrical situation that forces the approximation to fail. Namely, when the guiding system has a hyperbolic rest point with unstable dimension one, exact and approximate solutions can be split apart by the stable manifold and sent in opposite directions along the unstable manifold. Since it takes time $\mathcal{O}(1/\varepsilon)$ for the solution (and its approximation) to approach the rest point, the breakdown takes place after this amount of time. (In regular perturbations, the same thing happens after time $\mathcal{O}(1)$.)

This phenomenon is closely related to a more general one that happens near hyperbolic rest points of the guiding system regardless of their unstable dimension. Away from the rest point, solutions move at a relatively steady speed, covering a distance $\mathcal{O}(1)$ in time $\mathcal{O}(1/\varepsilon)$ in the averaging case, or in time $\mathcal{O}(1)$ in the regular case. In other words, away from a rest point approximate solutions are valid for as long as it takes to cross a compact set. But near a rest point solutions slow down, and close to the rest point they become arbitrarily slow, so that the error estimates no longer remain valid across a compact set. This difficulty is due to the fact that the approximation theorems we have proved are for initial value problems; by posing a suitable boundary value problem instead, the difficulty can be overcome. A full resolution requires considering solutions that pass several rest points, but this is postponed to later sections. Here we deal only with a (small, but fixed) neighborhood of a single hyperbolic rest point.

In applied mathematics, approximate solutions are often used to understand the possible behaviors of a system. These approximate solutions are often "exact solutions of approximate equations"; that is, we simplify the equation by dropping small terms, or by averaging, and use an exact solution of the simplified equation as an approximate solution of the original equation. But which solution of the simplified equation should we choose? Consider an exact solution of the original equation, and a narrow tube around this solution. A good approximate solution should be one that remains within the tube for a long time. It is usual to choose for an approximate solution a solution of the simplified equation that has a point (usually the initial point) in common with the exact solution, on the assumption that this choice will stay in the tube longer than any other. But this assumption is often not correct. We will see that there often exists a solution of the simplified equation that remains within the tube for a very long time, sometimes even for all time, although it may not have any point in common with the exact solution. An approximate solution of this type is said to **shadow** the exact solution. A formal definition of shadowing can be given along the following lines, although the specifics vary with the setting (and will be clear in each theorem).

Definition 6.4.1. *Let* $\dot{\boldsymbol{u}} = \mathbf{f}(\boldsymbol{u}, \varepsilon)$ *and* $\dot{\boldsymbol{v}} = \mathbf{g}(\boldsymbol{v}, \varepsilon)$ *be two systems of differential equations on* $\mathbb{R}^n$*, and let* Ω *be an open subset of* $\mathbb{R}^n$*. Then the* $\boldsymbol{v}$ *system has the* $\mathcal{O}(\varepsilon^k)$ **shadowing property** *with respect to the* $\boldsymbol{u}$ *system in* Ω *if there is a constant* $c > 0$ *such that for every family* $\mathbf{u}(t, \varepsilon)$ *of solutions of the* $\boldsymbol{u}$*-equation with* $\mathbf{u}(0,0) \in \Omega$ *there exists a family* $\mathbf{v}(t, \varepsilon)$ *of solutions of the* $\boldsymbol{v}$*-equation such that* $\|\mathbf{u}(t, \varepsilon) - \mathbf{v}(t, \varepsilon)\| < c\varepsilon^k$ *for as long as* $\mathbf{u}(t, \varepsilon)$ *remains in* Ω*. Here "as long as" means that the estimate holds for both the future* $(t > 0)$ *and the past* $(t < 0)$*, until* $\mathbf{u}(t, \varepsilon)$ *leaves* Ω *(and for all time if it does not). If* $\mathbf{u}(t, \varepsilon)$ *leaves* Ω *and later reenters, the estimate is not required to hold after reentry.*

It will always be the case, in our results, that shadowing is a two-way relationship, that is, every exact solution family is shadowed by an approximate one and vice versa. It is the "vice versa" that makes shadowing valuable for applied mathematics: *every* solution of the approximate equations is shadowed by an exact solution, and therefore illustrates a possible behavior of the exact system. This helps to remove the objection that we do not know which approximate solution to consider as an approximation to a particular exact solution (since we cannot use the initial condition to match them up). The solutions of the boundary value problems mentioned above have just this shadowing property. Numerical analysts will recognize the boundary value problem as one that is numerically stable where the initial value problem is not. Dynamical systems people will see that it corresponds to the stable and unstable fibrations used in the "tubular family" approach to structural stability arguments.

6.4.1 The Regular Case

Let (6.1.6) have a hyperbolic rest point $\boldsymbol{a}_\varepsilon$ for small ε. Then this rest point has stable and unstable manifolds, of dimension s and u respectively, with $s + u = n$. Stable and unstable manifolds are treated in [121] and [215], among many other references.

Suppose now that the unstable dimension of $\boldsymbol{a}_\varepsilon$ is $u = 1$. Then $s = n - 1$, and the stable manifold partitions $\mathbb{R}^n$ locally into two parts; solutions approaching $\boldsymbol{a}_\varepsilon$ close to the stable manifold, but on opposite sides, will split apart and travel away from each other in opposite directions along the unstable manifold. (See Figure 6.2.) Now consider a small nonzero ε. The stable manifold will typically have moved slightly from its unperturbed position, and there will be a narrow region of initial conditions that lie on one side of the

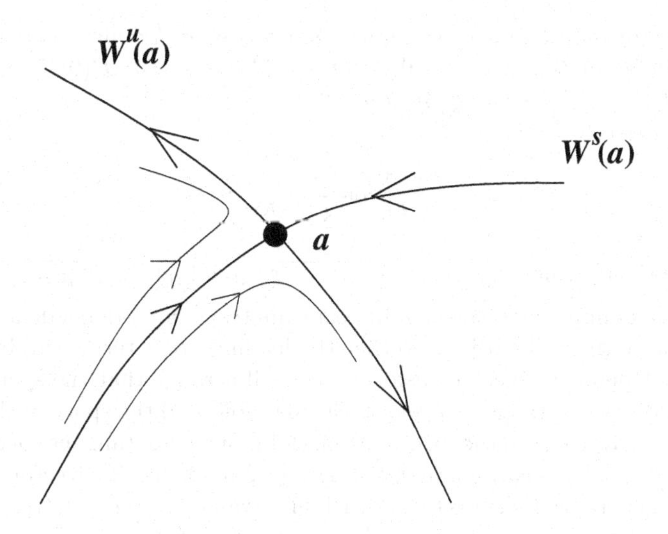

Fig. 6.2: Separation of nearby solutions by a hyperbolic rest point.

unperturbed stable manifold but on the other side of the perturbed one. But the unperturbed solution is the leading-order regular perturbation approximation to the perturbed one, so it is clear that these must separate after the solutions come close enough to the rest point, which happens in time $\mathcal{O}(1)$. (The idea is clear enough, and we forgo a more precise statement, involving $\mathcal{O}$ estimates that are uniform with respect to the initial conditions.)

To proceed further, we introduce special coordinates near $\boldsymbol{a}_\varepsilon$. If $s = 0$ or $u = 0$ then $\boldsymbol{\xi}$ or $\boldsymbol{\eta}$ will be absent in the following lemma. Euclidean norms are used in Item 4, rather than our usual norm (1.1.1), because it is important later that the spheres defined by these norms be smooth manifolds.

Lemma 6.4.2. *There exists an ε-dependent curvilinear coordinate system $(\boldsymbol{\xi}, \boldsymbol{\eta}) = (\xi_1, \dots, \xi_s, \eta_1, \dots, \eta_u)$ with the following properties:*

1. *The rest point $\boldsymbol{a}_\varepsilon$ is at the origin $(\boldsymbol{\xi}, \boldsymbol{\eta}) = (0, 0)$ for all small ε. (That is, the coordinate system moves as ε is varied, so that the origin "tracks" the rest point.)*
2. *The stable manifold, locally, coincides with the subspace $\boldsymbol{\eta} = 0$, and the unstable manifold with $\boldsymbol{\xi} = 0$. (That is, the coordinate system bends so that the subspaces "track" the manifold locally.)*
3. *The differential equations take the form*

$$\begin{bmatrix} \dot{\boldsymbol{\xi}} \\ \dot{\boldsymbol{\eta}} \end{bmatrix} = \begin{bmatrix} A\boldsymbol{\xi} + \boldsymbol{P}(\boldsymbol{\xi}, \boldsymbol{\eta}, \varepsilon) \\ B\boldsymbol{\eta} + \boldsymbol{Q}(\boldsymbol{\xi}, \boldsymbol{\eta}, \varepsilon) \end{bmatrix}. \tag{6.4.1}$$

Here A and B are constant matrices with their eigenvalues in the left and right half-planes respectively, and $\boldsymbol{P}$ and $\boldsymbol{Q}$ vanish at the origin for all

ε, contain only terms of quadratic and higher order for $\varepsilon = 0$ (but can contain linear terms for $\varepsilon \neq 0$), and in addition satisfy $\boldsymbol{P}(0, \boldsymbol{\eta}, \varepsilon) = 0$ and $\boldsymbol{Q}(\boldsymbol{\xi}, 0, \varepsilon) = 0$ for small $\boldsymbol{\xi}$, $\boldsymbol{\eta}$, and ε.

4. *There exists $K > 0$ such that*

$$\begin{aligned} \|e^{At}\boldsymbol{\xi}\|_s &\leq e^{-Kt}\|\boldsymbol{\xi}\|_s \\ \|e^{-Bt}\boldsymbol{\eta}\|_u &\leq e^{-Kt}\|\boldsymbol{\eta}\|_u \end{aligned} \tag{6.4.2}$$

for all $t > 0$, where $\|\boldsymbol{\xi}\|_s = \sqrt{\xi_1^2 + \cdots + \xi_s^2}$ and $\|\boldsymbol{\eta}\|_u = \sqrt{\eta_1^2 + \cdots + \eta_u^2}$.

Proof Beginning with the existing coordinates $\boldsymbol{x}$, an ε-dependent translation will place $\boldsymbol{a}_\varepsilon$ at the origin, so that the leading-order terms will be linear. Next an ε-dependent linear transformation will arrange that the space of the first s coordinates is tangent to the stable manifold and the space of the last u coordinates is tangent to the unstable manifold. Since the tangent space to the stable (respectively unstable) manifold is the span of the eigenvectors (and, if necessary, generalized eigenvectors) with eigenvalues in the left (respectively right) half-plane, the matrix of the linear part now takes the block diagonal form

$$\begin{bmatrix} A & \\ & B \end{bmatrix}.$$

Then a further linear coordinate change, preserving the block structure, will arrange that A and B satisfy (6.4.2). (An easy way to do this, at the risk of making the matrix complex, is to put A and B into modified Jordan form, with sufficiently small off-diagonal entries where the ones normally go. Since we require a real matrix, real canonical form should be used instead.) Let the coordinates obtained at this point be denoted by $(\widetilde{\boldsymbol{\xi}}, \widetilde{\boldsymbol{\eta}})$. Then the local stable manifold is the graph of a function $\widetilde{\boldsymbol{\eta}} = \mathbf{h}(\widetilde{\boldsymbol{\xi}}) = \mathcal{O}(\|\widetilde{\boldsymbol{\xi}}\|^2)$ and the local unstable manifold is the graph of a function $\widetilde{\boldsymbol{\xi}} = \mathbf{k}(\widetilde{\boldsymbol{\eta}}) = \mathcal{O}(\|\widetilde{\boldsymbol{\eta}}\|^2)$. Now the nonlinear change of coordinates $\boldsymbol{\xi} = \widetilde{\boldsymbol{\xi}} - \mathbf{k}(\widetilde{\boldsymbol{\eta}})$, $\boldsymbol{\eta} = \widetilde{\boldsymbol{\eta}} - \mathbf{h}(\widetilde{\boldsymbol{\xi}})$ will flatten the local stable and unstable manifolds without changing the linear terms. The resulting system will have the form (6.4.1) with all of the specified conditions. □

The norms used for $\boldsymbol{\xi}$ and $\boldsymbol{\eta}$ in (6.4.2) have the same form as the standard Euclidean norm, but are actually different because the mapping to curvilinear coordinates is not an isometry. However, it is a local diffeomorphism, so it and its inverse are Lipschitz. In the full n-dimensional space we choose the norm

$$|(\boldsymbol{\xi}, \boldsymbol{\eta})| = \max\{\|\boldsymbol{\xi}\|_s, \|\boldsymbol{\eta}\|_u\}. \tag{6.4.3}$$

Asymptotic ($\mathcal{O}$) estimates made in this norm will be equivalent to estimates in the Euclidean norm in the original variables $\boldsymbol{x}$, or in our usual norm (1.1.1). The equations (6.4.1) must be interpreted with some care because the coordinates are ε-dependent. For instance, the stable and unstable manifolds of (6.4.1) are independent of ε, although this is not true for the original system (6.1.6). The "splitting" situation discussed above (of an initial condition lying between the stable manifolds for $\varepsilon = 0$ and some $\varepsilon \neq 0$) would appear in these coordinates as two distinct initial conditions on opposite sides of $\boldsymbol{\eta} = 0$.

We now focus on a **box neighborhood** of the origin in the new coordinates, defined by

$$N = \{(\boldsymbol{\xi}, \boldsymbol{\eta}) : |(\boldsymbol{\xi}, \boldsymbol{\eta})| \leq \delta\}.$$

By choosing δ small enough, it can be guaranteed that the linear part of the flow dominates the nonlinear part in N. The neighborhood N is fixed in the $(\boldsymbol{\xi}, \boldsymbol{\eta})$ variables but depends on ε in the original x variables; to emphasize this, we sometimes write N_ε. By the definition of our norm, N_ε is a Cartesian product of closed Euclidean δ-balls $|\boldsymbol{\xi}|_s \leq \delta$ and $|\boldsymbol{\eta}|_u \leq \delta$, that is,

$$N_\varepsilon = \overline{B}_\delta^s \times \overline{B}_\delta^u.$$

The boundary of N_ε is therefore

$$\partial N_\varepsilon = (S_\delta^{s-1} \times \overline{B}_\delta^u) \cup (\overline{B}_\delta^s \times S_\delta^{u-1}).$$

Since the δ-sphere in a one-dimensional space is the two-point set $S_\delta^0 = \{+\delta, -\delta\}$ and the closed δ-ball is the interval $\overline{B}_\delta^1 = [-\delta, +\delta]$, N will be a square when $s = u = 1$. When $s = 1$ and $u = 2$, or vice versa, it is a cylinder. See Figure 6.3. If δ is taken small enough, an orbit entering N will do so through a point $(\boldsymbol{\alpha}, \boldsymbol{\eta}) \in S_\delta^{s-1} \times \overline{B}_\delta^u$, and an orbit leaving N will do so through a point $(\boldsymbol{\xi}, \boldsymbol{\beta}) \in \overline{B}_\delta^s \times S_\delta^{u-1}$. Most orbits will enter N at some time (which we usually take as $t = 0$) and leave after some finite time T. In this case we say that the orbit has **entrance data** $(\boldsymbol{\alpha}, \boldsymbol{\eta})$, **exit data** $(\boldsymbol{\xi}, \boldsymbol{\beta})$, and **box data** $(\boldsymbol{\alpha}, \boldsymbol{\beta}, T)$. Any of these forms of data will uniquely determine the orbit, if ε is also specified. (The point $\boldsymbol{\alpha} \in S^{s-1}$ is a vector in $\mathbb{R}^s$ subject to $\|\boldsymbol{\alpha}\| = \delta$, and as such has s components but only $s - 1$ independent variables. Similarly, $\boldsymbol{\beta}$ contains $u - 1$ independent variables. Therefore each form of data (entry, exit, or box) contains $n - 1$ independent variables, enough to determine an orbit without fixing the solution on the orbit, that is, the position when $t = 0$.) For use in a later section, we define the **entry and exit maps** by

$$\boldsymbol{\Phi}_\varepsilon(\boldsymbol{\alpha}, \boldsymbol{\beta}, T) = (\boldsymbol{\alpha}, \boldsymbol{\eta}), \qquad \boldsymbol{\Psi}_\varepsilon(\boldsymbol{\alpha}, \boldsymbol{\beta}, T) = (\boldsymbol{\xi}, \boldsymbol{\beta}), \tag{6.4.4}$$

where $(\boldsymbol{\alpha}, \boldsymbol{\eta})$ and $(\boldsymbol{\xi}, \boldsymbol{\beta})$ are the entrance and exit data corresponding to box data $(\boldsymbol{\alpha}, \boldsymbol{\beta}, T)$.

The orbits lying on the stable and unstable manifolds enter N but do not leave, or leave but do not enter, so they do not have box data in the sense defined so far. We include these by defining the **broken orbit** with box data $(\boldsymbol{\alpha}, \infty, \boldsymbol{\beta})$ to be the union of three orbits: the (unique) orbit on the stable manifold entering N_ε at $\boldsymbol{\alpha}$, the (unique) orbit on the unstable manifold exiting N_ε through $\boldsymbol{\beta}$, and the rest point itself. In the case of a source ($u = n$) or a sink ($u = 0$), all orbits have $T = \infty$ but they are not broken. In addition, for a source $\boldsymbol{\alpha}$ is empty, and for a sink $\boldsymbol{\beta}$ is empty, so in effect the only box data for a source is $\boldsymbol{\beta}$, and for a sink, $\boldsymbol{\alpha}$.

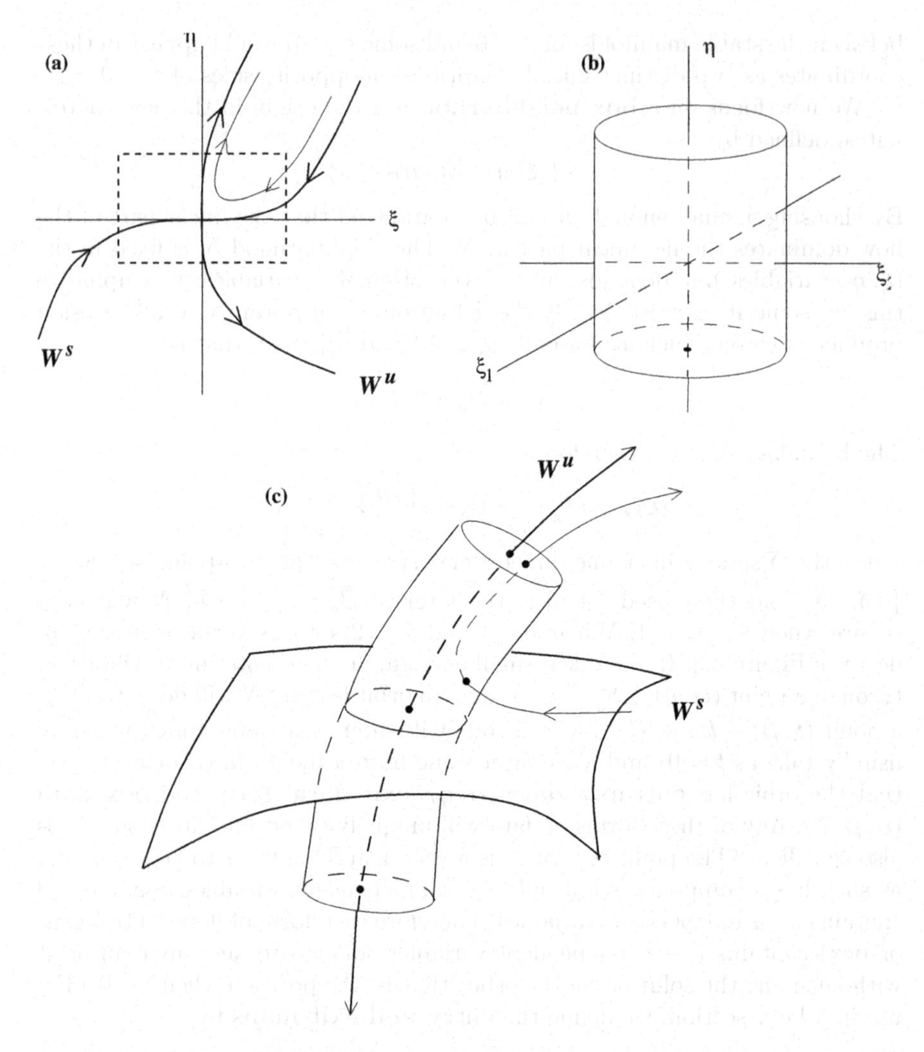

Fig. 6.3: (a) A box neighborhood when $s = u = 1$ showing the stable and unstable manifolds and a crossing solution. (b) A box neighborhood when $s = 2$, $u = 1$. (c) A crossing solution for the box in (b), drawn in the original coordinates.

There is a sense in which the **box data problem** is well posed, whereas the entrance and exit data problems are not. Of course, the entrance and exit data problems are initial value problems, and these are well posed in the sense that solutions exist, are unique, and depend continuously on the entrance or exit data *for each fixed t*. It follows that they also depend continuously on the entrance or exit data *uniformly for t in any finite interval.* But no finite interval of t suffices to handle all solutions passing through N for the full time that they remain in N; the box crossing time T approaches infinity as the orbit is moved closer to the stable and unstable manifolds. (This is equivalent to the fact mentioned earlier that the time of validity of regular perturbation estimates does not suffice to cross a compact set containing a rest point.) This difficulty does not exist for the box data problem: if the box data $(\boldsymbol{\alpha}, \boldsymbol{\beta}, T)$ is changed slightly, the solution is changed slightly, throughout its passage through N.

Theorem 6.4.3 (Box Data Theorem). *Let $\varepsilon_0 > 0$ and $\delta > 0$ be sufficiently small. Let N be the box neighborhood of size δ of a hyperbolic rest point, as described above. Then for every $\boldsymbol{\alpha} \in S_\delta^{s-1}$, $\boldsymbol{\beta} \in S_\delta^{u-1}$, $T > 0$, and ε satisfying $|\varepsilon| \leq \varepsilon_0$, there exists a unique solution*

$$\begin{aligned}\boldsymbol{\xi} &= \boldsymbol{\xi}(t; \boldsymbol{\alpha}, \boldsymbol{\beta}, T, \varepsilon),\\ \boldsymbol{\eta} &= \boldsymbol{\eta}(t; \boldsymbol{\alpha}, \boldsymbol{\beta}, T, \varepsilon),\end{aligned}$$

of (6.4.1) defined for t in some open interval including $0 \leq t \leq T$ and satisfying

$$\begin{aligned}\boldsymbol{\xi}(0; \boldsymbol{\alpha}, \boldsymbol{\beta}, T, \varepsilon) &= \boldsymbol{\alpha},\\ \boldsymbol{\eta}(T; \boldsymbol{\alpha}, \boldsymbol{\beta}, T, \varepsilon) &= \boldsymbol{\beta}.\end{aligned}$$

In addition, this solution satisfies $|(\boldsymbol{\xi}(t), \boldsymbol{\eta}(t))| \leq \delta$ for $0 \leq t \leq T$, and depends smoothly on $(\boldsymbol{\alpha}, \boldsymbol{\beta}, T, \varepsilon)$, with partial derivatives that are bounded even as $T \to \infty$; that is, they are bounded for $(\boldsymbol{\alpha}, \boldsymbol{\beta}, T, \varepsilon)$ in the noncompact set $S^{s-1} \times S^{u-1} \times [0, \infty) \times [-\varepsilon_0, \varepsilon_0]$, and for $0 \leq t \leq T$. As $T \to \infty$, the orbit $(\boldsymbol{\xi}(t), \boldsymbol{\eta}(t))$ approaches the broken orbit with box data $(\boldsymbol{\alpha}, \boldsymbol{\beta}, \infty, \varepsilon)$.

Proof One shows that the box data problem is equivalent to the system of integral equations

$$\begin{aligned}\boldsymbol{\xi}(t) &= e^{At}\boldsymbol{\alpha} + \int_0^t e^{A(t-s)} \boldsymbol{P}(\boldsymbol{\xi}(s), \boldsymbol{\eta}(s), \varepsilon)\, \mathrm{d}s,\\ \boldsymbol{\eta}(t) &= e^{B(t-T)}\boldsymbol{\beta} - \int_t^T e^{B(t-s)} \boldsymbol{Q}(\boldsymbol{\xi}(s), \boldsymbol{\eta}(s), \varepsilon)\, \mathrm{d}s.\end{aligned}$$

The theorem is then proved by a variation of the usual contraction mapping argument for solution of integral equations. For details see [199]. □

Recall that in the original $\boldsymbol{x}$ variables N depends on ε and is denoted by N_ε. Let $|\varepsilon| < \varepsilon_0$. We assign to each **guiding orbit** (an orbit or broken orbit of the guiding system $\dot{\boldsymbol{z}} = \mathbf{f}^0(\boldsymbol{z})$ in N_0) the **associated orbit** of $\dot{\boldsymbol{x}} = \mathbf{f}^{[0]}(\boldsymbol{x}, \varepsilon)$ in N_ε having the same box data $(\boldsymbol{\alpha}, \boldsymbol{\beta}, T)$. Notice that (unless $T = \infty$) the guiding orbit and associated orbit take the same time T to cross their respective box neighborhoods, so we can assign to each point $\boldsymbol{z}$ on a guiding orbit an associated point $\boldsymbol{x} = \mathbf{h}_\varepsilon(\boldsymbol{z})$ as follows: consider each orbit to enter its box at time 0; suppose the guiding orbit reaches $\boldsymbol{z}$ at time t with $0 \le t \le T$; let $\mathbf{h}_\varepsilon(\boldsymbol{z})$ be the point on the associated orbit at time t. A special definition of $\mathbf{h}_\varepsilon$ is required when $T = \infty$ and the orbits are broken; we match points on the entering segment by the time t from entrance (with $0 \le t < \infty$); we match the rest point to the rest point; and we match points on the exiting segment by time required to exit.

Theorem 6.4.4 (Local conjugacy and first-order shadowing, regular case). *The map $\mathbf{h}_\varepsilon : N_0 \to N_\varepsilon$ is a topological conjugacy of the guiding system in N_0 with the perturbed system in N_ε. This map satisfies $\|\mathbf{h}_\varepsilon(\boldsymbol{z}) - \boldsymbol{z}\| = \mathcal{O}(\varepsilon)$ uniformly for $\boldsymbol{z} \in N_0$. If $\mathbf{z}(t)$ is any solution of $\dot{\boldsymbol{z}} = \mathbf{f}^0(\boldsymbol{z})$ intersecting N_0, then there is a solution of $\dot{\boldsymbol{x}} = \mathbf{f}^{[0]}(\boldsymbol{x}, \varepsilon)$ that shadows $\mathbf{z}(t)$ with an error that is uniformly $\mathcal{O}(\varepsilon)$ throughout the time that $\mathbf{z}(t)$ remains in N_0. This shadowing solution is $\mathbf{x}(t, \varepsilon) = \mathbf{h}_\varepsilon(\mathbf{z}(t))$.*

Proof For $\mathbf{h}_\varepsilon$ to be a topological conjugacy means that it is a homeomorphism (that is, it is continuous and invertible, with a continuous inverse), and that it carries solutions of the guiding system to solutions of the perturbed system while preserving the time parameter. All of these statements follow from Theorem 5.3.1. In addition, since the solution of the box data problem is smooth in ε, the distance between perturbed and unperturbed solutions is $\mathcal{O}(\varepsilon)$, so $\mathbf{h}_\varepsilon$ moves points by a distance $\mathcal{O}(\varepsilon)$. Together, these facts imply that $\mathbf{h}_\varepsilon(\mathbf{z}(t))$ is a solution of the perturbed equation and shadows $\mathbf{z}(t)$ as claimed. As a side note, this construction is closely related to the one used in [215, Lemma 7.3] to prove Hartman's theorem using local tubular families. To relate the two constructions, take portions of the boundary of N_ε as the transverse disks to the stable and unstable manifolds needed in [215]. In our proof the box data theorem replaces the inclination lemma (lambda lemma) as the supporting analytical machinery. □

The shadowing part of this theorem is extended in [200] to show that any approximate solution of $\dot{\boldsymbol{x}} = \mathbf{f}^{[0]}(\boldsymbol{x}, \varepsilon)$ constructed by the kth-order regular perturbation method is $\mathcal{O}(\varepsilon^{k+1})$-shadowed by an exact solution within a box neighborhood of the hyperbolic rest point.

6.4.2 The Averaging Case

Assume now that the guiding system (6.1.4) for the averaging case has a hyperbolic rest point $\boldsymbol{a}_0$, and let N_0 be a sufficiently small box neighborhood. Then the suspended system (6.3.2) has a periodic solution $(\mathbf{z}, \theta) = (\mathbf{z}(\boldsymbol{a}, t, \varepsilon), t)$

with tubular neighborhood $N_0 \times S_1$. The conjugacy in the following theorem has the form $\mathbf{H}_\varepsilon(\boldsymbol{z},\theta) = (\boldsymbol{y},\theta)$, with θ preserved. Given a solution $\mathbf{z}(t,\varepsilon)$ of (6.1.3), a solution $\mathbf{x}(t,\varepsilon)$ of (6.1.1) that shadows $\mathbf{z}(t,\varepsilon)$ can be obtained from $\mathbf{H}_\varepsilon(\mathbf{z}(t,\varepsilon),t) = (\mathbf{x}(t,\varepsilon),t)$.

Theorem 6.4.5 (Local conjugacy and first-order shadowing, averaging case). *There is a homeomorphism $\mathbf{H}_\varepsilon$ carrying $N_0 \times S^1$ to a tubular neighborhood of the solution $(\mathbf{x},\theta) = (\mathbf{x}(\boldsymbol{a}_\varepsilon,t,\varepsilon),t)$ of (6.3.1) conjugating the solutions of (6.3.2) and (6.3.1). This homeomorphism depends smoothly on ε, moves points a distance $\mathcal{O}(\varepsilon)$, and assigns to each solution of the averaged equation an exact solution that shadows it with error $\mathcal{O}(\varepsilon)$ as long as the guiding solution remains in N_0.*

Proof It suffices to prove the corresponding statements for $\mathbf{y}(\boldsymbol{b}_\varepsilon,t,\varepsilon)$ in place of $\mathbf{x}(\boldsymbol{a}_\varepsilon,t,\varepsilon)$, because the averaging transformation $\mathbf{U}$ (see Lemma 2.8.4) provides a global smooth conjugacy between the $(\boldsymbol{x},\theta)$ and $(\boldsymbol{y},\theta)$ systems that moves points by a distance $\mathcal{O}(\varepsilon)$. A topological conjugacy of $(\boldsymbol{z},\theta)$ with $(\boldsymbol{y},\theta)$ can be composed with the smooth conjugacy $\mathbf{U}$ of $(\boldsymbol{y},\theta)$ to $(\boldsymbol{x},\theta)$ to produce the desired topological conjugacy of $(\boldsymbol{z},\theta)$ to $(\boldsymbol{x},\theta)$.

In the (suspended) $\boldsymbol{y}$ system

$$\dot{\boldsymbol{y}} = \varepsilon\overline{\mathbf{f}}^1(\boldsymbol{y}) + \varepsilon^2\mathbf{f}_\star^{[2]}(\boldsymbol{y},\theta,\varepsilon), \quad \dot\theta = 1, \tag{6.4.5}$$

we introduce ε-dependent coordinates $(\boldsymbol{\xi},\boldsymbol{\eta},\theta) \in \mathbb{R}^s \times \mathbb{R}^u \times S^1$ (θ is unchanged) in which the system takes the form

$$\begin{bmatrix}\dot{\boldsymbol{\xi}}\\ \dot{\boldsymbol{\eta}}\end{bmatrix} = \varepsilon\begin{bmatrix}A\boldsymbol{\xi} + \boldsymbol{P}(\boldsymbol{\xi},\boldsymbol{\eta},\theta,\varepsilon)\\ B\boldsymbol{\eta} + \boldsymbol{Q}(\boldsymbol{\xi},\boldsymbol{\eta},\theta,\varepsilon)\end{bmatrix}, \quad \dot\theta = 1. \tag{6.4.6}$$

As before, these coordinates are chosen so that the stable and unstable manifolds, locally, lie in the $\boldsymbol{\xi}$ and $\boldsymbol{\eta}$ coordinate spaces respectively. Construct the norm $|\ |$ as before, and take N_ε to be a neighborhood of the form $|(\boldsymbol{\xi},\boldsymbol{\eta})| \le \delta$ for sufficiently small δ. (The neighborhood does not depend on ε in the new coordinates, but does in the original ones.) If a solution family $(\boldsymbol{\xi}(t,\varepsilon),\boldsymbol{\eta}(t,\varepsilon),\theta(t,\varepsilon))$ enters $N_\varepsilon \times S^1$ at $t=0$ and leaves at time $T(\varepsilon)$, its *box data* is defined to be

$$(\boldsymbol{\alpha}_\varepsilon,\theta_0(\varepsilon),\boldsymbol{\beta}_\varepsilon,T(\varepsilon)), \tag{6.4.7}$$

where $\boldsymbol{\alpha}_\varepsilon = \boldsymbol{\xi}(0,\varepsilon)$, $\theta_0(\varepsilon) = \theta(0,\varepsilon)$, and $\boldsymbol{\beta}_\varepsilon = \boldsymbol{\eta}(T(\varepsilon),\varepsilon)$; note that the exiting value of θ will be $\theta(T(\varepsilon),\varepsilon) = \theta_0(\varepsilon) + T(\varepsilon)$. Special cases (for solutions on the stable and unstable manifolds) are treated as before. An integral equation argument shows that the box data problem is well posed. The z system corresponds to (6.4.6) with ε set equal to zero inside $\boldsymbol{P}$ and $\boldsymbol{Q}$ (but not in the initial coefficient). The map that pairs solutions of these two systems having the same box data is the required local conjugacy and defines the shadowing orbits. □

For additional details, and for extensions of the shadowing result to higher order, see [200]. (This paper does not address the conjugacy.)

6.5 Extended Error Estimate for Solutions Approaching an Attractor

The results of the last section can be used to give short proofs of some of the attraction results already proved in Chapter 5, in both regular and averaging cases. These results apply only when the hyperbolic rest point $\boldsymbol{a}_0$ of the guiding system is a sink, that is, has unstable dimension zero. The idea to use shadowing for this purpose is due to Robinson [224].

For the regular case, suppose that $\boldsymbol{a}_0$ is a sink for the unperturbed (or guiding) system (6.1.8), and let $\boldsymbol{b}$ be a point such that the solution $\mathbf{z}(\boldsymbol{b}, t)$ of (6.1.8) with initial point $\mathbf{z}(\boldsymbol{b}, t) = \boldsymbol{b}$ approaches $\boldsymbol{a}_0$ as $t \to \infty$. Let $\boldsymbol{a}_\varepsilon$ be the rest point of (6.1.6) that reduces to $\boldsymbol{a}_0$ when $\varepsilon = 0$, and let $\mathbf{x}(\boldsymbol{b}, t, \varepsilon)$ be the solution of (6.1.6) with $\mathbf{x}(\boldsymbol{b}, 0, \varepsilon) = \boldsymbol{b}$. It is clear that for small enough ε, $\mathbf{x}(\boldsymbol{b}, t, \varepsilon) \to \boldsymbol{a}_\varepsilon$ as $t \to \infty$. Although there are a few technical points, the idea of the following proof is extremely simple: the approximate and exact solutions remain close for the time needed to reach a small neighborhood of the rest point, and from this time on, there is a shadowing solution that remains close to both of them.

Theorem 6.5.1 (Eckhaus/Sanchez–Palencia). *Under these circumstances,*

$$\|\mathbf{x}(\boldsymbol{b}, t, \varepsilon) - \mathbf{z}(\boldsymbol{b}, t)\| = \mathcal{O}(\varepsilon)$$

for all $t \geq 0$.

Proof Let N_ε be a box neighborhood of $\boldsymbol{a}_\varepsilon$ as constructed in the previous section. (Since $\boldsymbol{a}_\varepsilon$ is a sink, there are no $\boldsymbol{\eta}$ variables, and N_ε is simply a ball around $\boldsymbol{a}_\varepsilon$, not a product of balls.) Since $\mathbf{z}(\boldsymbol{b}, t)$ approaches $\boldsymbol{b}$, there exists a time $L > 0$ such that $\mathbf{z}(\boldsymbol{b}, t)$ lies within N_0 at time $t = L$, and it follows that for small ε, $\mathbf{x}(\boldsymbol{b}, L, \varepsilon)$ lies in N_ε. By the usual error estimate for regular perturbations,

$$\|\mathbf{x}(\boldsymbol{b}, t, \varepsilon) - \mathbf{z}(\boldsymbol{b}, t)\| = \mathcal{O}(\varepsilon) \quad \text{for} \quad 0 \leq t \leq L. \tag{6.5.1}$$

By Theorem 6.4.4 there exists a family $\widetilde{\mathbf{x}}(t, \varepsilon)$ of (exact) solutions of (6.1.6) that shadows $\mathbf{z}(\boldsymbol{b}, t)$ in N_ε, and in particular,

$$\|\widetilde{\mathbf{x}}(t, \varepsilon) - \mathbf{z}(\boldsymbol{b}, t)\| = \mathcal{O}(\varepsilon) \quad \text{for} \quad t \geq L. \tag{6.5.2}$$

It follows from (6.5.1) and (6.5.2) that $\|\mathbf{x}(\boldsymbol{b}, L, \varepsilon) - \widetilde{\mathbf{x}}(L, \varepsilon)\| = \mathcal{O}(\varepsilon)$. Since $\mathbf{x}(\boldsymbol{b}, t, \varepsilon)$ and $\widetilde{\mathbf{x}}(t, \varepsilon)$ are exact solutions of a system that is contracting in $|\ |$ in N_ε, we have $|\mathbf{x}(\boldsymbol{b}, t, \varepsilon) - \widetilde{\mathbf{x}}(t, \varepsilon)| = \mathcal{O}(\varepsilon)$ for $t \geq L$. Since all norms are equivalent in a finite-dimensional space (and since the ε-dependent norm $|\ |$

is continuous in ε), the same estimate holds in $\| \ \|$ (even though the flow need not be contracting in this norm), and we have

$$\|\mathbf{x}(\boldsymbol{b},t,\varepsilon) - \widetilde{\mathbf{x}}(t,\varepsilon)\| = \mathcal{O}(\varepsilon) \quad \text{for} \quad t \geq L. \tag{6.5.3}$$

The desired result follows from

$$\|\mathbf{x}(\boldsymbol{b},t,\varepsilon) - \mathbf{z}(\boldsymbol{b},t)\| \leq \|\mathbf{x}(\boldsymbol{b},t,\varepsilon) - \widetilde{\mathbf{x}}(t,\varepsilon)\| + \|\widetilde{\mathbf{x}}(t,\varepsilon) - \mathbf{z}(\boldsymbol{b},t)\|,$$

combined with (6.5.1), (6.5.2), and (6.5.3). □

For the averaging case, suppose that $\boldsymbol{a}_0$ is a sink for (6.1.4). Let $\boldsymbol{b}$ be a point such that the solution $\mathbf{w}(\boldsymbol{b},\tau)$ of (6.1.4) approaches $\boldsymbol{a}_0$ as $\tau \to \infty$. Let $\mathbf{z}(\boldsymbol{b},t,\varepsilon) = \mathbf{w}(\boldsymbol{b},\varepsilon t)$ and $\mathbf{x}(\boldsymbol{b},t,\varepsilon)$ be the solutions of (6.1.3) and (6.1.1), respectively, with initial point $\boldsymbol{b}$.

Theorem 6.5.2. *Under these circumstances,*

$$\|\mathbf{x}(\boldsymbol{b},t,\varepsilon) - \mathbf{z}(\boldsymbol{b},t,\varepsilon)\| = \mathcal{O}(\varepsilon)$$

for all $t > 0$.

Proof Let N_0 be a box neighborhood of $\boldsymbol{a}_0$ and let L be such that $\mathbf{w}(\boldsymbol{b},\tau)$ lies in N_0 at $\tau = L$. Let $\mathbf{H}_\varepsilon(N_0 \times S^1)$ be the tubular neighborhood of the periodic solution of (6.3.1) constructed in Theorem 6.4.5. Then $(\mathbf{x}(\boldsymbol{b},t,\varepsilon),t)$ and $(\mathbf{z}(\boldsymbol{b},t,\varepsilon),t)$ will reach this neighborhood in time L/ε and will remain $\mathcal{O}(\varepsilon)$-close during that time. After this time the shadowing solution $\mathbf{H}_\varepsilon(\mathbf{z}(\boldsymbol{b},t,\varepsilon),t)$ remains close to both of the others. The details are almost identical to those of the last theorem and may be left to the reader. □

6.6 Conjugacy and Shadowing in a Dumbbell-Shaped Neighborhood

In Section 6.4 it was shown that the unperturbed system in a regular perturbation problem is topologically conjugate to the perturbed system near a hyperbolic rest point, and that the conjugacy moves points by a distance $\mathcal{O}(\varepsilon)$, resulting in local shadowing of approximate solutions by exact ones. Similar results were proved for first-order averaging. In this section we show that these conjugacy and shadowing results can be extended to a neighborhood of a heteroclinic orbit connecting two rest points. This set is called a **dumbbell neighborhood** because it consists of two boxes connected by a tube, and thus resembles a weightlifter's dumbbell. As usual in dynamical systems, the stable and unstable manifolds of a hyperbolic rest point $\mathbf{p}$ are denoted by $W^s(\mathbf{p})$ and $W^u(\mathbf{p})$, where s and u simply stand for "stable" and "unstable." We will also use s, u, s', and u' to denote the dimensions of such manifolds, but this should not cause confusion.

6.6.1 The Regular Case

Suppose that the guiding system $\dot{\boldsymbol{z}} = \mathbf{f}^0(\boldsymbol{z})$ has two hyperbolic rest points, $\boldsymbol{a}_0$ and $\boldsymbol{a}_0'$, with a connecting (or heteroclinic) orbit γ_0 from $\boldsymbol{a}_0$ to $\boldsymbol{a}_0'$. Then γ_0 belongs to $W^s(\boldsymbol{a}_0) \cap W^u(\boldsymbol{a}_0')$. We assume that the intersection of these stable and unstable manifolds is transverse; that is, at any point of the intersection, the tangent spaces to the two invariant manifolds together span the ambient space $\mathbb{R}^n$. Let the stable and unstable dimensions of $\boldsymbol{a}_0$ and $\boldsymbol{a}_0'$ be s, u and s', u' respectively. (See Figure 6.4 for the case $n = 3$, $u = 2$, $u' = 1$ and Figure 6.5 for $n = 2$, $u = 2$, $u' = 0$). Let N_0 and N_0' be box neighborhoods of $\boldsymbol{a}_0$ and $\boldsymbol{a}_0'$ with coordinate systems $(\boldsymbol{\xi}, \boldsymbol{\eta})$, $(\boldsymbol{\xi}', \boldsymbol{\eta}')$, chosen as in Section 6.4 such that the box data problems $(\boldsymbol{\alpha}, \boldsymbol{\beta}, T)$ and $(\boldsymbol{\alpha}', \boldsymbol{\beta}', T')$ are well posed by Theorem 6.4.3.

The flow along connecting orbits runs "downhill" in terms of the unstable dimension of the rest points; that is, we have

$$u' < u. \tag{6.6.1}$$

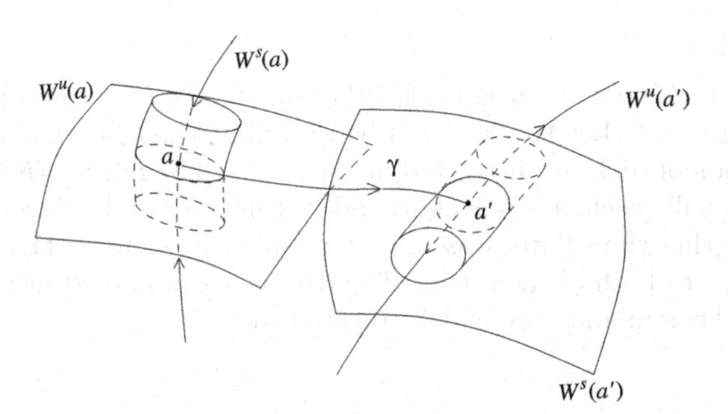

(a) A connecting orbit γ from rest point $\boldsymbol{a}$ ($s = 1$, $u = 2$) to rest point $\boldsymbol{a}'$ ($s' = 2$, $u' = 1$) showing transverse intersection of $W^u(\boldsymbol{a})$ with $W^s(\boldsymbol{a}')$.

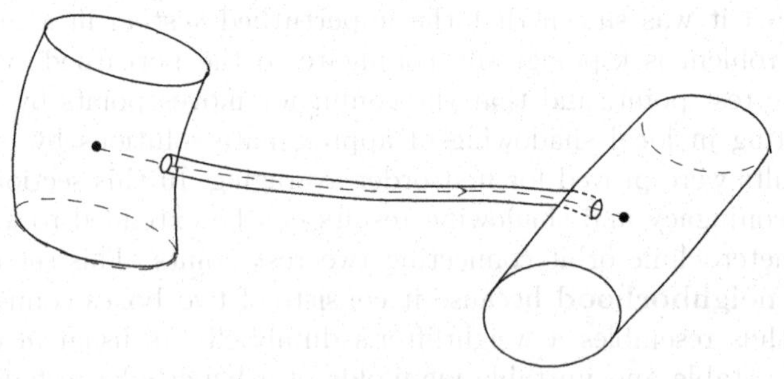

(b) A dumbbell neighborhood for γ .

Fig. 6.4: A connecting orbit.

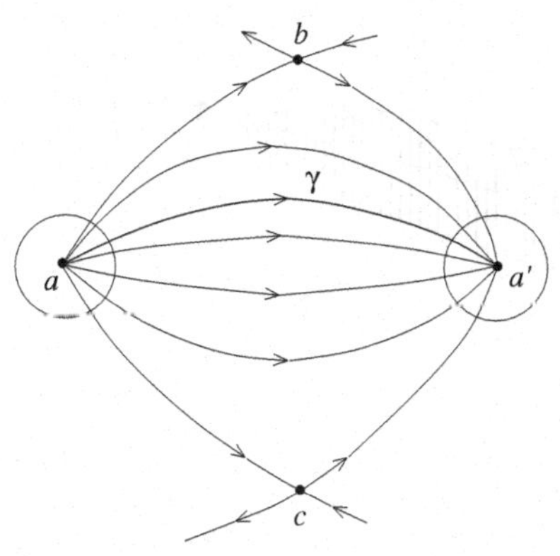

(a) A connecting orbit γ in the plane ($n = 2$) joining a source $\boldsymbol{a}$ ($u = 2$) with a sink $\boldsymbol{a}'$ ($u' = 0$). Since the dimension drop $u - u'$ is greater than one, there are many connecting orbits (bounded by the stable and unstable manifolds of saddle points b and c).

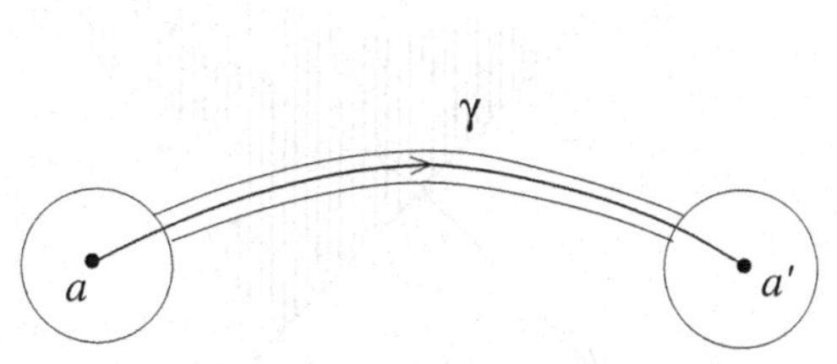

(b) A dumbbell neighborhood containing γ.

Fig. 6.5: A connecting orbit.

(The tangent spaces of $W^u(\boldsymbol{a}_0)$ and $W^s(\boldsymbol{a}_0')$ at a point of γ_0 must span $\mathbb{R}^n$, but they have at least one direction in common, namely the tangent direction to γ_0. Therefore $u + s' - 1 \geq n$. But $s' = n - u'$, so $u \geq u' + 1$.) If $u' = u - 1$, γ_0 is the only connecting orbit from $\boldsymbol{a}_0$ to $\boldsymbol{a}_0'$. If the dimension drop is greater than one, there is a continuum of connecting orbits, and the set of points $\boldsymbol{\beta}$ (on the exit sphere in $W^u(\boldsymbol{a}_0)$) that lie on heteroclinic orbits from $\boldsymbol{a}_0$ to $\boldsymbol{a}_0'$ is an embedded disk of dimension $u - u' - 1$; see Figure 6.5 where the disk is an arc. (*Disk* means the same as *ball*, but emphasizes that the dimension may be less than that of the ambient space.) But for the moment we continue to focus on a single connecting orbit γ_0.

The orbit γ_0 of the guiding system leaves N_0 with exit data of the form $(\boldsymbol{\xi}, \boldsymbol{\eta}) = (0, \boldsymbol{\beta}) \in \overline{B}_\delta^s \times S_\delta^{u-1}$; here $\boldsymbol{\xi} = 0$ because γ_0 lies on the unstable manifold. Choose a neighborhood U of $(0, \boldsymbol{\beta})$ on the boundary of N such that all orbits exiting N through U lie close enough to γ_0 that they enter N_0'; later an additional smallness condition will be imposed on U. Let M_0 be the region between N_0 and N_0' filled with the orbits passing through U. Let

$$D_0 = N_0 \cup M_0 \cup N_0'.$$

This is called a *dumbbell neighborhood* because of its shape. Let $\widetilde{D}_0$ be the portion of D_0 filled with orbits (and broken orbits) that enter N_0, pass through M_0 into N_0', and then exit N_0'. Usually the broken orbits will lie on the boundary of $\widetilde{D}_0$, which is therefore not generally open. Figure 6.6 shows D_0 and $\widetilde{D}_0$

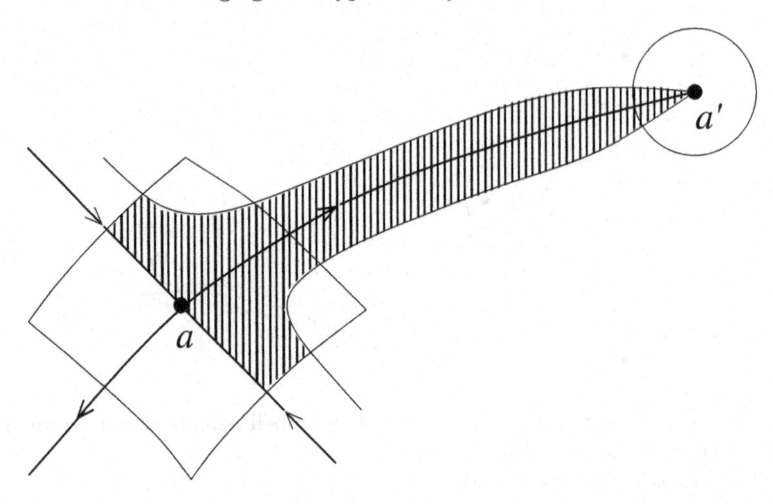

Fig. 6.6: A dumbbell neighborhood D of a connecting orbit γ from a saddle $\boldsymbol{a}$ ($u = 1$) to a sink $\boldsymbol{a}'$ ($u' = 0$) in the plane ($n = 2$). The shaded portion is the region $\widetilde{D}$ filled with orbits passing through the tube.

in the case in which γ_0 connects a saddle to a sink in the plane ($n = 2$, $u = 1$, $u' = 0$). In the next theorem, we construct a conjugacy of the guiding flow on $\widetilde{D}_0$ with the perturbed flow on a similar set $\widetilde{D}_\varepsilon$

Theorem 6.6.1 (Dumbbell conjugacy and first-order shadowing, regular case). *If U is taken small enough in the definition of $\widetilde{D}_0$, there is a homeomorphism $\boldsymbol{h}_\varepsilon : \widetilde{D}_0 \to \widetilde{D}_\varepsilon = \boldsymbol{h}_\varepsilon(\widetilde{D}_0)$ conjugating $\dot{\boldsymbol{z}} = \mathbf{f}^0(\boldsymbol{z})$ with $\dot{\boldsymbol{x}} = \mathbf{f}^{[0]}(\boldsymbol{x}, \varepsilon)$. The conjugacy depends smoothly on ε and satisfies $\|\mathbf{h}_\varepsilon(\boldsymbol{z}) - \boldsymbol{z}\| = \mathcal{O}(\varepsilon)$ uniformly for $\mathbf{x}(t) = \boldsymbol{h}_\varepsilon(\mathbf{z}(t)) \in D_0$. If $\mathbf{z}(t)$ is a solution of $\dot{\boldsymbol{z}} = \mathbf{f}^0(\boldsymbol{z})$ passing through $\widetilde{D}_0$, then $\mathbf{x}(t, \varepsilon) = \boldsymbol{h}_\varepsilon(\mathbf{z}(t))$ is a solution of $\dot{\boldsymbol{x}} = \mathbf{f}^{[0]}(\boldsymbol{x}, \varepsilon)$ that shadows $\mathbf{z}(t)$ with error $\mathcal{O}(\varepsilon)$ throughout the interval in which $\mathbf{z}(t)$ remains in $\widetilde{D}_0$.*

Proof In the local coordinates $(\boldsymbol{\xi}, \boldsymbol{\eta})$, the rest point $\boldsymbol{a}(\varepsilon)$ of $\dot{\boldsymbol{x}} = \mathbf{f}^{[0]}(\boldsymbol{x}, \varepsilon)$ is fixed at the origin, and the neighborhood $N = N_\varepsilon$ is also fixed (independent of ε). Therefore the set U is well defined even for $\varepsilon \neq 0$, and becomes a "moving" set U_ε in the x coordinates. The orbits of $\dot{\boldsymbol{x}} = \mathbf{f}^{[0]}(\boldsymbol{x}, \varepsilon)$ through U form a tube M_ε, and we set $D_\varepsilon = N_\varepsilon \cup M_\varepsilon \cup N'_\varepsilon$. Then we define $\widetilde{D}_\varepsilon$ to be the part of D_ε filled with orbits that pass through M_ε.

Whenever it is not confusing we suppress ε in the following discussion. The first step is to observe that the box data problem $(\boldsymbol{\alpha}, \boldsymbol{\beta}, T)$, which is well posed in N by Theorem 6.4.3, remains well posed in $N \cup M$. This is because the exit map $\boldsymbol{\Psi}$ defined in (6.4.4) provides initial data for a problem in M, which is crossed by all orbits in bounded time. (Initial value problems are *uniformly* well posed in compact sets with no rest point.) Now observe that as $T \to \infty$ (so that the orbit with box data $(\boldsymbol{\alpha}, \boldsymbol{\beta}, T)$ approaches γ_0), T is changing rapidly but the time taken to cross M approaches that of γ_0 and is therefore bounded. It follows that the total crossing time S is a smooth, monotonically increasing function of T (with $\boldsymbol{\alpha}$ and $\boldsymbol{\beta}$ held constant) with positive derivative for T sufficiently large. Therefore, if U is taken sufficiently small in the definition of M, the function $S(T)$ is invertible to give $T(S)$, and $(\boldsymbol{\alpha}, \boldsymbol{\beta}, T)$ can be replaced by $(\boldsymbol{\alpha}, \boldsymbol{\beta}, S)$. It is clear (by local applications of the one-variable implicit function theorem) that $S(T)$ is smooth for finite T and that $S = \infty$ if and only if $T = \infty$. Thus we have proved that the problem with modified box data $(\boldsymbol{\alpha}, \boldsymbol{\beta}, S)$ is uniformly well posed on $N \cup M$. It is now clear how to define a conjugacy of the unperturbed and perturbed flows on $N \cup M$: to each unperturbed ($\varepsilon = 0$) orbit we assign the perturbed orbit with the same data, and then match points along the two orbits by their time from entry (into N), or equivalently (since paired orbits have the same S), by their time until exit (from M into N'). As in the case of box neighborhoods, only one of these methods of matching points will work in the case of broken orbits, but continuity of the conjugacy at points of the broken orbits is clear.

Any orbit passing through $\tilde{D}$ has modified box data $(\boldsymbol{\alpha}, \boldsymbol{\beta}, S)$ in $N \cup M$ and (ordinary) box data $(\boldsymbol{\alpha}', \boldsymbol{\beta}', T')$ in N'. Conversely, given arbitrary $(\boldsymbol{\alpha}, \boldsymbol{\beta}, S)$ and $(\boldsymbol{\alpha}', \boldsymbol{\beta}', T')$, these will usually define orbits that do not connect to form a single orbit. We want to write down an equation stating the condition that these orbits do connect. This condition takes the form

$$\boldsymbol{F}(\boldsymbol{\alpha}, \boldsymbol{\beta}, S; \boldsymbol{\alpha}', \boldsymbol{\beta}', T'; \varepsilon) = \widehat{\boldsymbol{\Psi}}_\varepsilon(\boldsymbol{\alpha}, \boldsymbol{\beta}, S) - \boldsymbol{\Phi}'_\varepsilon(\boldsymbol{\alpha}', \boldsymbol{\beta}', T') = 0, \tag{6.6.2}$$

where $\boldsymbol{\Phi}'$ is the entry map for N' as defined in (6.4.4), and $\widehat{\boldsymbol{\Psi}}$ is the exit map for $N \cup M$, assigning to $(\boldsymbol{\alpha}, \boldsymbol{\beta}, S)$ the point where the corresponding orbit leaves M and enters N'. In this equation we now treat $\boldsymbol{\alpha}, \boldsymbol{\beta}, \boldsymbol{\alpha}', \boldsymbol{\beta}'$ as given in local coordinates on their respective spheres, so that (for instance) $\boldsymbol{\alpha}$ has $s - 1$ independent coordinates (rather than s components with a relation).

Now we specialize to $\varepsilon = 0$. The hypothesis of transversality of the intersection of stable and unstable manifolds along γ_0 is equivalent to the matrix of partial derivatives $[\boldsymbol{F}_{\boldsymbol{\beta}}, \boldsymbol{F}_{\boldsymbol{\alpha}'}]$ having maximal rank when $S = T' = \infty$ and $\varepsilon = 0$. (In greater detail, the columns of $F_{\boldsymbol{\beta}}$ span a $(u-1)$-dimensional space tangent to $W^u(\boldsymbol{a}_0)$ and transverse to γ_0 at the point where γ_0 enters N', while the columns of $F_{\boldsymbol{\alpha}'}$ span an $(s'-1)$-dimensional space tangent to $W^s(a_0')$ and transverse to γ_0 at the same point.) In the simplest case, when the drop in unstable dimensions is one ($u' = u - 1$), the matrix $[F_{\boldsymbol{\beta}}, F_{\boldsymbol{\alpha}'}]$ is square, and since it has maximal rank, it is invertible. Before considering the general case, we complete the proof of the theorem in this special case.

The idea is to show that for S and T' sufficiently large and ε sufficiently small, the *dumbbell data problem* with data $(\boldsymbol{\alpha}, S, T', \boldsymbol{\beta}')$ is well posed. Once this is established, the argument follows the now-familiar form: the unperturbed orbit having given dumbbell data is associated with the perturbed orbit having the same data, and (since both orbits take time $S + T'$ to cross the dumbbell) points on these orbits can be matched. All that is necessary, then, is to show that there exist unique smooth functions $\boldsymbol{\beta} = \boldsymbol{\beta}(\boldsymbol{\alpha}, S, \boldsymbol{\beta}', T', \varepsilon)$ and $\boldsymbol{\alpha}' = \boldsymbol{\alpha}'(\boldsymbol{\alpha}, S, \boldsymbol{\beta}', T', \varepsilon)$ such that $F(\boldsymbol{\alpha}, \boldsymbol{\beta}, S; \boldsymbol{\alpha}', \boldsymbol{\beta}', T'; \varepsilon) = 0$. Then the solution having data $(\boldsymbol{\alpha}, \boldsymbol{\beta}, S)$ in $N_\varepsilon \cup M_\varepsilon$ will connect with the solution having data $(\boldsymbol{\alpha}', \boldsymbol{\beta}', T')$ in N'_ε to form the desired unique solution of the box data problem. But the existence of the functions $\boldsymbol{\beta}$ and $\boldsymbol{\alpha}'$, for large S and T' and small ε, follows by the implicit function theorem from the invertibility of $[F_{\boldsymbol{\beta}}, F_{\boldsymbol{\alpha}'}]$ at $S = T' = \infty$ and $\varepsilon = 0$. (It can be checked, by a slight modification of any of the usual proofs, that the implicit function theorem is valid around $S = T' = \infty$. We need a version that is valid for $\boldsymbol{\alpha}, \boldsymbol{\beta}'$ in their spheres, which are compact sets.) The necessity of taking S and T' large is responsible for the final reduction in the size of U required in the statement of the theorem.

When the dimension drop $u - u'$ is greater than one, the proof requires a slight modification. In this case the matrix $[F_{\boldsymbol{\beta}}, F_{\boldsymbol{\alpha}'}]$ has more columns than rows, and can be made invertible by crossing out enough correctly chosen columns from $F_{\boldsymbol{\beta}}$. (The decision to cross out columns from $F_{\boldsymbol{\beta}}$ rather than from $F_{\boldsymbol{\alpha}'}$ is arbitrary, but it is never necessary to use both.) Let $\widehat{\boldsymbol{\beta}}$ be the part of $\boldsymbol{\beta}$ corresponding to columns crossed out. Then the implicit function theorem allows $F = 0$ to be solved for the rest of the components of $\boldsymbol{\beta}$, and all of $\boldsymbol{\alpha}'$, as functions of $(\boldsymbol{\alpha}, \widehat{\boldsymbol{\beta}}, S, T', \boldsymbol{\beta}', \varepsilon)$ for small ε and large S, T'. In other words, the dumbbell data must be expanded to $(\boldsymbol{\alpha}, \widehat{\boldsymbol{\beta}}, S, T', \boldsymbol{\beta}', \varepsilon)$ in order to obtain a well-posed problem, but other than this, the argument is the same.

Some additional details are contained in [199]. However, in that paper only the shadowing is proved and not the conjugacy, and the replacement of T by S was not made, so that associated orbits did not cross D in exactly the same time (but in a time that could differ by $\mathcal{O}(\varepsilon)$). □

Notice that the construction involves modifying only the "internal" variables $\boldsymbol{\beta}$ and $\boldsymbol{\alpha}'$, not the "external" variables $\boldsymbol{\alpha}$ and $\boldsymbol{\beta}'$. This makes it possible to extend the argument in a natural way to "multiple dumbbells." For instance, given connecting orbits from $\boldsymbol{a}$ to $\boldsymbol{a}'$ and from $\boldsymbol{a}'$ to $\boldsymbol{a}''$ (another hyperbolic rest point), we can create a neighborhood of the broken orbit from $\boldsymbol{a}$ to $\boldsymbol{a}''$ by joining three box neighborhoods with two tubes, and obtain shadowing and conjugacy results for the orbits that pass through both tubes. We will not state a theorem formally, but this will be used in Section 6.6.2.

6.6.2 The Averaging Case

It should be clear how to modify the proofs in this section for the averaging case, so we only outline the steps and state the result. First a dumbbell neighborhood is defined for the guiding system, exactly as in the regular case. Next the Cartesian product with S^1 is taken. Then the box data for the first box, $(\boldsymbol{\alpha}, \theta_0, \boldsymbol{\beta}, T)$ (see (6.4.7)), is replaced by box tube data $(\boldsymbol{\alpha}, \theta_0, \boldsymbol{\beta}, S)$ as before, noticing that the orbit exits the tube (and enters the second box) with $\theta = \theta_0 + S$; the same construction is repeated for system (6.4.5). Next a function $\mathbf{F}(\boldsymbol{\alpha}, \boldsymbol{\beta}, \theta_0, S; \boldsymbol{\alpha}', \boldsymbol{\beta}', \theta_0', T'; \varepsilon)$ is constructed such that $\mathbf{F} = 0$ if the orbit with data $(\boldsymbol{\alpha}, \boldsymbol{\beta}, \theta_0, S)$ in the box tube connects with the orbit having data $(\boldsymbol{\alpha}', \boldsymbol{\beta}', \theta_0', T')$ in the second box for parameter value ε. The vector equation $\mathbf{F} = 0$ will include $\theta_0 + S = \theta_0'$ as one entry. The assumption that the stable and unstable manifolds of the guiding system intersect transversely again implies that, beginning with data that match when $\varepsilon = 0$, $\boldsymbol{\beta}$ and $\boldsymbol{\alpha}'$ can be adjusted (smoothly in ε) so that the modified data match for ε near zero, provided (again) that the size of the tube may need to be reduced.

Theorem 6.6.2 (Dumbbell conjugacy and shadowing, averaging case). *Let $D_0 = N_0 \cup M_0 \cup N_0'$ be a dumbbell neighborhood for the guiding system (6.1.5) and let $\widetilde{D}_0$ be the union of the orbits in D_0 that pass through the tube M_0. Assume than M_0 is sufficiently narrow. Then there exists a homeomorphism $\mathbf{H}_\varepsilon : \widetilde{D}_0 \times S^1 \to \mathbf{H}_\varepsilon(\widetilde{D}_0 \times S^1)$ conjugating solutions of (6.3.2) with solutions of (6.3.1). The conjugacy depends smoothly on ε, moves points a distance $\mathcal{O}(\varepsilon)$, and maps approximate solutions to shadowing exact solutions.*

Shadowing for the unsuspended systems follows as discussed before Theorem 6.4.5. For an extension to higher-order averaging see [200].

6.7 Extension to Larger Compact Sets

It is now easy to prove that shadowing holds on large compact sets (which are closures $\overline{\Omega}$ of bounded open sets Ω). Establishing conjugacy in this context is much harder, and only a few indications will be given. Both regular and averaging cases will be treated simultaneously. We assume in either case that the guiding system is a gradientlike Morse–Smale system in the following sense.

Let Ω be a bounded open subset of $\mathbb{R}^n$ with smooth boundary. An autonomous system of differential equations defined on a neighborhood of $\overline{\Omega}$ is called a *gradientlike Morse–Smale system on $\overline{\Omega}$* provided that

1. Ω contains a finite collection of hyperbolic rest points $\boldsymbol{a}_1, \dots, \boldsymbol{a}_s$ for the system.
2. The stable and unstable manifolds of these rest points intersect transversely whenever they intersect in $\overline{\Omega}$.
3. Every orbit beginning in $\overline{\Omega}$ either approaches one of the rest points $\boldsymbol{a}_j$ as $t \to \infty$ or else leaves $\overline{\Omega}$ in finite time, and the same is true as $t \to -\infty$. (An orbit cannot approach *the same* rest point in both directions, because of equation (6.6.1).)

In order to state the results for the regular and averaging cases simultaneously, we write $\mathbf{z}(t, \varepsilon)$ for a solution of either (6.1.3) or (6.1.8), even though in the regular case $\mathbf{z}(t)$ does not depend on ε.

Theorem 6.7.1 (Shadowing on compact sets). *If the guiding system is a gradientlike Morse–Smale system on $\overline{\Omega}$, there exist constants $c > 0$ and*

$\varepsilon_0 > 0$ *such that for each approximate solution family* $\mathbf{z}(t,\varepsilon)$ *there is an exact solution family* $\mathbf{x}(t,\varepsilon)$ *satisfying*

$$\|\mathbf{z}(t,\varepsilon) - \mathbf{x}(t,\varepsilon)\| < c\varepsilon$$

as long as $\mathbf{z}(t,\varepsilon)$ *remains in* Ω, *for every* ε *such that* $0 \leq \varepsilon \leq \varepsilon_0$.

Proof In this proof "orbit" means "a connected component of the intersection of an orbit of the guiding system with $\overline{\Omega}$." (The reason for taking a connected component is that if an orbit leaves $\overline{\Omega}$ the error estimate ceases to hold, and is not recovered if the orbit reenters this set at a later time.) The idea of the proof is to construct a finite collection of open sets O_j and constants $c_j > 0$, $\varepsilon_j > 0$ for $j = 1, \ldots, r$ with the following properties:

1. The open sets $O_1, \ldots, O_r$ cover $\overline{\Omega}$.
2. Every approximate solution $\mathbf{z}(t,\varepsilon)$ that passes through O_j is shadowed by an exact solution $\mathbf{x}(t,\varepsilon)$ in the sense that $\|\mathbf{z}(t,\varepsilon) - \mathbf{x}(t,\varepsilon)\| < c_j\varepsilon$ as long as $\mathbf{z}(t,\varepsilon)$ remains in O_j, provided that $0 \leq \varepsilon \leq \varepsilon_j$.
3. Every orbit γ (in the sense defined above) is contained completely in (at least) one set O_j.

It is the last of these requirements that leads to the difficulties in the construction, explained below. After this construction is made, let c be the maximum of the c_j and ε_0 the minimum of the ε_j, and the theorem follows immediately.

To carry out the construction of the open cover, we begin by placing a box neighborhood around each sink in $\overline{\Omega}$. Next we consider the rest points of unstable dimension one; from each such point, two orbits leave and approach either a sink or the boundary of Ω. In the first case we cover the orbit by a dumbbell neighborhood suitable for shadowing. In the second we cover it by a box neighborhood and a flow tube leading to the boundary. (Shadowing follows in such a neighborhood exactly as for the first box and tube in a dumbbell.) Up to this point we have constructed only a finite number of open sets. Next we consider the rest points of unstable dimension two; for clarity, let $\boldsymbol{a}_2$ be such a point. From $\boldsymbol{a}_2$ there are an uncountable number of departing orbits. First we consider an orbit γ that leaves $\boldsymbol{a}_2$ and approaches a rest point $\boldsymbol{a}_1$ of unstable dimension one; we cover γ by a dumbbell (containing $\boldsymbol{a}_2$ and $\boldsymbol{a}_1$) suitable for shadowing and satisfying the additional narrowness condition that all orbits passing through the tube and not terminating at $\boldsymbol{a}_1$ pass through one of the tubes, already constructed, that leave $\boldsymbol{a}_1$ and end at a sink $\boldsymbol{a}_0$ or at the boundary. But the dumbbell from $\boldsymbol{a}_2$ to $\boldsymbol{a}_1$ does not become an open set in our cover; instead we use the double dumbbell containing $\boldsymbol{a}_2$, $\boldsymbol{a}_1$, and $\boldsymbol{a}_0$ (or the "one-and-a-half dumbbell" ending at the boundary), and recall the multiple dumbbell shadowing argument mentioned briefly in the last section. The construction guarantees that every orbit that is completely contained in the double dumbbell will be shadowed, thus satisfying condition 3 above. After covering orbits from $\boldsymbol{a}_2$ of this type, we cover the remaining orbits from $\boldsymbol{a}_2$ that connect directly to a sink or to the boundary by a single dumbbell as before. Finally, before going on to rest points of unstable dimension 3, we use the compactness of the exit sphere from $\boldsymbol{a}_2$ to select a finite subset of the open sets just constructed that still cover all orbits leaving $\boldsymbol{a}_2$. Now we continue in the same way. If $\boldsymbol{a}_3$ is a rest point of unstable dimension 3 and γ is an orbit leaving $\boldsymbol{a}_3$ and approaching $\boldsymbol{a}_2$, we make the tube from $\boldsymbol{a}_3$ to $\boldsymbol{a}_2$ narrow enough that all orbits passing through the tube (and not terminating at $\boldsymbol{a}_2$) pass through one of the (finite number of) tubes leading from $\boldsymbol{a}_2$ constructed at the previous step, and then we add the resulting multiple dumbbells to our cover. Finally, after covering all orbits beginning at the sources, we treat orbits entering across the boundary of $\overline{\Omega}$. These either approach a rest point (in which case we cover them with a flow tube narrow enough to feed into the subsequent tubes) or another point on the boundary (in which case we simply use any flow tube, and shadowing can be done via initial conditions since the orbits leave in finite time). Having used the compactness of the exit spheres to obtain finiteness at each stage, we make a final use of the compactness of $\overline{\Omega}$ to obtain finiteness at the end. □

The argument used to prove Theorem 6.7.1 cannot be used to prove conjugacy on $\overline{\Omega}$, because the local constructions for conjugacies in each O_j may not agree on the intersection of two such sets. (Shadowing orbits need not be unique, but a conjugacy must be a homeomorphism.) Constructing the local conjugacies so that they do agree requires careful coordination of the features that lead to ambiguity in the dumbbell conjugacies. (As presented above, these ambiguities result from the choice of coordinates for $\boldsymbol{\alpha}$ and $\boldsymbol{\beta}$ on the entry and exit spheres and the choices of $\widehat{\boldsymbol{\beta}}$. For a global conjugacy argument it is better to formulate things in a coordinate-free way, and then the ambiguity depends on the choices of certain transverse fibrations to certain smoothly embedded cells.) The details have not been carried out, but similar things have been done in other proofs of conjugacy for Morse–Smale systems. Completing the present argument would prove a new result, namely, that when a Morse–Smale vector field depends smoothly on a parameter, the conjugacy does also. (We have proved this for the local conjugacies here.) The Morse–Smale structural stability theorem is usually proved on a compact manifold without boundary (instead of on our $\overline{\Omega}$), but this change does not present any difficulties. (As a technical aside related to this, it is important that in our argument we allow the conjugacy to move the boundary of $\overline{\Omega}$; see [226]. Otherwise, tangencies of orbits with the boundary of Ω pose difficulties.)

One step in the conjugacy argument is both easy to prove and significant by itself in applications. The *diagram* of a Morse–Smale flow is a directed graph with vertices corresponding to the rest points (and/or periodic orbits) and a directed edge from one vertex to another if they are connected by a heteroclinic orbit.

Theorem 6.7.2 (Diagram stability). *If the guiding system is gradientlike Morse–Smale on $\overline{\Omega}$, then the diagram of the original system is the same as* *the diagram of the guiding system (with rest points replaced by periodic orbits in the averaging case).*

Proof For the averaging case see [198, Section 6]. The regular case, which is better known, can be handled similarly. □

6.8 Extensions and Degenerate Cases

Three central assumptions govern the results discussed in this chapter. First it was assumed that the rest points of the guiding system were simple (Theorems 6.3.1 and 6.3.2). Next it was added that they were hyperbolic (Theorems 6.3.1 and 6.3.3). Finally, it was required that the stable and unstable manifolds intersect transversely (Theorems 6.6.1, 6.6.2, 6.7.1, and 6.7.2). Now we briefly discuss what happens if these hypotheses are weakened. There are many open questions in this area.

If the guiding system has a rest point that is not simple, then the rest point is expected to bifurcate in some manner in the original system. That is, in the regular case there may be different numbers of rest points for $\varepsilon < 0$ and for $\varepsilon > 0$, all coming together at the nonsimple rest point when $\varepsilon = 0$. In the averaging case, there will typically be different numbers of periodic solutions on each side of $\varepsilon = 0$. Bifurcation theory is a vast topic, and we will not go into it here. Most treatments focus on the existence and stability of the bifurcating solutions without discussing their interconnections by heteroclinic orbits. There are actually two problems here, the connections between the rest points (or periodic orbits) in the bifurcating cluster, and the connections between these and other rest points (or periodic orbits) originating from other rest points of the guiding system. The first problem is local in the sense that it takes place near the nonsimple rest point, but is global in the sense that it involves intersections of stable and unstable manifolds, and often becomes a global problem in the usual sense after a rescaling of the variables; the rescaled problem is "transplanted" to a new "root" guiding problem, which sometimes has hyperbolic rest points and can be studied by the methods described here. But this leaves the second problem untouched, because the rescaling moves the other rest points to infinity.

Next we turn to the case that the guiding system has simple rest points, but these are not hyperbolic. In this case existence of the expected rest points (or periodic orbits) in the original system is assured, but their stability is unclear. Of particular interest is the case in which these actually are hyperbolic when $\varepsilon \neq 0$ (even though this hyperbolicity fails at $\varepsilon = 0$). A typical situation is that the guiding system has a rest point with a pair of conjugate pure imaginary eigenvalues, but these move off the imaginary axis when ε is varied. Of course, a Hopf bifurcation (which is not a bifurcation *of the rest point*) could occur here, but our first interest is simply the hyperbolicity of the rest point. (We speak in terms of the regular case, but the averaging case can be handled similarly; in place of "rest point" put "periodic orbit," and if there is a Hopf bifurcation, in place of "bifurcating periodic orbit" put "invariant torus.")

So suppose that (6.1.8) has a simple rest point $\boldsymbol{a}_0$ that gives rise (via Theorem 6.3.1) to a rest point $a(\varepsilon)$ for (6.1.6). Let

$$A(\varepsilon) = \mathsf{Df}^{[0]}(\boldsymbol{a}_\varepsilon, \varepsilon)$$

and suppose that this has Taylor expansion

$$A(\varepsilon) = A_0 + \varepsilon A_1 + \cdots + \varepsilon^k A_k + \cdots .$$

Suppose that the truncation (or k-jet) $A_0 + \cdots + \varepsilon^k A_k$ is hyperbolic for $0 < \varepsilon < \varepsilon_0$. Does it follow that $A(\varepsilon)$ itself is hyperbolic (with the same unstable dimension)? Not always, as the example in Section 5.9.1 already shows. But there are circumstances under which hyperbolicity of $A(\varepsilon)$ can be decided from its k-jet alone, and then we speak of *k-determined hyperbolicity.* Several criteria for k-determined hyperbolicity have been given in [206], [198, Section 5], and [203, Section 3.7]. The criterion given in the last reference is algorithmic in character (so that after a finite amount of calculation one has an answer).

But this is not the end of the story. We have seen (Theorem 6.4.4) that when A_0 is hyperbolic, there is a local conjugacy between the guiding and original systems near the rest point. Is there a similar result when $A(\varepsilon)$ has k-determined hyperbolicity? The best that can be done is to prove conjugacy on an ε-dependent neighborhood that shrinks as ε approaches zero at a rate depending on k. (It is clear that if there is a Hopf bifurcation, conjugacy cannot hold in a fixed neighborhood because the guiding system lacks the periodic orbit. A conjugacy theorem in the shrinking neighborhood is proved under certain conditions in [206] using rather different techniques from those used here; it is stated in a form suitable for mappings rather than flows. The result does imply a shadowing result in the shrinking neighborhood, although this is not stated.) The shrinking of the neighborhood makes it difficult to proceed with the rest of the program carried out in this chapter. For instance, it takes longer than time $1/\varepsilon$ for a solution starting at a finite distance from the rest point to reach the neighborhood of conjugacy, so neither the attraction argument (Theorem 6.5.1) nor the dumbbell argument (Theorem 6.6.1) can be carried out. No shadowing results have been proved outside the shrinking neighborhood. Nevertheless, it is sometimes possible to prove conjugacy results (without shadowing) on large compact sets. See [198, Section 8] and [227] for an example in which the existence of a Lyapunov function helps to bridge the gap between a fixed and a shrinking neighborhood.

Next we turn to the situation in which the guiding system has two hyperbolic rest points with a connecting orbit that is not a transverse intersection between the stable and unstable manifolds. Only one case has been studied, the two-dimensional case with a saddle connection. In higher dimensions the geometry of nontransverse intersections can be very complicated and there is probably no general result.

Consider a system of the form

$$\dot{\boldsymbol{x}} = \mathbf{f}^0(\boldsymbol{x}) + \varepsilon \mathbf{f}^1(\boldsymbol{x}) + \varepsilon^2 \mathbf{f}^{[2]}(\boldsymbol{x}, \varepsilon),$$

with $\boldsymbol{x} \in R^2$, and the associated system

$$\dot{\boldsymbol{z}} = \mathbf{f}^0(\boldsymbol{z}) + \varepsilon \mathbf{f}^1(\boldsymbol{z}),$$

from which the ε^2 terms have been omitted. Assume that when $\varepsilon = 0$, the $\boldsymbol{z}$ system has two saddle points $\boldsymbol{a}$ and $\boldsymbol{a}'$ with a saddle connection as in Figure 6.7, which is not transverse. (The unstable dimension of $\boldsymbol{a}'$ is not less than that of $\boldsymbol{a}$.) The saddle connection is assumed to split as shown for $\varepsilon > 0$, so that the nontransverse intersection exists only for the unperturbed system, and the splitting is caused by the term $\varepsilon \mathbf{f}^1(\boldsymbol{z})$, so that it is of the same topological type for both the $\boldsymbol{x}$ and $\boldsymbol{z}$ systems. Then every solution of the $\boldsymbol{z}$ system is shadowed in a dumbbell neighborhood by a solution of the $\boldsymbol{x}$ system, but the shadowing is (uniformly) only of order $\mathcal{O}(\varepsilon)$ and not $\mathcal{O}(\varepsilon^2)$ as one might hope. The proof hinges on the fact that no orbit can pass arbitrarily close to both rest points; there is a constant k such that every orbit remains a distance $k\varepsilon$ from at least one of the two rest points. Let $\mathbf{z}(t, \varepsilon)$ be an orbit that is bounded away from $\boldsymbol{a}'$ by distance $k\varepsilon$. Then $\mathbf{z}(t, \varepsilon)$ is shadowed with error $\mathcal{O}(\varepsilon^2)$, in a box neighborhood of $\boldsymbol{a}$, by the orbit $\mathbf{x}(t, \varepsilon)$ having the same box data. In the course of passing through the tube and second box of the dumbbell, it loses accuracy, but because it is bounded away from $\boldsymbol{a}'$ it retains accuracy $\mathcal{O}(\varepsilon)$. Orbits that are bounded away from $\boldsymbol{a}$ are shadowed by using box data near $\boldsymbol{a}'$. (See [131], but there is an error corrected in [205]: the $\mathcal{O}(\varepsilon^2)$ shadowing claimed in [131] is correct for single orbits, but not uniformly.)

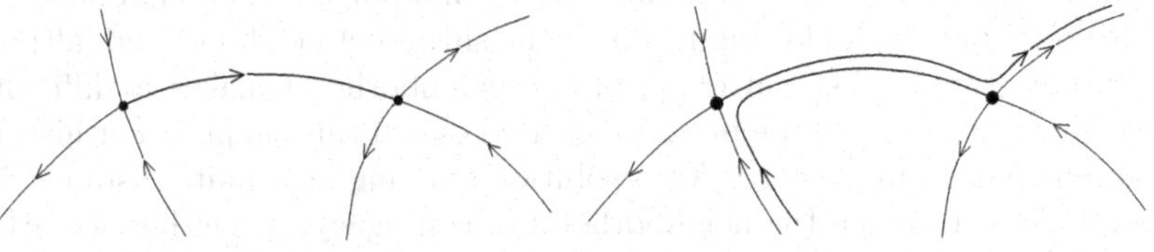

(a) A saddle connection in the plane for $\dot{\boldsymbol{z}} = \mathbf{f}^0(\boldsymbol{z})$.

(b) Breaking of the saddle connection for $\dot{\boldsymbol{z}} = \mathbf{f}^0(\boldsymbol{z}) + \varepsilon \mathbf{f}^1(\boldsymbol{z})$ wi th $\varepsilon > 0$.

Fig. 6.7: A saddle connection in the plane.

7

Averaging over Angles

7.1 Introduction

In this chapter we consider systems of the form

$$\begin{bmatrix} \dot{\boldsymbol{r}} \\ \dot{\boldsymbol{\theta}} \end{bmatrix} = \begin{bmatrix} 0 \\ \boldsymbol{\Omega}^0(\boldsymbol{r}) \end{bmatrix} + \varepsilon \begin{bmatrix} \mathbf{f}^{[1]}(\boldsymbol{r},\boldsymbol{\theta},\varepsilon) \\ \boldsymbol{\Omega}^{[1]}(\boldsymbol{r},\boldsymbol{\theta},\varepsilon) \end{bmatrix}, \tag{7.1.1}$$

where $\boldsymbol{r} \in \mathbb{R}^n$, $\boldsymbol{\theta} \in \mathbb{T}^m$, and ε is a small parameter. Here $\mathbb{T}^m$ is the m-torus, and to say that $\boldsymbol{\theta} \in \mathbb{T}^m$ merely means that $\boldsymbol{\theta} \in \mathbb{R}^m$ but the functions $\mathbf{f}^{[1]}$ and $\boldsymbol{\Omega}^{[1]}$ are 2π-periodic in each component of $\boldsymbol{\theta}$; we refer to components of $\boldsymbol{\theta}$ as angles. The radial variables $\boldsymbol{r}$ may be actual radii (in which case the coordinate system is valid only when each $r_i > 0$) or just real numbers (so that when $n = m = 1$ the state space may be a plane in polar coordinates, or a cylinder). The variable names may differ in the examples. Before turning to the (often rather intricate) details of specific examples, it is helpful to mention a few basic generalities about such systems that connect this chapter with the previous ones and with other mathematical and physical literature.

7.2 The Case of Constant Frequencies

The simplest case of (7.1.1) is the case in which $\boldsymbol{\Omega}^{[1]} = 0$ and $\boldsymbol{\Omega}^0(\boldsymbol{r}) = \boldsymbol{\omega}$ is constant:

$$\begin{bmatrix} \dot{\boldsymbol{r}} \\ \dot{\boldsymbol{\theta}} \end{bmatrix} = \begin{bmatrix} 0 \\ \boldsymbol{\omega} \end{bmatrix} + \varepsilon \begin{bmatrix} \mathbf{f}^{[1]}(\boldsymbol{r},\boldsymbol{\theta},\varepsilon) \\ 0 \end{bmatrix}. \tag{7.2.1}$$

In this case the angle equations can be solved with initial conditions $\boldsymbol{\theta}(0) = \boldsymbol{\beta}$ to give

$$\boldsymbol{\theta}(t) = \boldsymbol{\omega} t + \boldsymbol{\beta}, \tag{7.2.2}$$

and this can be substituted into the radial equations to give

$$\dot{\boldsymbol{r}} = \varepsilon \mathbf{f}^{[1]}(\boldsymbol{r}, \boldsymbol{\omega} t + \boldsymbol{\beta}, \varepsilon). \tag{7.2.3}$$

The right-hand side of this system is quasiperiodic in t, hence almost-periodic, and hence the system is a KBM system, so according to Chapter 4 it may be averaged to first order, giving

$$\dot{\boldsymbol{\rho}} = \varepsilon \overline{\mathbf{f}}^1(\boldsymbol{\rho}, \boldsymbol{\beta}), \tag{7.2.4}$$

where

$$\overline{\mathbf{f}}^1(\boldsymbol{\rho}, \boldsymbol{\beta}) = \lim_{T\to\infty} \frac{1}{T} \int_0^T \mathbf{f}^1(\boldsymbol{\rho}, \boldsymbol{\omega} t + \boldsymbol{\beta})\, \mathrm{d}t. \tag{7.2.5}$$

This remark alone is sufficient to justify some of the averaging arguments given in this chapter, although the error estimate can often be strengthened from that of Theorem 4.3.6 to $\mathcal{O}(\varepsilon)$ for time $\mathcal{O}(1/\varepsilon)$, as for periodic averaging (see below). The nature of the average defined by (7.2.5) depends strongly on the frequency vector $\boldsymbol{\omega}$, and it is very helpful to reframe the averaging process in a more geometrical way that clarifies the role of $\boldsymbol{\omega}$.

To this end, let $\boldsymbol{\nu}$ denote an integer vector and define

$$\boldsymbol{\omega}^\perp = \{\boldsymbol{\nu} : \boldsymbol{\nu} \cdot \boldsymbol{\omega} = \nu_1\omega_1 + \cdots + \nu_m\omega_m = 0\}. \tag{7.2.6}$$

The set $\boldsymbol{\omega}^\perp$ is closed under addition and under multiplication by integers, that is, it is a $\mathbb{Z}$-module, and is called the **annihilator module** of $\boldsymbol{\omega}$. For a formal definition of module, see Definition 11.2.1. In the case $\boldsymbol{\omega}^\perp = \{\mathbf{0}\}$, called the **nonresonant case**, the curves (7.2.2) are dense in $\mathbb{T}^m$ and, by a theorem of Kronecker and Weyl [250], are such that

$$\overline{\mathbf{f}}^1(\boldsymbol{\rho}, \boldsymbol{\beta}) = \frac{1}{(2\pi)^m} \int_{\mathbb{T}^m} \mathbf{f}^1(\boldsymbol{\rho}, \boldsymbol{\theta})\, \mathrm{d}\theta_1 \cdots \mathrm{d}\theta_m. \tag{7.2.7}$$

In particular, $\overline{\mathbf{f}}^1(\boldsymbol{\rho}, \boldsymbol{\beta})$ is independent of $\boldsymbol{\beta}$. Furthermore, if we write $\mathbf{f}^{[1]}$ as a (multiple) Fourier series

$$\mathbf{f}^{[1]}(\boldsymbol{r}, \boldsymbol{\theta}, \varepsilon) = \sum_{\boldsymbol{\nu}\in\mathbb{Z}^m} \mathbf{a}^{[1]}_{\boldsymbol{\nu}}(\boldsymbol{r}, \varepsilon) e^{i\boldsymbol{\nu}\boldsymbol{\theta}}, \tag{7.2.8}$$

then (in the same nonresonant case) we have

$$\overline{\mathbf{f}}^1(\boldsymbol{\rho}, \boldsymbol{\beta}) = \mathbf{a}^{[1]}_0(\boldsymbol{r}, 0) = \mathbf{a}^1_0(\boldsymbol{r}). \tag{7.2.9}$$

In the resonant case, the curves (7.2.2) are dense in some subtorus (depending on $\boldsymbol{\beta}$) embedded in $\mathbb{T}^m$, and $\overline{\mathbf{f}}^1(\boldsymbol{\rho}, \boldsymbol{\beta})$ is an average over this subtorus. In this case $\overline{\mathbf{f}}^1$ does depend on $\boldsymbol{\beta}$, but only through the subtorus that $\boldsymbol{\beta}$ belongs to. We have

$$\overline{\mathbf{f}}^1(\boldsymbol{\rho}, \boldsymbol{\beta}) = \sum_{\boldsymbol{\nu}\in\boldsymbol{\omega}^\perp} \mathbf{a}^1_{\boldsymbol{\nu}}(\boldsymbol{\rho}) e^{i\boldsymbol{\nu}\boldsymbol{\beta}}, \tag{7.2.10}$$

an equation that reduces to (7.2.9) in the nonresonant case. In order to see this, it is helpful to transform the angular variables into a form that reveals the invariant subtori of $\mathbb{T}^m$ more clearly. This is called *separating the fast and slow angles*. In fact, without doing this there is no convenient way to write an integral expression similar to (7.2.6) for $\overline{\mathbf{f}}^1$ in the resonant case.

Theorem 7.2.1. *Given a frequency vector $\boldsymbol{\omega} \in \mathbb{R}^m$, there exists a (unique) integer k, with $0 \le k < m$ and a (nonunique) unimodular matrix $S \in SL_m(\mathbb{Z})$ (that is, an integer matrix with determinant one, so that S^{-1} is also an integer matrix) such that*

$$S\boldsymbol{\omega} = (0, \dots, 0, \lambda_1, \dots, \lambda_k) \qquad \text{and} \qquad \boldsymbol{\lambda}^\perp = \{\mathbf{0}\}. \tag{7.2.11}$$

There are $m - k$ initial zeros in $S\boldsymbol{\omega}$.

Proof The example below will illustrate the ideas of this proof and the procedure for finding S. Let T be an $r \times m$ matrix whose rows are linearly independent and generate $\boldsymbol{\omega}^\perp$, and let $k = m - r$. (Such a basis is possible because $\boldsymbol{\omega}^\perp$ is not an arbitrary submodule of $\mathbb{Z}^m$ but is a **pure submodule**, that is, if an element of $\boldsymbol{\omega}^\perp$ is divisible by an integer, the quotient also belongs to $\boldsymbol{\omega}^\perp$. Pure submodules behave much like vector subspaces.) For any integer matrix T there exist unimodular matrices S $(m \times m)$ and U $(r \times r)$ such that UTS^{-1} has the following form, called **Smith normal form**:

$$UTS^{-1} = \begin{bmatrix} D & 0 \\ 0 & 0 \end{bmatrix},$$

where $D = \operatorname{diag}(\delta_1, \dots, \delta_k)$ where the δ_i are positive integers with δ_i dividing δ_{i+1} for each i. In our situation the zero rows at the bottom will not exist, and each $\delta_i = 1$, so the Smith normal form is

$$UTS^{-1} = \begin{bmatrix} I & 0 \end{bmatrix}.$$

(This again follows from the fact that $\boldsymbol{\omega}^\perp$ is a pure submodule.) The Smith normal form may be obtained by performing integer row and column operations on T; the matrix U is the product of the elementary matrices for the row operations, and S^{-1} is the product of the elementary matrices for the column operations. (Elementary operations over $\mathbb{Z}$ are interchanges, adding a multiple of one row or column to another, and multiplying a row or a column by ± 1, that is, by a unit of the ring.) The matrix U will not be used, but S is the matrix in the theorem. For Smith normal form see [212], for pure submodules (or subgroups) see [124], and for additional details about this application see [198] and [196]. □

Write

$$S\boldsymbol{\theta} = (\boldsymbol{\varphi}, \boldsymbol{\psi}) = (\varphi_1, \dots, \varphi_{m-k}, \psi_1, \dots, \psi_k) \tag{7.2.12}$$

(understood as a column vector). Since S is unimodular, this is a legitimate change of angle variables, in the sense that if any component of $\boldsymbol{\varphi}$ or $\boldsymbol{\psi}$ is shifted by 2π then the components of $\boldsymbol{\theta}$ are shifted by integer multiples of 2π. In the new coordinates, (7.2.1) can be written as

$$\begin{bmatrix} \dot{\boldsymbol{r}} \\ \dot{\boldsymbol{\varphi}} \\ \dot{\boldsymbol{\psi}} \end{bmatrix} = \begin{bmatrix} 0 \\ 0 \\ \boldsymbol{\lambda} \end{bmatrix} + \varepsilon \begin{bmatrix} \mathbf{F}^{[1]}(\boldsymbol{r}, \boldsymbol{\varphi}, \boldsymbol{\psi}, \varepsilon) \\ 0 \\ 0 \end{bmatrix}$$

$$= \begin{bmatrix} 0 \\ 0 \\ \boldsymbol{\lambda} \end{bmatrix} + \varepsilon \begin{bmatrix} \mathbf{f}^{[1]}(\boldsymbol{r}, S^{-1}(\boldsymbol{\varphi}, \boldsymbol{\psi}), \varepsilon) \\ 0 \\ 0 \end{bmatrix}. \tag{7.2.13}$$

The components of $\boldsymbol{\varphi}$ are called **slow angles** (and in fact in the present situation they are constant), while those of $\boldsymbol{\psi}$ are **fast angles**. Now (7.2.13) can be viewed as a new system of the form (7.2.1) with $(\boldsymbol{r}, \boldsymbol{\varphi})$ as $\boldsymbol{r}$ and $\boldsymbol{\psi}$ as $\boldsymbol{\theta}$; viewed in this way, (7.2.13) is nonresonant, because $\boldsymbol{\lambda}^{\perp} = \{0\}$. Therefore the average is obtained by averaging over the k-torus with variables $\boldsymbol{\psi}$. That is, the averaged system is

$$\dot{\boldsymbol{r}} = \varepsilon \overline{\mathbf{f}}^1(\boldsymbol{r}, \boldsymbol{\varphi}), \quad \dot{\boldsymbol{\varphi}} = 0,$$

with

$$\overline{\mathbf{f}}^1(\boldsymbol{r}, \boldsymbol{\varphi}) = \frac{1}{(2\pi)^k} \int_{T^k} \mathbf{f}^1(\boldsymbol{r}, S^{-1}(\boldsymbol{\varphi}, \boldsymbol{\psi}), 0)\, \mathrm{d}\psi_1 \cdots \mathrm{d}\psi_k.$$

Example 7.2.2. Suppose $\boldsymbol{\omega} = (\sqrt{2}, \sqrt{3}, \sqrt{2} - \sqrt{3}, 3\sqrt{2} + 2\sqrt{3})$. Then we may take

$$T = \begin{bmatrix} 1 & -1 & -1 & 0 \\ 3 & 2 & 0 & -1 \end{bmatrix}.$$

Adding the first column to the second and third gives

$$\begin{bmatrix} 1 & 0 & 0 & 0 \\ 3 & 5 & 3 & -1 \end{bmatrix}.$$

Subtracting three times the first row from the second gives

$$\begin{bmatrix} 1 & 0 & 0 & 0 \\ 0 & 5 & 3 & -1 \end{bmatrix}.$$

Now interchange the second and fourth columns, multiply the bottom row by -1, and add multiples of the (new) second column to the third and fourth to obtain

$$\begin{bmatrix} 1 & 0 & 0 & 0 \\ 0 & 1 & 0 & 0 \end{bmatrix}.$$

Ignoring the row operations (which only affect U in the proof of the theorem), we can multiply the elementary matrices producing the column operations and arrive at S^{-1}, but since inverting the elementary matrices is trivial, it is easy to multiply the inverses (in the reverse order) to obtain S directly:

$$S = \begin{bmatrix} 1 & -1 & -1 & 0 \\ 0 & -5 & -3 & 1 \\ 0 & 0 & 1 & 0 \\ 0 & 1 & 0 & 0 \end{bmatrix}.$$

Now

$$S\boldsymbol{\omega} = \begin{bmatrix} 0 \\ 0 \\ \sqrt{2}-\sqrt{3} \\ \sqrt{3} \end{bmatrix} = \begin{bmatrix} 0 \\ 0 \\ \lambda_1 \\ \lambda_2 \end{bmatrix}$$

and

$$S\boldsymbol{\theta} = \begin{bmatrix} \theta_1-\theta_2-\theta_3 \\ -5\theta_2-3\theta_1+\theta_4 \\ \theta_3 \\ \theta_2 \end{bmatrix} = \begin{bmatrix} \varphi_1 \\ \varphi_2 \\ \psi_1 \\ \psi_2 \end{bmatrix}.$$

◇

When $\boldsymbol{\Omega}^{[1]} \neq 0$ in (7.1.1), but $\boldsymbol{\Omega}^0(\boldsymbol{r})$ is still constant, separation of fast and slow angles carries

$$\begin{bmatrix} \dot{\boldsymbol{r}} \\ \dot{\boldsymbol{\theta}} \end{bmatrix} = \begin{bmatrix} 0 \\ \boldsymbol{\omega} \end{bmatrix} + \varepsilon \begin{bmatrix} \mathbf{f}^{[1]}(\boldsymbol{r},\boldsymbol{\theta},\varepsilon) \\ \boldsymbol{\Omega}^{[1]}(\boldsymbol{r},\boldsymbol{\theta},\varepsilon) \end{bmatrix} \tag{7.2.14}$$

into

$$\begin{bmatrix} \dot{\boldsymbol{r}} \\ \dot{\boldsymbol{\phi}} \\ \dot{\boldsymbol{\psi}} \end{bmatrix} = \begin{bmatrix} 0 \\ 0 \\ \boldsymbol{\lambda} \end{bmatrix} + \varepsilon \begin{bmatrix} \mathbf{F}^{[1]}(\boldsymbol{r},\boldsymbol{\phi},\boldsymbol{\psi},\varepsilon) \\ \mathbf{G}^{[1]}(\boldsymbol{r},\boldsymbol{\phi},\boldsymbol{\psi},\varepsilon) \\ \mathbf{H}^{[1]}(\boldsymbol{r},\boldsymbol{\phi},\boldsymbol{\psi},\varepsilon) \end{bmatrix}. \tag{7.2.15}$$

The slow angles $\boldsymbol{\phi}$ are no longer constant, but move slowly compared to $\boldsymbol{\psi}$.

Now we turn briefly to the question of improving the error estimate for first-order averaging from the one given by almost-periodic averaging of (7.2.3) to $\mathcal{O}(\varepsilon)$ for time $\mathcal{O}(1/\varepsilon)$. At the same time we address the case of (7.2.14), which does not reduce to (7.2.3). The first observation is that it suffices (for this purpose) to study (7.2.14) with nonresonant $\boldsymbol{\omega}$. (If $\boldsymbol{\omega}$ is resonant, we simply pass to (7.2.15) and then absorb $\boldsymbol{\phi}$ into $\boldsymbol{r}$ and rename $\boldsymbol{\psi}$ as $\boldsymbol{\theta}$, obtaining a new system of the form (7.2.14) that is nonresonant.) We define $\overline{\mathbf{f}}^1(\boldsymbol{r})$ as in (7.2.5), noticing that it does not depend on $\boldsymbol{\beta}$, and define $\overline{\boldsymbol{\Omega}}^1(\boldsymbol{r})$ similarly. The idea is to imitate the classical proof of first-order averaging for periodic systems, Theorem 2.8.8. Thus we consider a change of variables from $(\boldsymbol{r},\boldsymbol{\theta})$ to $(\boldsymbol{\rho},\boldsymbol{\eta})$ having the form

$$\begin{bmatrix} \boldsymbol{r} \\ \boldsymbol{\theta} \end{bmatrix} = \begin{bmatrix} \boldsymbol{\rho} \\ \boldsymbol{\eta} \end{bmatrix} + \varepsilon \begin{bmatrix} \mathbf{u}^1(\boldsymbol{\rho},\boldsymbol{\eta}) \\ \mathbf{v}^1(\boldsymbol{\rho},\boldsymbol{\eta}) \end{bmatrix}, \tag{7.2.16}$$

where $\mathbf{u}^1$ and $\mathbf{v}^1$ are 2π-periodic in each component of $\boldsymbol{\eta}$. This will carry (7.2.14) into

$$\begin{bmatrix} \dot{\boldsymbol{\rho}} \\ \dot{\boldsymbol{\eta}} \end{bmatrix} = \begin{bmatrix} 0 \\ \boldsymbol{\omega} \end{bmatrix} + \varepsilon \begin{bmatrix} \overline{\mathbf{f}}^1(\boldsymbol{\rho}) \\ \overline{\boldsymbol{\Omega}}^1(\boldsymbol{r}) \end{bmatrix} + \varepsilon^2 \begin{bmatrix} \mathbf{f}_\star^{[2]}(\boldsymbol{\rho},\boldsymbol{\eta},\varepsilon) \\ \boldsymbol{\Omega}_\star^{[2]}(\boldsymbol{\rho},\boldsymbol{\eta},\varepsilon) \end{bmatrix}, \tag{7.2.17}$$

provided that $\mathbf{u}^1$ and $\mathbf{v}^1$ satisfy the homological equations (which are now partial differential equations)

$$\begin{bmatrix} \omega_1 \frac{\partial \mathbf{u}^1}{\partial \eta_1} + \cdots + \omega_m \frac{\partial \mathbf{u}^1}{\partial \eta_m} \\ \omega_1 \frac{\partial \mathbf{v}^1}{\partial \eta_1} + \cdots + \omega_m \frac{\partial \mathbf{v}^1}{\partial \eta_m} \end{bmatrix} = \begin{bmatrix} \mathbf{f}^1(\boldsymbol{\rho}, \boldsymbol{\eta}, 0) - \overline{\mathbf{f}}^1(\boldsymbol{\rho}) \\ \boldsymbol{\Omega}^1(\boldsymbol{\rho}, \boldsymbol{\eta}, 0) - \overline{\boldsymbol{\Omega}}^1(\boldsymbol{\rho}) \end{bmatrix}. \tag{7.2.18}$$

If these equations can be solved, it only remains to estimate the error due to deletion of $\mathbf{f}_\star^{[2]}$ and $\boldsymbol{\Omega}_\star^{[2]}$ from (7.2.17). This goes much as in Chapter 2 and will be omitted. So we turn our attention to (7.2.18), and in particular to the equation for $\mathbf{u}^1$, since the one for $\mathbf{v}^1$ is handled in the same way.

It is easy to write down a *formal* solution of (7.2.18) in view of (7.2.8):

$$\mathbf{u}^1(\boldsymbol{\rho}, \boldsymbol{\eta}) = \sum_{\boldsymbol{\nu} \neq 0} \frac{\mathbf{a}_{\boldsymbol{\nu}}^1(\boldsymbol{\rho})}{i\boldsymbol{\nu} \cdot \boldsymbol{\omega}} e^{i\boldsymbol{\nu}\boldsymbol{\eta}}. \tag{7.2.19}$$

This is obtained by subtracting the mean $\mathbf{a}_0^1$ and taking the zero-mean antiderivative of the remaining terms. Since this is a Fourier series, not a Taylor series, there is no such thing as *asymptotic validity*; the series (7.2.19) must be convergent if it is to have any meaning at all. Since $\boldsymbol{\omega}$ is nonresonant, the denominators $i\boldsymbol{\nu} \cdot \boldsymbol{\omega}$ (with $\boldsymbol{\nu} \neq \mathbf{0}$) are never zero, so the coefficients are well defined. On the other hand, if $m > 1$, then for any $\boldsymbol{\omega}$, there will be values of $\boldsymbol{\nu}$ for which $\boldsymbol{\nu} \cdot \boldsymbol{\omega}$ is arbitrarily small; this is the famous **small divisor problem**, or more precisely, the **easy small divisor problem** (since related problems that are much harder to handle arise in connection with the Kolmogorov–Arnol'd–Moser theorem). Unless the corresponding values of $\mathbf{a}_{\boldsymbol{\nu}}^1$ are sufficiently small, they may be magnified by the effect of the small divisors so that (7.2.19) diverges even though (7.2.8) converges. In this case one cannot achieve the error bound $\mathcal{O}(\varepsilon)$ for time $\mathcal{O}(1/\varepsilon)$, and must be content with the weaker bound from Chapter 4. But whenever (7.2.19) converges, the stronger bound holds. The case $m = 1$ is quite special here; small divisors cannot occur, and (7.2.19) always converges.

The simplest case is when the series in (7.2.8) is finite. In this case there is no difficulty at all, because (7.2.19) is also finite. (This case also falls under Lemma 4.6.5, at least when $\boldsymbol{\Omega}^{[1]} = 0$.) Another important case is when there exist constants $\alpha > 0$ and $\gamma > 0$ such that

$$|\boldsymbol{\nu} \cdot \boldsymbol{\omega}| \geq \frac{\gamma}{|\boldsymbol{\nu}|^\alpha} \tag{7.2.20}$$

for all $\boldsymbol{\nu}$, where $|\boldsymbol{\nu}| = |\nu_1| + \cdots + |\nu_m|$. In this case the components of $\boldsymbol{\omega}$ are said to be **badly incommensurable**; this is a strong form of nonresonance. In this case, if f and g are real analytic, (7.2.19) converges. Details of the averaging proof in this case, including the higher-order case, are given in [217, Section 5].

7.3 Total Resonances

When the integer k in Theorem 7.2.1 is one, so that there is only one fast angle, the frequency vector $\boldsymbol{\omega}$ is called **totally resonant**, or a **total resonance**. (It is common to speak loosely of any resonant ω as "a resonance.") Averaging over one fast angle is easy, as there can be no small divisors. In this section we prove some technical lemmas about total resonances that will be used in Chapter 10 for Hamiltonian systems. This may be omitted on a first reading.

Lemma 7.3.1. *If ω is totally resonant, there is a real number μ such that $\mu\omega \in \mathbb{Z}^m$.*

Proof The matrix S from Theorem 7.2.1 satisfies

$$S\omega = \begin{bmatrix} 0 \\ \vdots \\ 0 \\ \lambda \end{bmatrix}. \tag{7.3.1}$$

Let $\mu = 1/\lambda$. Then

$$\mu\omega = S^{-1} \begin{bmatrix} 0 \\ \vdots \\ 0 \\ 1 \end{bmatrix} \in \mathbb{Z}^m,$$

since S^{-1} is an integer matrix (since $S \in SL_m(\mathbb{Z})$). □

In the sequel we assume that this scaling has been done, so that ω is already an integer vector and $\lambda = 1$. Writing

$$S = \begin{bmatrix} R \\ p \end{bmatrix} = \begin{bmatrix} r_{11} & r_{12} & \cdots & r_{1m} \\ \vdots & \vdots & & \vdots \\ r_{m-1,1} & r_{m-2,1} & \cdots & r_{m-1,m} \\ p_1 & p_2 & \cdots & p_m \end{bmatrix}, \tag{7.3.2}$$

we have $R\omega = \mathbf{0}$ and $p_1\omega_1 + \cdots + p_m\omega_m = 1$, which implies that the greatest common divisor of the integers ω_i is one. It is not necessary that R and p be obtained by the method of Smith normal forms; the rows of R can be any $m-1$ generators of the annihilator module of ω, and the existence of $\mathbf{p}$ follows by number theory from the gcd condition on ω. It will be convenient to choose R so as to minimize the integer N such that

$$|r_{i1}| + \cdots + |r_{im}| \le N \qquad \text{for} \qquad i = 1, \ldots, m-1. \tag{7.3.3}$$

The smallest such N can be said to measure the "order" of the resonance ω (although traditionally for Hamiltonian systems the order is $M = N - 2$, and this will be used in Chapter 10). The 1-norm of a vector is the sum of the absolute values of the components ($\|\boldsymbol{v}\|_1 = |v_1| + \cdots + |v_n|$), so the expressions occurring in (7.3.3) are the 1-norms of the rows of R.

Since S is invertible, the equation (7.3.1) can be solved for the components of ω by Cramer's rule, resulting in

$$\omega_j = \frac{\det R_j}{\det S}, \tag{7.3.4}$$

where R_j is obtained by deleting the jth column of R. We now use this solution to obtain estimates on the components of $\boldsymbol{\omega}$ in terms of N. These estimates will be best if N has been minimized (as discussed above), but are valid in any case. (The main ideas for the proofs that follow were suggested by P. Noordzij and F. Van Schagen in a private communication dated 1982.)

As a first step, we prove the following lemma, valid for any matrix (not necessarily an integer matrix).

Lemma 7.3.2. *Let K be an $n \times n$ matrix satisfying the bound*

$$|k_{i1}| + \cdots + |k_{in}| \leq L$$

on the 1-norm of the ith row for $i = 1, \ldots, n$. Then

$$|\det K| \leq L^n.$$

Proof The proof is by induction on n. For $n = 1$, the result is trivial. Suppose the result has been proved for matrices of size $n - 1$, and let K be of size n. Then the matrix K_{ij} obtained by deleting the ith row and jth column of K satisfies the same bound on the 1-norms of its rows as K, so by the inductive hypothesis, $|\det K_{ij}| \leq L^{n-1}$. Therefore by minoring on the top row,

$$\begin{aligned}
|\det K| &= |k_{11} \det K_{11} - k_{12} \det K_{12} + \cdots \pm k_{1n} \det K_{1n}| \\
&\leq |k_{11}| L^{n-1} + \cdots + |k_{1n}| L^{n-1} \\
&= (|k_{11}| + \cdots + |k_{1n}|) L^{n-1} \\
&\leq L^n.
\end{aligned}$$

Notice, for future use, that the same argument would work minoring on any row. □

Theorem 7.3.3. *Let $\boldsymbol{\omega} \in \mathbb{Z}^m$ be a total resonance, scaled so that its entries are integers with greatest common divisor one. Let R be an $(m-1) \times m$ integer matrix (as above) such that $R\boldsymbol{\omega} = 0$, and let N be an integer such that (7.3.3) holds. Then for each $j = 1, \ldots, m$ we have*

$$|\omega_j| \leq (N - 1)^{m-1}. \tag{7.3.5}$$

Since the denominator in (7.3.4) is a positive integer, it is ≥ 1, so it suffices to prove that

$$|\det R_j| \leq (N - 1)^{m-1}. \tag{7.3.6}$$

This will be done in a series of lemmas. The first deals with an easy special case, the case that R contains no zeros. (It will turn out, at the end of this section, that a much stronger estimate holds in this case. This is ultimately because an R with no zeros will not minimize N, so (7.3.5) will be a weak estimate.)

Lemma 7.3.4. *Equation (7.3.6) holds if all entries of R are nonzero.*

Proof Since the entries of R are nonzero integers, deleting the jth column reduces the 1-norm of each row by at least one, so R_j satisfies the conditions of Lemma 7.3.2 with $L = N - 1$ and $n = m - 1$. $\square$

Next we consider the case that R may have zero entries, but $\boldsymbol{\omega}$ does not. In this case, deleting the jth column does not necessarily reduce the 1-norm of every row, but only of those rows in which the jth column has a nonzero entry. We again use the repeated minoring strategy from the proof of Lemma 7.3.2, but now we must show that at each step it is possible to find a row with 1-norm $\leq N - 1$ on which to minor. At the first step, this is simple: The jth column must have a nonzero entry, because if all entries were zero, R_j would be nonsingular (since R has rank $m - 1$ by the definition of total resonance). From this it would follow that all but one of the entries of $\boldsymbol{\omega}$ are zero, contrary to hypothesis. The next lemma generalizes this remark.

Lemma 7.3.5. *If $\boldsymbol{\omega}$ has no zero entries, it is impossible for R to have ℓ columns (with $\ell < m$) which have nonzero entries only in the same $\ell - 1$ rows.*

Proof Suppose that the columns with indices $j_1, \ldots, j_\ell$ have zero entries outside of the rows with indices $i_1, \ldots, i_{\ell-1}$. Delete these columns and rows from R to obtain $\widehat{R}$, delete the entries $\omega_{j_1}, \ldots, \omega_{j_\ell}$ from $\boldsymbol{\omega}$ to obtain $\widehat{\boldsymbol{\omega}}$, and observe that $\widehat{R}\widehat{\boldsymbol{\omega}} = 0$. We will show in a moment that $\widehat{R}$ is nonsingular. It follows that $\widehat{\boldsymbol{\omega}} = 0$, contrary to the hypothesis that $\boldsymbol{\omega}$ has no zero entries.

Permute the columns of R so that $j_1, \ldots, j_\ell$ occur first and the other columns remain in their original order. Do the same with the rows to put $i_1, \ldots, i_{\ell-1}$ first. The resulting matrix has the form

$$\begin{bmatrix} P & Q \\ 0 & \widehat{R} \end{bmatrix},$$

and still has rank $m - 1$. It follows that $\widehat{R}$ has rank $m - \ell$, and is invertible. $\square$

Lemma 7.3.6. *Equation (7.3.6) holds if all entries of $\boldsymbol{\omega}$ are nonzero.*

Proof We claim that $\det R_j$ can be evaluated by repeated minoring on rows having 1-norm $\leq N - 1$. In the following argument, all rows and columns of matrices are identified by their original indices, even after various rows and columns have been deleted. Set $j_1 = j$. By Lemma 7.3.5 with $\ell = 1$, the j_1 column has at least one nonzero element. Let i_1 be the index for a row in which such an element occurs. Then the i_1 row in R_j has 1-norm $\leq N - 1$. We now delete this row from R_j to obtain an $(m-2) \times (m-1)$ matrix $R_{j_1 i_1}$. The cofactors of elements in the i_1 row of R_j are obtained by deleting a column j_2 from $R_{j_1 i_1}$ to obtain $R_{j_1 i_1 j_2}$, and then taking the determinant. We must show that for every choice of j_2 there is a row (say i_2) in $R_{j_1 i_1 j_2}$ that has 1-norm $\leq N - 1$. Suppose not. Then, in the original matrix R, columns j_1 and j_2 have nonzero entries only in row i_1. This is impossible by Lemma 7.3.5 with $\ell = 2$. Continuing in this way, the minoring can be completed with rows of 1-norm $\leq N - 1$, and (7.3.6) follows. $\square$

Lemma 7.3.7. *Equation (7.3.6) holds in the general case when $\boldsymbol{\omega}$ may have zero elements.*

Proof For the zero elements of $\boldsymbol{\omega}$, (7.3.5) is trivially true. We can delete these elements from $\boldsymbol{\omega}$ to obtain $\widehat{\boldsymbol{\omega}}$, delete the corresponding columns from R to obtain $\widehat{R}$, and still have $\widehat{R}\widehat{\boldsymbol{\omega}} = 0$. Some rows of $\widehat{R}$ will be linearly dependent on the rest, and these can be deleted. The remaining problem has the form treated in Lemma 7.3.6. □

It is clear that these estimates can be strengthened in special cases. For instance, in the case of Lemma 7.3.4, each successive column that is deleted will reduce the 1-norm of the remaining rows by at least one. Therefore $|\det R_j| \le (N-1)(N-2)\cdots(N-m+1)$.

Remark 7.3.8. Theorem 7.3.3 can be used to produce a list of all possible resonances with a given order $M = N - 2$ in m variables. A list of total first-order resonances appeared in [262] and of second-order resonances in [279], in both cases for Hamiltonian systems with 3 degrees of freedom. The estimate obtained in Theorem 7.3.3 is moreover sharp when N is minimized. For instance, with $\boldsymbol{\omega} = (1, m-1, (m-1)^2, \ldots, (m-1)^{m-1})$, it is easy to find an R with $N = m$. Then $\omega_m = (N-1)^{m-1}$, and the bound in the theorem is attained. ♡

7.4 The Case of Variable Frequencies

Turning to (7.1.1) when $\boldsymbol{\Omega}^0(\boldsymbol{r})$ is not constant, the first observation is that in a certain sense, the problem can be reduced locally to the constant frequency case after all. To see this, choose a fixed $\mathbf{r}_0$ and dilate the variable $\boldsymbol{r}$ around this value by setting

$$\boldsymbol{r} = \mathbf{r}_0 + \varepsilon^{1/2}\boldsymbol{\sigma}. \tag{7.4.1}$$

The result is

$$\begin{bmatrix} \dot{\boldsymbol{\sigma}} \\ \dot{\boldsymbol{\theta}} \end{bmatrix} = \begin{bmatrix} 0 \\ \boldsymbol{\Omega}^0(\mathbf{r}_0) \end{bmatrix} + \varepsilon^{1/2} \begin{bmatrix} \varepsilon^{1/2}\mathbf{f}^1(\mathbf{r}_0, \boldsymbol{\theta}) \\ \mathcal{O}(1) \end{bmatrix} + \mathcal{O}(\varepsilon). \tag{7.4.2}$$

This has the same form as (7.2.1), with $\boldsymbol{\omega} = \boldsymbol{\Omega}^0(\mathbf{r}_0)$ and with ε replaced by $\varepsilon^{1/2}$, and it may be averaged over the appropriate torus depending on the resonance module of $\boldsymbol{\Omega}^0(\mathbf{r}_0)$. Under suitable circumstances (if, for instance, there is only one fast angle, or the Fourier series for $\mathbf{f}^{[1]}$ and $\boldsymbol{\Omega}^{[1]}$ are finite, or if $\mathbf{f}^{[1]}$ and $\boldsymbol{\Omega}^{[1]}$ are real analytic and the frequencies of the fast angles are badly incommensurable) the results will have error $\mathcal{O}(\varepsilon^{1/2})$ for time $\mathcal{O}(1/\varepsilon^{1/2})$ as long as the solution remains in a compact set of $\boldsymbol{\sigma}$. Such a compact set corresponds under (7.4.1) to a shrinking neighborhood of $\mathbf{r}_0$ with radius $\mathcal{O}(\varepsilon^{1/2})$. The difficulty, then, is that for each such shrinking neighborhood around a different $\mathbf{r}_0$, the appropriate type of average to be taken may be different, depending on the resonances present. To study the global behavior of solutions, the first step is to find out what resonances are relevant in a given problem, and then to try to follow solutions as they pass from one type of resonance to another.

Much of this chapter and in particular, Chapter 8, is devoted to the simplest special case, $m = 1$, which is very special indeed. We already know that in this case, small divisors cannot arise. Moreover, in this case there is only one possibility for resonance, and that is $\Omega^0(\mathbf{r}_0) = 0$. In the typical (generic) case, then, the resonant values of $\mathbf{r}_0$ will occur on isolated hypersurfaces in $\mathbb{R}^n$ (or simply at points, if $n = 1$). Away from these resonant manifolds, one can average over the (single) angle θ, and near the resonance (within distance $\mathcal{O}(\varepsilon^{1/2})$ one cannot average at all, but regular perturbation approximations are possible. The problem becomes one of matching these approximations. This problem is treated in detail in Chapter 8.

When $m > 1$ things are much harder, because the resonance module typically changes with any change in $\boldsymbol{r}$, and the set of $\boldsymbol{r}$ for which $\boldsymbol{\Omega}^0(\boldsymbol{r})$ is resonant is dense. It now becomes important to distinguish between *engaged* and *disengaged* resonances. We say that $\boldsymbol{\Omega}^0(\boldsymbol{r})$ is an **engaged resonance** if its resonance module contains nonzero integer vectors $\boldsymbol{\nu}$ for which the corresponding Fourier coefficients of $\mathbf{f}^{[1]}$ and $\boldsymbol{\Omega}^{[1]}$ (such as $\mathbf{a}^1_{\boldsymbol{\nu}}$ in (7.2.8)) are nonzero. Resonances that are not engaged may be ignored, because the average appropriate to them coincides with the nonresonant average. (For instance, the only nonzero term in (7.2.10) will be $\mathbf{a}^1_0$.) In particular, if the Fourier series for $\mathbf{f}^{[1]}$ and $\boldsymbol{\Omega}^{[1]}$ are finite, there will be finitely many **resonance manifolds** corresponding to engaged resonance modules of dimension one, and the multiple resonance manifolds will just be intersections of these. So the resonance manifolds will still be isolated, and the case $m > 1$ will not be so different from $m = 1$. Away from the resonances, average over all angles. Near the resonances, average over particular angles. Then attempt to match.

If a dense set of resonance manifolds are engaged, then one tries to determine how many are *active*. A resonance is **active** if the correct local averaged system has rest points within the resonance band. In this case, some orbits will not pass through the resonance, and others will be delayed within the resonance band for long periods of time. In the opposite (passive) case, all solutions will pass through the resonance in time $\mathcal{O}(1/\varepsilon^{1/2})$. Arnol′d has shown that in this case, the resonance can be ignored, at the cost of a significant weakening in the error estimate. Neishstadt has shown that even in the active case, the resonance can be ignored (with a weakening of the error estimate) for most initial conditions (since most solutions still pass through sufficiently rapidly). The theorems are formulated in terms of the measure of the set of exceptional solutions for which the error estimate fails. Exact statements of these results are technical and will not be given here. A rather thorough exposition of this point of view has been given in [177, Chapters 3-6]. It should be clear that the goal of this *Russian* approach is to average over *all the angles* even when this leads to a weaker result. The goal of the method presented below is to get a stronger result by doing the *correct* averaging only over the fast angles and matching the resulting pieces. The reference [177] also contains other topics related to multi-frequency averaging that we do not address here, such as the Kolmogorov–Arnol′d–Moser and Nekhoroshev theorems (which apply to the Hamiltonian case) and adiabatic invariants.

7.5 Examples

In our analysis of slowly varying systems we have developed up till now a theory for equations in the standard form

$$\dot{\boldsymbol{x}} = \varepsilon \mathbf{f}^1(\boldsymbol{x}, t).$$

In Section 3.3.1 we studied an oscillator with slowly varying coefficients which could be brought into standard form after a rather special transformation of the time scale. Systems with slowly varying coefficients, in particular varying frequencies, arise often in applications and we have to develop a systematic theory for these problems. Systems with slowly varying frequencies have been studied by Mitropolsky [190]. An interesting example of passage through resonance has been considered by Kevorkian [146] using a two-time scale method. Our treatment of the asymptotic estimates in this chapter is based on Sanders [236, 238] and forms an extension of the averaging theory of the periodic case as treated in Chapters 1 and 2. We start by discussing a number of examples to see what the difficulties are. In Sections 7.8–7.10 we discuss the regular case which is relatively simple.

7.5.1 Einstein Pendulum

We consider a linear oscillator with slowly varying frequency

$$\ddot{x} + \omega^2(\varepsilon t)x = 0.$$

We put $\dot{x} = \omega y$. Differentiation produces $\ddot{x} = \dot{\omega} y + \omega \dot{y}$ and using the equation we obtain

$$\dot{y} = -\omega x - \frac{\dot{\omega}}{\omega} y.$$

We transform $(x, y) \mapsto (r, \phi)$ by

$$x = r\sin(\phi), \quad y = r\cos(\phi),$$

to obtain

$$\dot{r} = -\frac{\dot{\omega}}{\omega} r\cos^2(\phi),$$
$$\dot{\phi} = \omega + \frac{\dot{\omega}}{\omega} \sin(\phi)\cos(\phi).$$

Introducing $\tau = \varepsilon t$ we have the third-order system:

$$\begin{bmatrix} \dot{r} \\ \dot{\tau} \\ \dot{\phi} \end{bmatrix} = \begin{bmatrix} 0 \\ 0 \\ \omega \end{bmatrix} + \varepsilon \begin{bmatrix} -\frac{1}{\omega}\frac{d\omega}{d\tau} r\cos^2(\phi) \\ 1 \\ \frac{1}{\omega}\frac{d\omega}{d\tau}\sin(\phi)\cos(\phi) \end{bmatrix}.$$

Remark 7.5.1. This system is of the form

$$\begin{bmatrix} \dot{\boldsymbol{x}} \\ \dot{\phi} \end{bmatrix} = \begin{bmatrix} 0 \\ \Omega^0(\boldsymbol{x}) \end{bmatrix} + \varepsilon \begin{bmatrix} \mathbf{X}^1(\boldsymbol{x},\phi) \\ \Omega^1(\boldsymbol{x},\phi) \end{bmatrix}, \quad \boldsymbol{x} \in D \subset \mathbb{R}^2, \quad \phi \in S^1, \tag{7.5.1}$$

where $\boldsymbol{x} = (r,\tau)$ and ϕ is an angular variable which is defined on the circle S^1. ♡

Remark 7.5.2. One can remove the $\mathcal{O}(\varepsilon)$ terms in the equation for ϕ by a slightly different coordinate transformation. The price for this is an increase of the dimension of the system. Transform

$$x = r\sin(\phi+\psi), \quad y = r\cos(\phi+\psi),$$

with $\dot{\phi} = \omega$; we obtain

$$\begin{bmatrix} \dot{r} \\ \dot{\psi} \\ \dot{\tau} \\ \dot{\phi} \end{bmatrix} = \begin{bmatrix} 0 \\ 0 \\ 0 \\ \omega \end{bmatrix} + \varepsilon \begin{bmatrix} -\frac{1}{\omega}\frac{d\omega}{d\tau} r\cos^2(\phi+\psi) \\ \frac{1}{\omega}\frac{d\omega}{d\tau}\sin(\phi+\psi)\cos(\phi+\psi) \\ 1 \\ 0 \end{bmatrix}.$$

This form of the perturbation equations has some advantages in treating the passage through resonance problems of Chapter 8. For the sake of simplicity the theorems in this chapter concern system (7.5.1) with $\dot{\phi} = \Omega^0(\boldsymbol{x})$. ♡

Remark 7.5.3. Since $\phi \in S^1$ it seems natural to average the equation for $\boldsymbol{x}$ in system (7.5.1) over ϕ to obtain an approximation of $\boldsymbol{x}(t)$. It turns out that under certain conditions this procedure can be justified as we shall see later on. ♡

7.5.2 Nonlinear Oscillator

It is a simple exercise to formulate in the same way the case of a nonlinear equation with a frequency governed by an independent equation:

$$\begin{aligned} \ddot{x} + \omega^2 x &= \varepsilon f(x,\dot{x},\varepsilon t), \\ \dot{\omega} &= \varepsilon g(x,\dot{x},\varepsilon t). \end{aligned}$$

Put again $\dot{x} = \omega y$; by differentiation and using the equations we have

$$\dot{y} = -\omega x + \varepsilon\frac{f}{\omega} - \varepsilon y\frac{g}{\omega}.$$

Transforming

$$x = r\sin(\phi), \quad y = r\cos(\phi),$$

we obtain with $\tau = \varepsilon t$ the fourth-order system

$$\begin{aligned}
\dot{r} &= \frac{\varepsilon}{\omega}\cos(\phi)[f(r\sin(\phi), \omega r\cos(\phi), \tau) - r\cos(\phi)g(r\sin(\phi), \omega r\cos(\phi), \tau)],\\
\dot{\omega} &= \varepsilon g(r\sin(\phi), \omega r\cos(\phi), \tau),\\
\dot{\tau} &= \varepsilon,\\
\dot{\phi} &= \omega - \frac{\varepsilon}{\omega r}\sin(\phi)[f(r\sin(\phi), \omega r\cos(\phi), \tau) - r\cos(\phi)g(r\sin(\phi), \omega r\cos(\phi), \tau)].
\end{aligned}$$

Comparing with system (7.5.1) we have $\boldsymbol{x} = (r, \omega, \tau) \in \mathbb{R}^3$. We discuss now a problem in which two angles have to be used.

7.5.3 Oscillator Attached to a Flywheel

The equations for such an oscillator have been discussed by Goloskokow and Filippow [108, Chapter 8.3]. The frequency ω_0 of the oscillator is a constant in this case; we assume that the friction, the nonlinear restoring force of the oscillator and several other forces are small. The equations of motion are

$$\begin{aligned}
\ddot{x} + \omega_0^2 x &= \varepsilon F(\phi, \dot{\phi}, \ddot{\phi}, x, \dot{x}, \ddot{x}),\\
\ddot{\phi} &= \varepsilon G(\phi, \dot{\phi}, \ddot{\phi}, x, \dot{x}, \ddot{x}),
\end{aligned}$$

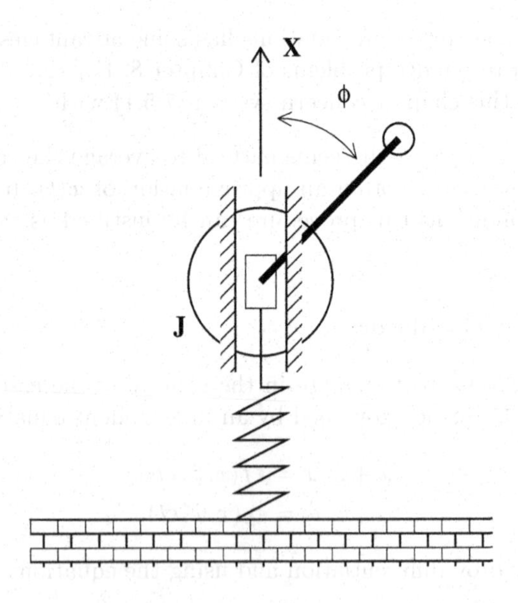

Fig. 7.1: Oscillator attached to a flywheel

where

$$F = \frac{1}{m}[-f(x) - \beta\dot{x} + q_1(\dot{\phi}^2\cos(\phi) + \ddot{\phi}\sin(\phi)],$$
$$G = \frac{1}{J_0}[M(\dot{\phi}) - M_w(\dot{\phi})] + q_2\sin(\phi)(\ddot{x} + g).$$

Here β, q_1 and q_2 are constants, g is the gravitational constant, J_0 is the moment of inertia of the rotor. $M(\dot{\phi})$ represents the known static characteristic of the motor, $M_w(\dot{\phi})$ stands for the damping of rotational motion. The equations of motion can be written as

$$\begin{aligned}
\dot{x} &= \omega_0 y\\
\dot{y} &= -\omega_0 x + \frac{\varepsilon}{\omega_0}F(\phi, \Omega, \dot{\Omega}, x, \omega_0 y, \omega_0\dot{y})\\
\dot{\Omega} &= \varepsilon G(\phi, \Omega, \dot{\Omega}, x, \omega_0 y, \omega_0\dot{y})\\
\dot{\phi} &= \Omega,
\end{aligned}$$

We put $\phi = \phi_1$. As in the preceding examples we can put $x = r\sin(\phi_2)$, $y = r\cos(\phi_2)$ to obtain

$$\begin{aligned}
\dot{r} &= \frac{\varepsilon}{\omega_0}\cos(\phi_2)F(\phi_1, \Omega, \dot{\Omega}, r\sin(\phi_2), \omega_0 r\cos(\phi_2), \omega_0\dot{y}),\\
\dot{\Omega} &= \varepsilon G(\phi_1, \Omega, \dot{\Omega}, r\sin(\phi_2), \omega_0 r\cos(\phi_2), \omega_0\dot{y}),\\
\dot{\phi}_2 &= \omega_0 - \frac{\varepsilon}{\omega_0 r}\sin(\phi_2)F(\phi_1, \Omega, \dot{\Omega}, r\sin(\phi_2), \omega_0 r\cos(\phi_2), \omega_0\dot{y})\\
\dot{\phi}_1 &= \Omega
\end{aligned}$$

$\dot{\Omega}$ and $\omega_0\dot{y}$ still have to be replaced using the equations of motion after which we can expand with respect to ε. The system is of the form (7.5.1) with higher-order terms added: $\boldsymbol{x} = (r, \Omega)$, $\boldsymbol{\phi} = (\phi_1, \phi_2)$. Again, it can be useful to simplify the equation for the angle ϕ_2. We achieve this by starting with the equations of motion and putting

$$\begin{aligned}
\phi &= \phi_1, & x &= r\sin(\phi_2 + \psi),\\
\phi_2 &= \omega_0 t, & y &= r\cos(\phi_2 + \psi).
\end{aligned}$$

The reader may want to verify that we obtain the fifth-order system

$$\begin{aligned}
\dot{r} &= \frac{\varepsilon}{\omega_0}\cos(\phi_2 + \psi)F(\phi_1, \Omega, \dot{\Omega}, r\sin(\phi_2 + \psi), \omega_0 r\cos(\phi_2 + \psi), \omega_0\dot{y}),\\
\dot{\psi} &= -\frac{\varepsilon}{\omega_0 r}\sin(\phi_2 + \psi)F(\phi_1, \Omega, \dot{\Omega}, r\sin(\phi_2 + \psi), \omega_0 r\cos(\phi_2 + \psi), \omega_0\dot{y}),\\
\dot{\Omega} &= \varepsilon G(\phi_1, \Omega, \dot{\Omega}, r\sin(\phi_2 + \psi), \omega_0 r\cos(\phi_2 + \psi), \omega_0\dot{y}),\\
\dot{\phi}_2 &= \omega_2\\
\dot{\phi}_1 &= \Omega,
\end{aligned}$$

where $\dot{\Omega}$ and $\omega_0 \dot{y}$ still have to be replaced using the equations of motion. We return to this example in Section 8.7.

Remark 7.5.4. Equations in the standard form $\dot{\boldsymbol{x}} = \varepsilon \mathbf{f}^1(\boldsymbol{x}, t)$, periodic in t, can be put in the form of system (7.5.1) in a trivial way. The equation is equivalent with

$$\dot{\boldsymbol{x}} = \varepsilon \mathbf{f}^1(\boldsymbol{x}, \phi), \quad \dot{\phi} = 1,$$

with the spatial variable $\phi \in S^1$. ♡

7.6 Secondary (Not Second Order) Averaging

It often happens that a system containing fast and slow angles can be averaged over the fast angles, producing a new system in which the remaining angles (formerly all slow) can again be separated into fast and slow angles on a different time scale as a result of **secondary resonances**. In this case one can begin again with a new (or secondary) first-order averaging. A famous instance of such a secondary resonance is the **critical inclination resonance** that arises in the so-called **oblate planet problem**, the study of artificial satellite motion around the Earth modeled as an oblate spheroid. In this section we discuss the general case with one fast angle in the original system; the oblate planet problem is treated in Appendix D.

Suppose that the original system has the form

$$\begin{bmatrix} \dot{\boldsymbol{r}} \\ \dot{\boldsymbol{\theta}} \\ \dot{\phi} \end{bmatrix} = \begin{bmatrix} 0 \\ 0 \\ \omega^0(\boldsymbol{r}) \end{bmatrix} + \varepsilon \begin{bmatrix} \mathbf{f}^1(\boldsymbol{r}, \boldsymbol{\theta}, \phi) \\ \mathbf{g}^1(\boldsymbol{r}, \boldsymbol{\theta}, \phi) \\ \omega^1(\boldsymbol{r}, \boldsymbol{\theta}, \phi) \end{bmatrix}, \tag{7.6.1}$$

with $\boldsymbol{r} \in \mathbb{R}^n$, $\boldsymbol{\theta} \in \mathbb{T}^m$, and $\phi \in S^1$. (We consider only scalar ϕ in order to avoid small divisor problems at the beginning.) Suppose that $\omega^0 \neq 0$ in the region under consideration, so that ϕ may be regarded as a fast angle. Then it is possible to average over ϕ to second-order by making a change of variables

$$\begin{bmatrix} \boldsymbol{R} \\ \boldsymbol{\Theta} \\ \Phi \end{bmatrix} = \begin{bmatrix} \boldsymbol{r} \\ \boldsymbol{\theta} \\ \phi \end{bmatrix} + \varepsilon \begin{bmatrix} \mathbf{u}^1(\boldsymbol{r}, \boldsymbol{\theta}, \phi) \\ \mathbf{v}^1(\boldsymbol{r}, \boldsymbol{\theta}, \phi) \\ \mathrm{w}^1(\boldsymbol{r}, \boldsymbol{\theta}, \phi) \end{bmatrix} + \varepsilon^2 \begin{bmatrix} \mathbf{u}^2(\boldsymbol{r}, \boldsymbol{\theta}, \phi) \\ \mathbf{v}^2(\boldsymbol{r}, \boldsymbol{\theta}, \phi) \\ \mathrm{w}^2(\boldsymbol{r}, \boldsymbol{\theta}, \phi) \end{bmatrix},$$

carrying (7.6.1) into

$$\begin{bmatrix} \dot{\boldsymbol{R}} \\ \dot{\boldsymbol{\Theta}} \\ \dot{\Phi} \end{bmatrix} = \begin{bmatrix} 0 \\ 0 \\ \omega^0(\boldsymbol{R}) \end{bmatrix} + \varepsilon \begin{bmatrix} \mathbf{F}^1(\boldsymbol{R}, \boldsymbol{\Theta}) \\ \mathbf{G}^1(\boldsymbol{R}, \boldsymbol{\Theta}) \\ \mathrm{H}^1(\boldsymbol{R}, \boldsymbol{\Theta}) \end{bmatrix} + \varepsilon^2 \begin{bmatrix} \mathbf{F}^2(\boldsymbol{R}, \boldsymbol{\Theta}) \\ \mathbf{G}^2(\boldsymbol{R}, \boldsymbol{\Theta}) \\ \mathrm{H}^2(\boldsymbol{R}, \boldsymbol{\Theta}) \end{bmatrix} + \cdots, \tag{7.6.2}$$

where the dotted terms will still depend on Φ. If the dotted terms are deleted, the $\boldsymbol{R}$ and $\boldsymbol{\Theta}$ equations decouple from the Φ equation. We now study the decoupled system under the assumptions that

$$\mathbf{F}^1(\boldsymbol{R},\boldsymbol{\Theta}) = 0, \quad \mathbf{G}^1(\boldsymbol{R},\boldsymbol{\Theta}) = \mathbf{G}^1(\boldsymbol{R}). \tag{7.6.3}$$

These assumptions may seem strong and unnatural, but in fact when the system is Hamiltonian (see Chapter 10) there is a single assumption on the Hamiltonian function, reflecting an underlying symmetry in the system, that implies (7.6.3) (and in addition implies that $H^1(R,\Theta) = H^1(R)$). Therefore the conditions (7.6.3) do arise naturally in actual examples such as the oblate planet problem.

So the problem to be considered is now

$$\begin{bmatrix} \dot{\boldsymbol{R}} \\ \dot{\boldsymbol{\Theta}} \end{bmatrix} = \varepsilon \begin{bmatrix} 0 \\ \mathbf{G}^1(\boldsymbol{R}) \end{bmatrix} + \varepsilon^2 \begin{bmatrix} \mathbf{F}^2(\boldsymbol{R},\boldsymbol{\Theta}) \\ \mathbf{G}^2(\boldsymbol{R},\boldsymbol{\Theta}) \end{bmatrix}. \tag{7.6.4}$$

Introducing slow time $\tau = \varepsilon t$ and putting $' = d/d\tau$, this becomes

$$\begin{bmatrix} \boldsymbol{R}' \\ \boldsymbol{\Theta}' \end{bmatrix} = \begin{bmatrix} 0 \\ \mathbf{G}^1(\boldsymbol{R}) \end{bmatrix} + \varepsilon \begin{bmatrix} \mathbf{F}^2(\boldsymbol{R},\boldsymbol{\Theta}) \\ \mathbf{G}^2(\boldsymbol{R},\boldsymbol{\Theta}) \end{bmatrix}, \tag{7.6.5}$$

which again has the form of (7.1.1). Therefore the procedure is to examine $\mathbf{G}^1(\boldsymbol{R}) = \boldsymbol{\Omega}^0(\boldsymbol{R})$ for (secondary) resonances, separate $\boldsymbol{\Theta}$ into fast and slow angles with respect to these resonances, and average again (this will be a **secondary first-order averaging**) over the fast angles. In the simplest case there will only be one fast angle, and small divisor problems will not arise. For a combined treatment of primary and secondary averaging in one step, see [135].

7.7 Formal Theory

We now begin a more detailed treatment of the case $m = 1$ (a single angle), using the notation of (7.5.1) rather than (7.1.1). To see what the difficulties are, we start with a formal presentation. We put

$$\boldsymbol{x} = \boldsymbol{y} + \varepsilon \mathbf{u}^1(\boldsymbol{y},\phi),$$

where $\mathbf{u}^1$ is to be an averaging transformation. So we have, using (7.5.1),

$$\dot{\boldsymbol{y}} + \varepsilon \frac{d\mathbf{u}^1}{dt} = \varepsilon \mathbf{X}^1(\boldsymbol{y} + \varepsilon \mathbf{u}^1, \phi)$$

or

$$\dot{\boldsymbol{y}} + \varepsilon \Omega^0(\boldsymbol{y}) \frac{\partial \mathbf{u}^1}{\partial \phi}(\boldsymbol{y} + \varepsilon \mathbf{u}^1) + \varepsilon \mathsf{D}_{\boldsymbol{y}} \mathbf{u}^1 \cdot \dot{\boldsymbol{y}} = \varepsilon \mathbf{X}^1(\boldsymbol{y} + \varepsilon \mathbf{u}^1, \phi).$$

Expansion with respect to ε yields

$$(I + \varepsilon \mathsf{D}_{\boldsymbol{y}} \mathbf{u}^1)\dot{\boldsymbol{y}} = \varepsilon \mathbf{X}^1(\boldsymbol{y},\phi) - \varepsilon \Omega^0(\boldsymbol{y}) \frac{\partial \mathbf{u}^1}{\partial \phi}(\boldsymbol{y},\phi) + \varepsilon^2 \cdots .$$

In the spirit of averaging it is natural to define

$$\mathbf{u}^1(\boldsymbol{y},\phi) = \frac{1}{\Omega^0(\boldsymbol{y})}\int^{\phi} (\mathbf{X}^1(\boldsymbol{y},\varphi) - \overline{\mathbf{X}}^1(\boldsymbol{y}))\,\mathrm{d}\varphi, \tag{7.7.1}$$

where $\overline{\mathbf{X}}^1$ is the 'ordinary' average of $\mathbf{X}^1$ over ϕ, i.e. $\overline{\mathbf{X}}^1(\cdot) = \frac{1}{2\pi}\int_0^{2\pi} \mathbf{X}^1(\cdot,\varphi)\,\mathrm{d}\varphi$. Notice that even when $\overline{\mathbf{X}}^1$ exists, the definition of $\mathbf{u}^1$ is purely formal, since we divide through $\Omega^0(\boldsymbol{y})$. In particular, even if $\mathbf{u}^1$ exists, we do not have an a priori bound on it, so the $\varepsilon^2\cdots$ terms can not be replaced by $\mathcal{O}(\varepsilon^2)$.

The equation becomes

$$\dot{\boldsymbol{y}} = \varepsilon\overline{\mathbf{X}}^1(\boldsymbol{y}) + \varepsilon^2\cdots$$

and we add

$$\dot{\phi} = \Omega^0(\boldsymbol{y}) + \varepsilon\cdots.$$

Remark 7.7.1. Before analyzing these equations we note that one of the motivations for this formulation is that it is easy to generalize this formal procedure to multi-frequency systems. Assume $\boldsymbol{\phi} = (\phi_1,\cdots,\phi_m)$ and let $\mathbf{X}^1(\boldsymbol{x},\boldsymbol{\phi})$ be written as

$$\mathbf{X}^1(\boldsymbol{x},\boldsymbol{\phi}) = \sum_{i=1}^{m} \mathbf{X}_i^1(\boldsymbol{x},\phi_i).$$

The equation for $\boldsymbol{\phi}$ in (7.5.1) consists now of m scalar equations of the form

$$\dot{\phi}_i = \Omega_i^0(\boldsymbol{x}).$$

In the transformation $\boldsymbol{x} = \boldsymbol{y} + \varepsilon\mathbf{u}^1(\boldsymbol{y},\boldsymbol{\phi})$ we put

$$\mathbf{u}^1(\boldsymbol{y},\boldsymbol{\phi}) = \sum_{i=1}^{m} \mathbf{u}_i^1(\boldsymbol{y},\phi_i),$$

with

$$\mathbf{u}_i^1(\boldsymbol{y},\phi_i) = \frac{1}{\Omega_i^0(\boldsymbol{y})}\int^{\phi_i} (\mathbf{X}_i^1(\boldsymbol{y},\varphi_i) - \overline{\mathbf{X}}_i^1(\boldsymbol{y}))\,\mathrm{d}\varphi_i.$$

The equation for $\boldsymbol{y}$ then becomes

$$\dot{\boldsymbol{y}} = \varepsilon\sum\nolimits_{i=1}^{m} \overline{\mathbf{X}}_i^1(\boldsymbol{y}) + \varepsilon^2\cdots. \tag{7.7.2}$$

One can obtain a formal approximation of the solutions of equation (7.7.2) by omitting the higher-order terms and integrating the system

$$\dot{\boldsymbol{z}} = \varepsilon\sum\nolimits_{i=1}^{m} \overline{\mathbf{X}}_i^1(\boldsymbol{z}), \quad \dot{\boldsymbol{\psi}} = \boldsymbol{\Omega}^0(\boldsymbol{z}).$$

To obtain in this way an asymptotic approximation of the solution of equation (7.5.1) we have to show that $\mathbf{u}^1$ is bounded. Then however, we have to know a priori that each Ω_i^0 is bounded away from zero (cf. equation (7.7.1)). The following simple example illustrates the difficulty. ♡

Example 7.7.2 (Arnol'd [7])). Consider the scalar equations

$$\begin{aligned}\dot{x} &= \varepsilon(1 - 2\cos(\phi)) \ , \ x(0) = x_0 \ ; \ x \in \mathbb{R},\\ \dot{\phi} &= x, \quad \phi(0) = \phi_0 \ ; \quad \phi \in S^1\end{aligned}$$

(written as a second-order equation for ϕ the system becomes the familiar looking equation $\ddot{\phi} + 2\varepsilon\cos(\phi) = \varepsilon$). The averaged equation (7.7.2) takes the form

$$\begin{aligned}\dot{y} &= \varepsilon + \varepsilon^2 \cdots ,\\ \dot{\phi} &= y + \varepsilon \cdots .\end{aligned}$$

We would like to approximate (x, ϕ) by

$$(x_0 + \varepsilon t, \phi_0 + x_0 t + \frac{1}{2}\varepsilon t^2).$$

The original equation has stationary solutions $(0, \pi/3)$ and $(0, 5\pi/3)$ so if we put for instance $(x_0, \phi_0) = (0, \pi/3)$, the error grows as $(\varepsilon t, \varepsilon t^2/2)$. Note that the averaged equations contain no singularities; it can be shown however that the higher-order terms do. In the following we shall discuss the approximate character of the formal solutions in the simple case that Ω^0 is bounded away from zero; this will be called the regular case. ◇

7.8 Systems with Slowly Varying Frequency in the Regular Case; the Einstein Pendulum

The following assumption will be a blanket assumption till the end of this chapter.

Assumption 7.8.1 *Suppose* $0 < m \leq \inf_{x \in D} |\Omega^0(\boldsymbol{x})| \leq \sup_{x \in D} |\Omega^0(\boldsymbol{x})| \leq M < \infty$ *where* m *and* M *are* ε*-independent constants.*

We formulate and prove the following lemma which provides a useful perturbation scheme for system (7.5.1).

Lemma 7.8.2. *Consider the equation with* C^1*-right-hand sides*

$$\begin{bmatrix}\dot{\boldsymbol{x}}\\ \dot{\phi}\end{bmatrix} = \begin{bmatrix}0\\ \Omega^0(\boldsymbol{x})\end{bmatrix} + \varepsilon\begin{bmatrix}\mathbf{X}^1(\boldsymbol{x}, \phi)\\ 0\end{bmatrix}, \quad \begin{bmatrix}\mathbf{x}\\ \phi\end{bmatrix}(0) = \begin{bmatrix}\boldsymbol{x}_0\\ \phi_0\end{bmatrix}, \quad \begin{array}{l}\boldsymbol{x} \in D \subset \mathbb{R}^n,\\ \phi \in S^1.\end{array} \tag{7.8.1}$$

We transform

$$\begin{bmatrix}\boldsymbol{x}\\ \phi\end{bmatrix} = \begin{bmatrix}\boldsymbol{y}\\ \psi\end{bmatrix} + \varepsilon\begin{bmatrix}\mathbf{u}^1(\boldsymbol{y}, \psi)\\ 0\end{bmatrix}, \tag{7.8.2}$$

with $(\mathbf{y}, \psi)$ *the solution of*

$$\begin{bmatrix} \dot{\boldsymbol{y}} \\ \dot{\psi} \end{bmatrix} = \begin{bmatrix} 0 \\ \Omega^0(\boldsymbol{y}) \end{bmatrix} + \varepsilon \begin{bmatrix} \overline{\mathbf{X}}^1(\boldsymbol{y}) \\ \Omega_\star^{[1]}(\boldsymbol{y}, \psi, \varepsilon) \end{bmatrix} + \varepsilon^2 \begin{bmatrix} \mathbf{X}_\star^{[2]}(\boldsymbol{y}, \psi, \varepsilon) \\ 0 \end{bmatrix}. \tag{7.8.3}$$

Here $\Omega_\star^{[1]}$ *and* $\mathbf{X}_\star^{[2]}$ *are to be constructed later on and* $\mathbf{y}(0), \psi(0)$ *are determined by (7.8.2). One defines*

$$\overline{\mathbf{X}}^1(\boldsymbol{y}) = \frac{1}{2\pi} \int_0^{2\pi} \mathbf{X}^1(\boldsymbol{y}, \varphi)\, \mathrm{d}\varphi$$

and

$$\mathbf{u}^1(\boldsymbol{y}, \phi) = \frac{1}{\Omega^0(\boldsymbol{y})} \int^{\phi} (\mathbf{X}^1(\boldsymbol{y}, \varphi) - \overline{\mathbf{X}}^1(\boldsymbol{y}))\, \mathrm{d}\varphi. \tag{7.8.4}$$

We choose the integration constant such that

$$\int_0^{2\pi} \mathbf{u}^1(\boldsymbol{y}, \varphi)\, \mathrm{d}\varphi = 0.$$

Then $\mathbf{u}^1, \Omega_\star^{[1]}$ *and* $\mathbf{X}_\star^{[2]}$ *are uniformly bounded.*

Proof Here $\mathbf{u}^1$ has been defined explicitly and is uniformly bounded because of the two-sided estimate for Ω^0 and the integrand in (7.8.4) having zero average. $\Omega_\star^{[1]}$ and $\mathbf{X}_\star^{[2]}$ have been defined implicitly and will now be determined, at least to zeroth order in ε. We differentiate the relations (7.8.2) and substitute the vector field (7.8.1). The ϕ-component

$$\dot{\phi} = \Omega^0(\boldsymbol{x})$$

is transformed to

$$\dot{\psi} = \Omega^0(\boldsymbol{y} + \varepsilon \mathbf{u}^1(\boldsymbol{y}, \psi)) = \Omega(\boldsymbol{y}) + \varepsilon \Omega_\star^{[1]}(\boldsymbol{y}, \psi, \varepsilon),$$

with

$$\varepsilon \Omega_\star^{[1]}(\boldsymbol{y}, \psi, \varepsilon) = \Omega^0(\boldsymbol{y} + \varepsilon \mathbf{u}^1(\boldsymbol{y}, \psi)) - \Omega^0(\boldsymbol{y}).$$

For $\varepsilon \downarrow 0$, $\Omega_\star^{[1]}$ approaches

$$\Omega_\star^1(\boldsymbol{y}, \psi) = \mathsf{D}_{\boldsymbol{y}} \Omega^0(\boldsymbol{y}) \cdot \mathbf{u}^1(\boldsymbol{y}, \psi).$$

With the implicit function theorem we establish the existence and uniform boundedness of $\Omega_\star^{[1]}$ for $\varepsilon \in (0, \varepsilon_0]$. For the $\boldsymbol{x}$-component we have the following relations:

$$\dot{\boldsymbol{x}} = \varepsilon \mathbf{X}^1(\boldsymbol{x}, \phi) = \varepsilon \mathbf{X}^1(\boldsymbol{y} + \varepsilon \mathbf{u}^1(\boldsymbol{y}, \psi), \psi)$$

and

$$
\begin{aligned}
\dot{\boldsymbol{x}} &= \dot{\boldsymbol{y}} + \varepsilon \frac{\partial \mathbf{u}^1}{\partial \psi}\dot{\psi} + \varepsilon \mathsf{D}\mathbf{u}^1 \cdot \dot{\boldsymbol{y}} \\
&= \varepsilon \overline{\mathbf{X}}^1(\boldsymbol{y}) + \varepsilon^2 \mathbf{X}_\star^{[2]}(\boldsymbol{y},\psi,\varepsilon) + \varepsilon \frac{\partial \mathbf{u}^1}{\partial \psi}(\Omega^0(\boldsymbol{y}) + \varepsilon \Omega_\star^{[1]}(\boldsymbol{y},\psi,\varepsilon)) \\
&\quad + \varepsilon \mathsf{D}\mathbf{u}^1 \cdot \left(\varepsilon \overline{\mathbf{X}}^1(\boldsymbol{y}) + \varepsilon^2 \mathbf{X}_\star^{[2]}(\boldsymbol{y},\psi,\varepsilon)\right) \\
&= \varepsilon \overline{\mathbf{X}}^1(\boldsymbol{y}) + \varepsilon^2 \mathbf{X}_\star^{[2]}(\boldsymbol{y},\psi,\varepsilon) + \frac{\varepsilon}{\Omega^0(\boldsymbol{y})}(\mathbf{X}^1(\boldsymbol{y},\psi) - \overline{\mathbf{X}}^1(\boldsymbol{y}))(\Omega^0(\boldsymbol{y}) \\
&\quad + \varepsilon \Omega_\star^{[1]}(\boldsymbol{y},\psi,\varepsilon)) + \varepsilon \mathsf{D}\mathbf{u}^1 \cdot \left(\varepsilon \overline{\mathbf{X}}^1(\boldsymbol{y}) + \varepsilon^2 \mathbf{X}_\star^{[2]}(\boldsymbol{y},\psi,\varepsilon)\right) \\
&= \varepsilon \overline{\mathbf{X}}^1(\boldsymbol{y}) + \varepsilon^2 \mathbf{X}_\star^{[2]}(\boldsymbol{y},\psi,\varepsilon) + \varepsilon \mathbf{X}^1(\boldsymbol{y},\psi) - \varepsilon \overline{\mathbf{X}}^1(\boldsymbol{y}) \\
&\quad + \frac{\varepsilon^2}{\Omega^0(\boldsymbol{y})}\Omega_\star^{[1]}(\boldsymbol{y},\psi,\varepsilon)\left(X(\boldsymbol{y},\psi,\varepsilon) - \overline{\mathbf{X}}^1(\boldsymbol{y})\right) \\
&\quad + \varepsilon \mathsf{D}\mathbf{u}^1 \cdot \left(\varepsilon X^o(\boldsymbol{y}) + \varepsilon^2 \mathbf{X}_\star^{[2]}(\boldsymbol{y},\psi,\varepsilon)\right) \\
&= \varepsilon \mathbf{X}^1(\boldsymbol{y},\psi) + \varepsilon^2 \mathbf{X}_\star^{[2]}(\boldsymbol{y},\psi,\varepsilon) + \varepsilon^2 \Omega_\star^{[1]}(\boldsymbol{y},\psi,\varepsilon)\frac{\partial \mathbf{u}^1}{\partial \psi}(\boldsymbol{y},\psi) \\
&\quad + \varepsilon \mathsf{D}\mathbf{u}^1 \cdot \left(\varepsilon \overline{\mathbf{X}}^1(\boldsymbol{y}) + \varepsilon^2 \mathbf{X}_\star^{[2]}(\boldsymbol{y},\psi,\varepsilon)\right).
\end{aligned}
$$

This gives the equation

$$
\varepsilon(I + \varepsilon \mathsf{D}\mathbf{u}^1)\mathbf{X}_\star^{[2]} = \mathbf{X}^1(\boldsymbol{y} + \varepsilon \mathbf{u}^1(\boldsymbol{y},\psi),\psi) - \mathbf{X}^1(\boldsymbol{y},\psi) - \varepsilon \Omega_\star^{[1]} \frac{\partial \mathbf{u}^1}{\partial \psi} - \varepsilon \mathsf{D}\mathbf{u}^1 \overline{\mathbf{X}}^1.
$$

In the limit $\varepsilon \downarrow 0$, we can solve this:

$$
\mathbf{X}_\star^2 = \mathsf{D}\mathbf{X}^1 \cdot \mathbf{u}^1 - \mathsf{D}\mathbf{u}^1 \cdot \overline{\mathbf{X}}^1 - \Omega_\star^1 \frac{\partial \mathbf{u}^1}{\partial \psi}.
$$

Using again the implicit function theorem we obtain the existence and uniform boundedness of $\mathbf{X}_\star^{[2]}$. □

Transformation (7.8.2) has produced (7.8.3); later on, in Chapter 13, we shall call this calculation a normalization process. We truncate (7.8.3) and we shall first prove the validity of the solution of the resulting equation as an asymptotic approximation to the solution of the nontruncated equation (7.8.3).

Lemma 7.8.3. *Consider (7.8.3) in Lemma 7.8.2 with the same conditions and solution $(\psi, \boldsymbol{y})$. Let $(\zeta, \mathbf{z})$ be the solution of*

$$
\begin{bmatrix} \dot{\boldsymbol{z}} \\ \dot{\zeta} \end{bmatrix} = \begin{bmatrix} 0 \\ \Omega^0(\boldsymbol{z}) \end{bmatrix} + \varepsilon \begin{bmatrix} \overline{\mathbf{X}}^1(\boldsymbol{z}) \\ 0 \end{bmatrix}, \quad \begin{bmatrix} \mathbf{z} \\ \zeta \end{bmatrix}(0) = \begin{bmatrix} z_0 \\ \zeta_0 \end{bmatrix}, \quad \boldsymbol{z} \in D.
$$

Remark that the initial values of both systems need not be the same. Then

$$
\| \mathbf{y} - \mathbf{z} \| \le (\| y_0 - z_0 \| + \varepsilon^2 t \| \mathbf{X}_\star^{[2]} \|) e^{\varepsilon \lambda_{\overline{\mathbf{X}}^1} t},
$$

where

$$\| \mathbf{X}_\star^{[2]} \| = \sup_{(\boldsymbol{y},\psi,\varepsilon)\in S^1\times D\times(0,\varepsilon_0]} |\mathbf{X}_\star^{[2]}(\boldsymbol{y},\psi,\varepsilon)|.$$

If $\boldsymbol{x}_0 = \boldsymbol{y}_0 + \mathcal{O}(\varepsilon)$ *this implies*

$$\boldsymbol{x}(t) = \boldsymbol{z}(t) + \mathcal{O}(\varepsilon)$$

on the time scale $1/\varepsilon$.

Proof The proof is standard. We write

$$\begin{aligned}\mathbf{y}(t) - \mathbf{z}(t) = \boldsymbol{y}_0 + \varepsilon \int_0^t \overline{\mathbf{X}}^1(\mathbf{y}(\tau))\,\mathrm{d}\tau \\ +\varepsilon^2 \int_0^t \mathbf{X}_\star^{[2]}(\mathbf{y}(\tau),\psi(\tau),\varepsilon)\,\mathrm{d}\tau - \boldsymbol{z}_0 - \varepsilon\int_0^t \overline{\mathbf{X}}^1(\mathbf{z}(\tau))\,\mathrm{d}\tau\end{aligned}$$

or

$$\| \mathbf{y}(t) - \mathbf{z}(t) \| \le \| \boldsymbol{y}_0 - \boldsymbol{z}_0 \| + \varepsilon \int_0^t \overline{\mathbf{X}}^1(\mathbf{y}(\tau)) - \overline{\mathbf{X}}^1(\mathbf{z}(\tau)) \|\,\mathrm{d}\tau + \varepsilon^2 \| \mathbf{X}_\star^{[2]} \|\, t.$$

Noting that $\| \overline{\mathbf{X}}^1(\boldsymbol{y}) - \overline{\mathbf{X}}^1(\boldsymbol{z}) \| \le \lambda_{\overline{\mathbf{X}}^1} \| \boldsymbol{y} - \boldsymbol{z} \|$ and applying the Gronwall Lemma 1.3.3 produces the desired result. □

From a combination of the two lemmas we obtain an averaging theorem:

Theorem 7.8.4. *Consider the equations with initial values*

$$\begin{bmatrix}\dot{\boldsymbol{x}}\\ \dot{\phi}\end{bmatrix} = \begin{bmatrix}0\\ \Omega^0(\boldsymbol{x})\end{bmatrix} + \varepsilon\begin{bmatrix}\mathbf{X}^1(\boldsymbol{x},\phi)\\ 0\end{bmatrix}, \quad \begin{bmatrix}\mathbf{x}\\ \phi\end{bmatrix}(0) = \begin{bmatrix}\boldsymbol{x}_0\\ \phi_0\end{bmatrix}, \quad \begin{matrix}\boldsymbol{x}\in D\subset\mathbb{R}^n\\ \phi\in S^1\end{matrix} \tag{7.8.5}$$

Let $(\mathbf{z},\zeta)$ *be the solution of*

$$\begin{aligned}\dot{\boldsymbol{z}} &= \varepsilon\overline{\mathbf{X}}^1(\boldsymbol{z}), \quad \mathbf{z}(0) = \boldsymbol{z}_0\ ,\ \boldsymbol{z}\in D,\\ \dot{\zeta} &= \Omega^0(\boldsymbol{z}), \quad \zeta(0) = \zeta_0,\end{aligned}$$

where

$$\overline{\mathbf{X}}^1(\cdot) = \frac{1}{2\pi}\int_0^{2\pi} \mathbf{X}^1(\cdot,\varphi)\,\mathrm{d}\varphi.$$

Then, if $\mathbf{z}(t)$ *remains in* $D^o \subset D$

$$\mathbf{x}(t) = \mathbf{z}(t) + \mathcal{O}(\varepsilon)$$

on the time scale $1/\varepsilon$. *Furthermore* $\phi(t) = \zeta(t) + \mathcal{O}(\varepsilon t e^{\varepsilon t})$.

Proof Transform equations (7.8.5) with Lemma 7.8.2, $(\boldsymbol{x},\phi)\mapsto(\boldsymbol{y},\psi)$ and apply Lemma 7.8.3. □

7.8.1 Einstein Pendulum

Consider the equation with slowly varying frequency

$$\ddot{x} + \omega^2(\varepsilon t)x = 0,$$

with initial conditions given. In Section 7.5 we obtained in this case the perturbation equations

$$\begin{aligned}
\dot{r} &= -\frac{\varepsilon}{\omega}\frac{d\omega}{d\tau}\, r\cos^2(\phi),\\
\dot{\tau} &= \varepsilon,\\
\dot{\phi} &= \omega + \frac{\varepsilon}{\omega}\frac{d\omega}{d\tau}\,\sin(\phi)\cos(\phi).
\end{aligned}$$

Averaging over ϕ we obtain the equations

$$\begin{aligned}
\dot{\overline{r}} &= -\frac{\varepsilon}{2\omega}\frac{d\omega}{d\tau}\,\overline{r},\\
\dot{\overline{\tau}} &= \varepsilon.
\end{aligned}$$

After integration we obtain

$$\overline{r}(t)\omega^{\frac{1}{2}}(\varepsilon t) = r_0\omega^{\frac{1}{2}}(0)$$

and $r(t) = \overline{r}(t) + \mathcal{O}(\varepsilon)$ on the time scale $1/\varepsilon$. In the original coordinates we may write

$$\omega(\varepsilon t)x^2 + \frac{\dot{x}^2}{\omega(\varepsilon t)} = \text{constant } + \mathcal{O}(\varepsilon)$$

on the time scale $1/\varepsilon$, which is a well-known adiabatic invariant of the system (the energy of the system changes linearly with the frequency). Note that in Section 3.3.1 we have treated these problems using a special time-like variable; the advantage here is that there is no need to find such special transformations.

7.9 Higher Order Approximation in the Regular Case

The estimates obtained in the preceding lemmas and in Theorem 7.8.4 can be improved. this is particularly useful in the case of the angle ϕ for which only an $\mathcal{O}(1)$ estimate has been obtained on the time scale $1/\varepsilon$. First we have a second-order version of Lemma 7.8.2:

Lemma 7.9.1. *Consider the equation*

$$\begin{bmatrix}\dot{\boldsymbol{x}}\\ \dot{\phi}\end{bmatrix} = \begin{bmatrix}0\\ \Omega^0(\boldsymbol{x})\end{bmatrix} + \varepsilon\begin{bmatrix}\mathbf{X}^1(\boldsymbol{x},\phi)\\ 0\end{bmatrix}, \quad \begin{bmatrix}\boldsymbol{x}\\ \phi\end{bmatrix}(0) = \begin{bmatrix}\boldsymbol{x}_0\\ \phi_0\end{bmatrix}, \quad \begin{array}{l}\boldsymbol{x}\in D\subset\mathbb{R}^n,\\ \phi\in S^1,\end{array} \tag{7.9.1}$$

and assume the conditions of Lemma 7.8.2. For the solutions $\boldsymbol{x}, \phi$ of equation (7.9.1) we can write

$$\begin{bmatrix} \mathbf{x}(t) \\ \phi(t) \end{bmatrix} = \begin{bmatrix} \mathbf{y}(t) \\ \psi(t) \end{bmatrix} + \varepsilon \begin{bmatrix} \mathbf{u}^1(\mathbf{y}(t), \psi(t)) \\ \mathrm{v}^1(\mathbf{y}(t), \psi(t)) \end{bmatrix} + \varepsilon^2 \begin{bmatrix} \mathbf{u}^2(\mathbf{y}(t), \psi(t)) \\ 0 \end{bmatrix}, \tag{7.9.2}$$

where $\mathbf{y}$ *and* ψ *are solutions of*

$$\begin{bmatrix} \dot{\boldsymbol{y}} \\ \dot{\psi} \end{bmatrix} = \begin{bmatrix} 0 \\ \Omega^0(\boldsymbol{y}) \end{bmatrix} + \varepsilon \begin{bmatrix} \overline{\mathbf{X}}^1(\boldsymbol{y}) \\ 0 \end{bmatrix} + \varepsilon^2 \begin{bmatrix} \overline{\mathbf{X}}^2_\star(\boldsymbol{y}) \\ \boldsymbol{\Omega}^{[2]}_\star(\boldsymbol{y}, \psi, \varepsilon) \end{bmatrix} + \varepsilon^3 \begin{bmatrix} \mathbf{X}^{[3]}_\star(\boldsymbol{y}, \psi, \varepsilon) \\ 0 \end{bmatrix}, \tag{7.9.3}$$

with $\mathbf{y}(0) = \boldsymbol{y}_0$ *and* $\psi(0) = \psi_0$*. Here* $\overline{\mathbf{X}}^2_\star$ *is defined by*

$$\overline{\mathbf{X}}^2_\star(\boldsymbol{y}) = \frac{1}{2\pi} \int_0^{2\pi} \left(\mathsf{D}\mathbf{X}^1 \cdot \mathbf{u}^1 - \frac{\mathsf{D}\Omega^0 \cdot \mathbf{u}^1}{\Omega^0} \mathbf{X}^1 \right) \mathrm{d}\varphi.$$

and $\mathbf{u}^1(\psi, \boldsymbol{y})$ *is defined as in Lemma 7.8.2, equation (7.8.4),* $\overline{\mathbf{X}}^1(\boldsymbol{y})$ *as in Lemma 7.8.2.*

Proof We present the formal computation and we shall not give all the technical details as in the proof of Lemma 7.8.2. From equations (7.9.1-7.9.2) we have

$$\dot{\phi} = \Omega^0(\boldsymbol{y} + \varepsilon \mathbf{u}^1 + \varepsilon^2 \mathbf{u}^2) = \Omega^0(\boldsymbol{y}) + \varepsilon \mathsf{D}\Omega^0 \cdot \mathbf{u}^1 + \mathcal{O}(\varepsilon^2).$$

On the other hand, differentiating the second part of transformation (7.9.2) yields

$$\dot{\phi} = \dot{\psi} + \varepsilon \frac{\partial \mathrm{v}^1}{\partial \psi} \dot{\psi} + \varepsilon \mathsf{D}\mathrm{v}^1 \cdot \dot{\boldsymbol{y}} = \Omega^0(\boldsymbol{y}) + \varepsilon \frac{\partial \mathrm{v}^1}{\partial \psi} \Omega^0 + \mathcal{O}(\varepsilon^2),$$

using (7.9.3). Comparing the two expressions for $\dot{\phi}$ we define

$$\mathrm{v}^1(\boldsymbol{y}, \psi) = \frac{1}{\Omega^0(\boldsymbol{y})} \int^{\psi} \mathsf{D}\Omega^0(\boldsymbol{y}) \cdot \mathbf{u}^1(\boldsymbol{y}, \varphi) \, \mathrm{d}\varphi.$$

In the same way we have from equation (7.9.1) with transformation (7.9.2)

$$\begin{aligned} \dot{\boldsymbol{x}} &= \varepsilon \mathbf{X}^1(\boldsymbol{y} + \varepsilon \mathbf{u}^1 + \varepsilon^2 \mathbf{u}^2, \psi + \varepsilon \mathrm{v}^1) \\ &= \varepsilon \mathbf{X}^1(\boldsymbol{y}, \psi) + \varepsilon^2 \mathrm{v}^1 \frac{\partial \mathbf{X}^1}{\partial \psi} + \varepsilon^2 \mathsf{D}\mathbf{X}^1 \cdot \mathbf{u}^1 + \mathcal{O}(\varepsilon^3). \end{aligned}$$

Differentiating the first part of transformation (7.9.2) yields

$$\dot{\boldsymbol{x}} = \dot{\boldsymbol{y}} + \varepsilon \dot{\psi} \frac{\partial \mathbf{u}^1}{\partial \psi} + \varepsilon \mathsf{D}\mathbf{u}^1 \cdot \dot{\boldsymbol{y}} + \varepsilon^2 \dot{\psi} \frac{\partial \mathbf{u}^2}{\partial \psi} + \varepsilon^2 \mathsf{D}\mathbf{u}^2 \cdot \dot{\boldsymbol{y}}$$

and, with (7.9.3),

$$= \varepsilon \overline{\mathbf{X}}^1 + \varepsilon^2 \mathbf{X}^2_\star + \varepsilon \frac{\partial \mathbf{u}^1}{\partial \psi} \Omega + \varepsilon^2 \mathsf{D}\mathbf{u}^1 \cdot \overline{\mathbf{X}}^1 + \varepsilon^2 \Omega^0 \frac{\partial \mathbf{u}^2}{\partial \psi} + \mathcal{O}(\varepsilon^3).$$

Comparing the two expressions for $\dot{\boldsymbol{x}}$ we have indeed

$$\mathbf{u}^1(\boldsymbol{y},\phi) = \frac{1}{\Omega^0(\boldsymbol{y})}\int^{\phi}(\mathbf{X}^1(\boldsymbol{y},\varphi) - \overline{\mathbf{X}}^1(\boldsymbol{y}))\,\mathrm{d}\varphi$$

and moreover we obtain

$$\mathbf{u}^2 = \frac{1}{\Omega^0}\int^{\phi}\left(\mathrm{v}^1\frac{\partial \mathbf{X}^1}{\partial\varphi} + \mathsf{D}\mathbf{X}^1\cdot\mathbf{u}^1 - \mathsf{D}\mathbf{u}^1\cdot\overline{\mathbf{X}}^1 - \overline{\mathbf{X}}^2_\star\right)\mathrm{d}\varphi.$$

There is no need to compute v^1 explicitly at this stage; we obtain, requiring $\mathbf{u}^2$ to have zero average

$$\begin{aligned}\overline{\mathbf{X}}^2_\star &= \frac{1}{2\pi}\int_0^{2\pi}\left(\mathrm{v}^1\frac{\partial \mathbf{X}^1}{\partial\varphi} + \mathsf{D}\mathbf{X}^1\mathbf{u}^1 - \mathsf{D}\mathbf{u}^1\cdot\overline{\mathbf{X}}^1\right)\mathrm{d}\varphi\\ &= \frac{1}{2\pi}\int_0^{2\pi}\left(-\frac{\partial \mathrm{v}^1}{\partial\varphi}\mathbf{X}^1 + \mathsf{D}\mathbf{X}^1\mathbf{u}^1\right)\mathrm{d}\varphi\\ &= \frac{1}{2\pi}\int_0^{2\pi}\left(\mathsf{D}\mathbf{X}^1 - \mathbf{X}^1\frac{\mathsf{D}\Omega^0}{\Omega^0}\right)\cdot\mathbf{u}^1\,\mathrm{d}\varphi,\end{aligned}$$

and we have proved the lemma. □

Following the same reasoning as in Section 7.8 we first approximate the solutions of equation (7.9.3).

Lemma 7.9.2. *Consider equation (7.9.3) with initial values*

$$\begin{aligned}\dot{\boldsymbol{y}} &= \varepsilon\overline{\mathbf{X}}^1(\boldsymbol{y}) + \varepsilon^2\overline{\mathbf{X}}^2_\star(\boldsymbol{y}) + \varepsilon^3\mathbf{X}^{[3]}_\star(\boldsymbol{y},\psi,\varepsilon), \quad \boldsymbol{y}(0) = \boldsymbol{y}_0\\ \dot{\psi} &= \Omega^0(\boldsymbol{y}) + \varepsilon^2\Omega^{[2]}_\star(\boldsymbol{y},\psi,\varepsilon), \quad \psi(0) = \psi_0.\end{aligned}$$

Let $(\boldsymbol{z},\zeta)$ be the solution of the truncated system

$$\begin{aligned}\dot{\boldsymbol{z}} &= \varepsilon\overline{\mathbf{X}}^1(\boldsymbol{z}) + \varepsilon^2\overline{\mathbf{X}}^2_\star(\boldsymbol{z}), \quad \boldsymbol{z}(0) = \boldsymbol{z}_0,\\ \dot{\zeta} &= \Omega^0(\boldsymbol{y}), \quad \zeta(0) = \zeta_0,\end{aligned}$$

then

$$\parallel \boldsymbol{z}(t) - \boldsymbol{y}(t) \parallel \le (\parallel \boldsymbol{z}_0 - \boldsymbol{y}_0 \parallel + \varepsilon^3 t \parallel \mathbf{X}^{[3]}_\star \parallel) e^{\varepsilon\lambda_{\overline{\mathbf{X}}^1 + \varepsilon\overline{\mathbf{X}}^2_\star} t}.$$

If $\boldsymbol{z}_0 = \boldsymbol{y}_0 + \mathcal{O}(\varepsilon^2)$, this implies that

$$\boldsymbol{y}(t) = \boldsymbol{z}(t) + \mathcal{O}(\varepsilon^2)$$

on the time scale $1/\varepsilon$. Furthermore

$$|\psi(t) - \zeta(t)| \le |\psi_0 - \zeta_0| + \parallel \mathsf{D}\Omega^0 \parallel \parallel \boldsymbol{z}(t) - \boldsymbol{y}(t) \parallel t + \varepsilon^2 t \parallel \Omega^{[1]}_\star \parallel.$$

If $\boldsymbol{z}_0 = \boldsymbol{y}_0 + \mathcal{O}(\varepsilon^2)$ and $\zeta_0 = \psi_0 + \mathcal{O}(\varepsilon)$ this produces an $\mathcal{O}(\varepsilon)$-estimate on the time scale $1/\varepsilon$ for the angular variable ψ.

Proof The proof is standard and runs along precisely the same lines as the proof of Lemma 7.8.3. □

We are now able to approximate the solutions of the original equation (7.9.1) to a higher-order precision.

Theorem 7.9.3 (Second-Order Averaging). *Consider equation (7.9.1)*

$$\begin{aligned} \dot{\boldsymbol{x}} &= \varepsilon \mathbf{X}^1(\boldsymbol{x}, \phi), \quad \boldsymbol{x}(0) = \boldsymbol{x}_0, \quad \boldsymbol{x} \in D \subset \mathbb{R}^n \\ \dot{\phi} &= \Omega^0(\boldsymbol{x}), \quad \phi(0) = \phi_0\ , \quad \phi \in S^1, \end{aligned}$$

and assume the conditions of Lemma 7.9.1. Following Lemma 7.9.2 we define $(\zeta, \boldsymbol{z})$ *as the solution of*

$$\begin{aligned} \dot{\boldsymbol{z}} &= \varepsilon \overline{\mathbf{X}}^1(\boldsymbol{z}) + \varepsilon^2 \overline{\mathbf{X}}^2_\star(\boldsymbol{z}), \quad \boldsymbol{z}(0) = \boldsymbol{x}_0 - \varepsilon \mathbf{u}^1(\boldsymbol{x}_0, \phi_0) \\ \dot{\zeta} &= \Omega^0(\boldsymbol{x}), \quad \zeta(0) = \phi_0. \end{aligned}$$

Then, on the time scale $1/\varepsilon$,

$$\begin{aligned} \boldsymbol{x}(t) &= \boldsymbol{z}(t) + \varepsilon \mathbf{u}^1(\boldsymbol{z}(t), \zeta(t)) + \mathcal{O}(\varepsilon^2) \\ \phi(t) &= \zeta(t) + \mathcal{O}(\varepsilon). \end{aligned}$$

Proof If $(\boldsymbol{y}, \psi)$ is defined as in Lemma 7.9.2, then

$$\boldsymbol{x}_0 = \boldsymbol{y}_0 + \varepsilon \mathbf{u}^1(\boldsymbol{y}_0, \psi_0) + \mathcal{O}(\varepsilon^2),$$

so

$$\begin{aligned} \boldsymbol{z}_0 - \boldsymbol{y}_0 &= \boldsymbol{x}_0 - \varepsilon \mathbf{u}^1(\boldsymbol{x}_0, \phi_0) - \boldsymbol{y}_0 \\ &= \varepsilon \mathbf{u}^1(\boldsymbol{y}_0, \psi_0) - \varepsilon \mathbf{u}^1(\boldsymbol{x}_0, \phi_0) + \mathcal{O}(\varepsilon^2) = \mathcal{O}(\varepsilon^2), \\ \zeta_0 - \psi_0 &= \mathcal{O}(\varepsilon). \end{aligned}$$

Applying Lemma 7.9.2 we have on the time scale $1/\varepsilon$

$$\begin{aligned} \boldsymbol{y}(t) &= \boldsymbol{z}(t) + \mathcal{O}(\varepsilon^2) \\ \psi(t) &= \zeta(t) + \mathcal{O}(\varepsilon). \end{aligned}$$

Since $\boldsymbol{x}(t) = \boldsymbol{y}(t) + \varepsilon \mathbf{u}^1(\boldsymbol{y}(t), \psi(t)) + \mathcal{O}(\varepsilon^2)$, we obtain the estimate of the theorem. Note that we also have

$$\boldsymbol{x}(t) = \boldsymbol{z}(t) + \mathcal{O}(\varepsilon)$$

on the time scale $1/\varepsilon$. □

7.10 Generalization of the Regular Case; an Example from Celestial Mechanics

In a number of problems in mechanics one encounters equations in which Ω^0 in equation (7.9.1) also depends on the angle ϕ. For instance

$$\dot{\boldsymbol{x}} = \varepsilon \mathbf{X}^1(\boldsymbol{x}, \phi),$$
$$\dot{\phi} = \Omega^0(\boldsymbol{x}, \phi).$$

We shall show here how to obtain a first-order approximation for $\boldsymbol{x}(t)$. The right-hand sides may also depend explicitly on t. The computations in that case become rather complicated and we shall not explore such problems here. Note however, that if the dependence on t is periodic we can interpret t as an angle θ while adding the equation $\dot{\theta} = 1$.

One might wonder, as in Section 7.8, if Ω^0 is bounded away from zero, why not divide the equation for $\dot{\boldsymbol{x}}$ by $\dot{\phi}$ and simply average over ϕ. The answer is first, that one would have have then an approximation in the time-like variable ϕ as independent variable; the estimate would still have to be extended to the behavior in t. More importantly, it is not clear how by such a simple approach one can generalize the procedure to multi-frequency systems ($\boldsymbol{\phi} = (\phi_1, \cdots, \phi_m)$) and to cases where the right-hand side depends on t. In the following we shall not repeat all the technical details of Section 7.8 but we shall try to convey that the general ideas of Section 7.8 apply in this case.

Lemma 7.10.1. *Consider the equation*

$$\begin{bmatrix} \dot{\boldsymbol{x}} \\ \dot{\phi} \end{bmatrix} = \begin{bmatrix} 0 \\ \Omega^0(\boldsymbol{x}, \phi) \end{bmatrix} + \varepsilon \begin{bmatrix} \mathbf{X}^1(\boldsymbol{x}, \phi) \\ 0 \end{bmatrix}, \quad \begin{bmatrix} \mathbf{x} \\ \phi \end{bmatrix}(0) = \begin{bmatrix} \boldsymbol{x}_0 \\ \phi_0 \end{bmatrix}, \quad \begin{matrix} \boldsymbol{x} \in D \subset \mathbb{R}^n, \\ \phi \in S^1. \end{matrix} \tag{7.10.1}$$

Transform

$$\begin{bmatrix} \boldsymbol{x} \\ \phi \end{bmatrix} = \begin{bmatrix} \boldsymbol{y} \\ \psi \end{bmatrix} + \varepsilon \begin{bmatrix} \mathbf{u}^1(\boldsymbol{y}, \psi) \\ 0 \end{bmatrix}. \tag{7.10.2}$$

Let $(\mathbf{y}, \psi)$ *be the solution of*

$$\begin{bmatrix} \dot{\boldsymbol{y}} \\ \dot{\psi} \end{bmatrix} = \begin{bmatrix} 0 \\ \Omega^0(\boldsymbol{y}, \psi) \end{bmatrix} + \varepsilon \begin{bmatrix} \overline{\mathbf{X}}^1(\boldsymbol{y}) \\ \Omega^{[1]}(\boldsymbol{y}, \psi, \varepsilon) \end{bmatrix} + \varepsilon^2 \begin{bmatrix} \mathbf{X}_\star^{[2]}(\boldsymbol{y}, \psi, \varepsilon) \\ 0 \end{bmatrix}. \tag{7.10.3}$$

Define

$$\overline{\mathbf{X}}^1(\boldsymbol{y}) = \frac{\frac{1}{2\pi}\int_0^{2\pi} \frac{\mathbf{X}^1(\boldsymbol{y},\varphi)}{\Omega^0(\boldsymbol{y},\varphi)}\, \mathrm{d}\varphi}{\frac{1}{2\pi}\int_0^{2\pi} \frac{1}{\Omega^0(\boldsymbol{y},\varphi)}\, \mathrm{d}\varphi} \tag{7.10.4}$$

and

$$\mathbf{u}^1(\phi, \boldsymbol{y}) = \int^{\phi} \frac{1}{\Omega^0(\varphi, \boldsymbol{y})} \left(\mathbf{X}^1(\varphi, \boldsymbol{y}) - \overline{\mathbf{X}}^1(\boldsymbol{y})\right) \mathrm{d}\varphi, \tag{7.10.5}$$

where the integration constant is such that

$$\frac{1}{2\pi}\int_0^{2\pi} \mathbf{u}^1(\varphi, \boldsymbol{y})\, \mathrm{d}\varphi = 0;$$

then $\mathbf{u}^1$, $\Omega^{[1]}$ *and* $\mathbf{X}_\star^{[2]}$ *are uniformly bounded.*

Proof The transformation generator $\mathbf{u}^1$ has been defined explicitly by equation (7.10.5) and is uniformly bounded because of the two-sided estimate for Ω^0 and the integrand in (7.10.5) having zero average. Differentiation of the relation between $\boldsymbol{x}$ and $\boldsymbol{y}$ and substitution of the vector field in (7.10.3) produces

$$\begin{aligned}\dot{\boldsymbol{y}} &+ \varepsilon\Omega^0(\boldsymbol{y}+\varepsilon\mathbf{u}^1,\phi)\frac{\partial\mathbf{u}^1}{\partial\phi}(\boldsymbol{y}+\varepsilon\mathbf{u}^1,\phi)+\varepsilon\mathsf{D}\mathbf{u}^1(\boldsymbol{y}+\varepsilon\mathbf{u}^1,\phi)\cdot\dot{\boldsymbol{y}}\\ &= \varepsilon\mathbf{X}^1(\boldsymbol{y}+\varepsilon\mathbf{u}^1,\phi).\end{aligned}$$

Expanding with respect to ε and using (7.10.4-7.10.5) yields

$$\dot{\boldsymbol{y}} = \varepsilon\overline{\mathbf{X}}^1(\boldsymbol{y}) + \varepsilon^2\mathbf{X}_\star^{[2]}(\boldsymbol{y},\psi,\varepsilon).$$

In the same way

$$\dot{\psi} = \Omega^0(\boldsymbol{y}+\varepsilon\mathbf{u}^1,\psi) = \Omega^0(\boldsymbol{y},\psi) + \varepsilon\Omega_\star^{[1]}(\boldsymbol{y},\psi,\varepsilon).$$

The existence and uniform boundedness of $\Omega_\star^{[1]}$ and $\mathbf{X}_\star^{[2]}$ re established as in Lemma 7.8.2. □

We now formulate an analogous version of Lemma 7.8.3 for equation (7.10.3).

Lemma 7.10.2. *Consider equation (7.10.3) with initial conditions*

$$\begin{bmatrix}\dot{\boldsymbol{y}}\\ \dot{\psi}\end{bmatrix} = \begin{bmatrix}0\\ \Omega^0(\boldsymbol{y},\psi)\end{bmatrix} + \varepsilon\begin{bmatrix}\overline{\mathbf{X}}^1(\boldsymbol{y})\\ \Omega_\star^{[1]}(\boldsymbol{y},\psi,\varepsilon)\end{bmatrix} + \varepsilon^2\begin{bmatrix}\mathbf{X}_\star^{[2]}(\boldsymbol{y},\psi,\varepsilon)\\ 0\end{bmatrix}, \quad \begin{bmatrix}\mathbf{y}\\ \psi\end{bmatrix}(0) = \begin{bmatrix}\boldsymbol{y}_0\\ \psi_0\end{bmatrix}.$$

Let $(\mathbf{z},\zeta)$ *be the solution of the truncated system*

$$\begin{bmatrix}\dot{\boldsymbol{z}}\\ \dot{\zeta}\end{bmatrix} = \begin{bmatrix}0\\ \Omega^0(\boldsymbol{z},\zeta)\end{bmatrix} + \varepsilon\begin{bmatrix}\overline{\mathbf{X}}^1(\boldsymbol{z})\\ 0\end{bmatrix}, \quad \begin{bmatrix}\mathbf{z}\\ \zeta\end{bmatrix}(0) = \begin{bmatrix}\boldsymbol{z}_0\\ \zeta_0\end{bmatrix}. \tag{7.10.6}$$

Then

$$\| \boldsymbol{y}(t) - \boldsymbol{z}(t) \| \le (\| \boldsymbol{y}_0 - \boldsymbol{z}_0 \| + \varepsilon^2 t \| \mathbf{X}_\star^{[2]} \|) e^{\lambda_{\overline{\mathbf{X}}^1}\varepsilon t}.$$

If $\boldsymbol{z}_0 = \boldsymbol{y}_0 + \mathcal{O}(\varepsilon)$ *this implies*

$$\boldsymbol{y}(t) = \boldsymbol{z}(t) + \mathcal{O}(\varepsilon)$$

on the time scale $1/\varepsilon$. *If, moreover,* $\overline{\mathbf{X}}^1 = 0$, *one has* $\lambda_{\overline{\mathbf{X}}^1} = 0$ *and*

$$\| \boldsymbol{y}(t) - \boldsymbol{z}(t) \| \le \| \boldsymbol{y}_0 - \boldsymbol{z}_0 \| + \varepsilon^2 t \| \mathbf{X}_\star^{[2]} \|, \tag{7.10.7}$$

which implies the possibility of extension of the time scale of validity.

Remark 7.10.3. On estimating $|\psi(t) - \zeta(t)|$ one obtains an $\mathcal{O}(1)$ estimate on the time scale 1, which result is even worse than the one in Lemma 7.8.3. However, though the error grows faster in this case, one should realize that both ψ and ζ are in S^1 so that the error never exceeds $\mathcal{O}(1)$. ♡

Proof The proof runs along the same lines as for Lemma 7.8.3. We note that if $\overline{\mathbf{X}}^1 = 0$ we put

$$\boldsymbol{y}(t) - \boldsymbol{z}(t) = \boldsymbol{y}_0 - \boldsymbol{z}_0 + \varepsilon^2 \int_0^t \mathbf{X}_\star^{[2]}(\psi(\boldsymbol{y}(\tau), \tau, \varepsilon)\, \mathrm{d}\tau,$$

which directly produces (7.10.7). □

Apart from the expressions (7.10.4-7.10.5) no new results have been obtained thus far. However we had to formulate Lemmas 7.10.1 and 7.10.2 to obtain the following theorem.

Theorem 7.10.4. *Consider the equations with initial values*

$$\begin{bmatrix} \dot{\boldsymbol{y}} \\ \dot{\phi} \end{bmatrix} = \begin{bmatrix} 0 \\ \Omega^0(\boldsymbol{y}, \phi) \end{bmatrix} + \varepsilon \begin{bmatrix} \overline{\mathbf{X}}^1(\boldsymbol{y}) \\ \Omega_\star^{[1]}(\boldsymbol{y}, \phi, \varepsilon) \end{bmatrix} + \varepsilon^2 \begin{bmatrix} \mathbf{X}_\star^{[2]}(\boldsymbol{y}, \phi, \varepsilon) \\ 0 \end{bmatrix}, \quad \begin{bmatrix} \mathbf{y} \\ \phi \end{bmatrix}(0) = \begin{bmatrix} \boldsymbol{y}_0 \\ \phi_0 \end{bmatrix}, \tag{7.10.8}$$

with $\boldsymbol{y} \in D \subset \mathbb{R}^n$ *and* $\phi \in S^1$. *Let* $(\mathbf{z}, \zeta)$ *be the solution of the truncated system*

$$\begin{bmatrix} \dot{\boldsymbol{z}} \\ \dot{\zeta} \end{bmatrix} = \begin{bmatrix} 0 \\ \Omega^0(\boldsymbol{z}, \zeta) \end{bmatrix} + \varepsilon \begin{bmatrix} \overline{\mathbf{X}}^1(\boldsymbol{z}) \\ 0 \end{bmatrix}, \quad \begin{bmatrix} \mathbf{z} \\ \zeta \end{bmatrix}(0) = \begin{bmatrix} \boldsymbol{z}_0 \\ \zeta_0 \end{bmatrix},$$

where

$$\overline{\mathbf{X}}^1(\boldsymbol{x}) = \frac{\int_0^{2\pi} \frac{\mathbf{X}^1(\boldsymbol{x},\varphi)}{\Omega^0(\boldsymbol{x},\varphi)}\, \mathrm{d}\varphi}{\int_0^{2\pi} \frac{1}{\Omega^0(\boldsymbol{x},\varphi)}\, \mathrm{d}\varphi}.$$

Then if $\mathbf{z}(t)$ *remains in* D^o

$$\mathbf{x}(t) = \mathbf{z}(t) + \mathcal{O}(\varepsilon)$$

on the time scale $1/\varepsilon$. *If* $\overline{\mathbf{X}}^1 = 0$, $\mathbf{z}(t) = \boldsymbol{z}_0$ *and we have, if we put* $\boldsymbol{z}_0 = \boldsymbol{x}_0$,

$$\mathbf{x}(t) = \mathbf{x}_0 + \mathcal{O}(\varepsilon) + \mathcal{O}(\varepsilon^2 t).$$

Proof Apply Lemmas 7.10.1 and 7.10.2. □

To illustrate the preceding theory we shall discuss an example from celestial mechanics. The equations contain time t explicitly but this causes no complications as this is in the form of slow time εt.

7.10.1 Two-Body Problem with Variable Mass

Consider the Newtonian two-body problem in which the total mass m decreases monotonically and slowly with time. If the loss of mass is isotropic and is removed instantly from the system, the equation of motion in polar coordinates θ, r is

$$\ddot{r} = -\frac{Gm}{r^2} + \frac{c^2}{r^3},$$

with angular momentum integral

$$r^2\dot{\theta} = c.$$

G is the gravitational constant. To express the slow variation with time we put $m = m(\tau)$ with $\tau = \varepsilon t$. Hadjidemetriou [117] derived the perturbation equations for the orbital elements e (eccentricity), E (eccentric anomaly, a phase angle) and ω (angle indicating the direction of the line of apsides); an alternative derivation has been given by Verhulst [274]. We have

$$\frac{de}{dt} = -\varepsilon\frac{(1-e^2)\cos(E)}{1-e\cos(E)}\frac{1}{m}\frac{dm}{d\tau}, \tag{7.10.9a}$$

$$\frac{d\omega}{dt} = -\varepsilon\frac{(1-e^2)\sin(E)}{e(1-e\cos(E))}\frac{1}{m}\frac{dm}{d\tau}, \tag{7.10.9b}$$

$$\frac{d\tau}{dt} = \varepsilon, \tag{7.10.9c}$$

$$\frac{dE}{dt} = \frac{(1-e^2)^{\frac{3}{2}}}{1-e\cos(E)}\frac{G^2}{c^3}m^2 + \varepsilon\frac{\sin(E)}{e(1-e\cos(E))}\frac{1}{m}\frac{dm}{d\tau}. \tag{7.10.9d}$$

Here E plays the part of the angle ϕ in the standard system (7.10.1); note that here $n = 3$. To apply the preceding theory the first term on the right-hand side of equation (7.10.9d) must be bounded away from zero. This means that for perturbed elliptic orbits we have the restriction $0 < \alpha < e < \beta < 1$ with α, β independent of ε. Then we can apply Theorem 7.10.4 with $\boldsymbol{x} = (e, \omega, \tau)$; in fact (7.10.9d) is somewhat more complicated than the equation for ϕ in Theorem 7.10.4 but this does not affect the first-order computation. We obtain

$$\int_0^{2\pi} \frac{X(\boldsymbol{x}, \varphi)}{\Omega(\boldsymbol{x}, \varphi)}\, d\varphi = 0$$

for the first two equations and $\dot{\tau} = \varepsilon$ as it should be. So the eccentricity e and the position of the line of apsides ω are constant with error $\mathcal{O}(\varepsilon)$ on the time scale $1/\varepsilon$. In other words: we have proved that the quantities e and ω are adiabatic invariants for these perturbed elliptic orbits if we exclude the nearly-circular and nearly-parabolic cases. To obtain nontrivial behavior one has to calculate higher-order approximations in ε or first-order approximations on a longer time scale or one has to study the excluded domains in e: $[0, \alpha]$ and $[\beta, 1]$. This work has been carried out and for further details we refer the reader to [274]; see also Appendix D.

8

Passage Through Resonance

8.1 Introduction

In Chapter 7 we met a difficulty while applying straightforward averaging techniques to the problem at hand. We studied the case where this difficulty could not happen, that is, $\Omega^0(\boldsymbol{x})$ does not vanish, calling this the regular case. We now return to the problem where Ω^0 can have zeros or can be small. We cannot present a complete theory as such a theory is not available, so we rather aim at introducing the reader to the relevant concepts. This may serve as an introduction to the literature. In this context we mention [176, 125]; for the literature on passage of separatrices see [281]. To be more concrete, we will study the following equations.

$$\begin{bmatrix}\dot{\boldsymbol{x}}\\ \dot{\phi}\end{bmatrix} = \begin{bmatrix}0\\ \Omega^0(\boldsymbol{x})\end{bmatrix} + \varepsilon\begin{bmatrix}\mathbf{X}^1(\boldsymbol{x},\phi)\\ 0\end{bmatrix}, \quad \begin{bmatrix}\mathbf{x}\\ \phi\end{bmatrix}(0) = \begin{bmatrix}\boldsymbol{x}_0\\ \phi_0\end{bmatrix}, \quad \begin{matrix}\boldsymbol{x}\in D\subset\mathbb{R}^n,\\ \phi\in S^1.\end{matrix} \tag{8.1.1}$$

If we try to average this equation in the sense of Section 7.8, our averaging transformation becomes singular at the zeros of Ω^0. If Ω^0 is (near) zero, we say that the system is in resonance. This terminology derives from the fact that in many applications the angle ϕ is in reality the difference between two angles: to say that $\Omega^0 \approx 0$ is equivalent to saying that the frequencies of the two angles are about equal, or that they are in resonance. We shall meet this point of view again in Chapter 10. Well, this vanishing of Ω^0 is certainly a problem, but, as is so often the case, it has a local character. And if the problem is local, we can use this information to simplify our equations by Taylor expansion. To put it more formally, we define the **resonance manifold**M as follows.

$$\mathsf{M} = \{(\boldsymbol{x},\phi)\in D\times S^1 \subset \mathbb{R}^n\times S^1 | \Omega^0(\boldsymbol{x}) = 0\}.$$

Here N is a manifold only in the very original sense of the word, that is of the solution set of an equation. N is in general *not* invariant under the flow of the differential equation; in general this flow is transversal to N in a sense which will be made clear in the sequel.

To study the behavior of the equations locally we need local variables, a concept originating from boundary layer theory; for an introduction to these concepts see [82].

8.2 The Inner Expansion

Assume that $\Omega^0(\mathbf{0}) = 0$. A local variable $\boldsymbol{\xi}$ is obtained by scaling $\boldsymbol{x}$ near $\mathbf{0}$.

$$\boldsymbol{x} = \delta(\varepsilon)\boldsymbol{\xi},$$

where $\delta(\varepsilon)$ is an order function with $\lim_{\varepsilon\downarrow 0}\delta(\varepsilon) = 0$ and $\boldsymbol{\xi}$ is supposed to describe the **inner region** or **boundary layer**. In the local variable the equations read

$$\begin{aligned}
\delta(\varepsilon)\dot{\boldsymbol{\xi}} &= \varepsilon\mathbf{X}^1(\mathbf{0},\phi) + \mathcal{O}(\varepsilon\delta) \\
\dot{\phi} &= \Omega^0(\delta(\varepsilon)\boldsymbol{\xi}) = \Omega^0(\mathbf{0}) + \delta(\varepsilon)\mathsf{D}\Omega^0(\mathbf{0})\cdot\boldsymbol{\xi} + \mathcal{O}(\delta^2) \\
&= \delta(\varepsilon)\mathsf{D}\Omega^0(\mathbf{0})\cdot\boldsymbol{\xi} + \mathcal{O}(\delta^2).
\end{aligned}$$

Truncating these equations we obtain the **inner vector field** on the right-hand sides:

$$\dot{\boldsymbol{\xi}} = \delta^{-1}(\varepsilon)\varepsilon\mathbf{X}^1(\mathbf{0},\phi), \quad \dot{\phi} = \delta(\varepsilon)\mathsf{D}\Omega^0(\mathbf{0})\cdot\boldsymbol{\xi}.$$

Solutions of this last set of equations are called **formal inner expansions** or **formal local expansions** (If the asymptotic validity has been shown we leave out the *formal*). The corresponding second-order equation is

$$\ddot{\phi} = \varepsilon\mathsf{D}\Omega^0(\mathbf{0})\cdot\mathbf{X}^1(\mathbf{0},\phi)$$

so that the natural time scale of the inner equation is $1/\sqrt{\varepsilon}$. A consistent choice for $\delta(\varepsilon)$ might then be $\delta(\varepsilon) = \sqrt{\varepsilon}$, but we shall see that there is some need to take the size of the boundary layer around N somewhat larger.

Example 8.2.1. Consider the two-dimensional system determined by

$$\Omega^0(x) = x, \quad \mathrm{X}^1(x,\phi) = \alpha(x) - \beta(x)\sin(\phi).$$

Expanding near the resonance manifold $x = 0$ we have the inner equation

$$\ddot{\phi} + \varepsilon\beta(0)\sin(\phi) = \varepsilon\alpha(0).$$

For $\beta(0) > 0$ and $|\alpha(0)/\beta(0)| < 1$ the phase flow is sketched in Figure 8.1. Note that N corresponds to $\phi = 0$. If the solution enters the boundary layer near the stable manifold of the saddle point it might stay for a long while in the inner region, in fact much longer than the natural time scale. In such cases we will be in trouble as the theory discussed so far does not extend beyond the **natural time scales**. ♢

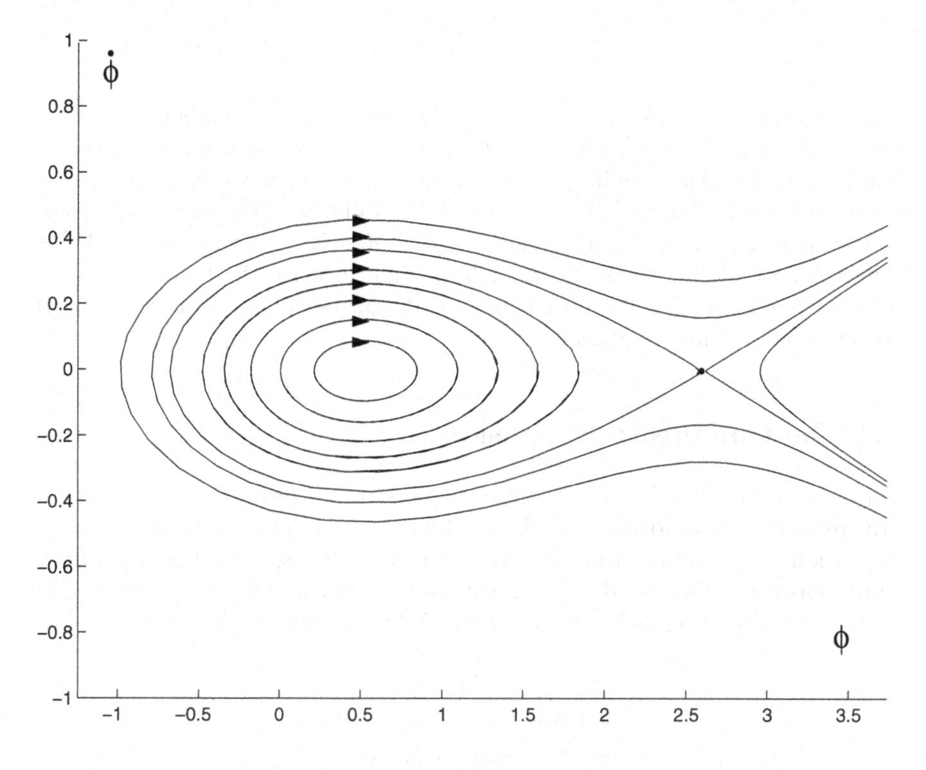

Fig. 8.1: Phase flow of $\ddot{\phi} + \varepsilon\beta(0)\sin(\phi) = \varepsilon\alpha(0)$

8.3 The Outer Expansion

Away from the resonance manifold and its neighborhood where $\Omega^0(\boldsymbol{x}) \approx 0$ we have the **outer region** and corresponding **outer expansion** of the solution. In fact we have already seen this outer expansion as it can be taken to be the solution of the averaged equation in the sense of Section 7.8. The averaging process provides us with valid answers if one keeps fair distance from the resonance manifold. This explains the term outer, since one is always looking out from the singularity. We shall see however, that if we try to extend the validity of the outer expansion in the direction of the resonance manifold we need to make the following assumption, which will be a blanket assumption for the outer domain.

Assumption 8.3.1

$$\frac{\sqrt{\varepsilon}}{\Omega^0(\boldsymbol{x})} = o(1). \tag{8.3.1}$$

This means that if $d(\boldsymbol{x}, \mathsf{N})$ is the distance of $\boldsymbol{x}$ to the resonance manifold N *we have*

$$\frac{\varepsilon}{d^2(\boldsymbol{x},\mathsf{N})} = o(1).$$

So, if we take $\delta(\varepsilon) = \sqrt{\varepsilon}$, as suggested in the discussion of the inner expansion, we cannot extend the averaging results to the boundary of the inner domain. Thus we shall consider an inner region of size δ, somewhat larger than $\sqrt{\varepsilon}$, but this poses the problem of how to extend the validity of the inner expansion to the time scale $1/\delta(\varepsilon)$ in the region $c_1\sqrt{\varepsilon} \leq d(\boldsymbol{x},\mathsf{N}) \leq c_2\delta(\varepsilon)$. We need this longer time scale for the solution to have time to leave the boundary layer. This problem, which is by no means trivial, can be solved using the special structure of the inner equations.

8.4 The Composite Expansion

Once we obtain the inner and outer expansion we proceed to construct a **composite expansion**. To do this we add the inner expansion to the outer expansion while subtracting the common part, the so called **inner-outer expansion**. For the foundations of this process of composite expansions and matching we refer again to [82, Chapter 3]. In formula this becomes

$$\boldsymbol{x}_C = \boldsymbol{x}_I + \boldsymbol{x}_O - \boldsymbol{x}_{IO},$$

$\boldsymbol{x}_C$: the composite expansion,

$\boldsymbol{x}_I$: the inner expansion,

$\boldsymbol{x}_O$: the outer expansion,

$\boldsymbol{x}_{IO}$: the inner-outer expansion.

In the inner region, $\boldsymbol{x}_C$ has to look like $\boldsymbol{x}_I$, so $\boldsymbol{x}_{IO}$ should look like $\boldsymbol{x}_O$ to cancel the outer expansion; this means that $\boldsymbol{x}_{IO}$ should be the inner expansion of the outer expansion, i.e. the outer expansion reexpanded in the inner variables. Analogous reasoning applies to the composite expansion in the outer region. This type of expansion procedure can be carried out for vector fields. We shall define the **inner-outer vector field** as the averaged inner expansion or, equivalently, the averaged equation expanded around N. That the averaged equation can be expanded at all near N may surprise us but it turns out to be possible at least to first order. The second-order averaged equation might be singular at N. The solution of the inner-outer vector field is then the inner-outer expansion. From the validity of the averaging method and the expansion method, which we shall prove, follows the validity of the composite expansion method, that is we can write the original solution $\boldsymbol{x}$ as

$$\boldsymbol{x} = \boldsymbol{x}_C + \mathcal{O}(\eta(\varepsilon)) \quad \eta = o(1).$$

If the solution enters the inner domain and leaves it at the other side, then we speak of **passage through resonance**. We can only describe this asymptotically if the inner vector field is transversal to N. Otherwise the asymptotic solution, i.e. the composite expansion, cannot pass through the resonance, even though the real solution might be able to do this (be it on a much longer time scale than the natural time scale of the inner expansion).

8.5 Remarks on Higher-Dimensional Problems

8.5.1 Introduction

In our discussion thus far we have scaled $\boldsymbol{x}$ uniformly in all directions, but if the dimension of the spatial variable n is larger than one, one could choose a minimal number of coordinates transversal to N (measuring the distance to N) and split $\mathbb{R}^n$ accordingly. For instance if $n = 2$ and

$$\dot{\phi} = x_1(1 + {x_2}^2),$$

the dimension of the system is three. N is determined by $x_1 = 0$ and x_1 can be used as a transversal coordinate; x_2 plays no essential part. The remaining coordinates, such as x_2 in this example, remain unscaled in the inner region and have variations of $\mathcal{O}(\sqrt{\varepsilon})$ on the time scale $1/\sqrt{\varepsilon}$. They play no part in the asymptotic analysis of the resonance to first order and we can concentrate on the lower dimensional problem, in this case with dimension 2.

8.5.2 The Case of More Than One Angle

If $\boldsymbol{\phi} \in \mathbb{T}^m, m > 1$, the situation is complicated and the theory is far from complete. We outline the problems for the case $m = 2$ and n spatial variables:

$$\begin{bmatrix} \dot{\boldsymbol{x}} \\ \dot{\boldsymbol{\phi}} \end{bmatrix} = \begin{bmatrix} 0 \\ \boldsymbol{\Omega}^0(\boldsymbol{x}) \end{bmatrix} + \varepsilon \begin{bmatrix} \mathbf{X}^1(\boldsymbol{x}, \boldsymbol{\phi}) \\ 0 \end{bmatrix}, \quad \begin{array}{l} \boldsymbol{x} \in \mathbb{R}^n, \\ \boldsymbol{\phi} \in \mathbb{T}^2. \end{array}$$

The right-hand side of the equation for $\boldsymbol{x}$ is expanded in a complex Fourier series; we have

$$\dot{\boldsymbol{x}} = \varepsilon \sum_{k,l=-\infty}^{\infty} \mathbf{c}_{kl}^1(\boldsymbol{x}) e^{i(k\phi_1 + l\phi_2)}.$$

Averaging over the angles *outside* the resonances

$$k\Omega_1^0 + l\Omega_2^0 = 0$$

leads to the averaged equation

$$\dot{\boldsymbol{y}} = \varepsilon \mathbf{c}_{00}^1(\boldsymbol{y}).$$

A resonance arises for instance if

$$k\Omega_1^0(\boldsymbol{x}) + l\Omega_2^0(\boldsymbol{x}) = 0, \quad k, l \in \mathbb{Z}. \tag{8.5.1}$$

In the part of $\mathbb{R}^n$ where (8.5.1) is satisfied it is natural to introduce two independent linear combination angles for ϕ_1 and ϕ_2. If we take $\psi = k\phi_1 + l\phi_2$ as one of them, the equation for ψ is varying slowly in a neighborhood of the domain where (8.5.1) holds. Here we cannot average over ψ.

This resonance condition has as a consequence that, in principle, an infinite number of resonance domains can be found. In each of these domains we have to localize around the resonance manifold given by (8.5.1). Locally we construct an expansion with respect to the resonant variable $\phi_{kl} = k\phi_1 + l\phi_2$ while averaging over all the other combinations. Note that we have assumed that the resonance domains are disjunct. So, locally we have again a problem with one angle ϕ_{kl} and what remains is the problem of obtaining a global approximation to the solution. These ideas have already been discussed in Section 7.4. Before treating some simple examples we mention a case which occurs quite often in practice. Suppose we have *one* angle ϕ and a perturbation which is also a periodic function of t, period 1.

$$\dot{\boldsymbol{x}} = \varepsilon \mathbf{X}^{[1]}(\boldsymbol{x}, \phi, t, \varepsilon)$$
$$\dot{\phi} = \Omega^{[0]}(\boldsymbol{x}, \varepsilon).$$

It is natural to introduce now two angles $\phi_1 = \phi$, $\phi_2 = t$ and adding the equation

$$\dot{\phi}_2 = 1.$$

The resonance condition (8.5.1) becomes in this case

$$k\Omega(x) + l = 0.$$

So each rational value assumed by $\Omega_1^0(\boldsymbol{x})$ corresponds to a resonance domain provided that the resonance is engaged (cf. Section 7.4), that is, the corresponding k, l-coefficient arises in the Fourier expansion of $\mathbf{X}^1$. If $\phi \in \mathbb{T}^m, m > 1$, the analysis and estimates are much more difficult. It is surprising that we can still describe the flow in the resonance manifold to first order.

We conclude this discussion with two simple examples given by Arnol′d [7]; in both cases $n = 2$, $m = 2$.

8.5.3 Example of Resonance Locking

The equations are

$$\begin{aligned}
\dot{x}_1 &= \varepsilon, \\
\dot{x}_2 &= \varepsilon \cos(\phi_1 - \phi_2), \\
\dot{\phi}_1 &= x_1, \\
\dot{\phi}_2 &= x_2.
\end{aligned}$$

The resonance condition (8.5.1) reduces to the case $k = 1$, $l = -1$:

$$x_1 = x_2.$$

There are two cases to consider: First suppose we start in the resonance manifold, so $\mathrm{x}_1(0) = \mathrm{x}_2(0)$, and let $\phi_1(0) = \phi_2(0)$. Then the solutions are easily seen to be

$$\begin{aligned}
\mathrm{x}_1(t) &= \mathrm{x}_2(t) = \mathrm{x}_1(0) + \varepsilon t,\\
\phi_1(t) &= \phi_1(0) + \mathrm{x}_1(0)t + \frac{1}{2}\varepsilon t^2,\\
\phi_2(t) &= \phi_1(t).
\end{aligned}$$

The solutions are locked into resonance, due to the special choice of initial conditions. The second case arises when we start outside the resonance domain: $\mathrm{x}_1(0) - \mathrm{x}_2(0) = a \neq 0$ with a independent of ε. Averaging over $\phi_1 - \phi_2$ produces the equations

$$\begin{aligned}
\dot{y}_1 &= \varepsilon,\\
\dot{y}_2 &= 0,\\
\dot{\psi}_1 &= y_1,\\
\dot{\psi}_2 &= y_2,
\end{aligned}$$

with the solutions

$$\begin{aligned}
\mathrm{y}_1(t) &= \mathrm{x}_1(0) + \varepsilon t,\\
\mathrm{y}_2(t) &= \mathrm{x}_2(0),\\
\psi_1(t) &= \phi_1(0) + \mathrm{x}_1(0)t + \frac{1}{2}\varepsilon t^2,\\
\psi_2(t) &= \phi_2(0) + \mathrm{x}_2(0)t.
\end{aligned}$$

To establish the asymptotic character of these formal approximations, note first that

$$\mathrm{x}_1(t) = \mathrm{y}_1(t).$$

Furthermore we introduce $x = x_1 - x_2$, $\phi = \phi_1 - \phi_2$ to obtain

$$\begin{aligned}
\dot{x} &= \varepsilon(1 - \cos(\phi)),\\
\dot{\phi} &= x,
\end{aligned}$$

which has the integral

$$x^2 = a^2 + 2\varepsilon\phi - 2\varepsilon\sin(\phi).$$

At the same time we have

$$(y_1 - y_2)^2 = a^2 + 2a\varepsilon t + \varepsilon^2 t^2 = a^2 + 2\varepsilon(\psi_1 - \psi_2).$$

This expression agrees with the integral to $\mathcal{O}(\varepsilon)$ for all time. Although the approximate integral constitutes a valid approximation of the integral which exists for the system, this is still not enough to characterize the individual orbits on the integral manifold. We omit the technical discussion for this detailed characterization. The (x, ϕ)-phase flow resembles closely the flow depicted in Figure 8.1.

8.5.4 Example of Forced Passage through Resonance

The equations are

$$\begin{aligned}
\dot{x}_1 &= \varepsilon, \\
\dot{x}_2 &= \varepsilon \cos(\phi_1 - \phi_2), \\
\dot{\phi}_1 &= x_1 + x_2, \\
\dot{\phi}_2 &= x_2.
\end{aligned}$$

The resonance condition reduces to

$$x_1 = 0.$$

There are two cases to consider:

The Case $a = \mathrm{x}_1(0) > 0$

1. Let $b = \phi_1(0) - \phi_2(0)$. Since $\dot{x}_1 > 0$ we have $\mathrm{x}_1(t) > 0$ for all $t > 0$. Thus there will be no resonance. Using averaging, we can show that $\mathrm{x}_2(t) = \mathrm{x}_2(0) + \mathcal{O}(\varepsilon)$, but we can also see this from solving the original equations:

$$\begin{aligned}
\mathrm{x}_1(t) &= \mathrm{x}_1(0) + \varepsilon t, \\
\mathrm{x}_2(t) &= \mathrm{x}_2(0) + \varepsilon \int_0^t \cos(b + a\tau + \frac{1}{2}\varepsilon\tau^2)\, d\tau \qquad (8.5.2)\\
&= \mathrm{x}_2(0) + \mathcal{O}(\varepsilon), \\
\phi_1(t) - \phi_2(t) &= b + \mathrm{x}_1(0)t + \frac{1}{2}\varepsilon t^2.
\end{aligned}$$

The estimate of the integral is valid for all time and can be obtained by transforming $\sigma = b + a\tau + \frac{1}{2}\varepsilon\tau^2$, followed by partial integration.

Exercise 8.5.1. Where exactly do we use the fact that $a > 0$?

The Case $a < 0$.

2. Using formula (8.5.2) we can take the limit for $t \to \infty$ to compute the change in x_2 induced by the passage through the resonance. The integral is a well known Fresnel integral and the result is

$$\lim_{t\to\infty} \mathrm{x}_2(t) = \mathrm{x}_2(0) + \sqrt{\frac{1}{2}\pi\varepsilon} \cos(b - \frac{a^2}{2\varepsilon} + \frac{\pi}{4}).$$

The $\sqrt{\varepsilon}$-contribution of the resonance agrees with the analysis given in Section 8.5.1.

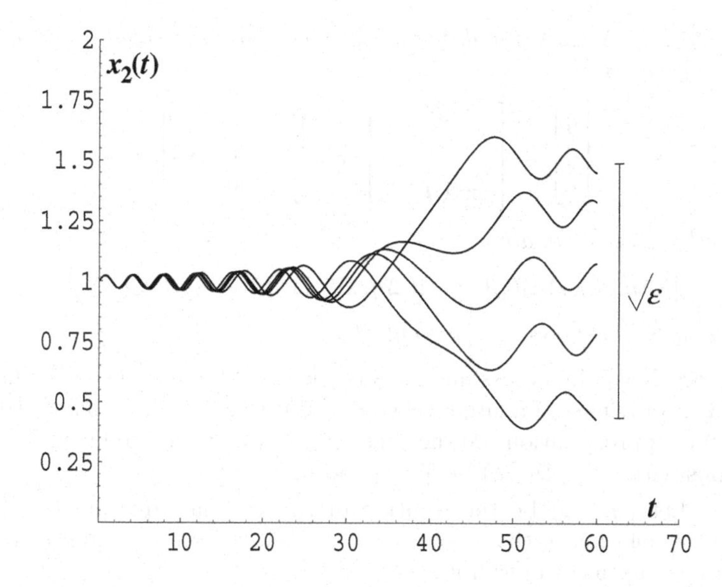

Fig. 8.2: Solutions $\mathrm{x} = \mathrm{x}_2(t)$ based on equation (8.5.2) with $b = 0$, $\varepsilon = 0.1$ and $\mathcal{O}(\varepsilon)$ variations of a around -2.

Observe that the dependency of the shift in x_2 is determined by the orbit, as it should be, since changing the initial point in time should have no influence on this matter.

8.6 Analysis of the Inner and Outer Expansion; Passage through Resonance

After the intuitive reasoning in the preceding sections we shall now develop the asymptotic analysis of the expansions outside and in the resonance region. We shall also discuss the rather intricate problem of matching these expansions and thus the phenomenon of passage through resonance.

Lemma 8.6.1. *Suppose that we can find local coordinates for equation (8.1.1) such that one can split $\boldsymbol{x}$ into $(\boldsymbol{\eta}, \xi) \in \mathbb{R}^{n-1} \times \mathbb{R}$ such that $\Omega^0(\boldsymbol{\eta}, 0) = 0$. Denote the $\boldsymbol{\eta}$-component of $\mathbf{X}^1$ by $\mathbf{X}^1_{\parallel}$ and the ξ-component by $\mathrm{X}^1_{\perp}$. Then consider the equation*

$$\begin{bmatrix} \dot{\boldsymbol{\eta}} \\ \dot{\xi} \\ \dot{\phi} \end{bmatrix} = \begin{bmatrix} 0 \\ 0 \\ \Omega^0(\boldsymbol{\eta}, \xi) \end{bmatrix} + \varepsilon \begin{bmatrix} \mathbf{X}^1_{\parallel}(\boldsymbol{\eta}, \xi, \phi) \\ \mathrm{X}^1_{\perp}(\boldsymbol{\eta}, \xi, \phi) \\ 0 \end{bmatrix} \tag{8.6.1}$$

in a $\sqrt{\varepsilon}$-neighborhood of $\xi = 0$. Introduce the following norm $|\ |_\varepsilon$:

$$|((\boldsymbol{\eta}, \xi, \phi)|_\varepsilon = \| (\boldsymbol{\eta}, \phi) \| + \frac{1}{\sqrt{\varepsilon}} \| \xi \| .$$

Denote by $(\Delta\boldsymbol{\eta}, \Delta\xi, \Delta\phi)$ the difference of the solution of equation (8.6.1) and the inner equation:

$$\begin{bmatrix} \dot{\boldsymbol{\eta}} \\ \dot{\xi} \\ \dot{\phi} \end{bmatrix} = \begin{bmatrix} 0 \\ 0 \\ \frac{\partial \Omega^0}{\partial \xi}(\boldsymbol{\eta}, 0)\xi \end{bmatrix} + \varepsilon \begin{bmatrix} 0 \\ \mathrm{X}^1_{\perp}(\boldsymbol{\eta}, 0, \phi) \\ 0 \end{bmatrix} . \tag{8.6.2}$$

Then we have the estimate

$$|(\Delta\boldsymbol{\eta}, \Delta\xi, \Delta\phi)(t)|_\varepsilon \leq (|(\Delta\boldsymbol{\eta}, \Delta\xi, \Delta\phi)(0)|_\varepsilon + R\varepsilon t) e^{K\sqrt{\varepsilon} t}.$$

with R and K constants independent of ε.

Remark 8.6.2. We do not assume any knowledge of the initial conditions here, since this is only part of a larger scheme; this estimate indicates however that considering approximations on the time scale $1/\sqrt{\varepsilon}$ one has to know the initial conditions with error $\mathcal{O}(\sqrt{\varepsilon})$ in the $|\ |_\varepsilon$-norm. ♡

Proof Let $(\boldsymbol{\eta}, \xi, \phi)$ be the solution of the original equation (8.6.1) and $(\overline{\boldsymbol{\eta}}, \overline{\xi}, \overline{\phi})$ the solution of the inner equation (8.6.2). Then we can estimate the difference using the Gronwall lemma:

$$\begin{aligned}
&|(\Delta\boldsymbol{\eta}, \Delta\xi, \Delta\phi)(t)|_\varepsilon \\
&\leq |(\Delta\boldsymbol{\eta}, \Delta\xi, \Delta\phi)(0)|_\varepsilon + \int_0^t \Big(\| \Omega^0(\boldsymbol{\eta}, \xi) - \frac{\partial \Omega^0}{\partial \xi}(\overline{\boldsymbol{\eta}}, 0)\overline{\xi} \| \\
&\quad + \varepsilon^{\frac{1}{2}} \| \mathrm{X}^1_{\perp}(\boldsymbol{\eta}, \xi, \phi) - \mathrm{X}^1_{\perp}(\overline{\boldsymbol{\eta}}, 0, \overline{\phi}) \| + \varepsilon \| \mathbf{X}^1_{\parallel}(\boldsymbol{\eta}, \xi, \phi) \| \Big) \, \mathrm{d}\tau \\
&\leq |(\Delta\boldsymbol{\eta}, \Delta\xi, \Delta\phi)(0)|_\varepsilon \\
&\quad + \int_0^t \Big(\| \frac{\partial \Omega}{\partial \xi} \| \| \xi - \overline{\xi} \| + C\sqrt{\varepsilon}(\| \phi - \overline{\phi} \| + \| \boldsymbol{\eta} - \overline{\boldsymbol{\eta}} \|) + R\varepsilon \Big) \, \mathrm{d}\tau \\
&\leq |(\Delta\boldsymbol{\eta}, \Delta\xi, \Delta\phi)(0)|_\varepsilon + K\varepsilon^{\frac{1}{2}} \int_0^t |(\Delta\boldsymbol{\eta}, \Delta\xi, \Delta\phi)(\tau)|_\varepsilon \, \mathrm{d}\tau + R \int_0^t \varepsilon \, \mathrm{d}\tau,
\end{aligned}$$

and this implies

$$|(\Delta\boldsymbol{\eta}, \Delta\xi, \Delta\phi)(t)|_\varepsilon \leq (|(\Delta\boldsymbol{\eta}, \Delta\xi, \Delta\phi)(0)|_\varepsilon + R\varepsilon t) e^{K\sqrt{\varepsilon} t},$$

as desired. □

Exercise 8.6.3. Generalize this lemma to the case where $\xi = \mathcal{O}(\delta(\varepsilon))$. Is it possible to get estimates on a larger time scale than $1/\sqrt{\varepsilon}$?

In the next lemma we shall generalize Lemma 7.9.1; the method of proof is the same, but we are more careful about the inverse powers of Ω^0 appearing in the perturbation scheme.

Lemma 8.6.4. *Consider the equation*

$$\begin{bmatrix}\dot{\boldsymbol{x}}\\ \dot{\phi}\end{bmatrix} = \begin{bmatrix}0\\ \Omega^0(\boldsymbol{x})\end{bmatrix} + \varepsilon\begin{bmatrix}\mathbf{X}^1(\boldsymbol{x},\phi)\\ 0\end{bmatrix}, \quad \begin{bmatrix}\mathbf{x}\\ \phi\end{bmatrix}(0) = \begin{bmatrix}\boldsymbol{x}_0\\ \phi_0\end{bmatrix}, \quad \begin{matrix}\boldsymbol{x}\in D\subset\mathbb{R}^n,\\ \phi\in S^1.\end{matrix}$$

Then we can write the solution of this equation $(\boldsymbol{x},\phi)$ as follows:

$$\begin{bmatrix}\mathbf{x}\\ \phi\end{bmatrix}(t) = \begin{bmatrix}\mathbf{y}\\ \psi\end{bmatrix}(t) + \varepsilon\begin{bmatrix}\mathbf{u}^1\\ \mathrm{v}^1\end{bmatrix}(\mathbf{y}(t),\psi(t)) + \varepsilon^2\begin{bmatrix}\mathbf{u}^2\\ 0\end{bmatrix}(t)(\mathbf{y}(t),\psi(t)), \qquad (8.6.3)$$

where $(\mathbf{y},\psi)$ is the solution of

$$\begin{bmatrix}\dot{\boldsymbol{y}}\\ \dot{\psi}\end{bmatrix} = \begin{bmatrix}0\\ \Omega^0(\boldsymbol{y})\end{bmatrix} + \varepsilon\begin{bmatrix}\overline{\mathbf{X}}^1(\boldsymbol{y})\\ 0\end{bmatrix} + \varepsilon^2\begin{bmatrix}\overline{\mathbf{X}}^2_\star(\boldsymbol{y})\\ \Omega^{[2]}_\star(\boldsymbol{y},\psi,\varepsilon)\end{bmatrix} + \varepsilon^3\begin{bmatrix}\mathbf{X}^{[3]}_\star(\boldsymbol{y},\psi,\varepsilon)\\ 0\end{bmatrix}. \quad (8.6.4)$$

with $\mathbf{y}(0) = \boldsymbol{y}_0, \boldsymbol{y}\in D\subset\mathbb{R}^n$ and $\psi(0) = \psi_0, \psi\in S^1$. Here $(\boldsymbol{y}_0,\psi_0)$ is the solution of the equation

$$\begin{bmatrix}\boldsymbol{x}_0\\ \phi_0\end{bmatrix} = \begin{bmatrix}\boldsymbol{y}_0\\ \psi_0+)\end{bmatrix} + \varepsilon\begin{bmatrix}\mathbf{u}^1(\boldsymbol{y}_0,\psi_0)\\ \mathrm{v}^1(\boldsymbol{y}_0,\psi_0)\end{bmatrix} + \varepsilon^2\begin{bmatrix}\mathbf{u}^2(\boldsymbol{y}_0,\psi_0)\\ 0\end{bmatrix}.$$

On $D\times S^1\times(0,\varepsilon_0]$ we have the following (nonuniform) estimates for $\Omega^{[2]}_\star$ and $\mathbf{X}^{[3]}_\star$:

$$\mathbf{X}^{[3]}_\star = \mathcal{O}(\frac{1}{\Omega^0(\boldsymbol{y})^4}), \quad \Omega^{[2]}_\star = \mathcal{O}(\frac{1}{\Omega^0(\boldsymbol{y})^3}),$$

and v^1, $\mathbf{u}^1$ and $\mathbf{u}^2$ can be estimated by

$$\mathbf{u}^1 = \mathcal{O}(\frac{1}{\Omega^0(\boldsymbol{y})}), \quad \mathrm{v}^1 = \mathcal{O}(\frac{1}{\Omega^0(\boldsymbol{y})^2}), \quad \mathbf{u}^2 = \mathcal{O}(\frac{1}{\Omega^0(\boldsymbol{y})^2}).$$

Here $\overline{\mathbf{X}}^1$ and $\overline{\mathbf{X}}^2_\star$ are defined as follows:

$$\overline{\mathbf{X}}^1(\cdot) = \frac{1}{2\pi}\int_0^{2\pi}\mathbf{X}^1(\cdot,\varphi)\,\mathrm{d}\varphi,$$

$$\Omega^0(\cdot)\mathbf{u}^1(\cdot,\phi) = \int^{\phi}(\mathbf{X}^1(\cdot,\varphi) - \overline{\mathbf{X}}^1(\cdot))\,\mathrm{d}\varphi,$$

$$\overline{\mathbf{X}}^2_\star(\cdot) = \frac{1}{2\pi}\int_0^{2\pi}\mathsf{D}\mathbf{X}^1(\cdot,\varphi)\cdot\mathbf{u}^1(\cdot,\varphi) - \mathbf{X}^1(\cdot,\varphi)\cdot\frac{\mathsf{D}\Omega^0(\cdot)}{\Omega^0(\cdot)}\mathbf{u}^1(\cdot,\varphi)\,\mathrm{d}\varphi.$$

It follows that

$$\overline{\mathbf{X}}^1(\boldsymbol{y}) = \mathcal{O}(1), \quad \overline{\mathbf{X}}^2_\star(\boldsymbol{y}) = \mathcal{O}(\frac{1}{\Omega^0(\boldsymbol{y})^2}).$$

Of course, one has to choose $(\boldsymbol{x}_0, \phi_0)$ *well outside the inner domain, since otherwise it may not be possible to solve the equation for the initial conditions* $(\boldsymbol{y}_0, \psi_0)$. *In the same sense the estimates are nonuniform. For the proof to work, one has to require that Assumption 8.3.1 holds, that is,*

$$\frac{\varepsilon}{\Omega^0(\boldsymbol{y})^2} = o(1) \text{ as } \varepsilon \downarrow 0.$$

That is to say, y *should be outside a* $\sqrt{\varepsilon}$*-neighborhood of the resonance manifold.*

Proof First we differentiate the relation (8.6.3) along the vector field:

$$\begin{bmatrix} \dot{\boldsymbol{x}} \\ \dot{\phi} \end{bmatrix} = \begin{bmatrix} 1+\varepsilon \mathsf{D}\mathbf{u}^1 + \varepsilon^2 \mathsf{D}\mathbf{u}^2 & \varepsilon\frac{\partial \mathbf{u}^1}{\partial\psi} + \varepsilon^2\frac{\partial \mathbf{u}^2}{\partial\psi} \\ \varepsilon \mathsf{D}\mathrm{v}^1 & 1+\varepsilon\frac{\partial \mathrm{v}^1}{\partial\psi} \end{bmatrix} \begin{bmatrix} \dot{\boldsymbol{y}} \\ \dot{\psi} \end{bmatrix}$$
$$= \begin{bmatrix} 1+\varepsilon \mathsf{D}\mathbf{u}^1 + \varepsilon^2 \mathsf{D}\mathbf{u}^2 & \varepsilon\frac{\partial \mathbf{u}^1}{\partial\psi} + \varepsilon^2\frac{\partial \mathbf{u}^2}{\partial\psi} \\ \varepsilon \mathsf{D}\mathrm{v}^1 & 1+\varepsilon\frac{\partial \mathrm{v}^1}{\partial\psi} \end{bmatrix} \begin{bmatrix} \varepsilon\overline{\mathbf{X}}^1 + \varepsilon^2\overline{\mathbf{X}}^2_\star + \varepsilon^3\mathbf{X}^{[3]}_\star \\ \Omega^0 + \varepsilon^2\Omega^{[2]}_\star \end{bmatrix}$$
$$= \begin{bmatrix} \mathbf{0} \\ \Omega^0 \end{bmatrix} + \varepsilon \begin{bmatrix} \Omega^0\frac{\partial \mathbf{u}^1}{\partial\psi} + \overline{\mathbf{X}}^1 \\ \Omega^0\frac{\partial \mathrm{v}^1}{\partial\psi} \end{bmatrix} + \varepsilon^2 \begin{bmatrix} \Omega^0\frac{\partial \mathbf{u}^2}{\partial\psi} + \overline{\mathbf{X}}^2_\star + \mathsf{D}\mathbf{u}^1\cdot\overline{\mathbf{X}}^1 \\ \Omega^{[2]}_\star + \mathsf{D}\mathrm{v}^1\cdot\overline{\mathbf{X}}^1 \end{bmatrix}$$
$$+\varepsilon^3 \begin{bmatrix} \Omega^{[2]}_\star\frac{\partial \mathbf{u}^1}{\partial\psi} + \mathsf{D}\mathbf{u}^2\cdot\overline{\mathbf{X}}^1 + \mathsf{D}\mathbf{u}^1\cdot\overline{\mathbf{X}}^2_\star + \mathbf{X}^{[3]}_\star \\ \cdots \end{bmatrix}$$

Then we use (8.6.3) to replace $(\boldsymbol{x}, \phi)$ by $(\boldsymbol{y}, \psi)$ in the original differential equation:

$$\begin{bmatrix} \dot{\boldsymbol{x}} \\ \dot{\phi} \end{bmatrix} = \begin{bmatrix} \varepsilon\mathbf{X}^1(\boldsymbol{x},\phi) \\ \Omega^0(\boldsymbol{x}) \end{bmatrix} = \begin{bmatrix} \varepsilon\mathbf{X}^1(\boldsymbol{y}+\varepsilon\mathbf{u}^1+\varepsilon^2\mathbf{u}^2, \psi+\varepsilon\mathrm{v}^1) \\ \Omega^0(\boldsymbol{y}+\varepsilon\mathbf{u}^1+\varepsilon^2\mathbf{u}^2) \end{bmatrix}$$
$$= \begin{bmatrix} \varepsilon\mathbf{X}^1(\boldsymbol{y},\psi) + \varepsilon^2\frac{\partial\mathbf{X}^1}{\partial\psi}\mathrm{v}^1 + \varepsilon^2\mathsf{D}\mathbf{X}^1\mathbf{u}^1 + \mathcal{O}(\varepsilon^3(\|\mathbf{u}^1\|^2 + \|\mathbf{u}^2\|)) \\ \Omega^0(\boldsymbol{y}) + \varepsilon\mathsf{D}\Omega^0\mathbf{u}^1 + \mathcal{O}(\varepsilon^2\|\mathbf{u}^1\|^2 + \varepsilon^2\|\mathbf{u}^2\|) \end{bmatrix}$$
$$= \begin{bmatrix} \mathbf{0} \\ \Omega^0(\boldsymbol{y}) \end{bmatrix} + \varepsilon \begin{bmatrix} \mathbf{X}^1(\boldsymbol{y},\psi) \\ \mathsf{D}\Omega^0\cdot\mathbf{u}^1 \end{bmatrix} + \varepsilon^2 \begin{bmatrix} \mathrm{v}^1\frac{\partial\mathbf{X}^1}{\partial\psi} + \mathsf{D}\mathbf{X}^1\cdot\mathbf{u}^1 \\ \mathcal{O}((\|\mathbf{u}^1\|^2 + \|\mathbf{u}^2\|) \end{bmatrix}$$
$$+\varepsilon^3 \begin{bmatrix} \mathcal{O}(\|\mathbf{u}^1\|^2 + \|\mathbf{u}^2\|) \\ 0 \end{bmatrix}.$$

Equating powers of ε, we obtain the following relations:

$$\Omega^0 \begin{bmatrix} \frac{\partial\mathbf{u}^1}{\partial\psi} \\ \frac{\partial\mathrm{v}^1}{\partial\psi} \\ \frac{\partial\mathbf{u}^2}{\partial\psi} \end{bmatrix} = \begin{bmatrix} \mathbf{X}^1 - \overline{\mathbf{X}}^1 \\ \mathsf{D}\Omega^0\cdot\mathbf{u}^1 \\ \mathrm{v}^1\frac{\partial\mathbf{X}^1}{\partial\psi} + \mathsf{D}\mathbf{X}^1\cdot\mathbf{u}^1 - \mathsf{D}\mathbf{u}^1\cdot\overline{\mathbf{X}}^1 - \overline{\mathbf{X}}^2_\star \end{bmatrix}.$$

The second component of this equation is by now standard: Let $\mathbf{u}^1$ be defined by

$$\mathbf{u}^1(\boldsymbol{y},\phi) = \frac{1}{\Omega^0(\boldsymbol{y})}\int^{\phi} \left(\mathbf{X}^1(\boldsymbol{y},\varphi) - \overline{\mathbf{X}}^1(\boldsymbol{y})\right) \mathrm{d}\varphi$$

and

$$\int_0^{2\pi} \mathbf{u}^1(\boldsymbol{y},\varphi)\,\mathrm{d}\varphi = 0,$$

where as usual

$$\overline{\mathbf{X}}^1(\cdot) = \frac{1}{2\pi}\int_0^{2\pi} \mathbf{X}^1(\cdot,\varphi)\,\mathrm{d}\varphi.$$

Then we can also solve the first component (if the average of $\mathbf{u}^1$ had not been zero, the averaged vector field would have been different):

$$\mathrm{v}^1(\boldsymbol{y},\phi) = \frac{1}{\Omega^0(\boldsymbol{y})}\int^{\phi} \mathsf{D}\Omega^0(\boldsymbol{y})\cdot\mathbf{u}^1(\boldsymbol{y},\varphi)\,\mathrm{d}\varphi$$

and again

$$\int_0^{2\pi} \mathrm{v}^1(\boldsymbol{y},\varphi)\,\mathrm{d}\varphi = 0.$$

Thus we have that $\mathbf{u}^1 = \mathcal{O}(1/\Omega^0(\boldsymbol{y}))$ and $\mathrm{v}^1 = \mathcal{O}(1/\Omega^0(\boldsymbol{y})^2)$. We are now ready to solve the third component:

$$\mathbf{u}^2(\phi) = \frac{1}{\Omega^0}\int^{\phi} \left(\mathrm{v}^1(\varphi)\frac{\partial \mathbf{X}^1}{\partial\varphi}(\varphi) + \mathsf{D}\mathbf{X}^1(\varphi)\cdot\mathbf{u}^1(\varphi) - \mathsf{D}\mathbf{u}^1(\varphi)\cdot\overline{\mathbf{X}}^1 - \overline{\mathbf{X}}^2_\star\right) \mathrm{d}\varphi.$$

This is a bounded expression if we take

$$\begin{aligned}
&\overline{\mathbf{X}}^2(\boldsymbol{y})\\
&= \frac{1}{2\pi}\int_0^{2\pi} \left(\mathrm{v}^1(\boldsymbol{y},\varphi)\frac{\partial \mathbf{X}^1}{\partial\varphi} + \mathsf{D}\mathbf{X}^1(\boldsymbol{y},\varphi)\cdot\mathbf{u}^1(\boldsymbol{y},\varphi) - \mathsf{D}\mathbf{u}^1(\boldsymbol{y},\varphi)\cdot\overline{\mathbf{X}}^1(\boldsymbol{y})\right)\mathrm{d}\varphi\\
&= \frac{1}{2\pi}\int_0^{2\pi} \left(\mathrm{v}^1(\boldsymbol{y},\varphi)\frac{\partial \mathbf{X}^1}{\partial\varphi} + \mathsf{D}\mathbf{X}^1(\boldsymbol{y},\varphi)\cdot\mathbf{u}^1(\boldsymbol{y},\varphi)\right)\mathrm{d}\varphi\\
&= \frac{1}{2\pi}\int_0^{2\pi} \left(\mathsf{D}\mathbf{X}^1(\boldsymbol{y},\varphi)\mathbf{u}^1(\boldsymbol{y},\varphi) - \frac{\partial \mathrm{v}^1}{\partial\varphi}(\boldsymbol{y},\varphi)\mathbf{X}^1(\boldsymbol{y},\varphi)\right)\mathrm{d}\varphi\\
&= \frac{1}{2\pi}\int_0^{2\pi} \left(\mathsf{D}\mathbf{X}^1(\boldsymbol{y},\varphi)\cdot\mathbf{u}^1(\boldsymbol{y},\varphi) - \frac{\mathsf{D}\Omega^0(\boldsymbol{y})\cdot\mathbf{u}^1(\boldsymbol{y},\varphi)}{\Omega^0(\boldsymbol{y})}\mathbf{X}^1(\boldsymbol{y},\varphi)\right)\mathrm{d}\varphi.
\end{aligned}$$

From this last expression it follows that we do not have to compute either v^1 or $\mathbf{u}^2$, in order to compute $\overline{\mathbf{X}}^1$ and $\overline{\mathbf{X}}^2_\star$. It also follows that $\overline{\mathbf{X}}^2_\star = \mathcal{O}(\frac{1}{\Omega^0(\boldsymbol{y})^2})$. We can solve $\Omega^{[2]}_\star$ and $\mathbf{X}^{[3]}_\star$ from the equations, and we find that $\Omega^{[2]}_\star = \mathcal{O}(\frac{1}{\Omega^0(\boldsymbol{y})^3})$ and $\mathbf{X}^{[3]}_\star = \mathcal{O}(\frac{1}{\Omega^0(\boldsymbol{y})^4})$. In each expansion we have implicitly assumed that $\varepsilon/\Omega^0(\boldsymbol{y})^2 = o(1)$ as $\varepsilon \downarrow 0$ as in our blanket Assumption 8.3.1. □

Exercise 8.6.5. Formulate and prove the analogue of Lemma 7.9.2, for the situation described in Lemma 8.6.4.

In Lemma 7.8.4, one assumption is that Ω^0 be bounded away from zero in a uniform way. We shall have to drop this restriction, and assume for instance that the distance of the boundary of the outer domain to the resonance manifold is at least of order $\delta(\varepsilon)$, where $\delta(\varepsilon)$ is somewhere between $\sqrt{\varepsilon}$ and 1.

Then there is the time scale. If there is no passage through resonance, i.e. if in Lemma 8.6.1 the average of $\mathbf{X}_\perp^1$ vanishes, there is no immediate need to prove anything on a longer time scale than $1/\sqrt{\varepsilon}$, since that is the natural time scale in the inner region (this, of course, is a matter of taste; one might actually need longer time scale estimates in the outer region: the reader may want to try to formulate a lemma on this problem, cf. [236]. On the other hand, if there is a passage through resonance, then we can use this as follows: In our estimate, we meet expressions of the form

$$\int \frac{\varepsilon}{\Omega^0(\mathbf{z}(s))^k}\, \mathrm{d}s,$$

where $\mathbf{z}$ is the solution of the outer equation

$$\dot{\boldsymbol{z}} = \varepsilon \overline{\mathbf{X}}^1(\boldsymbol{z}) + \varepsilon^2 \overline{\mathbf{X}}^2(\boldsymbol{z}).$$

Let us take a simple example:

$$\dot{z} = \varepsilon\alpha, \quad \dot{\phi} = z.$$

Then $\mathrm{z}(t) = z_0 + \varepsilon\alpha t$ and the integral is

$$\begin{aligned}\int_{t_1}^{t_2} \frac{\varepsilon\, \mathrm{d}s}{(z_0+\varepsilon\alpha s)^k} &= \frac{1}{\alpha}\int_{\mathrm{z}(t_1)}^{\mathrm{z}(t_2)} \frac{\mathrm{d}\xi}{\xi^k} = \frac{1}{\alpha(k-1)}\left(\frac{1}{\mathrm{z}^{k-1}(t_1)} - \frac{1}{\mathrm{z}^{k-1}(t_2)}\right) \\ &= \frac{1}{\alpha(k-1)}\left(\frac{1}{\Omega^0(\mathrm{z}(t_1))^{k-1}} - \frac{1}{\Omega^0(\mathrm{z}(t_2))^{k-1}}\right), \ k \geq 2.\end{aligned}$$

This leads to the following

Assumption 8.6.6 *In the sequel we shall assume that for $k \geq 2$*

$$\int_{t_1}^{t_2} \frac{\varepsilon\, \mathrm{d}s}{\Omega^0(\mathbf{z}(s))^k} = \mathcal{O}\left(\frac{1}{\Omega^0(\mathbf{z}(t_1))^{k-1}}\right) + \mathcal{O}\left(\frac{1}{\Omega^0(\mathbf{z}(t_2))^{k-1}}\right),$$

as long as $\mathbf{z}(t)$ stays in the outer domain on $[t_1, t_2]$.

This is an assumption that can be checked, at least in principle, since it involves only the averaged equation, which we are supposed to be able to solve (Actually, solving may not even be necessary; all we need is some nice estimates on $\Omega^0(\mathbf{z}(t))$). One might wish to prove this assumption from basic facts about the vector field, but this turns out to be difficult, since this estimate incorporates rather subtly both the velocity of the $\boldsymbol{z}$-component, and the dependence of Ω^0 on $\boldsymbol{z}$.

Lemma 8.6.7. *Consider equation (8.6.4), introduced in Lemma 8.6.4,*

$$\begin{bmatrix}\dot{y}\\ \dot{\psi}\end{bmatrix} = \begin{bmatrix}0\\ \Omega^0(y)\end{bmatrix} + \varepsilon\begin{bmatrix}\overline{\mathbf{X}}^1(y)\\ 0\end{bmatrix} + \varepsilon^2\begin{bmatrix}\overline{\mathbf{X}}^2_\star\\ \Omega^{[2]}_\star(y,\psi,\varepsilon)\end{bmatrix} + \varepsilon^3\begin{bmatrix}\mathbf{X}^{[3]}_\star(y,\psi,\varepsilon)\\ 0\end{bmatrix},$$

where we have the following estimates

$$\Omega^{[2]}_\star = \mathcal{O}(\frac{1}{\Omega^0(y)^3}),\quad \mathbf{X}^{[3]}_\star = \mathcal{O}(\frac{1}{\Omega^0(y)^4}),\quad \text{and } \overline{\mathbf{X}}^2_\star = \mathcal{O}(\frac{1}{\Omega^0(y)^2}).$$

Let $(\mathbf{y},\psi)$ be the solution of this equation. Let $(\mathbf{z},\zeta)$ be the solution of the truncated system

$$\begin{bmatrix}\dot{z}\\ \dot{\zeta}\end{bmatrix} = \begin{bmatrix}0\\ \Omega^0(z)\end{bmatrix} + \varepsilon\begin{bmatrix}\overline{\mathbf{X}}^1(z)\\ 0\end{bmatrix} + \varepsilon^2\begin{bmatrix}\overline{\mathbf{X}}^2_\star(z)\\ 0\end{bmatrix},\quad \begin{bmatrix}\mathbf{z}\\ \zeta\end{bmatrix}(0) = \begin{bmatrix}z_0\\ \zeta_0\end{bmatrix},\quad \begin{matrix}z \in D^o \subset D,\\ \zeta \in S^1.\end{matrix}$$

Then, with Assumption 8.6.6, $(\mathbf{z},\zeta)$ is an approximation of $(\mathbf{y},\psi)$ in the following sense: Let δ be such that $\varepsilon/\delta^2(\varepsilon) = o(1)$ and $|\Omega^0(\mathbf{z}(t))| \geq C\delta(\varepsilon)$ for all $t \in [0, L/\varepsilon)$; then on $0 \leq \varepsilon t \leq L$

$$\| \mathbf{y}(t) - \mathbf{z}(t) \| = \mathcal{O}(\| y_0 - z_0 \|) + \mathcal{O}(\frac{\varepsilon^2}{\delta^3(\varepsilon)}),$$

$$|\psi(t) - \zeta(t)| = \mathcal{O}(|\psi_0 - \zeta_0|) + \mathcal{O}(\frac{\| y_0 - z_0 \|}{\varepsilon}) + \mathcal{O}(\frac{\varepsilon}{\delta^2(\varepsilon)}).$$

Proof In the following estimates, we shall not go into all technical details; the reader is invited to plug the visible holes. Using the differential equations, we obtain the following estimate for the difference between $\mathbf{y}$ and $\mathbf{z}$:

$$\begin{aligned}\| \mathbf{y}(t) - \mathbf{z}(t) \| &\leq \| y_0 - z_0 \| + \varepsilon\int_0^t \| \overline{\mathbf{X}}^1(\mathbf{y}(s)) - \overline{\mathbf{X}}^1(\mathbf{z}(s)) \| \,\mathrm{d}s\\ &+\varepsilon^2\int_0^t \| \overline{\mathbf{X}}^2_\star(\mathbf{y}(s)) - \overline{\mathbf{X}}^2_\star(\mathbf{z}(s)) \| \,\mathrm{d}s + \varepsilon^3\int_0^t \| \mathbf{X}^{[3]}_\star(\mathbf{y}(s),\psi(s),\varepsilon) \| \,\mathrm{d}s\\ &\leq \| y_0 - z_0 \| + \varepsilon\int_0^t C\left(1 + \frac{\varepsilon}{\Omega^0(\mathbf{z}(s))^3} + \frac{\varepsilon^2}{\Omega^0(\mathbf{z}(s))^5}\right) \| \mathbf{y}(s) - \mathbf{z}(s) \| \,\mathrm{d}s\\ &+\varepsilon^3\int_0^t \frac{C}{\Omega^0(\mathbf{z}(s))^4}\,\mathrm{d}s\end{aligned}$$

(For odd powers of Ω^0, we take, of course, the power of the absolute value). Using the Gronwall lemma, this implies

$$\begin{aligned}
\| \mathbf{y}(t) - \mathbf{z}(t) \| &\le \| y_0 - z_0 \| \, e^{\varepsilon \int_0^t C\left(1+\frac{\varepsilon}{\Omega^0(\mathbf{z}(s))^3}+\frac{\varepsilon^2}{\Omega^0(\mathbf{z}(s))^5}\right) \mathrm{d}s} \\
&\quad + \int_0^t \frac{C\varepsilon^3}{\Omega^0(\mathbf{z}(s))^4} e^{\varepsilon \int_0^t C\left(1+\frac{\varepsilon}{\Omega^0(\mathbf{z}(\sigma))^3}+\frac{\varepsilon^2}{\Omega^0(\mathbf{z}(\sigma))^5}\right]) \mathrm{d}\sigma} \mathrm{d}s \\
&\le (\| y_0 - z_0 \| + C(\frac{\varepsilon^2}{\Omega^0(\mathbf{z}(t))^3} + \frac{\varepsilon^2}{\Omega^0(\mathbf{z}(0))^3})) \\
&\quad e^{C(\varepsilon t + \frac{\varepsilon}{\Omega^0(\mathbf{z}(t))^2} + \frac{\varepsilon}{\Omega^0(\mathbf{z}(0))^2} + \frac{\varepsilon^2}{\Omega^0(\mathbf{z}(t))^4} + \frac{\varepsilon^2}{\Omega^0(\mathbf{z}(0))^4})} \\
&= \mathcal{O}(\| y_0 - z_0 \|) + \mathcal{O}(\frac{\varepsilon^2}{\delta^3}) \text{ on } 0 \le \varepsilon t \le L.
\end{aligned}$$

For the angular variables, the estimate is now very easy:

$$\begin{aligned}
\| \psi(t) - \zeta(t) \| &\le \| \psi_0 - \zeta_0 \| + C \int_0^t \| \mathbf{z}(s) - \mathbf{y}(s) \| \, \mathrm{d}s \\
&\quad + C\varepsilon^2 \int_0^t \frac{1}{\Omega^0(\mathbf{z}(s))^3} \mathrm{d}s \\
&= \mathcal{O}(\| \psi_0 - \zeta_0 \|) + \mathcal{O}(\frac{\| y_0 - z_0 \|}{\varepsilon}) + \mathcal{O}(\frac{\varepsilon}{\delta^2}) \text{ on } 0 \le \varepsilon t \le L
\end{aligned}$$

if $d(z_0, \mathsf{N}) = \mathcal{O}_\sharp(1)$; otherwise one has to include a term $\mathcal{O}(\varepsilon/\Omega^0(z_0)^2)$; this last term presents difficulties if one wishes to obtain estimates for the full passage through resonance, at least for the angular variables. The estimate implies that one can average as long as $\varepsilon/\delta^2 = o(1)$. On the other hand, the estimates for the inner region are valid only in a $\sqrt{\varepsilon}$-neighborhood of the resonance manifold. So there is a gap. □

As we have already pointed out however, it is possible to bridge this gap by using a time scale extension argument for one-dimensional monotone vector fields. The full proof of this statement is rather complicated and has been given in [238]. Here we would like to give only the essence of the argument in the form of a lemma.

Lemma 8.6.8 (Eckhaus). *Consider the vector field*

$$\dot{x} = \mathrm{f}^0(x) + \varepsilon \mathrm{f}^1(x,t) \ , \ \mathrm{x}(0) = x_0 \ , \ x \in D \subset \mathbb{R},$$

with $0 < m \le \mathrm{f}^0(x) \le M < \infty$, $\| R \| \le C$ *for* $x \in D$. *Then* y, *the solution of*

$$\dot{y} = \mathrm{f}^0(y), \quad \mathrm{y}(0) = x_0,$$

is an $\mathcal{O}(\varepsilon t)$*-approximation of* x *(as compared to the usual* $\mathcal{O}(\varepsilon t e^{\varepsilon t})$ *Gronwall estimate).*

Proof Let y be the solution of

$$\dot{y} = \mathrm{f}^0(y), \quad \mathrm{y}(0) = x_0$$

and let t^* be the solution of

$$\dot{t}^\star = 1 + \varepsilon \frac{\mathrm{f}^1(\mathrm{y}(t^\star), t)}{\mathrm{f}^0(\mathrm{y}(t^\star))}, \quad \mathrm{t}^\star(0) = 0.$$

This equation is well defined and we have

$$|\mathrm{y}(\mathrm{t}^\star(t)) - \mathrm{y}(t)| \le \int_t^{t^\star(t)} |\dot{\mathrm{y}}(s)| \, \mathrm{d}s \le M|\mathrm{t}^\star(t) - t|,$$

while on the other hand

$$|\mathrm{t}^\star(t) - t| \le \int_0^t \varepsilon \left|\frac{\mathrm{f}^1(\mathrm{y}(\mathrm{t}^\star(s)), s)}{\mathrm{f}^0(\mathrm{y}(\mathrm{t}^\star(s)))}\right| \mathrm{d}s \le \frac{C}{m}\varepsilon t.$$

Therefore $|\mathrm{y}(\mathrm{t}^\star(t)) - \mathrm{y}(t)| \le C\frac{M}{m}\varepsilon t$. Let

$$\tilde{\mathrm{x}}(t) = \mathrm{y}(\mathrm{t}^\star(t)).$$

Then $\tilde{\mathrm{x}}(0) = \mathrm{y}(\mathrm{t}^\star(0)) = \mathrm{y}(0) = x_0$ and

$$\begin{aligned}
\dot{\tilde{\mathrm{x}}}(t) = \dot{\mathrm{t}}^\star \dot{\mathrm{y}} &= (1 + \varepsilon \frac{\mathrm{f}^1(\mathrm{y}(\mathrm{t}^\star(t)), t)}{\mathrm{f}^0(\mathrm{y}(\mathrm{t}^\star(t))})\mathrm{f}^0(\mathrm{y}(\mathrm{t}^\star(t))) \\
&= \mathrm{f}^0(\mathrm{y}(\mathrm{t}^\star(t))) + \varepsilon \mathrm{f}^1(\mathrm{y}(\mathrm{t}^\star(t)), t) \\
&= \mathrm{f}^0(\tilde{x}(t)) + \varepsilon \mathrm{f}^1(\tilde{\mathrm{x}}(t), t).
\end{aligned}$$

By uniqueness, $\tilde{\mathrm{x}} = \mathrm{x}$, the solution of the original equation, and therefore $\mathrm{x}(t) - \mathrm{y}(t) = \mathcal{O}(\varepsilon t)$. □

Although we have a higher dimensional problem, the inner equation is only two-dimensional and integrable. This makes it possible to apply a variant of this lemma in our situation. The nice thing about this lemma is that it gives explicit order estimates and we do not have to rely on abstract extension principles giving only $o(1)$-estimates.

Concluding Remarks

Although we have neither given the full theorem on passage through resonance, nor an adequate discussion of the technical difficulties, we have given here the main ideas and concepts that are needed to do this. Note that from the point of view of asymptotics passage through resonance and locking in resonance is still an open problem. The main difficulty arises as follows: In the case of locking the solution in the inner domain enters a ball in which all solutions attract toward a critical point or periodic solution. During the start of this process the solution has to pass fairly close to the saddle point (since the inner equation is conservative to first order, all attraction has to be small, and so is the splitting up of the stable and unstable manifold). While passing the saddle point, errors grow as $e^{\sqrt{\varepsilon}t}$ ($1/\sqrt{\varepsilon}$ being the natural time scale in the inner domain); the attraction in the ball on the other hand takes place on the time scale larger than $1/\varepsilon$ so that we get into trouble with the asymptotic estimates in the case of attraction (see Chapter 5). Finally it should be mentioned that the method of multiple time scales provides us with the same formal results but equally fails to describe the full asymptotic problem rigorously.

8.7 Two Examples

8.7.1 The Forced Mathematical Pendulum

We shall sketch the treatment of the perturbed pendulum equation

$$\ddot{\phi} + \sin(\phi) = \varepsilon F(\phi, \dot{\phi}, t)$$

while specifying the results in the case $F = \sin(t)$. The treatment is rather technical, involving elliptic integrals. The notation of elliptic integrals and a number of basic results are taken from [51]. This calculation has been inspired by [114] and [53] and some conversations with S.-N. Chow. Putting $\varepsilon = 0$ the equation has the energy integral

$$\frac{1}{2}\dot{\phi}^2 - \cos(\phi) = c.$$

It is convenient to introduce $\phi = 2\theta$ and $k^2 = 2/(1+c)$; then

$$\dot{\theta} = \pm\frac{1}{k}(1 - k^2\sin^2(\theta))^{\frac{1}{2}}.$$

Instead of t we introduce the time-like variable u by

$$t = k\int^{\theta} \frac{d\tau}{(1 - k^2\sin^2\tau)^{\frac{1}{2}}} = ku.$$

This implies $\sin(\theta) = \mathrm{sn}(ku, k)$ and

$$\dot{\theta} = \pm\frac{1}{k}(1 - k^2\mathrm{sn}^2(ku,k))^{\frac{1}{2}} = \pm\frac{1}{k}\mathrm{dn}(ku,k).$$

In the spirit of the method of variation of constants we introduce for the perturbed problem the transformation $(\theta, \dot{\theta}) \mapsto (k, u)$.

$$\theta = \mathrm{am}(ku,k), \quad \dot{\theta} = \frac{1}{k}\mathrm{dn}(ku,k).$$

After some manipulation of elliptic functions we obtain

$$\begin{aligned}
\dot{k} &= -\frac{\varepsilon}{2}k^2\mathrm{dn}(ku,k)F,\\
\dot{u} &= \frac{1}{k^2} + \frac{\varepsilon}{2}\frac{F}{1-k^2}(-E(ku)\mathrm{dn}(ku,k) + k^2\mathrm{sn}(ku,k)\mathrm{cn}(ku,k)).
\end{aligned}$$

One can demonstrate that

$$\dot{u} = \frac{1}{k^2} + \mathcal{O}(\varepsilon)$$

uniform in k. If F depends explicitly and periodically on time, the system of equations for k and u constitutes an example which can be handled by averaging over two angles t and u, except that one has to be careful with the infinite series. We take

$$F = \sin(t).$$

Fourier expansion of the right-hand side of the equation for k can be written down using the elliptic integral $K(k)$:

$$\dot{k} = \frac{\varepsilon k^2 \pi}{4K(k)} \sum_{m=-\infty}^{\infty} \frac{1}{\cosh \frac{m\pi K'(k)}{K(k)}} \sin(\frac{m\pi k}{K(k)} u - t).$$

It seems evident that we should introduce the angles

$$\psi_m = \frac{m\pi k}{K(k)} u - t,$$

with corresponding equation

$$\dot{\psi}_m = \frac{m\pi}{kK(k)} - 1 + \mathcal{O}(\varepsilon).$$

A resonance arises if for some $m = m_r$, and certain k

$$\frac{m_r \pi}{kK(k)} = 1.$$

We call the resonant angle ψ_r. The analysis until here runs along the lines pointed out in Section 8.5; there are only technical complications owing to the use of elliptic functions. We have to use an averaging transformation which is a direct extension of the case with one angle. It takes the form

$$x = y + \varepsilon \sum_{m \neq m_r} u_m(\psi_m, y)$$

As a result of the calculations we have the following vector field in the m_r-th resonance domain

$$\begin{aligned} \dot{k} &= \varepsilon \frac{k^2}{4K(k)} [\cosh(\frac{m_r \pi K'}{K})]^{-1} \sin(\psi_r), \\ \dot{\psi}_r &= \frac{m_r \pi}{kK(k)} - 1 + \mathcal{O}(\varepsilon). \end{aligned}$$

This means that for k corresponding to the resonance given by $m = m_r$ there are two periodic solutions, given by $\psi_r = 0, \pi$, one elliptic, one hyperbolic.

Note that we can formally take the limit for $m_r \to \infty$ while staying in resonance:

$$\lim_{\substack{m_r \to \infty \\ m_r \pi = kK(k)}} \dot{k} = \frac{\varepsilon \pi}{4K(k)\cosh(\frac{\pi}{2})} \sin(\psi_r)$$

One should compare this answer with the Melnikov function for this particular problem. (See also [240]).

8.7.2 An Oscillator Attached to a Fly-Wheel

We return to the example in Section 7.5.3, which describes a flywheel mounted on an elastic support; the description was taken from [108, Chapter 8.3]. The same model has been described in [88, Chapter 3.3], where one studies a motor with a slightly eccentric rotor, which interacts with its elastic support. First we obtain the equations of motion for x, the oscillator, and ϕ, the rotational motion, separately. We abbreviate $M_1(\dot{\phi}) = M(\dot{\phi}) - M_w(\dot{\phi})$, obtaining

$$\ddot{x} + \omega_0^2 x = \frac{\varepsilon}{m} \frac{-f(x) - \beta\dot{x} + q_1\dot{\phi}^2\cos(\phi)}{1 - \varepsilon^2 \frac{q_1 q_2}{m}\sin^2\phi} + \varepsilon^2 R_1,$$

$$\ddot{\phi} = \frac{\frac{\varepsilon}{J_0} M_1(\dot{\phi}) + \varepsilon q_2 g \sin(\phi) - \varepsilon q_2 \omega_0^2 x \sin(\phi)}{1 - \varepsilon^2 \frac{q_1 q_2}{m}\sin^2(\phi)} + \varepsilon^2 R_2,$$

with

$$R_1 = \frac{-\frac{q_2\omega_0^2}{m} x \sin^2(\phi) + \frac{q_1}{m}\sin(\phi)(\frac{1}{J_0}M_1(\dot{\phi}) + q_2 g \sin(\phi)}{1 - \varepsilon^2 \frac{q_1 q_2}{m}\sin^2(\phi)},$$

$$R_2 = \frac{\frac{q_2}{m}\sin(\phi)(-f(x) - \beta\dot{x} + q_1\dot{\phi}^2\cos(\phi))}{1 - \varepsilon^2 \frac{q_1 q_2}{m}\sin^2(\phi)}.$$

Using the transformation from Section 7.5, $x = r\sin(\phi_2)$, $\dot{x} = \omega_0 r\cos(\phi_2)$, $\phi = \phi_1$, $\dot{\phi} = \Omega$ we obtain to first order (modulo $\mathcal{O}(\varepsilon^2)$-terms)

$$\begin{bmatrix} \dot{r} \\ \dot{\Omega} \\ \dot{\phi}_1 \\ \dot{\phi}_2 \end{bmatrix} = \begin{bmatrix} 0 \\ 0 \\ \Omega \\ \omega_0 \end{bmatrix} + \varepsilon \begin{bmatrix} \frac{\cos(\phi_2)}{m\omega_0}(-f(r\sin(\phi_2)) - \beta\omega_0 r\cos(\phi_2) + q_1\Omega^2\cos(\phi_1)) \\ \frac{1}{J_0}M_1(\Omega) + q_2 g\sin(\phi_1) - q_2\omega_0^2 r\sin(\phi_1)\sin(\phi_2) \\ 0 \\ \mathcal{O}(1) \end{bmatrix}, \tag{8.7.1}$$

The right-hand sides of equation (8.7.1) can be written as separate functions of ϕ_2, $\phi_1 - \phi_2$ and $\phi_1 + \phi_2$ (cf. Remark 7.7.1). Assuming Ω to be positive we have only one resonance manifold given by

$$\Omega = \omega_0.$$

Outside the resonance we average over the angles to obtain

$$\begin{bmatrix} \dot{r} \\ \dot{\Omega} \end{bmatrix} = \varepsilon \begin{bmatrix} -\frac{\beta}{2m} r \\ \frac{1}{J_0} M_1(\Omega) \end{bmatrix} + \mathcal{O}(\varepsilon^2). \tag{8.7.2}$$

Depending on the choice of the motor characteristic $M_1(\Omega)$, the initial value of Ω and the eigenfrequency ω_0 the system will move into resonance or stay outside the resonance domain. In the resonance domain near $\Omega = \omega_0$ averaging over the angles ϕ_2 and $\phi_1 + \phi_2$ produces

$$\begin{bmatrix} \dot{r} \\ \dot{\Omega} \\ \dot{\phi}_1 - \dot{\phi}_2 \end{bmatrix} = \begin{bmatrix} 0 \\ 0 \\ \Omega - \omega_0 \end{bmatrix} + \varepsilon \begin{bmatrix} -\frac{\beta}{2m} r + \varepsilon \frac{q_1}{2m\omega_0} \Omega^2 \cos(\phi_1 - \phi_2) \\ \frac{1}{J_0} M_1(\Omega) - \varepsilon \frac{q_2 {\omega_0}^2}{2} r \cos(\phi_1 - \phi_2) \\ \mathcal{O}(1) \end{bmatrix} + \mathcal{O}(\varepsilon^2). \tag{8.7.3}$$

Putting $\chi = \phi_1 - \phi_2$ we can derive the inner equation

$$\dot{r} = 0, \quad \ddot{\chi} = \frac{\varepsilon}{J_0} M_1(\Omega) - \varepsilon \frac{q_2 {\omega_0}^2}{2} r \cos(\chi)$$

To study the possibility of locking into resonance we have to analyze the equilibrium solutions of the equation for r, Ω and χ in the resonance domain. If we find stability we should realize that we cannot expect the equilibrium solutions to be globally attracting. Some solutions will be attracted into the resonance domain and stay there, others will pass through the resonance. The equilibrium solutions are given by

$$\frac{\beta}{2m} r_\star = \frac{q_1}{2m\omega_0} {\Omega_\star}^2 \cos(\chi_\star), \quad \frac{1}{J_0} M_1(\Omega_\star) = \frac{q_2 \omega_0^2}{2} r_\star \cos(\chi_\star), \quad \Omega_\star = \omega_0.$$

The analysis produces three small eigenvalues containing terms of $\mathcal{O}(\varepsilon^{\frac{1}{2}})$, $\mathcal{O}(\varepsilon)$ and higher order. A second-order approximation of the equations of motion and the eigenvalues may be advisable but we do not perform this computation here (Note that in [108, page 319], a force $P(x) = -cx - \gamma x^3$ is used which introduces a mixture of first- and second-order terms; from the point of asymptotics this is not quite satisfactory). We conclude our study of the first-order calculation by choosing the constants and $M_1(\Omega)$ explicitly; this enables us to compare against numerical results. Choose $m = \omega_0 = \beta = q_1 = q_2 = J_0 = g = 1$; a linear representation is suitable for the motor characteristic:

$$M_1(\Omega) = \frac{1}{4}(2 - \Omega).$$

The equilibrium solutions are then given by

$$r_\star = \cos(\chi_\star), \quad r_\star \cos(\chi_\star) = \frac{1}{2}, \quad \Omega_\star = 1$$

so that $r_\star = 1/\sqrt{2}$ and $\chi_\star = \pm\pi/4$. The calculation thus far suggests that locally, in the resonance manifold $r = 1/\sqrt{2}$, $\Omega = 1$, stable attracting solutions may exist corresponding to the phenomenon of locking in resonance. Starting with initial conditions $r(0) > 1/\sqrt{2}$, $\Omega(0) < 1$ equations (8.7.2) tell us that we move into resonance; some of the solutions, depending on the initial value of χ, will be caught in the resonance domain; other solutions will pass through resonance, and, again according to equation (8.7.2), will move to a region where r is near zero and Ω is near 2.

9

From Averaging to Normal Forms

9.1 Classical, or First-Level, Normal Forms

The essence of the method of averaging is to use near-identity coordinate changes to simplify a system of differential equations. (This is clearly seen, for instance, in Section 2.9, where the original system is periodic in time and the simplified system is autonomous up to some order k.) The idea of simplification by near-identity transformations is useful in other circumstances as well. In the remaining chapters of this book we turn to the topic of normal forms for systems of differential equations near an equilibrium point (or rest point). This topic has much in common with the method of averaging. A slow and detailed treatment of normal forms, with full proofs, may be found in [203]. These proofs will not be repeated here, and they are not needed in concrete examples. Instead, we will survey the theory without proofs in this chapter, and then, in later chapters, turn to topics that are not covered in [203]. These include a detailed treatment of normal forms for Hamiltonian resonances and recent developments in the theory of higher-level normal forms. (By *higher-level normal forms* we mean what various authors call *higher-order normal forms*, *hypernormal forms*, *simplest normal forms*, and *unique normal forms*.)

The starting point for normal form theory is a smooth autonomous system in $\mathbb{R}^n$ with a rest point at the origin, expanded in a formal (that is, not necessarily convergent) Taylor series

$$\dot{\boldsymbol{x}} = \mathbf{f}^{[0]}(\boldsymbol{x}) = A\boldsymbol{x} + \mathbf{f}^1(\boldsymbol{x}) + \mathbf{f}^2(\boldsymbol{x}) + \cdots , \tag{9.1.1}$$

where A is an $n \times n$ matrix and $\mathbf{f}^j$ contains only terms of degree $j+1$. For clarity we introduce the following terminology.

Definition 9.1.1. *A vector field $\mathbf{f}^j$ on $\mathbb{R}^n$ containing only terms of degree $j+1$, that is,*

$$\mathbf{f}^j(\boldsymbol{x}) = \sum_{\mu_1+\cdots+\mu_n=j+1} \mathbf{a}_m x_1^{\mu_1} \cdots x_n^{\mu_n},$$

will be called a **homogeneous vector polynomial of grade** j. *The (finite dimensional) vector space of all homogeneous vector polynomials of grade* j *is denoted* $\mathcal{V}^j$. *The (infinite dimensional) vector space of all formal power series of the form (9.1.1) is denoted*

$$\mathcal{V} = \prod_{j=0}^{\infty} \mathcal{V}^j.$$

(The elements of $\mathcal{V}$ *are written as sums, as in (9.1.1), but since we allow infinitely many nonzero terms,* $\mathcal{V}$ *is technically the direct product, rather than the direct sum, of the* $\mathcal{V}^j$.*)*

It is often convenient to *dilate* the coordinate system x around the origin by putting $\boldsymbol{x} = \varepsilon\boldsymbol{\xi}$, where ε is a small parameter. Then (9.1.1) becomes $\varepsilon\dot{\boldsymbol{\xi}} = \mathbf{f}^{[0]}(\varepsilon\boldsymbol{\xi})$, and after canceling a factor of ε we have

$$\dot{\boldsymbol{\xi}} = A\boldsymbol{\xi} + \varepsilon\mathbf{f}^1(\boldsymbol{\xi}) + \varepsilon^2\mathbf{f}^2(\boldsymbol{\xi}) + \cdots. \tag{9.1.2}$$

In this form, the notation conforms to our usual conventions explained in Notation 1.5.2, which justifies the use of grade rather than degree. (Another justification is given below.) Written with a remainder term, (9.1.2) becomes

$$\dot{\boldsymbol{\xi}} = A\boldsymbol{\xi} + \varepsilon\mathbf{f}^1(\boldsymbol{\xi}) + \varepsilon^2\mathbf{f}^2(\boldsymbol{\xi}) + \cdots + \varepsilon^k\mathbf{f}^k(\boldsymbol{\xi}) + \frac{1}{\varepsilon}\mathbf{f}^{[k+1]}(\varepsilon\boldsymbol{\xi}), \tag{9.1.3}$$

in which the remainder term is $\mathcal{O}(\varepsilon^{k+1})$ on the ball $\|\xi\| \leq 1$. In the original (undilated) coordinates, $\mathbf{f}^{[k+1]}(\boldsymbol{x})$ is $\mathcal{O}(\varepsilon^{k+1})$ in the small neighborhood $\|\boldsymbol{x}\| \leq \varepsilon$.

9.1.1 Differential Operators Associated with a Vector Field

There are two differential operators associated with any vector field $\mathbf{v}$ on $\mathbb{R}^n$ that will appear repeatedly in the sequel. (These have already been introduced briefly in Section 3.2). The first operator is

$$\mathcal{D}_{\mathbf{v}} = \mathrm{v}_1(\boldsymbol{x})\frac{\partial}{\partial x_1} + \cdots + \mathrm{v}_n(\boldsymbol{x})\frac{\partial}{\partial x_n}. \tag{9.1.4}$$

This operator is classically written $\mathbf{v} \cdot \nabla$, and in differential geometry is often identified with the vector field $\mathbf{v}$ itself (so that vector fields simply *are* differential operators). Applied to a scalar field $f : \mathbb{R}^n \to \mathbb{R}$, it produces a new scalar field $\mathcal{D}_{\mathbf{v}} f$ which may be written as

$$(\mathcal{D}_{\mathbf{v}} f)(x) = \mathsf{D}f(\boldsymbol{x})\mathbf{v}(\boldsymbol{x})$$

(remembering that $\mathsf{D}f(\boldsymbol{x})$ is a row and $\mathbf{v}(\boldsymbol{x})$ is a column). This is the rate of change of f along the flow of $\mathbf{v}$ at $\boldsymbol{x}$. Applied to a vector field,

$$(\mathcal{D}_{\mathbf{v}}\mathbf{w})(\boldsymbol{x}) = \mathsf{D}\mathbf{w}(\boldsymbol{x})\mathbf{v}(\boldsymbol{x}) = \begin{bmatrix} \mathcal{D}_{\mathbf{v}} w_1 \\ \vdots \\ \mathcal{D}_{\mathbf{v}} w_n \end{bmatrix}.$$

The second operator associated with $\mathbf{v}$ is the **Lie derivative** operator $\mathcal{L}_{\mathbf{v}}$ is a differential operator that can be applied only to vector fields. It is defined by

$$\mathcal{L}_{\mathbf{v}}\mathbf{w} = \mathcal{D}_{\mathbf{w}}\mathbf{v} - \mathcal{D}_{\mathbf{v}}\mathbf{w}, \tag{9.1.5}$$

or equivalently,

$$(\mathcal{L}_{\mathbf{v}}\mathbf{w})(\boldsymbol{x}) = [\mathbf{v}, \mathbf{w}] = \mathsf{D}\mathbf{w}(\boldsymbol{x})\mathbf{v}(\boldsymbol{x}) - \mathsf{D}\mathbf{v}(\boldsymbol{x})\mathbf{w}(\boldsymbol{x}). \tag{9.1.6}$$

The symbol $[\mathbf{v}, \mathbf{w}]$ is called the **Lie bracket** of $\mathbf{v}$ and $\mathbf{w}$. For the special case of linear vector fields $\mathbf{v}(\boldsymbol{x}) = A\boldsymbol{x}$ we write $\mathcal{D}_A$ and $\mathcal{L}_A$ for $\mathcal{D}_{\mathbf{v}}$ and $\mathcal{L}_{\mathbf{v}}$.

Remark 9.1.2. For some purposes it is better to replace $\mathcal{D}_{\mathbf{v}}$ with the operator

$$\nabla_{\mathbf{v}} = \frac{\partial}{\partial t} + \mathcal{D}_{\mathbf{v}} = \frac{\partial}{\partial t} + \mathrm{v}_1(x)\frac{\partial}{\partial x_1} + \cdots + \mathrm{v}_n(x)\frac{\partial}{\partial x_n}. \tag{9.1.7}$$

This is the appropriate generalization of $\mathcal{D}_{\mathbf{v}}$ for application to a time-dependent scalar field $f(\boldsymbol{x}, t)$. When f does not depend on t, the $\partial/\partial t$ term has no effect and the operator is the same as $\mathcal{D}_{\mathbf{v}}$. The operator (9.1.7) can be used even when the vector field $\mathbf{v}$ is time-dependent. For time-periodic vector fields, this operator has already appeared in (3.2.11). In Section 11.2, time-periodic vector fields will be identified with operators of this form. ♡

Definition 9.1.3. *Let P be a point in a set, and* $\mathbf{a}$, $\mathbf{b}$ *vectors. Then we define an* **affine space** *as consisting of points* $\mathbf{x} = P + \mathbf{a}$ *and* $\mathbf{y} = P + \mathbf{b}$, *with addition and scalar multiplication defined by*

$$\lambda\mathbf{x} + \mu\mathbf{y} = P + \lambda\mathbf{a} + \mu\mathbf{b}.$$

Notice that the $\nabla_{\mathbf{v}}$ form an affine space (with $P = \frac{\partial}{\partial t}$), with addition and multiplication by parameters redefined by

$$\mu(\boldsymbol{x})\nabla_{\mathbf{v}} + \lambda(\boldsymbol{x})\nabla_{\mathbf{w}} = \nabla_{\mu(\boldsymbol{x})\mathbf{v}+\lambda(\boldsymbol{x})\mathbf{w}}, \quad \lambda, \mu \in P[\mathbb{R}^n].$$

The rule for the application of $\nabla_{\mathbf{v}}$ on $\lambda(\boldsymbol{x})\mathbf{w}$ is the usual

$$\nabla_{\mathbf{v}}\lambda(\boldsymbol{x})\mathbf{w} = (\mathcal{D}_{\mathbf{v}}\lambda(\boldsymbol{x}))\mathbf{w} + \lambda(\boldsymbol{x})\nabla_{\mathbf{v}}\mathbf{w}.$$

Notice that

$$\nabla_{\mathbf{w}}\mathcal{D}_{\mathbf{v}} - \mathcal{D}_{\mathbf{v}}\nabla_{\mathbf{w}} = \mathcal{D}_{\mathbf{v}_t} + \mathcal{D}_{[\mathbf{w},\mathbf{v}]}.$$

This is also denoted by $\mathcal{L}_{\mathbf{v}}\hat{\mathbf{w}}$, and mimics the effect of a transformation generator $\mathbf{v}$ on a vector field $\mathbf{w}$.

Remark 9.1.4. The Lie bracket is sometimes (for instance, in [203]) defined to be the negative of our bracket, but the Lie derivative is always defined as in (9.1.5). Our version of the Lie bracket has the advantage that

$$\mathcal{D}_{[\mathbf{v},\mathbf{w}]} = [\mathcal{D}_{\mathbf{v}}, \mathcal{D}_{\mathbf{w}}] = \mathcal{D}_{\mathbf{v}}\mathcal{D}_{\mathbf{w}} - \mathcal{D}_{\mathbf{w}}\mathcal{D}_{\mathbf{v}},$$

where the bracket of linear operators $\mathcal{D}_{\mathbf{v}}$ and $\mathcal{D}_{\mathbf{w}}$ is their commutator. But from the point of view of group representation theory there are advantages to the other version. Vector fields form a Lie algebra under the bracket operation. In any Lie algebra, one writes

$$\mathrm{ad}(\mathbf{v})\mathbf{w} = [\mathbf{v}, \mathbf{w}].$$

Therefore with our choice of bracket, $\mathcal{L}_{\mathbf{v}} = \mathrm{ad}(\mathbf{v})$ and with the opposite choice $\mathcal{L}_{\mathbf{v}} = -\mathrm{ad}(\mathbf{v})$. Although this can be confusing in comparing books, in normal forms we are most often concerned only with the kernel and the image of $\mathcal{L}_{\mathbf{v}}$, which are the same as the kernel and image of $\mathrm{ad}(\mathbf{v})$ under either convention. If $\mathbf{f}^i \in \mathcal{V}^i$ and $\mathbf{f}^j \in \mathcal{V}^j$, then $[\mathbf{f}^i, \mathbf{f}^j] \in \mathcal{V}^{i+j}$, meaning that $\mathcal{V}$ is a **graded Lie algebra**. This is the promised second justification for using grade rather than degree. ♡

9.1.2 Lie Theory

The main idea of normal forms is to change coordinates in (9.1.1) from x to y by a transformation of the form

$$\boldsymbol{x} = \mathbf{U}(\boldsymbol{y}) = \boldsymbol{y} + \mathbf{u}^1(\boldsymbol{y}) + \mathbf{u}^2(\boldsymbol{y}) + \cdots, \tag{9.1.8}$$

with $\mathbf{u}^j \in \mathcal{V}^j$, to obtain a system

$$\dot{\boldsymbol{y}} = \mathbf{g}^{[0]}(\boldsymbol{y}) = A\boldsymbol{y} + \mathbf{g}^1(\boldsymbol{y}) + \mathbf{g}^2(\boldsymbol{y}) + \cdots \tag{9.1.9}$$

that is in some manner "simpler" than (9.1.1). If we work in dilated coordinates (with $\boldsymbol{x} = \varepsilon\boldsymbol{\xi}$ and $\boldsymbol{y} = \varepsilon\boldsymbol{\eta}$), the transformation (9.1.8) will take the form

$$\boldsymbol{\xi} = \boldsymbol{\eta} + \varepsilon\mathbf{u}^1(\boldsymbol{\eta}) + \varepsilon^2\mathbf{u}^2(\boldsymbol{\eta}) + \cdots.$$

In Section 3.2.2 we have seen that it is very helpful to express such a transformation in terms of a vector field called the **generator** of the transformation. Rather than work in dilated coordinates, we restate Theorem 3.2.1 in a manner suitable for (9.1.8).

Theorem 9.1.5. *Let $\mathbf{w}^{[1]}$ be a vector field (called a* **generator***) of the form*

$$\mathbf{w}^{[1]}(\boldsymbol{x}) = \mathbf{w}^1(\boldsymbol{x}) + \mathbf{w}^2(\boldsymbol{x}) + \cdots, \tag{9.1.10}$$

with $\mathbf{w}^j \in \mathcal{V}^j$. Then the transformation

$$x = \mathbf{U}(y) = e^{\mathcal{D}_{\mathbf{w}[1]}} y = y + \mathbf{u}^1(y) + \mathbf{u}^2(y) + \cdots \tag{9.1.11}$$

transforms the system

$$\dot{x} = \mathbf{f}^{[0]}(x) = Ax + \mathbf{f}^1(x) + \mathbf{f}^2(x) + \cdots \tag{9.1.12}$$

into the system

$$\dot{y} = \mathbf{g}^{[0]}(y) = e^{\mathcal{L}_{\mathbf{w}[1]}} \mathbf{f}^{[0]}(y) = Ay + \mathbf{g}^1(y) + \mathbf{g}^2(y) + \cdots . \tag{9.1.13}$$

For each j, $\mathbf{w}^j$ and $\mathbf{g}^j$ satisfy a homological equation *of the form*

$$\mathcal{L}_A \mathbf{w}^j(y) = \mathbf{K}^j(y) - \mathbf{g}^j(y), \tag{9.1.14}$$

where $\mathbf{K}^1 = \mathbf{f}^1$ and for $j > 1$, $\mathbf{K}^j$ equals $\mathbf{f}^j$ plus correction terms depending only on $\mathbf{f}^1, \dots, \mathbf{f}^{j-1}$ and $\mathbf{w}^1, \dots, \mathbf{w}^{j-1}$.

Remark 9.1.6. We usually assume that the transformation is carried out only to some finite order k, because this is all that can be done in finite time. However, in the usual sense of mathematical existence we can say that any formal power series for the generator determines a formal power series for the transformed system. Further, the Borel–Ritt theorem (see [203, Theorem A.3.2]) states that every formal power series is the Taylor series of some smooth function (which is not unique, but is determined only up to a "flat" function having zero Taylor series). Using this result, it follows that there does exist a smooth transformation taking the original system into a smooth system transformed to all orders. While it is convenient to speak of systems normalized to all orders for theoretical purposes, they are not computable in practice. The only way to achieve a system that is entirely in normal form is to normalize to some order k and truncate there, which, of course, introduces some error, discussed below in the semisimple case. ♡

9.1.3 Normal Form Styles

In order to use Theorem 9.1.5 to put a system (9.1.12) into a normal form (9.1.13), it is necessary to choose $\mathbf{g}^j$ (at each stage $j = 1, 2, \dots$) so that $\mathbf{K}^j - \mathbf{g}^j$ belongs to the image of $\mathcal{L}_A$ (regarded as a map of $\mathcal{V}^j$ to itself). Then the homological equation (9.1.14) will be solvable for $\mathbf{w}^j$, so it will be possible to construct a generator leading to the desired normal form. So the requirement that $\mathbf{K}^j - \mathbf{g}^j \in \operatorname{im} \mathcal{L}_A$ imposes a limitation on what kinds of normal forms are achievable. (It is, for instance, not always possible to choose $\mathbf{g}^j = 0$, because $\mathbf{K}^j$ will usually not belong to im $\mathcal{L}_A$.)

More precisely, let $(\operatorname{im} \mathcal{L}_A)^j$ denote the image of the map $\mathcal{L}_A : \mathcal{V}^j \to \mathcal{V}^j$ and let $\mathcal{N}^j$ be any complement to this image, so that

$$\mathcal{V}^j = (\operatorname{im} \mathcal{L}_A)^j \oplus \mathcal{N}^j. \tag{9.1.15}$$

Let

$$P : \mathcal{V}^j \to \mathcal{N}^j \tag{9.1.16}$$

be the projection into $\mathcal{N}^j$ associated with this direct sum. (That is, any $\mathbf{v} \in \mathcal{V}^j$ can be written uniquely as $\mathbf{v} = (\mathbf{v} - P\mathbf{v}) + P\mathbf{v}$ with $\mathbf{v} - P\mathbf{v} \in (\text{im } \mathcal{L}_A)^j$ and $P\mathbf{v} \in \mathcal{N}^j$.) Then we may take

$$\mathbf{g}^j = P\mathbf{K}^j \tag{9.1.17}$$

and the homological equation will be solvable.

So the question of what is the most desirable, or simplest, form for $\mathbf{g}^j$ reduces to the choice of the most desirable complement $\mathcal{N}^j$ to the image of $\mathcal{L}_A$ in each grade. Such a choice of $\mathcal{N}^j$ is called a **normal form style**.

9.1.4 The Semisimple Case

The simplest case is the case in which A is semisimple, meaning that it is diagonalizable over the complex numbers. In this case, $\mathcal{L}_A$ is also semisimple ([203, Lemma 4.5.2]). For any semisimple operator, the space on which the operator acts is the direct sum of the image and kernel of the operator. Therefore we can take

$$\mathcal{N}^j = (\ker \mathcal{L}_A)^j, \tag{9.1.18}$$

that is, the kernel of the map $\mathcal{L}_A : \mathcal{V}^j \to \mathcal{V}^j$. In other words, for the system (9.1.13) in normal form we will have

$$\mathcal{L}_A \mathbf{g}^j = 0. \tag{9.1.19}$$

This fact is commonly stated as "the nonlinear terms (in normal form) commute with the linear term." (If two vector fields $\mathbf{v}$ and $\mathbf{w}$ satisfy $[\mathbf{v}, \mathbf{w}] = 0$, they are said to commute. This implies that their flows ϕ^s and ψ^t commute in the sense that $\phi^s \circ \psi^t = \psi^t \circ \phi^s$ for all t and s.)

The consequences of (9.1.19) are quite profound, sufficiently so that no other choice of $\mathcal{N}^j$ is ever used in the semisimple case. Therefore this choice is called the **semisimple normal form style**. Some of these consequences are geometric: When the equation is normalized to any finite grade k and truncated at that grade, it will have symmetries that were not present (or were present but only in a hidden way) in the original system. These symmetries make it easy to locate the stable, unstable, and center manifolds of the rest point at the origin, and to determine preserved fibrations and foliations that reflect facts about the dynamics of the system near the origin. (See [203, Section 5.1].) Another consequence allows us to estimate the error due to truncation of the normal form, on the center manifold, in a manner very similar to the proof of the asymptotic estimate for higher-order periodic averaging. For simplicity we state the result for the case of pure imaginary eigenvalues (so that the center manifold is the whole space).

Theorem 9.1.7. *Suppose that A is semisimple and has only pure imaginary eigenvalues. Suppose that the system*

$$\dot{\boldsymbol{y}} = A\boldsymbol{y} + \mathbf{g}^1(\boldsymbol{y}) + \cdots + \mathbf{g}^k(\boldsymbol{y}) + \mathbf{g}^{[k+1]}(\boldsymbol{y})$$

is in semisimple normal form through grade k, so that $\mathcal{L}_A\mathbf{g}^j = 0$ for $j = 1,\dots,k$. Let $\varepsilon > 0$ and let $\mathbf{y}(t)$ be a solution with $\|\mathbf{y}(0)\| < \varepsilon$. Let $\mathbf{z}(t)$ be a solution of the truncated equation

$$\dot{\boldsymbol{z}} = A\boldsymbol{z} + \mathbf{g}^1(\boldsymbol{z}) + \cdots + \mathbf{g}^k(\boldsymbol{z})$$

with $\mathbf{z}(0) = \mathbf{y}(0)$. Then $\|\mathbf{y}(t) - \mathbf{z}(t)\| = \mathcal{O}(\varepsilon^k)$ for time $\mathcal{O}(1/\varepsilon)$.

Proof The idea of the proof is to make a change of variables $\boldsymbol{y} = e^{At}\boldsymbol{u}$. The linear term is removed, and because of (9.1.19) the nonlinear terms through grade k are not affected, resulting in

$$\dot{\boldsymbol{u}} = \mathbf{g}^1(\boldsymbol{u}) + \cdots + \mathbf{g}^k(\boldsymbol{u}) + e^{-At}\mathbf{g}^{[k+1]}(e^{At}\boldsymbol{u}) \tag{9.1.20}$$

The last term is bounded for all time because A is semisimple with imaginary eigenvalues, and the effect of truncating it may be estimated by the Gronwall inequality. For details see [203, Lemma 5.3.6]. □

In fact, with one additional condition on A, the semisimple normal form can actually be computed by the method of averaging. Suppose that A is semisimple with eigenvalues $\pm i\omega_j \neq 0$ for $j = 1,\dots,n/2$. (The dimension must be even.) Then the change of variables $\boldsymbol{\xi} = e^{At}\boldsymbol{v}$, applied to the dilated equation (9.1.2), results in

$$\dot{\boldsymbol{v}} = e^{-At}(\varepsilon\mathbf{f}^1(e^{At}\boldsymbol{x}) + \varepsilon^2\mathbf{f}^2(e^{At}\boldsymbol{x}) + \cdots)$$

If there is a common integer multiple T of the periods $2\pi/\omega_j$, this equation will be periodic of period T and in standard form for averaging. In this case, averaging to order k will eliminate the time dependence from the terms through order k. This will produce exactly the dilated form of (9.1.20). That is, the semisimple normal form coefficients $\mathbf{g}^1,\dots,\mathbf{g}^k$ can be computed by averaging.

9.1.5 The Nonsemisimple Case

When A is not semisimple, there is no obvious best choice of a normal form style. The simplest (and oldest) approach is to decompose A into semisimple and nilpotent parts

$$A = S + N \tag{9.1.21}$$

such that S and N commute ($SN = NS$, or $[S,N] = 0$), and then to normalize the higher order terms with respect to S only, so that the system (9.1.13) in normal form satisfies

$$\mathcal{L}_S\mathbf{g}^j = 0. \tag{9.1.22}$$

This is always possible, and the result is often called a **Poincaré–Dulac normal form.** (In [203] it is called an **extended semisimple normal form.**) Poincaré–Dulac normal forms have many of the advantages of semisimple normal forms. For instance, the normalized vector field (normalized to grade k and truncated there) commutes with S and therefore inherits symmetries from S (expressible as preserved fibrations and foliations).

Remark 9.1.8. The most common way to achieve the decomposition (9.1.21) is to put A into Jordan canonical form (which may require complex numbers) or real canonical form (which does not). When A is in Jordan form, S is the diagonal part of A and N is the off-diagonal part. However, there are algorithms to perform the decomposition without putting A into canonical form first, and these require less work if the canonical form is not required, see Algorithm 11.1. ♡

However, the Poincaré–Dulac normal form, by itself, is not a true normal form style as defined above. That is, ker $\mathcal{L}_S$ is not a complement to im $\mathcal{L}_A$ and we have only $\mathcal{V}^j = (\text{im } \mathcal{L}_A)^j + (\text{ker } \mathcal{L}_S)^j$, not $\mathcal{V}^j = (\text{im } \mathcal{L}_A)^j \oplus (\text{ker } \mathcal{L}_S)^j$. This means that ker $\mathcal{L}_S$ is too large a subspace: The Poincaré–Dulac normal form is capable of further simplification (while still remaining within the notion of a classical, or first-level, normal form).

So the goal is to define a normal form style $\mathcal{N}^j$ that is a true complement to im $\mathcal{L}_A$ but also satisfied $\mathcal{N}^j \subset \text{ker } \mathcal{L}_S$, so that the advantages of a Poincaré–Dulac normal form are not lost. There are two ways of doing this that are in common use, the **transpose normal form style** and the $\mathfrak{sl}_2$ **normal form style.** We now describe these briefly.

9.1.6 The Transpose or Inner Product Normal Form Style

The transpose normal form style is defined by

$$\mathcal{N}^j = (\text{ker } \mathcal{L}_{A^*})^j, \tag{9.1.23}$$

where A^* is the transpose of A (or the conjugate transpose, if complex numbers are allowed). This is always a complement to im $\mathcal{L}_A$, but *it is not always a subset of* ker $\mathcal{L}_S$. That is, a system in transpose normal form is not always in Poincaré–Dulac normal form. However, the transpose normal form is a Poincaré–Dulac normal form when A is in Jordan or real canonical form, and more generally, whenever S commutes with S^* (see [203, Lemma 4.6.10]).

The name **inner product normal form** comes from the way that (9.1.23) is proved to be a normal form style. One introduces an inner product on each space $\mathcal{V}^j$ in such a way that $(\mathcal{L}_A)^*$ (the adjoint of $\mathcal{L}_A$ with respect to the new inner product) coincides with $\mathcal{L}_{A^*}$. Since the Fredholm alternative theorem implies that $\mathcal{V}^j = (\text{im } \mathcal{L}_A)^j \oplus (\text{ker } (\mathcal{L}_A)^*)^j$, the proof is then complete. (In fact this direct sum decomposition is orthogonal.) The required inner product on $\mathcal{V}^j$ was first used for this purpose by Belitskii ([25]) but was rediscovered and popularized by [85]. See [203, Section 4.6] for a complete treatment. Notice that the transpose normal form can be computed and used from its definition (9.1.23) without any mention of the inner product.

There are other inner product normal forms besides the transpose normal form. In fact, any inner product defined on each $\mathcal{V}^j$ leads to a normal form style, by the simple declaration

$$\mathcal{N}^j = (\operatorname{im} \mathcal{L}_A)^\perp. \tag{9.1.24}$$

As noted above, the transpose normal form is of this kind. Inner product normal forms (in the general sense) will usually not be Poincaré–Dulac.

9.1.7 The $\mathfrak{sl}_2$ Normal Form

The other accepted way to achieve a normal form style consistent with the Poincaré–Dulac requirement is called the $\mathfrak{sl}_2$ normal form style, because the justification of the style is based on the representation theory of the Lie algebra called $\mathfrak{sl}_2$. Briefly, given the decomposition (9.1.21), it is possible to find matrices M and H such that the following conditions on commutator brackets hold:

$$[N, M] = H, \quad [H, N] = 2N, \quad [H, M] = -2M, \quad [M, S] = 0. \tag{9.1.25}$$

(Then M is nilpotent, H is semisimple, and N, M, and H span a three-dimensional vector space, closed under commutator bracket, which is isomorphic as a Lie algebra to $\mathfrak{sl}_2$.) The $\mathfrak{sl}_2$ normal form style is now defined by

$$\mathcal{N}^j = \ker \mathcal{L}_M. \tag{9.1.26}$$

This normal form style is Poincaré–Dulac in all cases, with no restrictions on A.

Remark 9.1.9. A detailed treatment of the $\mathfrak{sl}_2$ normal form is given in the starred sections of [203, Sections 2.5–2.7, 3.5, and 4.8]. In this reference, M is called the **pseudotranspose** of N, because when N is in Jordan form, M looks like the transpose of N with the off-diagonal ones replaced by other positive integer constants. In fact, when A is in Jordan form, a system in $\mathfrak{sl}_2$ normal form will look much like the same system in transpose normal form, except for the appearance of certain constant numerical factors. There are several advantages to the $\mathfrak{sl}_2$ style over the transpose style. There exists a certain computational algorithm, for use with symbolic processors, to compute the projection (9.1.16) for the $\mathfrak{sl}_2$ normal form; there is no equivalent algorithm for the transpose normal form. Also, the set of all systems in normal form with a given A has an algebraic structure (as a module of equivariants over a ring of invariants) that is best studied by the use of $\mathfrak{sl}_2$ methods (even if one is interested in the transpose style, [203, Section 4.7]). ♡

9.2 Higher Level Normal Forms

There are two questions that might occur naturally to anyone thinking about classical normal forms.

1. In the classical normal form, the generator coefficients $\mathbf{w}^1, \mathbf{w}^2, \ldots$ are chosen to satisfy the homological equations (9.1.14), but the solutions of these equations are not unique. Is it possible to achieve additional simplifications of the normal form by making judicious choices of the solutions to the homological equations?
2. In the classical normal form, the higher order terms are normalized "with respect to the linear term." That is, a system is in normal form if it satisfies a condition defined using only the matrix A. For instance, in the semisimple case a system is in normal form if the nonlinear terms satisfy (9.1.19). Would it not be reasonable to normalize the quadratic term with respect to the linear term, then to normalize the cubic term with respect to the sum of the linear and quadratic terms, and so forth? Would this lead to a more complete normalization?

The answer to both questions is yes, and the answers turn out to be the same. That is, making judicious choices of the solutions to the homological equations amounts to exactly the same thing as normalizing each term with respect to the sum of the preceding terms.

To see some of the difficulties of this problem, let us consider only normalization through the cubic term. According to Theorem 9.1.5, a generator of the form $\mathbf{w}^1 + \cdots$ carries the vector field $\mathbf{f}^0 + \mathbf{f}^1 + \mathbf{f}^2 + \cdots$ (with $\mathbf{f}^0(\boldsymbol{x}) = A\boldsymbol{x}$) into $\mathbf{g}^0 + \mathbf{g}^1 + \mathbf{g}^2 + \cdots$, where

$$\begin{pmatrix} \mathbf{g}^0 \\ \mathbf{g}^1 \\ \mathbf{g}^2 \end{pmatrix} = \begin{pmatrix} \mathbf{f}^0 \\ \mathbf{f}^1 + \mathcal{L}_{\mathbf{w}^1}\mathbf{f}^0 \\ \mathbf{f}^2 + \mathcal{L}_{\mathbf{w}^1}\mathbf{f}^1 + \mathcal{L}_{\mathbf{w}^2}\mathbf{f}^0 + \frac{1}{2}\mathcal{L}^2_{\mathbf{w}^1}\mathbf{f}^0 \end{pmatrix}. \tag{9.2.1}$$

(Be sure to notice that the sum of all the indices in a term is constant, equal to the grade, throughout each equation; for instance in the cubic $\mathbf{g}^2$ equation the indices sum to 2 in each term. In making this count we regard $\mathcal{L}^2_{\mathbf{w}^1}$ as $\mathcal{L}_{\mathbf{w}^1}\mathcal{L}_{\mathbf{w}^1}$.) The second equation in (9.2.1) is the same as the first homological equation $\mathcal{L}_A\mathbf{w}^1 = \mathbf{f}^1 - \mathbf{g}^1$; as usual, we choose a normal form style $\mathcal{N}^1$ and put $\mathbf{g}^1 = P\mathbf{f}^1$ (the projection of $\mathbf{f}^1$ into $\mathcal{N}^1$), then solve the homological equation for $\mathbf{w}^1$. But if $\mathbf{w}^1$ is any such solution, then $\mathbf{w}^1 + \boldsymbol{\kappa}^1$ is another, where $\boldsymbol{\kappa}^1 \in \ker \mathcal{L}_A$. (The notation $\boldsymbol{\kappa}^1$ is chosen to reflect the similar situation in averaging, discussed in Section 3.4.)

At this point we may proceed in two ways. Replacing $\mathbf{w}^1$ by $\mathbf{w}^1 + \boldsymbol{\kappa}^1$ in the third equation of (9.2.1), we may try to choose $\boldsymbol{\kappa}^1$ and $\mathbf{w}^2$ to simplify $\mathbf{g}^2$. This approach fits with the idea of question 1 at the beginning of this section. But the equations become quite messy, and it is better to follow the idea of question 2. To do this, we first ignore $\boldsymbol{\kappa}^1$ and apply only the generator

$\mathbf{w}^1$ (which, we recall, was any particular solution to the first homological equation), with $\mathbf{w}^2 = 0$. This changes $\mathbf{f}^1$ into its desired form $\mathbf{g}^1$, and changes $\mathbf{f}^2$ into $\mathbf{f}^2 + \mathcal{L}_{\mathbf{w}^1}\mathbf{f}^1 + \frac{1}{2}\mathcal{L}^2_{\mathbf{w}^1}\mathbf{f}^0$, which is simply an uncontrolled change with no particular improvement. We now rename this vector field as $\mathbf{f}^0 + \mathbf{f}^1 + \mathbf{f}^2 + \cdots$, and apply a second generator $\mathbf{w}^1 + \mathbf{w}^2 + \cdots$, using (9.2.1) once again. This time, since $\mathbf{f}^1$ is already in the desired form, we restrict $\mathbf{w}^1$ to lie in $\ker \mathcal{L}_{\mathbf{f}^0} = \ker \mathcal{L}_A$, so that $\mathcal{L}_{\mathbf{w}^1}\mathbf{f}^0 = 0$ and no change is produced in $\mathbf{f}^1$. (Notice that $\mathbf{w}^1$ now has exactly the same freedom as $\boldsymbol{\kappa}^1$ did in the first approach.) It also follows that $\mathcal{L}^2_{\mathbf{w}^1}\mathbf{f}^0 = 0$, so the third equation of (9.2.1) simplifies, and it can be rewritten in the form

$$\mathcal{L}_{\mathbf{f}^1}\mathbf{w}^1 + \mathcal{L}_{\mathbf{f}^0}\mathbf{w}^2 = \mathbf{f}^2 - \mathbf{g}^2. \tag{9.2.2}$$

(Warning: do not forget that $\mathbf{f}^1$ and $\mathbf{f}^2$ are the "new" functions after the first stage of normalization.) Equation (9.2.1) is an example of a **generalized homological equation**. It may be approached in a similar way to ordinary homological equations. The subspace of $\mathcal{V}^2$ consisting of all values of $\mathcal{L}_{\mathbf{f}^1}\mathbf{w}^1 + \mathcal{L}_{\mathbf{f}^0}\mathbf{w}^2$, as $\mathbf{w}^1$ ranges over $\ker \mathcal{L}_A$ and $\mathbf{w}^2$ ranges over $\mathcal{V}^2$, is called the **removable space** in $\mathcal{V}^2$. It is larger than the removable space $(\operatorname{im} \mathcal{L}_A)^2$ of classical normal form theory, since this is just the part coming from $\mathcal{L}_{\mathbf{f}^0}\mathbf{w}^2$. Letting $\mathcal{N}^2$ be a complement to the removable space, we can let $\mathbf{g}^2$ be the projection of $\mathbf{f}^2$ into this complement and then solve (9.2.2) for both $\mathbf{w}^1$ and $\mathbf{w}^2$. Since both $\mathbf{f}^0$ and $\mathbf{f}^1$ play a role in (9.2.2), we say that $\mathbf{g}^2$ is normalized with respect to $\mathbf{f}^0 + \mathbf{f}^1$.

It is clear that the calculations will become quite complicated at later stages, and some method of organizing them must be adopted. Recently, two closely related new ways of doing this have been found. One is to use ideas from spectral sequence theory (which arises in algebraic topology and homological algebra). This method will be described in detail in Chapter 13 below. The second is developed in [204], and will not be presented here. The main idea is to break the calculations into still smaller steps. For the cubic term discussed above, we would first apply a generator $\mathbf{w}^1$ to normalize the linear term (as before). Next we would apply a generator $\mathbf{w}^2$ to bring the quadratic term into its em classical normal form (with respect to the linear term $A\boldsymbol{x}$). Finally, we apply a third generator $\mathbf{w}^1 + \mathbf{w}^2$, with $\mathbf{w}^1 \in \ker \mathcal{L}_A$ as before and with an additional condition on $\mathbf{w}^2$ guaranteeing that the final change to the quadratic term does not take it out of classical normal form. That is, $\mathbf{w}^1 + \mathbf{w}^2$ has exactly the freedom necessary to improve the classical normalization without losing it. For the quartic term, there would be three steps: normalize first with respect to the linear term, then with respect to the quadratic term (without losing the first normalization), then finally with respect to the cubic term (without losing the previous gains). Algorithms are given in [204] to determine the spaces of generators for each of these substeps by row reduction methods, with the minimum amount of calculation. It is also shown there that the spectral sequence method for handling the calculations can be justified without using homological algebra.

Finally, a few words about the history and terminology of "higher-level" normal forms. The idea was first introduced by Belitskii ([25]). It was rediscovered by Baider, who (with coworkers) developed a fairly complete theory ([14, 12, 16]). Several other authors and teams have contributed; an annotated bibliography is given at the end of [203, Section 4.10]. It is still the case that only a small number of examples of higher-level normal forms have been successfully computed.

In the early days the phrase **normalizing beyond the normal form** was often used in regard to Baider's work. The term **unique normal form** came to be used to describe the situation in which each term is fully normalized with respect to all the preceding terms, but it must be understood that unique normal forms are not completely unique, since a style choice is still involved. However, once a style (a complement to the removable space in each grade) is chosen, there is no further flexibility in a unique normal form (as there is in a classical one). Other authors have used the terms **hypernormal form**, **simplest normal form**, and **higher-order normal form**. Since the term "order" is ambiguous, and could be taken to refer to the degree or grade, we have chosen to use **higher-level normal form** in this book. A first-level normal form is a classical normal form, normalized with respect to the linear term. Second-level means normalized (up to some grade k) with respect to the linear and quadratic terms, and so on.

10

Hamiltonian Normal Form Theory

10.1 Introduction

After introducing some concepts of Hamiltonian systems, we will discuss normalization in a Hamiltonian context. The applications will be directed at the basic resonances of two and three degree of freedom systems.

10.1.1 The Hamiltonian Formalism

Let M be a manifold and $\mathsf{T}^\star\mathsf{M}$ its cotangent bundle. On $\mathsf{T}^\star\mathsf{M}$ 'lives' a canonical symplectic form ω. That means that ω is a closed two-form (that is, $\mathbf{d}\omega = 0$), antisymmetric and nondegenerate. There are local coordinates $(\boldsymbol{q}, \boldsymbol{p})$ such that ω looks like

$$\omega = \sum_{i=1}^{n} \mathbf{d}q_i \wedge \mathbf{d}p_i$$

This can also be considered as the definition of ω, especially in the case $\mathsf{M} = \mathbb{R}^n$. Every $\mathrm{H} : \mathsf{T}^\star\mathsf{M} \to \mathbb{R}$ defines a vector field X_{H} on $\mathsf{T}^\star\mathsf{M}$ by the relation

$$\iota_{X_{\mathrm{H}}}\omega = \mathbf{d}\mathrm{H}$$

(There is considerable confusion in the literature due to a sign choice in ω. One should always be very careful with the application of formulas, and check to see whether this choice has been made in a consistent way).

The vector field X_{H} is called the **Hamilton equation**, and H the **Hamiltonian**. In local coordinates this looks like:

$$\begin{aligned}
\iota_{X_{\mathrm{H}}}\omega &= \iota_{\sum_{i=1}^{n}(X_{q_i}\frac{\partial}{\partial q_i} + X_{p_i}\frac{\partial}{\partial p_i})} \sum_{j=1}^{n} \mathbf{d}q_j \wedge \mathbf{d}p_j \\
&= \sum_{i=1}^{n} (X_{q_i}\mathbf{d}p_i - X_{p_i}\mathbf{d}q_i) \\
\mathbf{d}\mathrm{H} &= \sum_{i=1}^{n} (\frac{\partial \mathrm{H}}{\partial q_i}\mathbf{d}q_i + \frac{\partial \mathrm{H}}{\partial p_i}\mathbf{d}p_i)
\end{aligned}$$

or

$$\dot{q}_i = X_{q_i} = \frac{\partial \mathrm{H}}{\partial p_i}$$
$$\dot{p}_i = X_{p_i} = -\frac{\partial \mathrm{H}}{\partial q_i}$$

Let $(\mathsf{T}^\star\mathsf{M}, \omega_1)$ and $(\mathsf{T}^\star\mathsf{N}, \omega_2)$ be two symplectic manifolds and ϕ a diffeomorphism between them. We say that ϕ is symplectic if $\phi^\star\omega_1 = \omega_2$. Symplectic diffeomorphisms leave the Hamilton equation invariant:

$$\iota_{X_{\phi^\star \mathrm{H}}}\omega_2 = \mathbf{d}\phi^\star \mathrm{H} = \phi^\star \iota_{X_\mathrm{H}}\omega_1 = \iota_{\phi^\star X_\mathrm{H}}\phi^\star\omega_1 = \iota_{\phi\star X_\mathrm{H}}\omega_2$$

or $X_{\phi^\star \mathrm{H}} = \phi^\star X_\mathrm{H}$. Here, we used some results that can be found in [1]. Let $\boldsymbol{x}_o \in \mathsf{T}^\star\mathsf{M}$. We say that $\boldsymbol{x}_o$ is an **equilibrium point** of H if $d\mathrm{H}(\boldsymbol{x}_o) = 0$. We call $\dim \mathsf{M}$ the number of **degrees of freedom** of the system X_H. We say that a function or a differential form α is an **integral of motion** for the vector field X if

$$\mathcal{L}_X\alpha = 0$$

where $\mathcal{L}_X$ is defined as

$$\mathcal{L}_X\alpha = \iota_X\mathbf{d}\alpha + \mathbf{d}i_X\alpha.$$

If α is a function, this reduces to

$$\mathcal{L}_X\alpha = \iota_X\mathbf{d}\alpha.$$

It follows that H itself is an integral of motion of X_H, since

$$\mathcal{L}_{X_\mathrm{H}} = \iota_{X_\mathrm{H}}\mathbf{d}\mathrm{H} = \iota_{X_\mathrm{H}}\iota_{X_\mathrm{H}}\omega = 0$$

(ω is antisymmetric).

The **Poisson bracket** $\{\ ,\ \}$ is defined on functions (on the cotangent bundle) as follows:

$$\{\mathrm{F}, \mathrm{G}\} = -\iota_{X_\mathrm{F}}\iota_{X_\mathrm{G}}\omega.$$

Since ω is antisymmetric, $\{\mathrm{F}, \mathrm{G}\} = -\{\mathrm{G}, \mathrm{F}\}$. Note that

$$\{\mathrm{F}, \mathrm{G}\} = -\iota_{X_\mathrm{F}}\mathbf{d}\mathrm{G} = -\mathcal{L}_{X_\mathrm{F}}\mathrm{G}$$

and therefore G is invariant with respect to F iff $\{\mathrm{F}, \mathrm{G}\} = 0$. We call M the configuration space and $\boldsymbol{p}_m \in \mathsf{T}^\star_m\mathsf{M}$ is called the **momentum**. If M is the torus, we call the local coordinates **action-angle** variables. Action refers to the momentum and angle to the configuration space coordinates.

10.1.2 Local Expansions and Rescaling

In Hamiltonian mechanics, the small parameter necessary to do asymptotics is usually obtained by localizing the system around some well-known solution, e.g. an equilibrium or periodic solution; this involves the *dilation*, discussed in the preceding chapter. The quantity ε^2 is a measure for the energy with respect to equilibrium (or periodic solution). If the Hamiltonian is in polynomial form and starts with quadratic terms, we usually divide by ε^2. This implies that the grade of a Hamiltonian term is degree minus two. In most (but not all) cases, putting $\varepsilon = 0$, the equations of motion will reduce to linear decoupled oscillators. Hamiltonian mechanics represents a rich and important subject. In this chapter we take the narrow but useful view of how to obtain asymptotic approximations for Hamiltonian dynamics.

In the literature the emphasis is usually on the low-order resonances, such as $1:2$ or $1:1$, for the obvious reason that in these cases there is interesting dynamics while the number of nonlinear terms to be retained in the analysis is minimal. This emphasis is also found in applications, see for instance [210] for examples from mechanical engineering. We will restrict ourselves to semisimple cases. A low-order resonance such as $1:-1$ is a nonsemisimple example, arising for instance in problems of celestial mechanics. However, treatment of nonsemisimple problems is even more technical and takes too long. Note also, that in practice, higher-order resonance will occur more often than the lower-order cases, so we shall also consider such problems. In the various resonance cases which we shall discuss, the asymptotic estimates take different forms; this follows from the theory developed in the preceding chapters with, of course, special extensions for the Hamiltonian context.

There are quite a number of global results that one has to bear in mind while doing a local, asymptotic analysis. An old but useful introduction to the qualitative aspects can be found in [30]. See also [1], the books by Arnol′d, [8] and [9], and the series on Dynamical Systems edited by Anosov, Arnol′d, Kozlov et al., in particular [10]. A good introduction to resonance problems in dynamical systems, including Hamiltonian systems, is [125]. A useful reprint selection of seminal papers in the field is [181].

10.1.3 Basic Ingredients of the Flow

Following Poincaré, the main interest has been to obtain qualitative insight, i.e. to determine the basic ingredients *equilibria, periodic orbits, and invariant manifolds*; in the last case emphasis is often placed on invariant tori, which are covered by quasiperiodic orbits.

Equilibria constitute in general no problem, since they can be obtained by solving a set of algebraic or transcendental equations. Although this may be a far from trivial task in practice, we shall always consider it done.

To obtain periodic orbits is another matter. A basic theorem is due to Lyapunov [179] for analytic Hamiltonians with n degrees of freedom: if the eigenfrequencies of the linearized Hamiltonian near stable equilibrium are independent over $\mathbb{Z}$, there exist n families of periodic solutions filling up smooth 2-dimensional manifolds emanating from the equilibrium point. Fixing the energy level near an equilibrium point, one finds from these families n periodic solutions. These are usually called the normal modes of the system. The Lyapunov periodic solutions can be considered as a continuation of the n families of periodic solutions that one finds for the linearized equations of motion.

The assumption of nonresonance (the eigenfrequencies independent over $\mathbb{Z}$) has been dropped in a basic theorem by Weinstein [286]: for an n degrees of freedom Hamiltonian system near stable equilibrium, there exist (at least) n short-periodic solutions for fixed energy. Some of these solutions may be a continuation of linear normal mode families but some of the other ones are clearly not obtained as a continuation. As we shall see, in certain resonance cases the linear normal modes *can not* be continued. For instance, in the $1:2$ resonance case, we have general position periodic orbits with multi-frequency $(1, 2)$; we refer to such solutions as short-periodic.

Since the paper by Weinstein several results appeared in which all these periodic solutions have been indiscriminately referred to as normal modes. This is clearly confusing terminology; in our view a normal mode will be a periodic solution 'restricted' to a two-dimensional invariant subspace of the linearized system or an ε-close continuation of such a solution.

It is important to realize that the Lyapunov–Weinstein estimates of the number of periodic (families) of solutions are *lower* bounds. For instance in the case of two degrees of freedom, 2 short-periodic solutions are guaranteed to exist by the Weinstein theorem. But in the $1:2$ resonance case one finds generically 3 short-periodic solutions for each (small) value of the energy. One of these is a continuation of a linear normal mode, the other two are not. For higher-order resonances such as $3:7$ or $2:11$, there exist for an open set of parameters 4 short-periodic solutions of which two are continuations of the normal modes. Of course, symmetry and special Hamiltonian examples may change this picture drastically. For instance in the case of the famous Hénon–Heiles Hamiltonian

$$H(p,q) = \frac{1}{2}(p_1^2 + q_1^2 + p_2^2 + q_2^2) + \frac{1}{3}q_1^3 - q_1 q_2^2,$$

because of symmetry, there are 8 short-periodic solutions.

The existence of invariant tori around the periodic solutions is predicted by the Kolmogorov–Arnol'd–Moser theorem (or KAM-theorem) which is a collection of statements the first proofs of which have been provided by Arnol'd and Moser; see [9, 195] and [166]. and further references therein. According to this theorem, under rather general assumptions, the energy manifold ($(2n-1)$-dimensional in a n degrees of freedom system) contains an infinite number of n-dimensional tori, invariant under the flow. In a neighborhood of an equilibrium point in phase space, most of the orbits are located on these tori, or somewhat more precise: as $\varepsilon \downarrow 0$, the measure of orbits between the tori tends to zero. If we find only regular behavior by the asymptotic approximations, this can be interpreted as a further quantitative specification of the KAM-theorem. For instance, if we describe phase space by regular orbits with error of $\mathcal{O}(\varepsilon^k)$ on the time scale ε^{-m} $(k, m > 0)$ we have clearly an upper bound on how wild solutions can be on this time scale. However, in general one should keep in mind that the normal form can already be nonintegrable.

It may improve our insight into the possible richness of the phase-flow to enumerate some dimensions: The tori around periodic orbits in general posi-

degrees of freedom:	2	3	n
dimension of phase space:	4	6	2n
dimension of energy manifold:	3	5	2n-1
dimension invariant tori:	2	3	n

Table 10.1: Various dimensions.

tion are described by keeping the n actions fixed and varying the n angles, which makes them n-dimensional. The tori are embedded in the energy manifold and there is clearly no escape possible from between the tori if $n = 2$. For $n \geq 3$, orbits can escape between tori, see Table 10.1; for this process, called Arnol'd diffusion, see [6], Nekhoroshev [211] has shown that it takes place on at least an exponential time scale of the order $\varepsilon^{-a} e^{1/\varepsilon^b}$ where a, b are positive constants and ε^2 is a measure for the energy with respect to an equilibrium position.

Symmetries play an essential part in studying the theory and applications of dynamical systems. In the classical literature, say up to 1960, attention was usually paid to the relation between symmetry and the existence of first integrals. In general, one has that each one-parameter group of symmetries of a Hamiltonian system corresponds to a conserved quantity (Noether's theorem, [213]). One can think of translational invariance or rotational symmetries. The usual formulation is to derive the Jacobi identity from the Poisson bracket, together with some other simple properties, and to associate the system with a Lie algebra; see the introductory literature, for instance [1] and [67].

Recently the relation between symmetry and resonance, in particular its influence on normal forms has been explored using equivariant bifurcation and singularity theory, see [109] or [43], see also [276] for references. For symmetry in the context of Hamiltonian systems see [152] and [282]. It turns out, that symmetry assumptions often produce a hierarchy of resonances that can be very different from the generic cases.

10.2 Normalization of Hamiltonians around Equilibria

Normalization procedures contain a certain freedom of formulation. In application to Hamiltonian systems this freedom is used to meet and conserve typical aspects of these systems.

10.2.1 The Generating Function

The equilibria of a Hamiltonian vector field coincide with the critical points of the Hamiltonian. Suppose we have obtained such a critical point and consider it as the origin for local symplectic coordinates around the equilibrium. Since the value of the Hamiltonian at the critical point is not important, we take it to be zero, and we expand the Hamiltonian in the local coordinates in a Taylor expansion:

$$\mathrm{H}^{[0]} = \mathrm{H}^0 + \varepsilon \mathrm{H}^1 + \varepsilon^2 \mathrm{H}^2 + \cdots,$$

where H^k is homogeneous of degree $k+2$ and ε is a scaling factor (dilation), related to the magnitude of the neighborhood that we take around the equilibrium.

Assumption 10.2.1 *We shall assume* H^0 *to be in the following standard form*

$$\mathrm{H}^0 = \frac{1}{2}\sum_{j=1}^{n} \omega_j (q_j^2 + p_j^2)$$

(When some of the eigenvalues of $\mathrm{d}^2\mathrm{H}^{[0]}$ *have equal magnitude, this standard form may not be obtainable. In two degrees of freedom this makes the* 1 : 1- *and the* 1 : −1*-resonance exceptional (cf. Cushman [63] and Van der Meer [267]). We call* $\boldsymbol{\omega} = (\omega_1, \ldots, \omega_n)$ *the* frequency-vector.

In the following discussion of resonances, it turns out that it is convenient to have two other coordinate systems at our disposal, i.e. *complex coordinates* and *action-angle variables.* We shall introduce these first. Let

$$x_j = q_j - ip_j,$$
$$y_j = q_j + ip_j.$$

Then

$$\mathbf{d}x_j \wedge \mathbf{d}y_j = 2i\mathbf{d}q_j \wedge \mathbf{d}p_j.$$

In order to obtain the same vector field, and keeping in mind the definition

$$\iota_{X_{\mathrm{H}}}\omega = \mathbf{d}\mathrm{H},$$

(with two-form ω), we have to multiply the new Hamiltonian by $2i$ after the substitution of the new $\boldsymbol{x}$ and $\boldsymbol{y}$ coordinates in the old H. Thus

$$\mathrm{H}^{[0]} = \mathrm{H}^0 + \varepsilon \mathrm{H}^1 + \cdots ,$$

where

$$\mathrm{H}^0 = i \textstyle\sum_{j=1}^{n} \omega_j x_j y_j$$

(and H^k is again homogeneous of degree $k+2$, this time in $\boldsymbol{x}$ and $\boldsymbol{y}$).

The next transformation will be to action-angles variables. One should take care with this transformation since it is singular when a pair of coordinates vanishes. We put

$$\begin{aligned} x_j &= \sqrt{2\tau_j} e^{i\phi_j}, \quad \phi_j \in S^1, \\ y_j &= \sqrt{2\tau_j} e^{-i\phi_j}, \quad \tau_j \in (0, \infty). \end{aligned}$$

Then

$$\mathbf{d} x_j \wedge \mathbf{d} y_j = 2i \mathbf{d}\phi_j \wedge \mathbf{d}\tau_j .$$

Thus we have to divide the new Hamiltonian by the scaling factor $2i$ after substitution of the action-angle variables. We obtain

$$\mathrm{H}^0 = \textstyle\sum_{j=1}^{n} \omega_j \tau_j$$

and

$$\mathrm{H}^l = \sum_{\|\mathbf{m}\|_1 \le l+2} \mathrm{h}^l_{\mathbf{m}}(\boldsymbol{\tau}) e^{i\langle \mathbf{m}, \boldsymbol{\phi} \rangle},$$

where $\langle \mathbf{m}, \boldsymbol{\phi} \rangle = \sum_{j=1}^{n} m_j \phi_j$, $m_j \in \mathbb{Z}$ and $\| \mathbf{m} \|_1 = \sum_{j=1}^{n} |m_j|$. The $h^l_{\mathbf{m}}$ are homogeneous of degree $1 + l/2$ in $\boldsymbol{\tau}$. Applying the same transformation to the generating function of the normalizing transformation K, we can write

$$\mathrm{K}^{[1]} = \varepsilon \mathrm{K}^1 + \varepsilon^2 \mathrm{K}^2 + \cdots ,$$

where

$$\mathrm{K}^l = \sum_{\|\mathbf{m}\|_1 \le l+2} \mathrm{k}^l_{\mathbf{m}}(\boldsymbol{\tau}) e^{i\langle \mathbf{m}, \boldsymbol{\phi} \rangle}.$$

The term 'generating function' can be a cause of confusion. In Hamiltonian mechanics, the term is reserved for an auxiliary function, usually indicated by S (see [8]). In normal form theory, as developed in Chapter 11, the term is associated with the Hilbert–Poincaré series producing the normal forms. To avoid confusion, we will indicate the generating function in this more general sense by Hilbert–Poincaré series $P(t)$ (usually abbreviated 'Poincaré series'). The variable t stands for vector fields and invariants. This series predicts the terms which may be present in a normal form because of the resonances involved; it refers to the complete algebra of invariants. In the Hamiltonian context of this chapter, the normal form transformation conserves the symplectic structure of the system. The general formulation for generating functions and their computation in the framework of spectral sequences is given in Chapter 11 with examples in the Sections 13.4 and 13.5.

In the formulation of our problem the normal form equation is

$$\{\mathrm{H}^0, \mathrm{K}^1\} = \mathrm{H}^1 - \overline{\mathrm{H}}^1, \quad \{\overline{\mathrm{H}}^1, \mathrm{H}^0\} = 0,$$
$$\{\mathrm{K}^1, \mathrm{H}^0\} = \sum_{\|\mathbf{m}\|_1 \leq 3} \langle \mathbf{m}, \boldsymbol{\omega} \rangle \mathrm{k}^1_{\mathbf{m}}(\tau) e^{i\langle \mathbf{m}, \boldsymbol{\phi} \rangle}.$$

We can solve the normal form equation:

$$\mathrm{k}^1_{\mathbf{m}} = \begin{cases} \frac{1}{\langle \mathbf{m}, \boldsymbol{\omega} \rangle} \mathrm{h}^1_{\mathbf{m}}(\boldsymbol{\tau}) & 0 \\ \langle m, \omega \rangle \neq 0 & \langle m, \omega \rangle = 0. \end{cases}$$

Then

$$\overline{\mathrm{H}}^1 = \sum_{\substack{\langle \mathbf{m}, \boldsymbol{\omega} \rangle = 0 \\ \| \mathbf{m}\|_1 \leq 3}} \mathrm{h}^1_{\mathbf{m}}(\boldsymbol{\tau}) e^{i\langle \mathbf{m}, \boldsymbol{\phi} \rangle}$$

and $\overline{\mathrm{H}}^1$ commutes with H^0, i.e. $\{\mathrm{H}^0, \overline{\mathrm{H}}^1\} = 0$. For $\langle \mathbf{m}, \boldsymbol{\omega} \rangle$ nonzero, but very small, this introduces large terms in the asymptotic expansion *(small divisors)*. In that case it might be better to treat $\langle \mathbf{m}, \boldsymbol{\omega} \rangle$ as zero, and split of the part of H^0 that gives exactly zero, and consider this as the unperturbed problem. Suppose $\langle \mathbf{m}, \boldsymbol{\omega} \rangle = \delta$ and $\langle \mathbf{m}, \boldsymbol{\omega}^\sharp \rangle = 0$, where $\boldsymbol{\omega}$ and $\boldsymbol{\omega}^\sharp$ are close, then

$$\mathrm{H}^0 = \sum \omega_j \tau_j = \sum \omega^\sharp_j \tau_j + \sum (\omega_j - \omega_j{}^\sharp) \tau_j.$$

We say that $\mathbf{m} \in \mathbb{Z}^n$ is an *annihilator* of $\boldsymbol{\omega}^\sharp$ if

$$\langle \mathbf{m}, \boldsymbol{\omega}^\sharp \rangle = 0.$$

For the annihilator we use again the norm $\| \mathbf{m}\|_1 = \sum_{j=1}^n |m_j|$.

Definition 10.2.2. *If the annihilators with norm less than or equal to $\mu + 2$, span a codimension 1 sublattice of $\mathbb{Z}^n$, and $\nu \in \mathbb{N}$ is minimal, then we say that ω^o defines a* **genuine νth-order resonance**. *(ν is minimal means that ν represents the lowest natural number corresponding to genuine resonance).*

When normalizing step by step, using normal form polynomials as in the next section, the annihilators will determine the form of the polynomials, the norm of the annihilator determines their position in the normal form expansion.

Example 10.2.3. Some examples of genuine resonances are:

For $n = 2$

- $\boldsymbol{\omega}^\sharp = (k, l)$, with $\mathbf{m}^1 = (-l, k)$ and $k + l > 2$.

For $n = 3$

- $\omega^\sharp = (1,2,1)$, with $\mathbf{m}^1 = (2,-1,0)$ and $\mathbf{m}^2 = (0,-1,2)$,
- $\omega^\sharp = (1,2,2)$, with $\mathbf{m}^1 = (2,-1,0)$ and $\mathbf{m}^2 = (2,0,-1)$,
- $\omega^\sharp = (1,2,3)$, with $\mathbf{m}^1 = (2,-1,0)$ and $\mathbf{m}^2 = (1,1,-1)$,
- $\omega^\sharp = (1,2,4)$, with $\mathbf{m}^1 = (2,-1,0)$ and $\mathbf{m}^2 = (0,2,-1)$.

In all these examples we have $\| \mathbf{m}^i \|_1 = 3$ for $i = 1, 2$. ◇

10.2.2 Normal Form Polynomials

In the next sections we shall not carry out all details of normalizing concrete Hamiltonians, but we shall often assume that the Hamiltonian is already in normal form or point out the necessary steps, formulating Poincaré series and invariants. The idea is to study the general normal form, and determine its properties depending on the free parameters. This program has been carried out until now for two and, to some extent, for three degrees of freedom systems. For more than three degrees of freedom only some special cases were studied. The relevant free parameters can be computed in concrete problems by the normalization procedure. We shall first determine which polynomials are in normal form with respect to

$$\mathrm{H}^0 = \frac{1}{2}\sum_{j=1}^n \omega_j (q_j^2 + p_j^2).$$

Changing to complex coordinates, and introducing a multi-index notation, we can write a general polynomial term, derived from a real one, as

$$\mathrm{P}^\sigma = i(D\mathbf{x}^\mathbf{k}\mathbf{y}^\mathbf{l} + \overline{D}\mathbf{x}^\mathbf{l}\mathbf{y}^\mathbf{k}), D \in \mathbb{C},$$

where $\mathbf{x}^\mathbf{k} = x_1^{k_1} \cdots x_n^{k_n}$, $\mathbf{y}^\mathbf{l} = {y_1}^{l_n} \cdots y_n^{l_n}$, $k_i, l_i \geq 0$, $i = 1, \ldots, n$ and $\sigma = \| \mathbf{k} \|_1 + \| \mathbf{l} \|_1 - 2$. Since

$$\mathrm{H}^0 = i\sum_{j=1}^n \omega_j x_j y_j$$

we obtain

$$\begin{aligned}
\{\mathrm{H}^0, \mathrm{P}^\sigma\} &= \sum_{j=1}^n \left(\frac{\partial \mathrm{H}^0}{\partial x_j}\frac{\partial \mathrm{P}^\sigma}{\partial y_j} - \frac{\partial \mathrm{P}^\sigma}{\partial x_j}\frac{\partial \mathrm{H}^0}{\partial y_j}\right) \\
&= \sum_j \omega_j \left(x_j \frac{\partial}{\partial x_j} - y_j \frac{\partial}{\partial y_j}\right)(D\mathbf{x}^\mathbf{k}\mathbf{y}^\mathbf{l} + \overline{D}\mathbf{x}^\mathbf{l}\mathbf{y}^\mathbf{k}) \\
&= (D\mathbf{x}^\mathbf{k}\mathbf{y}^\mathbf{l} + \overline{D}\mathbf{x}^\mathbf{l}\mathbf{y}^\mathbf{k})\langle \omega, \mathbf{k} - \mathbf{l} \rangle.
\end{aligned}$$

So $\mathrm{P}^\sigma \in \ker(\mathrm{ad}\mathrm{H}^0)$ is equivalent with $\langle \boldsymbol{\omega}, \mathbf{k} - \mathbf{l} \rangle = 0$, where $\mathrm{ad}(\mathrm{H})\mathrm{K} = \{\mathrm{H}, \mathrm{K}\}$. Of course, this is nothing but the usual resonance relation. In action-angle variables, one only looks at the difference $\mathbf{k} - \mathbf{l}$, and the homogeneity condition puts a bound on this difference. The most important resonance term arises for $\| \mathbf{k} + \mathbf{l} \|_1$ minimal. Consider for instance two degrees of freedom, with ω_1 and $\omega_2 \in \mathbb{N}$ and relatively prime. The resonance term is

$$\mathrm{P}^{\omega_1+\omega_2-2} = D{x_1}^{\omega_2}{y_2}^{\omega_1} + \overline{D}{y_1}^{\omega_2}{x_2}^{\omega_1} .$$

Terms with $\mathbf{k} = \mathbf{l}$ are also resonant. They are called **self-interaction terms** and are polynomials in the variables $x_i y_i$ (or τ_i). Powers of $x_i y_i$ will arise as a basic part of the normal form. As the term "self-interaction" suggests, they do not produce dynamical interaction between the various degrees of freedom.

For a given number of degrees of freedom n and a given **resonant frequency vector** ω, we can calculate the normal form based on a *finite* set of monomials, the generators. The normal form is truncated at some degree, ideally as the qualitative results obtained are then robust with respect to higher-order perturbations. In practice, this robustness is often still a point of discussion.

10.3 Canonical Variables at Resonance

As we have seen, normal forms of Hamiltonians near equilibrium are characterized by combination angles $\langle \mathbf{m}, \boldsymbol{\phi} \rangle$. For instance in the case of two degrees of freedom and the 2 : 1-resonance, the normal form will contain the combination angle $\phi_1 - 2\phi_2$. It will often simplify and reduce the problem to take these angles as new variables, eliminating in the process one (fast) combination. For our notation we refer to Chapter 13. There we found that for a given resonant vector $\boldsymbol{\omega}$, there exists a matrix $\hat{M} \in GL_n(\mathbb{Z})$ such that

$$\hat{M}\boldsymbol{\omega} = \begin{bmatrix} 0 \\ 0 \\ \cdot \\ \cdot \\ 0 \\ 1 \end{bmatrix} + o(1),$$

where the o(1)-term represents a small detuning of the resonance. We drop this small term to keep the notation simple. Let

$$\boldsymbol{\psi} = \hat{M}\boldsymbol{\phi},$$

where $\boldsymbol{\phi}$ represents the angles we started out with. Then $\dot{\psi}_i = 0 + o(1)$, $i = 1, \ldots, n-1$ and $\dot{\psi}_n = 1 + o(1)$. The action variables are found by the dual definition:

$$\boldsymbol{\tau} = \hat{M}^\dagger \boldsymbol{r}$$

(where $\hat{M}^\dagger$ denotes the transpose of $\hat{M}$). This defines a symplectic change of coordinates from $(\boldsymbol{\phi}, \boldsymbol{\tau})$ to $(\boldsymbol{\psi}, \boldsymbol{r})$ variables, since

$$\sum_i \mathbf{d}\phi_i \wedge \mathbf{d}\tau_i = \sum_{ij} \hat{M}_{ji} \mathbf{d}\phi_i \wedge \mathbf{d}r_j = \sum_j \mathbf{d}\psi_j \wedge \mathbf{d}r_j.$$

We shall often denote r_n with E. Here E is the only variable independent of the extension from M to $\hat{M}$. Using the normalization process and introducing in this way $(\boldsymbol{\psi}, \boldsymbol{r})$, we will call the coordinates **canonical variables adapted to the resonance**. The resulting equations of motion will only produce interactions by the presence of combination angles, associated with the resonance vector ω. This will result in a reduction of the dimension of the system.

10.4 Periodic Solutions and Integrals

As we have seen in the introduction, Hamiltonian systems with n degrees of freedom have (at least) n families of periodic orbits in the neighborhood of a stable equilibrium, by Weinstein's 1973 theorem. However, this is the minimal number of periodic orbits for fixed energy, due to the resonance, the actual number of short-periodic solutions may be higher. The general normal form of a Hamiltonian near equilibrium depends on parameters, and the dimension of the parameter space depends on n, on the actual resonance, and on the order of truncation of the normal form one considers.

If we fix all these, we are interested in those values of the frequencies for which the normal form has more than n short-periodic (i.e. $\mathcal{O}(1)$) orbits for a given energy level. These frequency values are contained in the so-called **bifurcation set** of the resonance. For practical and theoretical reasons one is often interested in the dependency of the bifurcation set on the energy.

In Section 10.1.1 we introduced the Poisson bracket $\{,\}$. In time-independent Hamiltonian systems, the Hamiltonian itself, H, is an integral of motion, usually corresponding to the energy of the system. A functionally independent function $\mathrm{F}(\boldsymbol{q}, \boldsymbol{p})$ represents an independent integral of the system if it is in involution with the Hamiltonian, i.e.

$$\{\mathrm{F}, \mathrm{H}\} = 0.$$

If an n degrees of freedom time-independent Hamiltonian system has n independent integrals (including the Hamiltonian), it is called Liouville or completely integrable, or for short 'integrable'. In general, Hamiltonian systems with two or more degrees of freedom are not integrable. Although correct, this is a misleading or at least an incomplete statement. In actual physical systems, symmetries may add to the number of integrals, also in nonintegrable systems the chaos may be locally negligible. As we shall see, the possible integrability of the normal form near a solution (usually an equilibrium), produces information about such aspects of the dynamics.

In this respect we recall a basic result from Section 10.2. The normal form construction near equilibrium takes place with respect to the quadratic part of the Hamiltonian H^0. In each step of the normalization procedure we remove terms which are not in involution with H^0 with as a result that the resulting Hamiltonian in normal form $\overline{H}^{[0]}$ has H^0 as additional integral. The implication is that two degrees of freedom systems in normal form are integrable.

To determine whether a normal form of a Hamiltonian system with at least three degrees of freedom, is integrable or not is not easy. The earliest proofs are of a negative character, showing that integrals of a certain kind are not present. This is still a useful approach, for instance showing that algebraic integrals to a certain degree do not exist.

A problem is, that if a system is nonintegrable, one does not know what to expect. There will be an irregular component in the phase-flow, but we have no classification of irregularity, except in the case of two degrees of freedom (or symplectic two-dimensional maps). Ideally, any statement on the nonintegrability of a system should be followed up by a description of the geometry and the measure of the chaotic sets. One powerful and general criterion for chaos and nonintegrability is to show that a horseshoe map is embedded in the flow. The presence of a horseshoe involves locally an infinite number of unstable periodic solutions and sensitive dependence on initial values. This was exploited in [73] and used in [134].

Another approach is to locate and study certain singularities, often by analytic continuation of a suitable function. If the singularities are no worse than poles, the system is integrable; [39] is based on this. In the case that we have infinite branching for the singularity, we conclude that we have nonintegrability; this was used in [79].

We will return to these results in Section 10.7.

10.5 Two Degrees of Freedom, General Theory

Both in mathematical theory and with regards to applications in physics and engineering, the case of Hamiltonian systems with two degrees of freedom got most of the attention. We will discuss general results, important resonances and the case of symmetries. The first two subsections are elementary and can be skipped if necessary.

10.5.1 Introduction

A two-degrees of freedom system is characterized by a phase-flow that is four-dimensional, but restricted to the energy manifold. To visualize this flow is already complicated but to obtain a geometric picture is useful for the full understanding of a dynamical system, even if one is only interested in the asymptotics. It certainly helps to have a clear picture in mind of the linearized flow.

We give a list of possible ways of looking at a certain problem. To be specific, we assume that the Hamiltonian of the linearized system is positive definite, the flow is near stable equilibrium. The indefinite case is difficult from the asymptotic point of view, since solutions can grow without bounds, although one can still compute a normal form and use the results to find periodic orbits.

1. Fixing the energy we have, near a stable equilibrium, flow on a compact manifold, diffeomorphic to S^3. Since the energy manifold is compact, we have a priori bounds for the solutions. One should note that this sphere need not be a sphere in the (strict) geometric sense.
2. The Poincaré-mapping: One can take a plane, locally transversal to the flow on the energy manifold. This plane is mapped into itself under the flow, which defines a Poincaré-mapping. This map is easier to visualize than the full flow, since it is two-dimensional. Note that, due to its local character, the Poincaré map does not necessarily describe all orbits. In a situation with two normal modes for example, the map around one will produce this one as a fixed point, but the other normal mode will form the boundary of the map and has to be excluded because of the transversality assumption.
3. Projection into 'physical space': In the older literature one finds often a representation of the solutions by projection onto the base (or configuration) space, with coordinates q_1, q_2. In physical problems that is the space where one can see things happen. Under this projection periodic orbits typically look like algebraic curves, the order determined by the resonance. If they are stable and surrounded by tori, these tori project as tubes around these algebraic curves.
4. A visual representation that is also useful in systems with more than one degree of freedom is to plot the actions τ_i as functions of time; some authors prefer to use the amplitudes $\sqrt{2\tau_i}$ instead.
5. As we shall see, in two degrees of freedom systems, only one slowly varying combination angle ψ_1 plays a part. It is possible then, to plot the actions as functions of ψ_1; in this representation the periodic solutions show up as critical points of the τ, ψ_1-flow.
6. The picture of the periodic solution and their stability changes as the parameters of the Hamiltonian change. To illustrate these changes we draw bifurcation diagrams which illustrate the existence and stability of solutions and also the branching off and vanishing of periodic orbits. These bifurcation diagrams take various forms, see for instance Sections 10.6.1 and 10.8.1.

It is useful to have various means of illustration at our disposal as the complications of higher dimensional phase-flow are not so easy to grasp from only one type of illustration. It may also be useful for the reader to consult the pictures in [1] and [2].

10.5.2 The Linear Flow

Consider the linearized flow of a two degrees of freedom system. The Hamiltonian is

$$\mathrm{H} = \omega_1\tau_1 + \omega_2\tau_2$$

and the equations of motion

$$\dot{\phi}_i = \omega_i \ , \ \dot{\tau}_i = 0 \ , \ i = 1, 2,$$

corresponding to harmonic solutions

$$\begin{bmatrix} \boldsymbol{q} \\ \boldsymbol{p} \end{bmatrix} = \begin{bmatrix} \sqrt{2\tau_1(0)}\sin(\phi_1(0)+\omega_1 t) \\ \sqrt{2\tau_2(0)}\sin(\phi_2(0)+\omega_2 t) \\ \sqrt{2\tau_1(0)}\cos(\phi_1(0)+\omega_1 t) \\ \sqrt{2\tau_2(0)}\cos(\phi_2(0)+\omega_2 t) \end{bmatrix}.$$

If $\omega_1/\omega_2 \notin \mathbb{Q}$, there are two periodic solutions for each value of the energy, the normal modes given by $\tau_1 = 0$ and $\tau_2 = 0$. If $\omega_1/\omega_2 \in \mathbb{Q}$, all solutions are periodic. We fix the energy, choosing a positive constant E_o with

$$\omega_1\tau_1 + \omega_2\tau_2 = E_o.$$

In $\boldsymbol{q}, \boldsymbol{p}$-space this represents an ellipsoid which we identify with S^3. The energy manifold is invariant under the flow but also, in this linear case, τ_1 and τ_2 are conserved quantities, corresponding to invariant manifolds in S^3. The system is integrable. What do the invariant manifolds look like?
They are described by two equations

$$\omega_1\tau_1 + \omega_2\tau_2 = E_o$$

and

$$\omega_1\tau_1 = E_1 \text{ or } \omega_2\tau_2 = E_2.$$

E_1 and E_2 are both positive and their sum equals E_o. Choosing $E_1 = 0$ corresponds to a normal mode in the τ_2-component (all energy E_o in the second degree of freedom); as we know from harmonic solutions this is a circle lying in the q_2, p_2-plane. The same reasoning applies to the case $E_2 = 0$ with a normal mode in the τ_1-component. Consider one of these circles lying in S^3. The other circle passes through the center of the first one, because the center of the circle corresponds to a point where one of the actions τ is zero, which makes the other action maximal, and thus part of a normal mode; see Figure 10.1.

On the other hand, if we draw the second circle first, the picture must be the same, be it in another plane. This leads to Figure 10.2.

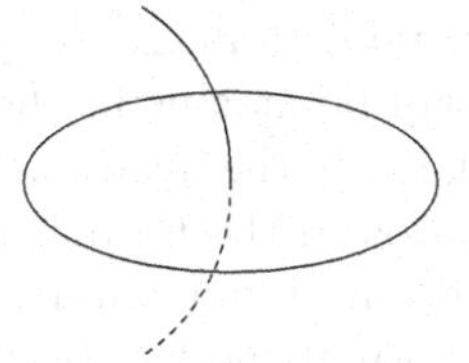

Fig. 10.1: One normal mode passes through the center of the second one.

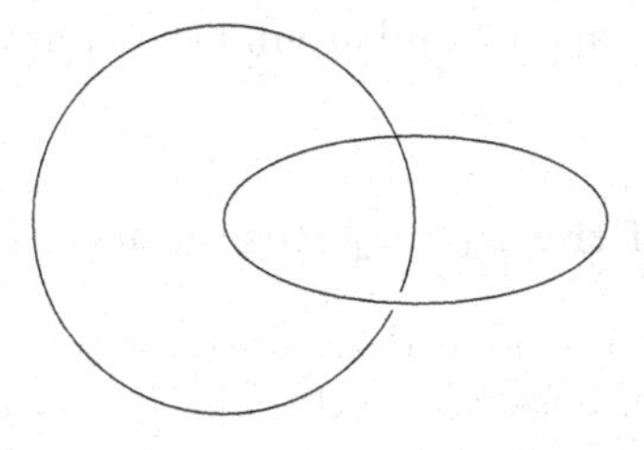

Fig. 10.2: The normal modes are linked.

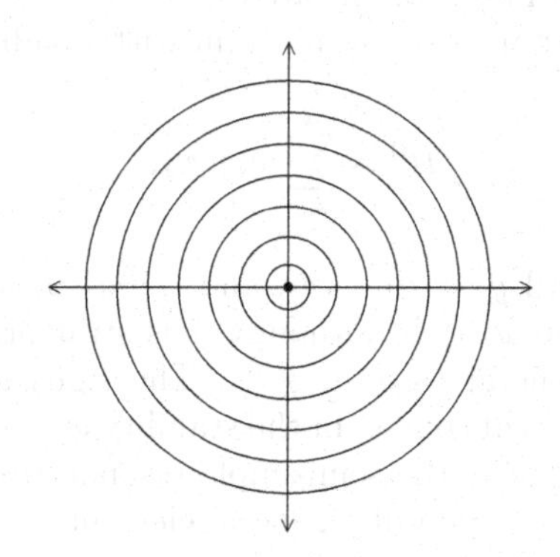

Fig. 10.3: Poincaré-map in the linear case.

The normal modes are linked and they form the extreme cases of the invariant manifolds $\omega_i \tau_i = E_i$, $i = 1, 2$. What do these manifolds look like when $E_1 E_2 > 0$? A Poincaré-mapping is easy to construct, it looks like a family of circles in the plane (Figure 10.3).

The center is a fixed point of the mapping corresponding to the normal mode in the second degree of freedom. The boundary represents the normal mode in the first degree of freedom; note that the normal mode does not belong to the Poincaré-mapping as the flow is not transversal here to the q_1, p_1-plane. Starting on one of the circles in the q_1, p_1-plane with $E_1, E_2 > 0$, the return map will produce another point on the circle. If $\omega_1/\omega_2 \notin \mathbb{Q}$ the starting point will never be reached again; If $\omega_1/\omega_2 \in \mathbb{Q}$ the orbits close after some time, i.e. the starting point will be attained. These orbits are called periodic orbits in **general position**. Clearly the invariant manifolds $\omega_1\tau_1+\omega_2\tau_2 = E_o$, $\omega_1\tau_1 = E_1$ are invariant tori around the normal modes.

We could have concluded this immediately, but with less detail of the geometric picture near the normal modes, by considering the action-angle variables $\boldsymbol{\tau}, \boldsymbol{\phi}$ and their equations of motion: if the τ_i are fixed, the ϕ_i are left to be varying and they describe the manifold we are looking for, the torus $\mathbb{T}^2$. Thus the energy surface has the following **foliation**: there are two invariant, linked circles, the normal modes and around these, invariant tori filling up the sphere.

10.5.3 Description of the $\omega_1 : \omega_2$-Resonance in Normal Form

In this section we give a general description of the description problem for Hamiltonian resonances, based on [241]. The two degrees of freedom case is rather simple, and it may seem that we use too much theory to formulate what we want. The theoretical discussion is aimed at the three degrees of freedom systems, which is much more complicated.

We consider Hamiltonians at equilibrium with quadratic term

$$H^0 = \sum_{j=1}^{2} \omega_j x_j y_j,$$

where $x_j = q_j + ip_j$ and $y_j = q_j - ip_j$, and the q_j, p_j are the real canonical coordinates. We assume $\omega_j \in \mathbb{N}$, although it is straightforward to apply the results in the more general case $\omega_j \in \mathbb{Z}$. The signs are important in the nonsemisimple case, and, of course, in the stability considerations. With these quadratic terms we speak of the semisimple resonant case. We now pose the problem to find the description of a general element

$$H = k[[x_1, y_1, x_2, y_2]]$$

such that $\{H^0, H\} = 0$ (see [203, Section 4.5]), with $k = \mathbb{R}$ or $\mathbb{C}$. Since the flow of H^0 defines a compact Lie group (S^1) action on $T^*\mathbb{R}^2$, we know beforehand that H can be written as a function of a finite number of invariants of the flow of H^0, that is, as

$$H = \sum_{k=1}^{q} F_k(\alpha_1, \cdots, \alpha_{p_k})\beta_k,$$

where $\{H^0, \alpha_\iota\} = \{H^0, \beta_\iota\} = 0$ for all relevant ι. If it follows from

$$\sum_{k=1}^{q} F_k(\alpha_1, \cdots, \alpha_{p_k})\beta_k = 0$$

that all the F_k are identically zero, we say that we have obtained a **Stanley decomposition** of the normal form. While the existence of the Stanley decomposition follows from the Hilbert finiteness theorem, it is general not unique: both $F(x)$ and $c+G(x)x$ are Stanley decompositions of general functions in one variable x. Notice that the number of primary variables α_ι is in principle variable, contrary to the case of Hironaka decompositions.

One can define the minimum number q in the Stanley decomposition as the **Stanley dimension**. In general one can only obtain upper estimates on this dimension by a smart choice of decomposition.

First of all, we see immediately that the elements $\tau_j = x_j y_j$ all Poisson commute with H^0. We let $\mathcal{I} = k[[\tau_1, \tau_2]]$. In principle, we work with real Hamiltonians as they are given by a physical problem, but it is easier to work with complex coordinates, so we take the coefficients to be complex too. In practice, one can forget the reality condition and work over $\mathbb{C}$. In the end, the complex dimension will be the same as the real one, after applying the reality condition.

Any monomial in ker ad(H^0) is an element of one of the spaces $\mathcal{I}$, $\mathcal{K} = \mathcal{I}[[y_1^{\omega_2} x_2^{\omega_1}]] y_1^{\omega_2} x_2^{\omega_1}$, $\bar{\mathcal{K}} = \mathcal{I}[[x_1^{\omega_2} y_2^{\omega_1}]] x_1^{\omega_2} y_2^{\omega_1}$. That is, the Stanley decomposition of the $\omega_1 : \omega_2$-resonance is

$$\mathcal{I} \oplus \mathcal{K} \oplus \bar{\mathcal{K}}.$$

We can simplify this formula to

$$\mathcal{I}[[y_1^{\omega_2} x_2^{\omega_1}]] \oplus \mathcal{I}[[x_1^{\omega_2} y_2^{\omega_1}]] x_1^{\omega_2} y_2^{\omega_1}.$$

The Hilbert–Poincaré series can be written as a rational function as follows:

$$P(t) = \frac{1 + t^{\omega_1+\omega_2}}{(1-t^2)^2(1-t^{\omega_1+\omega_2})}.$$

In the sequel we allow for detuning, that is, we no longer require the ω_i's to be integers. We assume that there exist integers k and l such that $\delta = l\omega_1 - k\omega_2$ is small. We then still call this a $k : l$-resonance.

10.5.4 General Aspects of the $k : l$-Resonance, $k \neq l$

We assume that $0 < k < l$, $(k, l) = 1$ but we will include detuning. In complex coordinates, the normal form of the $k : l$-resonance is according to Section 10.5.3

$$\begin{aligned} \mathrm{H} &= i(\omega_1 x_1 y_1 + \omega_2 x_2 y_2) + i\varepsilon^{k+l-2}(D {x_1}^l {y_2}^k + \overline{D} {y_1}^l {x_2}^k) \\ &\quad + i\varepsilon^2(\frac{1}{2}A(x_1 y_1)^2 + B(x_1 y_1)(x_2 y_2) + \frac{1}{2}C(x_2 y_2)^2) + \cdots, \end{aligned}$$

where $A, B, C \in \mathbb{R}$ and $D \in \mathbb{C}$. The terms depending on $x_1 y_1$, $x_2 y_2$ are terms in Birkhoff normal form; they are also included among the dots if $k + l > 4$. The term with coefficient ε^{k+l-2} is the first interaction term *between* the two degrees of freedom of the $k : l$-resonance. Of course, this describes the general case; as we shall see later, symmetries may shift the resonance interaction term to higher order.

It helps to use action-angle coordinates and then adapted resonance coordinates. Putting $D = |D|e^{i\alpha}$, we have in action-angle coordinates,

$$\begin{aligned} \mathrm{H} = \omega_1\tau_1 + \omega_2\tau_2 + \varepsilon^{k+l-2}|D|\sqrt{2\tau_1}(2\tau_2)^{\frac{k}{2}} \cos(l\phi_1 - k\phi_2 + \alpha) \\ +\varepsilon^2(A\tau_1^2 + 2B\tau_1\tau_2 + C\tau_2^2) + \cdots . \end{aligned}$$

In the sequel we shall drop the dots. Let $\delta = l\omega_1 - k\omega_2$ be the (small) detuning parameter. The resonance matrix can be taken as

$$\hat{M} = \begin{bmatrix} l & -k \\ k^\star & l^\star \end{bmatrix} \in SL_2(\mathbb{Z}).$$

Following Section 10.3, we introduce adapted resonance coordinates:

$$\begin{aligned} \psi_1 &= l\phi_1 - k\phi_2 + \alpha, \\ \psi_2 &= k^\star\phi_1 + l^\star\phi_2, \\ \tau_1 &= lr + k^\star E, \\ \tau_2 &= -kr + l^\star E. \end{aligned}$$

In the normal form given above, of the angles only the *combination angle* denoted by ψ_1 plays a part; we shall therefore replace ψ_1 by ψ in this section on two degrees of freedom systems. Then we have

$$\begin{aligned} \mathrm{H} = {} & (\omega_1 k^\star + \omega_2 l^\star)E + \delta r + \\ & \varepsilon^{k+l-2}|D|(2lr + 2k^\star E)^{\frac{l}{2}}(-2kr + 2l^\star E)^{\frac{k}{2}} \cos\psi + \\ & \varepsilon^2((Al^2 - 2Bkl + Ck^2)r^2 + 2(Alk^\star + B(ll^\star - kk^\star) - Ckl^\star)Er + \\ & (Ak^{\star 2} + 2Bk^\star l^\star + Cl^{\star 2})E^2). \end{aligned}$$

The angle ψ_2 is not present in the Hamiltonian, so that E is an integral of the equations of motion induced by the normal form ($\dot{E} = -\frac{\partial \mathrm{H}}{\partial \psi_2} = 0$). Since E corresponds to the H^0 part of the Hamiltonian and the energy manifold is bounded near stable equilibrium, the quantity E is *conserved* for the full problem to $\mathcal{O}(\varepsilon + \delta)$ *for all time*. Let

$$\Delta_1 = \begin{vmatrix} A & k \\ B & l \end{vmatrix}, \quad \Delta_2 = \begin{vmatrix} B & k \\ C & l \end{vmatrix},$$
$$\Delta_1^\star = \begin{vmatrix} A & k^\star \\ B & -l^\star \end{vmatrix}, \quad \Delta_2{}^\star = \begin{vmatrix} B & k^\star \\ C & -l^\star \end{vmatrix},$$

then

$$\begin{aligned} \mathrm{H} = (\omega_1 k^\star + \omega_2 l^\star)E + \delta r + \varepsilon^{k+l-2}|D|(2lr + 2k^\star E)^{\frac{l}{2}}(-2kr + 2l^\star E)^{\frac{k}{2}} \cos\psi \\ +\varepsilon^2((l\Delta_1 - k\Delta_2)r^2 + 2(k^\star\Delta_1 + l^\star\Delta_2)Er + (k^\star\Delta_1{}^\star + l^\star\Delta_2{}^\star)E^2). \end{aligned}$$

This leads to the reduced system of differential equations

$$\begin{aligned}
\dot{r} &= \varepsilon^{k+l-2}|D|(2k^\star E + 2lr)^{\frac{l}{2}}(2l^\star E - 2kr)^{\frac{k}{2}} \sin\psi,\\
\dot{\psi} &= \delta + \varepsilon^{k+l-2}|D|2(2k^\star E + 2lr)^{\frac{l}{2}-1}(2l^\star E - 2kr)^{\frac{k}{2}-1}\\
&\times((l^2 l^\star - k^2 k^\star)E - kl(k+l)r)\cos\psi +\\
&2\varepsilon^2((l\Delta_1 - k\Delta_2)r + (k^\star\Delta_1 + l^\star\Delta_2)E).
\end{aligned}$$

To complete the system we have to write down $\dot{E} = 0$ and the equation for ψ_2. We shall omit these equations in what follows.

In the beginning of this section, we characterized the normal modes by putting one of the actions equal to zero. For periodic orbits in general position we have $\tau_i \neq 0$ and constant, $i = 1, 2$. This implies that a periodic solution in general position corresponds to constant r during the motion, resulting in the condition

$$\sin(\psi) = 0, \text{ i.e., } \psi = 0, \pi,$$

during periodic motion. So we also have to look for stationary points of the equation for ψ, with the implication that δ must be of $\mathcal{O}(\varepsilon)$ if $k + l = 3$ or of $\mathcal{O}(\varepsilon^2)$ if $k + l \geq 4$. A good reference for the theory of periodic solutions for systems in resonance is [77].

10.6 Two Degrees of Freedom, Examples

10.6.1 The 1 : 2-Resonance

Although included in the $k : l$ case, the 1 : 2-resonance is so prominent in many applications that we discuss it separately, including the effect of detuning. For the normal form we have the general expression (see Section 10.5.3)

$$\overline{\mathrm{H}} = \mathrm{F}(x_1y_1, x_2y_2, x_1^2y_2) + x_2y_1^2\mathrm{G}(x_1y_1, x_2y_2, x_2y_1^2).$$

The 1 : 2-resonance is the only first-order resonance in two degrees of freedom systems. A convenient choice for the resonance matrix $\hat{M}$ turns out to be

$$\hat{M} = \begin{bmatrix} 2 & -1 \\ 1 & 0 \end{bmatrix},$$

producing the differential equations

$$\begin{aligned}
\dot{r} &= \varepsilon|D|\sqrt{-2r}(2E + 4r)\sin\psi,\\
\dot{\psi} &= \delta + 2\varepsilon\frac{|D|}{\sqrt{-2r}}(-E - 6r)\cos\psi,
\end{aligned}$$

where we ignore quartic and higher terms in the Hamiltonian. Periodic solutions in general position are obtained from the stationary points of this equation, leading to $\sin\psi = 0$ and $\cos\psi = \pm 1$. Moreover,

$$-r\delta^2 = 2\varepsilon^2|D|^2(E+6r)^2.$$

The normal modes, if present, are given by $\tau_1 = 0$ and $\tau_2 = 0$. Since

$$\tau_1 = 2r + E,$$
$$\tau_2 = -r,$$

this corresponds to $r = -\frac{1}{2}E$ and $r = 0$. The second one can only produce the trivial solution, but the first relation leads to

$$\frac{1}{2}E\delta^2 = 8\varepsilon^2|D|^2E^2$$

or

$$\delta = \pm 4\varepsilon|D|\sqrt{E} \quad \text{(bifurcation set)} .$$

The domain of resonance is defined by the inequality

$$|\delta| < 4\varepsilon|D|\sqrt{E}.$$

Strictly speaking, we have to add some reasoning about the existence of normal mode solutions, since the action-angle variables are singular at the normal modes. Therefore, we shall now analyze these cases somewhat more rigorously, using Morse theory. We put

$$x_j = q_j - ip_j,$$
$$y_j = q_j + ip_j.$$

In these real coordinates, the normal form is

$$\mathrm{H} = \mathrm{H}^0 + \varepsilon|D|(\cos\alpha((q_1^2 - p_1^2)q_2 + 2q_1p_1p_2) - \sin\alpha((q_1^2 - p_1^2)p_2 - 2q_1p_1q_2)).$$

We want to study the normal mode given by the equation $q_1 = p_1 = 0$. We put

$$p_2 = -\sqrt{2\tau}\sin\phi,$$
$$q_2 = \sqrt{2\tau}\cos\phi.$$

This symplectic transformation induces the new Hamiltonian

$$\mathrm{H} = \frac{1}{2}\omega_1(q_1^2 + p_1^2) + \omega_2\tau + \varepsilon|D|\sqrt{2\tau}(\cos(\phi-\alpha)(q_1^2 - p_1^2) - 2\sin(\phi-\alpha)q_1p_1).$$

We shall take as our unperturbed problem the Hamiltonian H^0, defined as

$$\mathrm{H}^0 = \frac{1}{2}(q_1^2 + p_1^2) + 2\tau.$$

To analyze the **normal mode**, we consider a periodic orbit as a critical orbit of H with respect to H^0. To show that the normal mode is indeed a critical orbit and to compute its stability type, we use **Lagrange multipliers**. The extended Hamiltonian H_e is defined as

$$H_e = \mu H^0 + H, \quad \mu \in \mathbb{R},$$

where we fix the energy level by $H^0 = E \in \mathbb{R}$. We obtain μ from

$$dH_e = 0$$

and substituting $q_1 = p_1 = 0$. Indeed,

$$dH_e = \begin{bmatrix} (\mu+\omega_1)q_1 \\ (\mu+\omega_1)p_1 \\ 0 \\ 2\mu+\omega_2 \end{bmatrix} + 2\varepsilon|D|(2\tau)^{\frac{1}{2}} \begin{bmatrix} \cos(\phi-\alpha)q_1 - \sin(\phi-\alpha)p_1 \\ -\cos(\phi-\alpha)p_1 - \sin(\phi-\alpha)q_1 \\ \mathcal{O}(q_1^2+p_1^2) \\ \mathcal{O}(q_1^2+p_1^2) \end{bmatrix}.$$

The critical orbit is given by the vector $(0,0,\phi,\frac{E}{2})$ and its tangent space is spanned by $(0,0,1,0)$. The kernel of $dH^0$is spanned by $(1,0,0,0)$, $(0,1,0,0)$ and $(0,0,1,0)$. This implies that the normal bundle of the critical orbit is spanned by $(1,0,0,0)$ and $(0,1,0,0)$. The second derivative of H_e, d^2H_e, is defined on this normal bundle and can easily be computed:

$$d^2H_e = (\mu+\omega_1)\begin{bmatrix} 1 & 0 \\ 0 & 1 \end{bmatrix} + 2\varepsilon|D|(2\tau)^{\frac{1}{2}} \begin{bmatrix} \cos(\phi-\alpha) & -\sin(\phi-\alpha) \\ -\sin(\phi-\alpha) & -\cos(\phi-\alpha) \end{bmatrix}.$$

It follows from $dH_e = 0$ that $\mu = -\frac{\omega_2}{2}$, so

$$d^2H_e = \begin{bmatrix} \frac{\delta}{2} + 2\varepsilon E^{\frac{1}{2}}|D|\cos(\phi-\alpha) & -2\varepsilon E^{\frac{1}{2}}|D|\sin(\phi-\alpha) \\ -2\varepsilon E^{\frac{1}{2}}|D|\sin(\phi-\alpha) & \frac{\delta}{2} - 2\varepsilon E^{\frac{1}{2}}|D|\cos(\phi-\alpha) \end{bmatrix}.$$

We obtain

$$\mathrm{tr}(d^2H_e) = \delta,$$
$$\det(d^2H_e) = \frac{\delta^2}{4} - 4\varepsilon^2 E|D|^2.$$

If $\det(d^2H_e) > 0$, the normal mode is elliptic since d^2H_e is definite; if $\det(d^2H_e) < 0$, the normal mode is hyperbolic. The bifurcation value is

$$\delta = \pm 4\varepsilon E^{\frac{1}{2}}|D|,$$

as we found before.

We can now give the picture of the periodic orbits versus detuning in Figure 10.4.

Note, that the crossing of the two elliptic orbits is not a bifurcation, since the solutions are π out of phase.

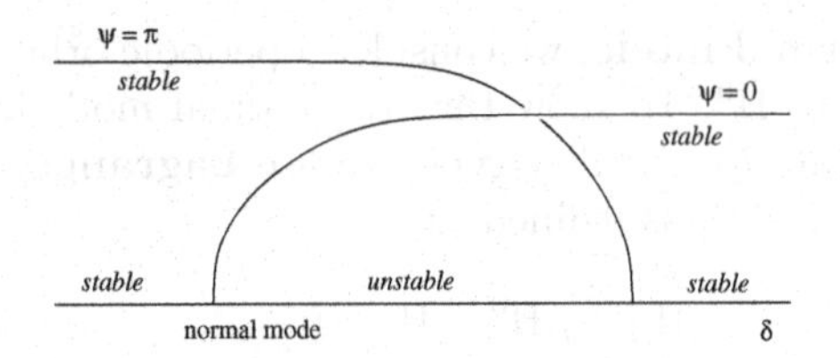

Fig. 10.4: Bifurcation diagram for the 1 : 2-resonance of periodic solutions in general position with the detuning as bifurcation parameter.

We conclude that there are two possibilities: either there are two elliptic periodic solutions (the minimum configuration), or there are two elliptic orbits and one hyperbolic periodic solution. The bifurcation of the elliptic periodic solution from the normal mode does not violate any index argument, because it is a **flip-orbit**:

In a Poincaré-section, Figure 10.5, there are four fixed points (apart from the normal mode), but this picture arises, because the frequency 2 causes the periodic solutions in general position to pass twice through the Poincaré-section.

As we have seen, the normalized Hamiltonian is of the form

$$\mathrm{H}^{[0]} = \mathrm{H}^0 + \varepsilon \overline{H}^1 + \mathcal{O}(\varepsilon^2)$$

in which $\overline{\mathrm{H}}^1$ stands for the normalized cubic part. We have found that H^0 corresponds to an integral of the normalized system, with $\mathcal{O}(\varepsilon)$-error with respect to the orbits of the original system and validity for all time. Of course,

Fig. 10.5: Poincaré section for the exact 1 : 2-resonance in normal form. The fixed point in the center corresponds to a hyperbolic normal mode; the four elliptic fixed points correspond to two stable periodic solutions.

the normalized Hamiltonian $\overline{H}$ is itself an integral of the normalized system, and we can take as two independent, Poisson commuting integrals H^0 and $\overline{\mathrm{H}}^1$.

This discussion generalizes to the n-degrees of freedom case, but for two degrees of freedom we found the minimal number of integrals to conclude that the normalized system is always completely integrable. This simplifies the analysis of two degrees of freedom systems considerably.

A useful concept is the **momentum map** which can be defined as follows:

$$M : T^\star\mathbb{R}^2 \to \mathbb{R}^2,$$
$$M(\boldsymbol{q},\boldsymbol{p}) = (\mathrm{H}^0(\boldsymbol{q},\boldsymbol{p}), \overline{\mathrm{H}}^1(\boldsymbol{q},\boldsymbol{p})).$$

Using this map, we can analyze the foliation induced by the two integrals. In general, $M^{-1}(x,y)$ will be a torus, or empty. For special values, the inverse image consists of a circle, which in this case is also a periodic orbit.

The asymptotic accuracy of the results in these computations follows from the estimates on the solutions of the differential equation. For the $1:2$-resonance we have $\mathcal{O}(\varepsilon)$-error on the time scale $1/\varepsilon$.

The normalization process can be chosen in various ways according to taste and efficiency notions. One approach is the reduction of the normal form Hamiltonian flow to a Poincaré map which, in the case of two degrees of freedom, is integrable. In [45] this is applied to the generic $1:2$-resonance which has a special feature: the planar Poincaré map has a central singularity, equivalent to a symmetric hyperbolic umbilic. The analysis of versal deformations and unfoldings leads to a general perturbation treatment which is applied to the spring-pendulum mechanical system (the spring-pendulum consists of a mass point on a spring with vertical motion only, to which a pendulum is attached). In addition to the $1:2$-resonance, the $2:2$-resonance is discussed in [47].

10.6.2 The Symmetric 1 : 1-Resonance

The general normal form expression for the $1:1$-resonance is (see Section 10.5.3)

$$\overline{\mathrm{H}} = \mathrm{F}(x_1y_1, x_2y_2, x_1y_2) + x_2y_1\mathrm{G}(x_1y_1, x_2y_2, x_2y_1).$$

We will include again detuning. The differential equations for the general $k:l$-resonance, $0<k<l$, do not apply to the $1:1$-resonance. We can, however, use them for what we shall call the **symmetric** $1:1$**-resonance**. The normal form for the general $1:1$-resonance is complicated, with many parameters. If we impose, however, mirror symmetry in each of the symplectic coordinates, that is invariance of the Hamiltonian under the four transformations

$$M_1 : (q_1,p_1,q_2,p_2) \mapsto (-q_1,p_1,q_2,p_2),$$
$$M_2 : (q_1,p_1,q_2,p_2) \mapsto (q_1,-p_1,q_2,p_2),$$
$$M_3 : (q_1,p_1,q_2,p_2) \mapsto (q_1,p_1,-q_2,p_2),$$
$$M_4 : (q_1,p_1,q_2,p_2) \mapsto (q_1,p_1,q_2,-p_2),$$

then this simplifies the normal form considerably. Since this symmetry assumption is natural in several applications (one must realize that the assumption need not be valid for the original Hamiltonian, only for the normal form), one could even say that the symmetric 1 : 1-resonance is more important than the general 1 : 1-resonance; it certainly merits a separate treatment.

Note that two normal modes exist in this case; we leave this to the reader. For the resonance matrix we take

$$\hat{M} = \begin{bmatrix} l & -k \\ k^\star & l^\star \end{bmatrix} = \begin{bmatrix} 2 & -2 \\ 2 & 2 \end{bmatrix}.$$

The differential equations are

$$\dot{r} = 16\varepsilon^2 |D|(E^2 - r^2)\sin\psi,$$
$$\dot{\psi} = \delta + 2\varepsilon^2 |D|(-16r)\cos\psi + 4\varepsilon^2((\Delta_1 - \Delta_2)r + (\Delta_1 + \Delta_2)E).$$

The stationary solutions, corresponding to periodic solutions in general position, are determined by

$$\sin\psi = 0 \quad \Rightarrow \quad \cos\psi = \pm 1,$$
$$\delta \pm 32\varepsilon^2 |D| r + 4\varepsilon^2(\Delta_1 - \Delta_2)r + 4\varepsilon^2(\Delta_1 + \Delta_2)E = 0.$$

Rescale

$$\delta = 4\varepsilon^2 E\Delta, \quad r = Ex,$$

then

$$\Delta \pm 8|D|x_\pm + (\Delta_1 - \Delta_2)x_\pm + (\Delta_1 + \Delta_2) = 0,$$

or

$$x_\pm = -\frac{\Delta + \Delta_1 + \Delta_2}{\Delta_1 - \Delta_2 \pm 8|D|},$$

with condition $|\Delta_1 - \Delta_2| \neq 8|D|$. Since, by definition, for orbits in general position,

$$\tau_1 = 2(E + r) > 0,$$
$$\tau_2 = 2(E - r) > 0,$$

we have $|x| < 1$ and this yields the following bifurcation equations:

$$\Delta = -(\Delta_1 + \Delta_2),$$
$$\Delta = -2(\Delta_1 \pm 4|D|).$$

The linearized equations at the stationary point (ψ_0, r_0) are

$$\dot{r} = 16\varepsilon^2|D|(E^2 - r_0^2)\cos(\psi_0)\psi,$$
$$\dot{\psi} = -32\varepsilon^2|D|\cos(\psi_0)r + 4\varepsilon^2(\Delta_1 - \Delta_2)r,$$

where $\cos\psi_0 = \pm 1$. The eigenvalues are given by

$$\lambda^2 - 16\varepsilon^2|D|(E^2 - r_0^2)\cos\psi_0(-32\varepsilon^2|D|\cos\psi_0 + 4\varepsilon^2(\Delta_1 - \Delta_2)) = 0.$$

The orbit is elliptic if

$$8|D| > \pm(\Delta_1 - \Delta_2)$$

and hyperbolic otherwise (excluding the bifurcation value). The bifurcation takes place when

$$8|D| = |\Delta_1 - \Delta_2|.$$

This is a so-called vertical bifurcation; for this ratio of the parameters, both normal modes bifurcate at the same moment, the equation for the stationary points is degenerate and in general one has to go to higher-order approximations to see what happens. Despite its degenerate character, this vertical bifurcation keeps turning up in applications, cf. [277] and [236].

10.6.3 The 1 : 3-Resonance

We will use the general results of Section 10.5.4. There are two second-order resonances in two degrees of freedom systems: the 1 : 1- and the 1 : 3-resonance. The latter has not been discussed very often in the literature. A reason for this might be, that mirror or discrete symmetry in one of the two degrees of freedom immediately causes degeneration of the normal form. In the case of for instance the 1 : 2-resonance, only mirror symmetry in the first degree of freedom causes degeneracy.

In general for n degrees of freedom, a low-order resonance with only odd resonance numbers, will be easily prone to degeneration.

Periodic Orbits in General Position

The Poincaré series and the normal form can be written down immediately as in the 1 : 2-case. For the resonance matrix we take

$$\hat{M} = \begin{bmatrix} 3 & -1 \\ 1 & 0 \end{bmatrix}.$$

The differential equations, derived from the normalized Hamiltonian, are

$$\dot{r} = \varepsilon^2|D|(2E + 6r)^{\frac{3}{2}}(-2r)^{\frac{1}{2}}\sin\psi,$$
$$\dot{\psi} = \delta + 2\varepsilon^2|D|(2E + 6r)^{\frac{1}{2}}(-2r)^{-\frac{1}{2}}(-E - 12r)\cos\psi +$$
$$2\varepsilon^2((3\Delta_1 - \Delta_2)r + \Delta_1 E).$$

This leads to the following equation for the stationary points

$$\sin \psi = 0,$$
$$(\delta + 2\varepsilon^2((3\Delta_1 - \Delta_2)r + \Delta_1 E))^2(-2r) = 4\varepsilon^4|D|^2(2E + 6r)(-E - 12r)^2.$$

This equation is cubic in r, there may be one or three real solutions. Let

$$r = Ex,$$
$$\delta = \Delta\varepsilon^2 E,$$

then

$$(\Delta + 2\Delta_1 + 2(3\Delta_1 - \Delta_2)x)^2(-2x) = 8|D|^2(1 + 3x)(1 + 12x)^2.$$

Put

$$\alpha = \Delta + 2\Delta_1,$$
$$\beta = 2(3\Delta_1 - \Delta_2),$$
$$\gamma = 2|D|.$$

Then we have

$$-(\alpha + \beta x)^2 x = \gamma^2(1 + 27x + 216x^2 + 432x^3),$$

or

$$(432\gamma^2 + \beta^2)x^3 + (216\gamma^2 + 2\alpha\beta)x^2 + (27\gamma^2 + \alpha^2)x + \gamma^2 = 0.$$

We shall not give the explicit solutions, but we are especially interested in the bifurcation set of this equation. First we transform to the standard form for cubic equations:

$$y^3 + uy + v = 0.$$

Let

$$ax^3 + bx^2 + cx + d = 0,$$

and put

$$y = x + \frac{b}{3a}.$$

Then we obtain

$$u = \frac{(ac - \frac{1}{3}b^2)}{a^2},$$
$$v = \frac{(da^2 - \frac{1}{3}abc + \frac{2}{27}b^3)}{a^3}.$$

The bifurcation set of this standard form is the well-known cusp equation

$$27v^2 + 4u^3 = 0.$$

After some extensive calculations, we find this to be equivalent to a homogeneous polynomial of degree 12 in α, β and γ. After factoring out, the bifurcation equation can be written as:

$$\alpha^4 + 54\alpha^2\gamma^2 - 243\gamma^4 - \frac{1}{3}\alpha^3\beta - 27\alpha\beta\gamma^2 + \frac{9}{4}\beta^2\gamma^2 = 0,$$

(we neglect here the isolated bifurcation plane $12\alpha = \beta$). Consider the curve $P = 0$, with

$$\begin{aligned} P(\alpha, \beta, \gamma) &= \alpha^4 + 54\alpha^2\gamma^2 - 243\gamma^4 - \frac{1}{3}\alpha^3\beta - 27\alpha\beta\gamma^2 + \frac{9}{4}\beta^2\gamma^2 \\ &= -\frac{1}{3}(\alpha^2 + 27\gamma^2)^2 + \frac{1}{3}\alpha^3(4\alpha - \beta) + \frac{9}{4}(4\alpha - \beta)(8\alpha - \beta)\gamma^2. \end{aligned}$$

This suggests the transformation

$$\begin{aligned} X &= \alpha, \\ Y &= (27)^{\frac{1}{2}}\gamma, \\ Z &= \frac{1}{2}(4\alpha - \beta). \end{aligned}$$

The resulting expression for P is

$$P^\star = -\frac{1}{3}\left(X^2 + Y^2 - XZ\right)^2 + \frac{1}{3}(X^2 + Y^2)Z^2.$$

Putting $Z = 1$, we have the equation for the **cardioid**:

$$\left(X^2 + Y^2 - X\right)^2 = (X^2 + Y^2).$$

Changing to polar coordinates

$$X = r\cos\theta, Y = r\sin\theta,$$

this takes the simple form

$$r = 1 + \cos\theta.$$

Another representation is obtained as follows. Intersecting the curve with the pencil of circles

$$X^2 + Y^2 - 2X = tY,$$

we obtain

$$(X + tY)^2 = (2X + tY).$$

This implies

$$(t^2 - 1)Y + 2tX = 0.$$

Substituting this in the equation for the circle bundle, we obtain

$$X = \frac{2(1 - t^2)}{(1 + t^2)^2},\ Y = \frac{4t}{(1 + t^2)^2},$$

so we have a rational parametrization of the bifurcation curve.

Normal Mode

With the same reasoning as in the 1 : 2 case we find only one normal mode for the 1 : 3-resonance. We analyze the normal form of the 1 : 3-resonance in real coordinates $\boldsymbol{q}$ and $\boldsymbol{p}$:

$$\begin{aligned}
\mathrm{H} &= \frac{1}{2}(\omega_1(q_1^2 + p_1^2) + \omega_2(q_2^2 + p_2^2)) + \\
&+\frac{1}{2}|D|\varepsilon^2((\cos\alpha + i\sin\alpha)(q_1 - ip_1)^3(q_2 + ip_2) + \\
&(\cos\alpha - i\sin\alpha)(q_1 + ip_1)^3(q_2 - ip_2) + \\
&\frac{1}{4}\varepsilon^2(A(q_1^2 + p_1^2)^2 + 2B(q_1^2 + p_1^2)(q_2^2 + p_2^2) + C(q_2^2 + p_2^2)^2) \\
&= \frac{1}{2}(\omega_1(q_1^2 + p_1^2) + \omega_2(q_2^2 + p_2^2)) + \\
&\varepsilon^2|D|(\cos\alpha((q_1^2 - 3p_1^2)q_1q_2 + (3q_1^2 - p_1^2)p_1p_2 - \\
&\sin\alpha((q_1{}^3 - 3p_1^2q_1)p_2 - (3q_1^2p_1 - p_1{}^3)q_2) + \\
&\frac{1}{4}\varepsilon^2(A(q_1^2 + p_1^2)^2 + 2B(q_1^2 + p_1^2)(q_2^2 + p_2^2) + C(q_2^2 + p_2^2)^2).
\end{aligned}$$

To study the normal mode $q_1 = p_1 = 0$, we put

$$p_2 = -(2\tau)^{\frac{1}{2}}\sin\phi,\ q_2 = (2\tau)^{\frac{1}{2}}\cos\phi,$$

obtaining

$$\begin{aligned}
\mathrm{H} &= \frac{1}{2}\omega_1(q_1^2 + p_1^2) + \omega_2\tau + \\
&\varepsilon^2|D|\sqrt{2\tau}(\cos(\phi - \alpha)(q_1^2 - 3p_1^2)q_1 - (3q_1^2 - p_1^2)p_1\sin(\phi - \alpha)) + \\
&\varepsilon^2(\frac{A}{4}(q_1^2 + p_1^2)^2 + B\tau(q_1^2 + p_1^2) + C\tau^2).
\end{aligned}$$

We introduce the extended Hamiltonian H_e and Lagrange multiplier μ as before by

$$\mathrm{H}_e = \mu \mathrm{H}^0 + \mathrm{H}, \quad \mathrm{H}^0 = 3E,$$

where

$$\mathrm{H}^0 = \frac{1}{2}(q_1^2 + p_1^2) + 3\tau.$$

Then

$$d\mathrm{H}_e = \begin{bmatrix} (\mu+\omega_1)q_1 \\ (\mu+\omega_1)p_1 \\ 0 \\ \omega_2 + 3\mu \end{bmatrix} + \varepsilon^2 \begin{bmatrix} \mathcal{O}(q_1^2+p_1^2) \\ \mathcal{O}(q_1^2+p_1^2) \\ \mathcal{O}(q_1^2+p_1^2) \\ \mathcal{O}(q_1^2+p_1^2) \end{bmatrix} + \varepsilon^2 \begin{bmatrix} 2B\tau q_1 \\ 2B\tau p_1 \\ 0 \\ 2C\tau + \mathcal{O}(q_1^2+p_1^2) \end{bmatrix}$$

and

$$d^2\mathrm{H}_e = \begin{bmatrix} \mu + \omega_1 + 2B\tau\varepsilon^2 & 0 \\ 0 & \mu + \omega_1 + 2B\tau\varepsilon^2 \end{bmatrix}.$$

Since $\omega_2 + 3\mu + 2C\varepsilon^2\tau = 0$ and $3\tau = E$,

$$\begin{aligned} d^2\mathrm{H}_e &= \begin{bmatrix} -\frac{\omega_2}{3} + \omega_1 + \frac{2}{3}(3B-C)E\varepsilon^2 & 0 \\ 0 & -\frac{\omega_2}{3} + \omega_1 + \frac{2}{3}(3B-C)E\varepsilon^2 \end{bmatrix} \\ &= \frac{1}{3}\begin{bmatrix} \delta + 2(3B-C)E\varepsilon^2 & 0 \\ 0 & \delta + 2(3B-C)E\varepsilon^2 \end{bmatrix}. \end{aligned}$$

This is a definite form, except if $\delta + 2(3B-C)E\varepsilon^2 = 0$, and the normal mode is elliptic. The bifurcation value, where $d^2\mathrm{H}_e = 0$, marks the 'flipping through' of a hyperbolic periodic orbit, in such a way that this orbit changes its phase with a factor π in the Poincaré section transversal to the normal mode.

10.6.4 Higher-order Resonances

After the low-order resonances of the preceding subsections, we will study *higher-order resonance* cases, starting with the general results of Section 10.5.4. This is the large group of resonances for which $k + l \geq 5$, allowing for detuning. In general we have again $(k, l) = 1$, but in the case of symmetries we have to relax this condition.

The differential equations in normal form have solutions, characterized by two different time scales; as we shall see, they are generated, by ε-terms of order (degree) 2, describing most of the flow in the Hamiltonian system, and of order $k + l - 2$, describing the flow in the so-called resonance domain. This particular structure of the normal form equations enables us to treat all the higher-order resonances at the same time. In contrast to the case of low-order resonances, we shall obtain the periodic orbits without making assumptions on k and l.

The discussion of the asymptotics of higher-order resonances is based on [237] and extensions in [257].

Periodic Orbits in General Position

The normal form equations are

$$\dot{r} = \varepsilon^{k+l-2}|D|(2k^\star E + 2lr)^{\frac{l}{2}}(2l^\star E - 2kr)^{\frac{k}{2}} \sin\psi,$$
$$\dot{\psi} = \delta + 2\varepsilon^2((l\Delta_1 - k\Delta_2)r + (k^\star\Delta_1 + l^\star\Delta_2)E) + \mathcal{O}(\varepsilon^{k+l-2}) + \mathcal{O}(\varepsilon^4).$$

As in Section 10.5.4, we have $D = |D|e^{i\alpha}, \psi = l\phi_1 - k\phi_2 + \alpha$.
To find periodic solutions in general position, we put $\sin\psi = 0$, producing an equation for r

$$\delta + 2\varepsilon^2((l\Delta_1 - k\Delta_2)r + (k^\star\Delta_1 + l^\star\Delta_2)E = \mathcal{O}(\varepsilon^{k+l-2}) + \mathcal{O}(\varepsilon^4).$$

It makes sense to choose $\delta = \mathcal{O}(\varepsilon^2)$. To find out whether the equation has a solution, we rescale

$$r = Ex,$$
$$\delta = 2E\varepsilon^2\Delta.$$

To $\mathcal{O}(\varepsilon^2)$ we have to solve

$$\Delta + k^\star\Delta_1 + l^\star\Delta_2 + (l\Delta_1 - k\Delta_2)x = 0, \tag{10.6.1}$$

with $-\frac{k^\star}{l} < x < \frac{l^\star}{k}$ (where $\frac{l^\star}{k} - (-\frac{k^\star}{l}) = \frac{1}{lk} > 0$ since $\hat{M} \in SL_2(\mathbb{Z})$). Equation (10.6.1) determines the so-called **resonance manifold**. Since

$$x = -\frac{(\Delta + k^\star\Delta_1 + l^\star\Delta_2)}{l\Delta_1 - k\Delta_2}, \quad l\Delta_1 - k\Delta_2 \neq 0,$$

the condition on the parameters for solvability becomes

$$-\frac{k^\star}{l} < -\frac{(\Delta + k^\star\Delta_1 + l^\star\Delta_2)}{l\Delta_1 - k\Delta_2} < \frac{l^\star}{k}.$$

This implies that the width of the parameter interval is given by

$$2\varepsilon^2 E\left|\frac{\Delta_1}{k} - \frac{\Delta_2}{l}\right|$$

Note, that the parameter that determines the presence of the resonance manifold is the rescaled detuning Δ.
In the resonance domain, if it exists, the condition $\sin\psi = 0$ results in two periodic orbits. Linearization produces easily that one is elliptic, the other hyperbolic. This conclusion holds for orbits in general position and not near the normal modes. A few examples were studied in [244] and are displayed in Figure 10.6.

Note that for the existence and location of the resonance manifold $\mathcal{O}(\varepsilon^2)$-terms of the normal form suffice, for the actual position of the periodic orbits we have to know α from the normal form to $\mathcal{O}(\varepsilon^{k+l-2})$.

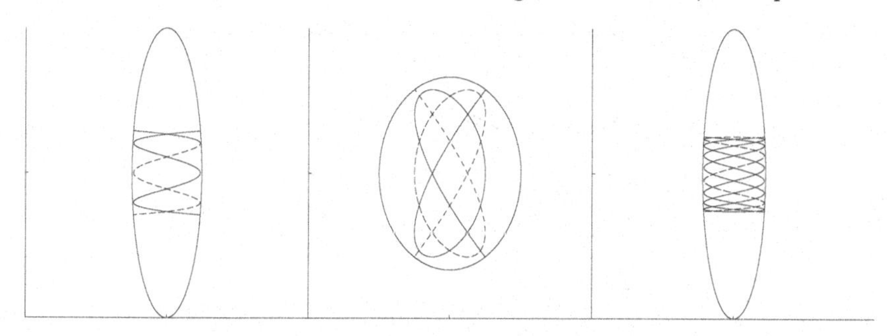

Fig. 10.6: Projections into base space for the resonances 4 : 1, 4 : 3 and 9 : 2; cf. Section 10.5.1, Option 2. The stable (full line) and unstable (− − −) periodic solutions are lying in the resonance manifold. The closed boundary is the curve of zero-velocity.

Asymptotic Estimates

The equations for amplitude and combination angle that we used in the preceding analysis are of the form

$$\dot{r} = \varepsilon^{k+l-2} f(r) \sin \psi + \cdots , \quad k + l - 2 \geq 3,$$
$$\dot{\psi} = \varepsilon^2 g(r) + \varepsilon^{k+l-2} h(r) + \cdots ;$$

$f(r)$, $g(r)$ and $h(r)$ are abbreviations for the expressions from the previous subsection. This system has to be supplemented by equations for E and ψ_2. The right-hand side of the equations starts with terms of $\mathcal{O}(\varepsilon^2)$ and, using the theory of Chapter 2, it is easy to obtain the estimate

$$r(t) = r(0) + \mathcal{O}(\varepsilon)$$

on the time scale $1/\varepsilon^2$. So, on this time scale, no appreciable change of the variable r takes place. To improve our insight in higher-order resonance we note that the right-hand side of the equation for r is $\mathcal{O}(\varepsilon^{k+l-2})$ with $k+l-2 \geq 3$ and of the equation for ψ is $\mathcal{O}(\varepsilon^2)$. In the spirit of Chapter 7, we can consider ψ to be rapidly varying with respect to the variable r and it is then natural to consider averaging the system over the angle ψ.

This procedure breaks down where ψ is not rapidly varying, i.e. in the domain where $g(r)$ is equal or near to zero. Note that the equation $g(r) = 0$, corresponding to equation (10.6.1) in the preceding section, defines the (so-called) resonance manifold M in phase space where the periodic orbits in general position are found.

For the asymptotic estimates we need to discern two domains in phase space.

- The **resonance domain** D_I, which is a neighborhood of the resonance manifold M. In terms of singular perturbations, this is the *inner* boundary layer.

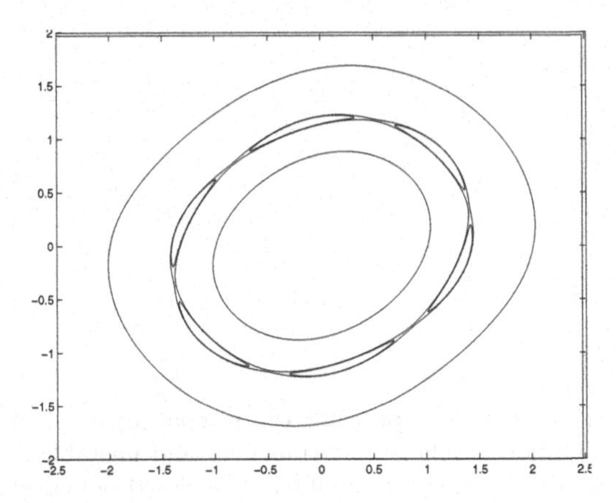

Fig. 10.7: The Poincaré map for the 1 : 6-resonance of the elastic pendulum ($\varepsilon = 0.75$, large for illustration purposes). The saddles are connected by heteroclinic cycles and inside the cycles are centers see [257], courtesy SIAP.

Introducing the distance $d(P, M)$ for a point P on the energy manifold to the manifold M we have

$$D_I = \{P | d(P, M) = \mathcal{O}(\varepsilon^{\frac{k+l-4}{2}}),\ \ k + l \geq 5.$$

- The remaining part of phase space, outside the resonance domain, is D_O, the *outer* domain. In the domain D_O, there is, to a certain approximation, no exchange of energy between the two degrees of freedom.

Following [257], the idea behind the estimate of the size of the resonance domain D_I, is as follows. In the Poincaré map, the periodic orbits in general position appear as $2k$ or $2l$ fixed points (excluding the origin) which are saddles and centers, corresponding to the unstable and stable periodic orbits in the resonance domain. Each two neighboring saddles are connected by a heteroclinic cycle. Inside each domain, bounded by these heteroclinic cycles, there is a center point. For an illustration, see Figure 10.7. We approximate the size of this domain by calculating the distance between the two intersection points of the heteroclinic cycle and a straight line connecting a center point to the origin. This leads to the estimate given above.

In the outer domain D_O, the flow can be described as a simple, nonlinear continuation of the linearized flow on a long time scale. This is expressed in terms of asymptotic estimates as follows:

Theorem 10.6.1. *Consider the equations for r, ψ and E with initial conditions in the outer domain D_O and the initial value problem*

$$\dot{\tilde{\psi}} = 2\varepsilon^2 E\Delta + 2\varepsilon^2[(l\Delta_1 - k\Delta_2)r + (k^\star\Delta_1 + l^\star\Delta_2)E]\ ,\ \tilde{\psi}(0) = \psi(0).$$

Then we have the estimates

$$r(t) - r(0), E(t) - E(0), \psi(t) - \tilde{\psi}(t) = \mathcal{O}(\varepsilon^{\frac{k+l-4}{6}})$$

on the time scale $\varepsilon^{-\frac{k+l}{2}}$.

Potential Problems

In a large number of problems, the Hamiltonian is characterized by quadratic momenta and a potential function for the positions:

$$H(p_1, p_2, q_1, q_2) = \frac{1}{2}(p_1^2 + p_2^2) + V(q_1, q_2). \tag{10.6.2}$$

Classical examples are the elastic pendulum and the generalized Hénon–Heiles Hamiltonian

$$H(p_1, p_2, q_1, q_2) = \frac{1}{2}(p_1^2 + p_2^2) + \frac{1}{2}(k^2 q_1^2 + l^2 q_2^2) - \varepsilon\left(\frac{1}{3}a_1 q_1^3 + a_2 q_1 q_2^2\right).$$

Resonance	$k+l-2$	d_ε	Interaction time scale
$1:4$	3	$\varepsilon^{1/2}$	$\varepsilon^{-5/2}$
$3:4$	5	$\varepsilon^{3/2}$	$\varepsilon^{-7/2}$
$1:6$	5	$\varepsilon^{3/2}$	$\varepsilon^{-7/2}$
$2:6$	6	ε^{2}	ε^{-4}
$1:8$	7	$\varepsilon^{5/2}$	$\varepsilon^{-9/2}$
$4:6$	8	ε^{3}	ε^{-5}

Table 10.2: The table presents the most prominent higher-order resonances of the elastic pendulum with lowest-order resonant terms $\mathcal{O}(\varepsilon^{k+l-2})$. The third column gives the size of the resonance domain in which the resonance manifold M is embedded, while in the fourth column we find the time scale of interaction in the resonance domain.

We have to normalize to $\mathcal{O}(\varepsilon^2)$ to locate the resonance manifold M by Eq. (10.6.1). However, as discussed in Section 10.6.4, for the position of the periodic orbits we have to normalize to $\mathcal{O}(\varepsilon^{k+l-2})$.

Fortunately, the answer is easy to obtain in the case of potential problems.

Lemma 10.6.2. *Consider the potential problem (10.6.2) where* $V(q_1, q_2)$ *has a Taylor expansion near* $(0,0)$ *which starts with* $\frac{1}{2}(k^2q_1^2 + l^2q_2^2)$*. Then the coefficient* D *of the normal form to* $\mathcal{O}(\varepsilon^{k+l-2})$ *is real or* $\alpha = 0$.

Proof The proof can be found in [257] with applications to the Hénon–Heiles Hamiltonian and the elastic pendulum. For the last case the Poincaré map for the 1 : 6-resonance is shown in Figure 10.7. □

It is interesting to consider the hierarchy of the first six higher-order resonances of the elastic pendulum in Table 10.2. Note that, because of symmetries, the 1 : 3-resonance is present as 2 : 6, the 2 : 3-resonance as 4 : 6.

The Double Eigenvalue Zero Case

An extreme kind of higher-order resonance is the case of widely separated frequencies. A typical Hamiltonian would be

$$H = \frac{1}{2}(p_1^2 + q_1^2) + \frac{1}{2}\varepsilon(p_2^2 + q_2^2) + \varepsilon \mathrm{H}^1 + \cdots .$$

In [44] these problems are discussed in the context of unfoldings of a singularity. Additional analysis and an application is given in [258], see also [125] for a discussion and applications.

10.7 Three Degrees of Freedom, General Theory

10.7.1 Introduction

In contrast with the case of two degrees of freedom systems, the literature on this subject is still growing. One of the reasons is doubtless the enormous increase in complexity of the expressions with the number of degrees of freedom; in the case of three degrees of freedom H^1 contains 56 terms, H^2 contains 126 terms. It is a question of considerable practical interest how to handle such longer expressions analytically. We shall find that by the process of normalization it is possible to obtain a drastic reduction of the size of these expressions.

One might wonder: are there new theoretical questions in systems with more than two degrees of freedom, are the questions not merely extensions of the same problems in a more complicated setting? To some extent this is true with respect to the analysis of periodic solutions of the normalized Hamiltonian. Note however that the question of stability of these solutions is essentially more difficult. In the case of two degrees of freedom the critical points of the equations for r and ψ (Section 10.5.4) will be elliptic or hyperbolic, characteristics which follow from a linear analysis. The existence of two-dimensional tori around these periodic solutions and the corresponding approximate integrals of motion which are valid for all time, then guarantee rigorously stability in the case of elliptic critical points of the reduced system. This property of rigorous results of a combined invariant tori/quasilinear analysis argument is lost in the case of three degrees of freedom. In this case we find again elliptic and hyperbolic orbits and there exist corresponding invariant tori around the elliptic orbits, but these are 3-dimensional in a 5-dimensional sphere, so the tori do not separate the sphere into distinct pieces, as in the lower-dimensional case. An easy way to see this is to consider only the actions. One can identify a torus with constant action-variables with a point on a $(n-1)$-simplex, where n is the number of degrees of freedom. For $n=2$ the point does divide the interval into two pieces, but for $n=3$ it does not divide the triangle into pieces. This topological fact gives rise to the so-called **Arnol′d diffusion** (for a discussion, see [174]) and other phenomena (see [125]). In the sequel we shall call periodic solutions corresponding to elliptic orbits again stable; note however that now we have stability only in a *formal* sense.

Another fundamental difference can be described as follows. In systems with two degrees of freedom we always find two integrals of the normalized Hamiltonian, providing us with a complete description of the phase flow. This is expressed by saying that the normalized Hamiltonian is (completely) integrable. In the case of three degrees of freedom we still have two integrals of the normal form, but we need three for the system to be integrable. To find a third integral is a nontrivial problem: in some cases it can be shown to exist, but there are also cases where it has been shown that a third analytic integral does not exist [79]. This makes the global description of the phase-flow of the normalized system essentially more difficult in the case of three degrees of freedom.

Another question that is only partially solved, is the asymptotic analysis of three degrees of freedom systems. In a number of cases, for instance for the genuine first-order resonances, the analytic difficulties can be overcome, and a complete analysis is possible of the periodic orbits and their formal stability. There are some results on second-order resonances and on higher-order resonances but the analysis is far from complete.

10.7.2 The Order of Resonance

For Hamiltonians near stable equilibrium and at exact resonance, we made the blanket Assumption 10.2.1 that

$$\mathrm{H}^0=\sum_{i=1}^{3}\frac{1}{2}\omega_i(q_i^2+p_i^2),\quad \omega_i\in\mathbb{N}\,,\quad i=1,2,3.$$

Following Section 10.2, we consider $\boldsymbol{k}\in\mathbb{Z}^3$ and $\boldsymbol{k}$-vectors such that $\sum_{i=1}^{3}\omega_i k_i=0$. We identify annihilation vectors $\boldsymbol{k}$ and $\boldsymbol{k}'$ if $\boldsymbol{k}+\boldsymbol{k}'=0$. The number $\kappa=\sum_{i=1}^{3}|k_i|$, the norm of $\boldsymbol{k}$, determines the order of normalization. However, to characterize the possible interactions between the three degrees of freedom on normalizing to H^κ, we need another quantity. Compare for example the resonances $1:2:3$ and $1:2:5$. On normalizing to H^1 ($\kappa=3$), we have for the $1:2:3$-resonance the annihilating vectors $(2,-1,0)$ and $(1,1,-1)$; for the $1:2:5$-resonance only $(2,-1,0)$. Up till H^1, or in the language of asymptotic approximations: up till an $\mathcal{O}(\varepsilon)$-approximation on the time scale $1/\varepsilon$, the $1:2:3$-resonance displays full interaction between all three degrees of freedom, the $1:2:5$-resonance decouples at this level to a two degrees of freedom system and a one degree of freedom system. The case of full interaction between all three degrees of freedom was called a **genuine resonance** in [262]. To indicate the number of annihilating vectors at a certain order κ, we introduce the interaction number σ_κ; intuitively, the larger σ_κ is, the more complex the analysis will appear to be. There are however no mathematical theorems to confirm this intuition and to measure exactly the complexity of any system in resonance. The same paper contains a list of genuine first-order resonances, and we reproduce it in Table 10.3, each resonance with its interaction number at order 3 and 4. (The reader may verify for instance that for the $1:2:1$- resonance annihilating k-vectors are $(2,-1,0)$, $(0,-1,2)$ and $(1,-1,1)$).

resonance	σ_3	σ_4
$1:2:1$	3	1
$1:2:2$	2	1
$1:2:3$	2	2
$1:2:4$	2	1

Table 10.3: The four genuine first-order resonances.

resonance	σ_3	σ_4
$1:1:1$	0	6
$1:1:3$	0	5
$1:2:5$	1	1
$1:2:6$	1	1
$1:3:3$	0	3
$1:3:4$	1	1
$1:3:5$	0	3
$1:3:6$	1	1
$1:3:7$	0	2
$1:3:9$	0	2
$2:3:4$	1	1
$2:3:6$	1	1

Table 10.4: The genuine second-order resonances.

For the sake of completeness we also list the 12 genuine second-order resonances with their interaction numbers in Table 10.4. The first two cases, $1:1:1$ and $1:1:3$, appear to be the most complicated, followed by the resonances $1:3:3$ and $1:3:5$. As noted before, resonances with odd annihilation numbers may degenerate easily and symmetry assumptions may change the dynamics and the complexity.

For the actual calculation of the normal forms, we can use an ad hoc approach, but a systematic treatment is based on the Poincaré series and the finite number of generators of the algebra of invariants. Such a program was initiated in [90] and we shall list a number of results.

With regards to the list of generators there are two remarks to keep in mind.

- The quadratic part of the normal form will always be H^0, but there can be more quadratic generators present in the case of a $1:1$- (or $\omega:\omega$-) subresonance. For instance in the case of the $1:2:1$-resonance, we have the generators x_1y_3, y_1x_3. In the corresponding normal form we have terms $x_1y_3P_1(\cdots), y_1x_3P_2(\cdots)$ with P_1, P_2 polynomials (without constant term) in the generators.
- In the same spirit, we will list generators of degree higher than two if they correspond to a subresonance. For instance in the case of the $1:2:5$-resonance, the $2:5$-subresonance will produce generators $x_2^5y_3^2$, $y_2^5x_3^2$, producing terms as products with the other generators of the $1:2:5$-resonance.
- A generator such as $y_1x_2y_3$ may be missing in the cubic part of the Hamiltonian normal form because of discrete symmetry. It will be present as $(y_1x_2y_3)^2$ at degree 6. We will omit such cases in the basic list of generators.

10.7.3 Periodic Orbits and Integrals

The quadratic part of the Hamiltonian is

$$\mathrm{H}^0 = \sum_{i=1}^{3} \frac{1}{2}\omega_i(q_i^2+p_i^2), \quad \omega_i \in \mathbb{N}\ ,\ i=1,2,3,$$

or in action-angle variables τ, ϕ

$$\mathrm{H}^0 = \sum_{i=1}^{3} \omega_i \tau_i,$$

and in complex variables

$$\mathrm{H}^0 = i\sum_{j=1}^{3} \omega_j x_j y_j.$$

Normalizing H^1 we find at most two linearly independent combinations of the three angles ϕ_i; we shall denote these combination angles by ψ_1 and ψ_2.

As discussed earlier, H^0 will be an approximate integral of the system, an exact integral of the normal form. In phase space $\mathrm{H}^0 =$ constant corresponds to S^5.

Periodic solutions are found as critical points of the normalized Hamiltonian reduced to H^0. In practice this involves the elimination of one action, the energy, leaving us with two action and two angle variables. The critical points are thus characterized by four eigenvalues; a pair of conjugate imaginary eigenvalues will be denoted in our pictures by E (elliptic), a pair of opposite real eigenvalues by H (hyperbolic), and the degenerate situation with zero eigenvalues by O. In Section 10.5.1 we discussed visual presentations of the phase-flow and periodic solutions. In the case of three degrees of freedom the following visualization (suggested by R. Cushman) is useful. We forget the angular variables and only plot the actions. For given energy, the set of allowable action values is a 2-simplex (triangular domain).

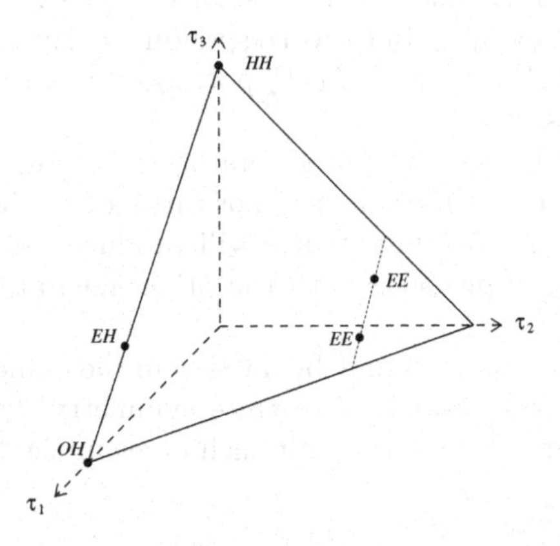

Fig. 10.8: Action simplex; dots indicate periodic solutions, normal modes are at the vertices. The stability characteristics are denoted by E, H and O.

The periodic solutions are points in this simplex since they have fixed actions. Note that according to [286] at least three periodic solutions exist for each energy value. To draw invariant surfaces is only possible in this representation if the angular variables do not play a part. The normal modes are the vertices of the simplex. The linear stability is indicated by two pairs of eigenvalues; for instance EE means two conjugate pairs of imaginary eigenvalues, OH means two eigenvalues zero and two real, HH means two real pairs, etc. In the next sections we will present results regarding the basic basic resonances of genuine first- and second-order; see also the paper [285] and a note on the $1:3:7$-resonance in [282]. The complete list of generators in each case, should enable the reader to compose normal forms of special interest. We leave out detuning and the subject of more than three degrees of freedom as the results here are still incidental. However, we mention two results of general interest. In [263] it is shown that the $1:2:\cdots:2$-resonance, normalized to H^1 is completely integrable. Another n degrees of freedom system, the famous Fermi–Pasta–Ulam problem is discussed by Rink [223] who demonstrates complete integrability of the system, normalized to H^2 and the presence of n-dimensional KAM tori.

10.7.4 The $\omega_1 : \omega_2 : \omega_3$-Resonance

We consider Hamiltonians at equilibrium with quadratic term

$$H^0 = \sum_{j=1}^{3} \omega_j x_j y_j,$$

where $x_j = q_j + ip_j$ and $y_j = q_j - ip_j$, and the q_j, p_j are the real canonical coordinates. We assume $\omega_j \in \mathbb{N}$, although it is straightforward to apply the results in the more general case $\omega_j \in \mathbb{Z}$. The signs are important in the nonsemisimple case, and, of course, in the stability considerations. With these quadratic terms we speak of the semisimple resonant case. We now pose the problem to find the description of a general element

$$H \in k[[x_1, y_1, x_2, y_2, x_3, y_3]]$$

such that $\{H^0, H\} = 0$ (see [203, Section 4.5]).

We show that if $M = \omega_1 + \omega_2 + \omega_3$, the Stanley dimension (see Section 10.5.3 of the ring of invariants of H^0 is bounded by $2M$.

We do this by giving an algorithm to compute a Stanley decomposition, and we illustrate this by giving the explicit formulae for the genuine zeroth, first and second-order resonances, that is, those resonances which have more than one generator of degree ≤ 4, not counting complex conjugates and $x_j y_j$'s. These resonances are the most important ones from the point of view of the asymptotic approximation of the solutions.

10.7.5 The Kernel of $\mathrm{ad}(H^0)$

First of all, we see immediately that the elements $\tau_j = x_j y_j$ all commute with H^0. We let $\mathcal{I} = k[[\tau_1, \tau_2, \tau_3]]$. In principle, we work with real Hamiltonians as they are given by a physical problem, but it is easier to work with complex coordinates, so we take the coefficients to be complex too. In practice, one can forget the reality condition and work over $\mathbb{C}$. In the end, the complex dimension will be the same as the real one, after applying the reality condition.

Any monomial in ker $\mathrm{ad}(H^0)$ is an element of one of the spaces

$$\mathcal{I}[[y_1^{n_1} x_2^{n_2} x_3^{n_3}]], \quad \mathcal{I}[[x_1^{n_1} y_2^{n_2} x_3^{n_3}]], \quad \mathcal{I}[[x_1^{n_1} x_2^{n_2} y_3^{n_3}]],$$

where $\mathbf{n} = (n_1, n_2, n_3)$ is a solution of $n_1\omega_1 = n_2\omega_2 + n_3\omega_3$, $n_2\omega_2 = n_1\omega_1 + n_3\omega_3$, $n_3\omega_3 = n_1\omega_1 + n_2\omega_2$, respectively, and all the $n_j \geq 0$.

In the equation $n_1\omega_1 = n_2\omega_2 + n_3\omega_3$ one cannot have a nontrivial solution of $n_1 = 0$, but if $n_1 > 0$, one can either have $n_2 = 0$ or $n_3 = 0$, but not both. We allow in the sequel n_2 to be zero, that is, we require $n_1 > 0$, $n_2 \geq 0$ and $n_3 > 0$.

We formulate this in general as follows. Consider the three equations

$$n_i\omega_i = n_{i+}\omega_{i+} + n_{i++}\omega_{i++}.$$

where the increment in the indices is in $\mathbb{Z}/3 = (1,2,3)$ (that is, $2^{++} \equiv 1$, etc.), where we allow n_{i+} to be zero, but n_i and n_{i++} are strictly positive.

We now solve for given $\mathbf{m}$ the equation $n_1\omega_1 = n_2\omega_2 + n_3\omega_3$, and then apply a cyclic permutation to the indices of $\mathbf{m}$.

Suppose that $\gcd(\omega_2, \omega_3) = g_1 > 1$. In that case, assuming $\mathbf{m}$ is primitive, we may conclude that $g_1 | n_1$. Let $n_1 = g_1\bar{n}_1$, $\omega_j = g_1\bar{\omega}_j, j = 2,3$. Then

$$\bar{n}_1\omega_1 = n_2\bar{\omega}_2 + n_3\bar{\omega}_3, \quad \gcd(\bar{\omega}_2, \bar{\omega}_3) = 1.$$

By cyclic permutation, we assume now that $\gcd(\bar{\omega}_i, \bar{\omega}_j) = 1$, and we call $\bar{\mathbf{m}}$ the **reduced resonance**. Observe that the Stanley dimension of the ring of invariants is the same for a resonance and its reduction.

Obviously, keeping track of the divisions by the gcd's, one can reconstruct the solution of the original resonance problem from the reduced one. Observe that in terms of the coordinates, the division is equivalent to taking a root, and this is not a symplectic transformation.

Dropping the bars, we again consider $n_1\omega_1 = n_2\omega_2 + n_3\omega_3$, but now we have $\gcd(\omega_2, \omega_3) = 1$.

If $\omega_1 = 1$, we are immediately done, since the solution is simply $n_1 = n_2\omega_2 + n_3\omega_3$, with arbitrary integers $n_2 \geq 0, n_3 > 0$.

So we assume $\omega_1 > 1$ and we calculate $\mod \omega_1$, keeping track of the positivity of our coefficients. Let $\omega_j = \bar{\omega}_j + k_j\omega_1, j = 2,3$, with $0 < \omega_j < \omega_1$ since $\gcd(\omega_j, \omega_1) = 1$. Let $\tilde{\omega}_3 = \omega_1 - \omega_3$, so again $0 < \tilde{\omega}_3 < \omega_1$. For $q = 0, \ldots, \omega_1 - 1$ let

$$\begin{aligned} n_2 &= q\tilde{\omega}_3 + l_2\omega_1 \\ n_3 &= q\bar{\omega}_2 + l_3\omega_1 \end{aligned}$$

with the condition that if $q = 0$, then $l_3 > 0$. Then

$$\begin{aligned} n_1\omega_1 &= (q\tilde{\omega}_3 + l_2\omega_1)\omega_2 + (q\bar{\omega}_2 + l_3\omega_1)\omega_3 \\ &= q\tilde{\omega}_3\omega_2 + q\bar{\omega}_2\omega_3 + \omega_1(l_2\omega_2 + l_3\omega_3) \\ &= q\tilde{\omega}_3(\bar{\omega}_2 + k_2\omega_1) + q\bar{\omega}_2(\bar{\omega}_3 + k_3\omega_1) + \omega_1(l_2\omega_2 + l_3\omega_3) \\ &= q\tilde{\omega}_3\bar{\omega}_2 + q\bar{\omega}_2\bar{\omega}_3 + \omega_1(q\tilde{\omega}_3 k_2 + q\bar{\omega}_2 k_3 + l_2\omega_2 + l_3\omega_3) \\ &= \omega_1(q(k_2\tilde{\omega}_3 + (1 + k_3)\bar{\omega}_2) + l_2\omega_2 + l_3\omega_3) \end{aligned}$$

or

$$n_1 = q(k_2\tilde{\omega}_3 + (1 + k_3)\bar{\omega}_2) + l_2\omega_2 + l_3\omega_3, \quad q = 0, \ldots, \omega_1 - 1.$$

This is the general solution of the equation $n_1 = n_2\omega_2 + n_3\omega_3$.

The solution is not necessarily giving us an irreducible monomial: it could be the product of several monomials in $\ker \operatorname{ad}(H^0)$. To analyze this we put

$$q\bar{\omega}_2 = \psi_2^q\omega_1 + \phi_2^q, 0 \leq \phi_2^q < \omega_1, \psi_2^q \geq 0$$

and

$$q\tilde{\omega}_3 = \psi_3^q\omega_1 + \phi_3^q, 0 \leq \phi_3^q < \omega_1, \psi_3^q \geq 0.$$

We now write $y_1^{n_1}x_2^{n_2}x_3^{n_3}$ as $\langle n_1, n_2, n_3\rangle$. Then

$$\begin{aligned}
&\langle n_1, n_2, n_3\rangle \\
&= \langle q(k_2\tilde{\omega}_3 + (1 + k_3)\bar{\omega}_2) + l_2\omega_2 + l_3\omega_3, q\tilde{\omega}_3 + l_2\omega_1, q\bar{\omega}_2 + l_3\omega_1\rangle \\
&= \langle \omega_2, \omega_1, 0\rangle^{l_2}\langle \omega_3, 0, \omega_1\rangle^{l_3}\langle q(k_2\tilde{\omega}_3 + (1 + k_3)\bar{\omega}_2), \psi_3^q\omega_1 + \phi_3^q, \psi_2^q\omega_1 + \phi_2^q\rangle.
\end{aligned}$$

Let $\phi_1^q = q(k_2\tilde{\omega}_3 + (1 + k_3)\bar{\omega}_2) - \psi_2^q\omega_3$. Then

$$\begin{aligned}
\phi_1^q &= q(k_2\tilde{\omega}_3 + (1 + k_3)\bar{\omega}_2) - \psi_2^q\omega_3 \\
&= k_2q\tilde{\omega}_3 + (1 + k_3)(\psi_2^q\omega_1 + \phi_2^q) - \psi_2^q(\bar{\omega}_3 + k_3\omega_1) \\
&= k_2q\tilde{\omega}_3 + (1 + k_3)\phi_2^q + \psi_2^q\omega_1 - \psi_2^q\bar{\omega}_3 \\
&= k_2q\tilde{\omega}_3 + (1 + k_3)\phi_2^q + \psi_2^q\tilde{\omega}_3 \geq 0.
\end{aligned}$$

We now write $\phi_1^q = \tilde{\psi}_3^q\omega_2 + \chi_1^q$, and we let $\hat{\psi}_3^q = \min(\tilde{\psi}_3^q, \psi_3^q)$. We have

$$\begin{aligned}
&\langle n_1, n_2, n_3\rangle \\
&= \langle \omega_2, \omega_1, 0\rangle^{l_2+\hat{\psi}_3}\langle \omega_3, 0, \omega_1\rangle^{l_3+\psi_2^q}\langle(\tilde{\psi}_3^q - \hat{\psi}_3^q)\omega_2 + \chi_1^q, (\psi_3^q - \hat{\psi}_3^q)\omega_1 + \phi_3^q, \phi_2^q\rangle.
\end{aligned}$$

We define

$$\begin{aligned}
\alpha_\iota &= \langle \omega_{\iota^+}, \omega_\iota, 0\rangle \\
\beta_\iota^0 &= \langle \omega_{\iota^{++}}, 0, \omega_\iota\rangle \\
\beta_\iota^q &= \langle(\tilde{\psi}_{\iota^{++}}^q - \hat{\psi}_{\iota^{++}}^q)\omega_{\iota^+} + \chi_\iota^q, (\psi_{\iota^{++}}^q - \hat{\psi}_{\iota^{++}}^q)\omega_\iota + \phi_{\iota^{++}}^q, \phi_{\iota^+}^q\rangle.
\end{aligned}$$

Thus

$$\langle n_1, n_2, n_3\rangle = \alpha_1^{l_2'}(\beta_1^0)^{l_3'}\beta_1^q, \quad l_2', l_3' \in \mathbb{N}, q = 0, \ldots, \omega_1 - 1,$$

or, in other words, $\langle n_1, n_2, n_3\rangle \in \mathcal{I}[[\alpha_1, \beta_1^0]]\beta_1^q$. This means that $\mathcal{I}[[\alpha_1, \beta_1^0]]\beta_1^q$ is the solution space of the resonance problem. Notice that by construction these spaces have only 0 intersection.

Let $\mathcal{K}$ be defined as $\bigoplus_{\iota\in\mathbb{Z}/3}\mathcal{K}_\iota$, where

$$\mathcal{K}_\iota = \bigoplus_{q=0}^{\omega_\iota - 1}\mathcal{I}[[\alpha_\iota, \beta_\iota^0]]\beta_\iota^q.$$

Then we have

Theorem 10.7.1. *Let $\bar{\mathcal{K}}$ denote the space of complex conjugates (that is, x_j and y_j interchanged) of the elements of $\mathcal{K}$. Then $\mathcal{I} \oplus \mathcal{K} \oplus \bar{\mathcal{K}}$ is a Stanley decomposition of the $\omega_1 : \omega_2 : \omega_3$-resonance.*

Corollary 10.7.2. *In each $\mathcal{K}_\iota$ there are ω_ι direct summands. Therefore there are $M = \omega_1 + \omega_2 + \omega_3$ direct summands in $\mathcal{K}$. This enables us to estimate the Stanley dimension from above by $1 + 2M$.*

Remark 10.7.3. The number of generators need not be minimal. In particular the β^q's can be generated by one or more elements. We conjecture that the $\beta^q, q = 1, \ldots, \omega_\iota - 1$, are generated as polynomials by at most two invariants. Furthermore, the β_ι^q's, are for $q > 0$ not algebraically independent of α_ι and β_ι^0. The relations among them constitute what we will call here the defining curve. Since the Stanley decomposition is the ring freely generated by the invariants divided out by the ideal of the defining curve, this gives us a description of the normal form that is independent of the choices made in writing down the Stanley decomposition.

Remark 10.7.4. The generating functions of the resonances that follow below were computed by A. Fekken [90]. They are the Hilbert–Poincaré series of the Stanley decomposition and can be computed by computing the Molien series [192] of the group action given by the flow of H^0, that is, by computing circle integrals (or residues).

Table 10.5–10.19 contain all the information to compute the Stanley decomposition for the lower-order resonances.

ι	α	β^0
1	y_1x_2	y_1x_3
2	y_2x_3	x_1y_2
3	x_1y_3	x_2y_3

Table 10.5: The $1:1:1$-resonance (Section 10.8.10)

ι	α	β^0
1	$y_1^2x_2$	$y_1^2x_3$
2	y_2x_3	$x_1^2y_2$
3	$x_1^2y_3$	x_2y_3

Table 10.6: The $1:2:2$-resonance (Section 10.8.3). This is derived from the $1:1:1$-resonance by squaring x_1 and y_1.

Remark 10.7.5. An obvious application of the given results is the computation of the nonsemisimple case. Nilpotent terms in H^0 are possible whenever there is a $1:1$-subresonance and show up in the tables as quadratic terms of type

ι	α	β^0
1	$y_1^3 x_2$	$y_1^3 x_3$
2	$y_2 x_3$	$x_1^3 y_2$
3	$x_1^3 y_3$	$x_2 y_3$

Table 10.7: The 1 : 3 : 3-resonance. This is derived from the 1 : 1 : 1-resonance by raising x_1 and y_1 to the third power.

ι	α	β^0	β^1
1	$y_1 x_2$	$y_1^2 x_3$	
2	$y_2^2 x_3$	$x_1 y_2$	
3	$x_1^2 y_3$	$x_2^2 y_3$	$x_1 x_2 y_3$

Table 10.8: The 1 : 1 : 2-resonance (Section 10.8.1). The defining curve is $((\beta_3^1)^2 - \alpha_3 \beta_3^0)$.

ι	α	β^0	β^1
1	$y_1^2 x_2$	$y_1^4 x_3$	
2	$y_2^2 x_3$	$x_1^2 y_2$	
3	$x_1^4 y_3$	$x_2^2 y_3$	$x_1^2 x_2 y_3$

Table 10.9: The 1 : 2 : 4-resonance (Section 10.8.7). This is derived from the 1 : 1 : 2-resonance by squaring x_1 and y_1.

ι	α	β^0	β^1
1	$y_1^3 x_2$	$y_1^6 x_3$	
2	$y_2^2 x_3$	$x_1^3 y_2$	
3	$x_1^6 y_3$	$x_2^2 y_3$	$x_1^3 x_2 y_3$

Table 10.10: The 1 : 3 : 6-resonance. This is derived from the 1 : 1 : 2-resonance by raising x_1 and y_1 to the third power.

ι	α	β^0	β^1	β^2
1	$y_1 x_2$	$y_1^3 x_3$		
2	$y_2^3 x_3$	$x_1^3 y_2$		
3	$x_1^3 y_3$	$x_2^3 y_3$	$x_1^2 x_2 y_3$	$x_1 x_2^2 x_3$

Table 10.11: The 1 : 1 : 3-resonance. The defining curve is $(\beta_3^1 \beta_3^2 - \alpha_3 \beta_3^0, (\beta_3^1)^2 - \alpha_3 \beta_3^2, (\beta_3^2)^2 - \beta_3^0 \beta_3^1)$.

ι	α	β^0	β^1	β^2
1	$y_1^2 x_2$	$y_1^6 x_3$		
2	$y_2^3 x_3$	$x_1^6 y_2$		
3	$x_1^6 y_3$	$x_2^3 y_3$	$x_1^4 x_2 y_3$	$x_1^2 x_2^2 y_3$

Table 10.12: The $1:2:6$-resonance (Section 10.8.10). This is derived from the $1:1:3$-resonance by squaring x_1 and y_1.

ι	α	β^0	β^1	β^2
1	$y_1^3 x_2$	$y_1^9 x_3$		
2	$y_2^3 x_3$	$x_1^9 y_2$		
3	$x_1^9 y_3$	$x_2^3 y_3$	$x_1^6 x_2 y_3$	$x_1^3 x_2^2 y_3$

Table 10.13: The $1:3:9$-resonance. This is derived from the $1:1:3$-resonance by raising x_1 and y_1 to the third power.

ι	α	β^0	β^1	β^2
1	$y_1^2 x_2$	$y_1^3 x_3$		
2	$y_2^3 x_3^2$	$x_1^2 y_2$	$x_1 y_2^2 x_3$	
3	$x_1^3 y_3$	$x_2^3 y_3^2$	$x_1 x_2 y_3$	$x_1^2 x_2^2 y_3^2$

Table 10.14: The $1:2:3$-resonance (Section 10.8.5). The defining curve is $((\beta_2^1)^2 - \alpha_2\beta_2^0, (\beta_3^1)^3 - \alpha_3\beta_3^0)$.

ι	α	β^0	β^1	β^2
1	$y_1^2 x_2$	$y_1^3 x_3^2$		
2	$y_2^3 x_3^4$	$x_1^2 y_2$	$x_1 y_2^2 x_3^2$	
3	$x_1^3 y_3^2$	$x_2^3 y_3^4$	$x_1 x_2 y_3^2$	$x_1^2 x_2^2 y_3^4$

Table 10.15: The $2:4:3$-resonance ([259, 161]). This is derived from the $1:2:3$-resonance by squaring x_3 and y_3.

ι	α	β^0	β^1	β^2	β^3	β^4
1	$y_1^2 x_2$	$y_1^5 x_3$				
2	$y_2^5 x_3^2$	$x_1^2 y_2$	$x_1 y_2^3 x_3$			
3	$x_1^5 y_3$	$x_2^5 y_3^2$	$x_1^3 x_2 y_3$	$x_1 x_2^2 y_3$	$x_1^4 x_2^3 y_3^2$	$x_1^2 x_2^4 y_3^2$

Table 10.16: The $1:2:5$-resonance ([260, 126, 125]). The defining curve is $((\beta_2^1)^2 - \alpha_2\beta_2^0, \beta_3^3 - \beta_3^1\beta_3^2, \beta_3^4 - (\beta_3^2)^2, (\beta_3^2)^3 - \beta_3^0\beta_3^1, (\beta_3^1)^2 - \alpha_3\beta_3^2, \beta_3^1(\beta_3^2)^2 - \alpha_3\beta_3^0)$.

ι	α	β^0	β^1	β^2	β^3
1	$y_1^3x_2$	$y_1^4x_3$			
2	$y_2^4x_3^3$	$x_1^3y_2$	$x_1y_2^3x_3^2$	$x_1^2y_2^2x_3$	
3	$x_1^4y_3$	$x_2^4y_3^3$	$x_1x_2y_3$	$x_1^2x_2^2y_3^2$	$x_1^3x_2^3y_3^3$

Table 10.17: The 1 : 3 : 4-resonance. The defining curve is $((\beta_2^2)^2 - \beta_2^0\beta_2^1, (\beta_2^1)^2 - \alpha_2\beta_2^2, \beta_2^1\beta_2^2 - \alpha_2\beta_2^0, (\beta_3^1)^4 - \alpha_3\beta_3^0)$.

ι	α	β^0	β^1	β^2	β^3	β^4
1	$y_1^3x_2$	$y_1^5x_3$				
2	$y_2^5x_3^3$	$x_1^2y_2$	$x_1^2y_2^4x_3^2$	$x_1y_2^2x_3$		
3	$x_1^5y_3$	$x_2^5y_3^3$	$x_1^2x_2y_3$	$x_1^4x_2^2y_3^2$	$x_1x_2^3y_3^2$	$x_1^3x_2^4y_3^3$

Table 10.18: The 1 : 3 : 5-resonance. The defining curve is $(\beta_2^1 - (\beta_2^2)^2, (\beta_2^2)^3 - \alpha_2\beta_2^0), \beta_3^4 - \beta_3^1\beta_3^3, (\beta_3^1)^3 - \alpha_3\beta_3^3, (\beta_3^3)^2 - \beta_3^0\beta_3^1, (\beta_3^1)^2\beta_3^3 - \alpha_3\beta_3^0)$.

ι	α	β^0	β^1	β^2	β^3	β^4	β^5	β^6
1	$y_1^3x_2$	$y_1^7x_3$						
2	$y_2^7x_3^3$	$x_1^3y_2$	$x_1y_2^5x_3^2$	$x_1^2y_2^3x_3$				
3	$x_1^7y_3$	$x_2^7y_3^3$	$x_1^4x_2y_3$	$x_1x_2^2y_3$	$x_1^5x_2^3y_3^2$	$x_1^2x_2^4y_3^2$	$x_1^6x_2^5y_3^3$	$x_1^3x_2^6y_3^3$

Table 10.19: The 1 : 3 : 7-resonance (Section 10.8.10). The defining curve is $((\beta_2^1)^2 - \alpha_2\beta_2^2, (\beta_2^2)^2 - \beta_2^0\beta_2^1, \beta_2^1\beta_2^2 - \alpha_2\beta_2^0, \beta_3^3 - \beta_3^1\beta_3^2, \beta_3^4 - (\beta_3^2)^2, \beta_3^5 - \beta_3^1(\beta_3^2)^2, \beta_3^6 - (\beta_3^2)^3, \beta_3^1)^2 - \alpha_3\beta_3^2, (\beta_3^2)^4 - \beta_3^0\beta_3^1, \beta_3^1(\beta_3^2)^3 - \alpha_3\beta_3^0)$.

x_iy_j. By computing the action of the nilpotent term on the other generators, one can then try and obtain the nonsemisimple normal form, see Section 11.5.3 and [203].

10.8 Three Degrees of Freedom, Examples

10.8.1 The 1 : 2 : 1-Resonance

The first study of this relatively complicated case was by Van der Aa and Sanders [262], for an improved version (there were some errors in the calculations) see [259]. It turns out that by normalizing, the 56 constants (parameters) of H^1 are reduced to 6 constants. For the normal form of the other first-order resonances, we find an even larger reduction. As stated before we assume that the three degrees of freedom systems are in exact resonance, avoiding the analytical difficulties which characterize the detuned problem.

As an example we present the normal form truncated at degree three:

$$\bar{H} = H^0 + x_2(a_1y_1^2 + a_2y_3^2 + a_3y_1y_3) + y_2(a_4x_1^2 + a_5x_3^2 + a_6x_1x_3).$$

When writing out the normal form, action-angle variables are more convenient for the analysis. The normal form to H^1 (degree three) is

$$\begin{aligned}\overline{H} &= \tau_1 + 2\tau_2 + \tau_3 + 2\varepsilon\sqrt{2\tau_2}[a_1\tau_1\cos(2\phi_1 - \phi_2 - a_2)\\ &+a_3\sqrt{\tau_1\tau_3}\cos(\phi_1 - \phi_2 + \phi_3 - a_4) + a_5\tau_3\cos(2\phi_3 - \phi_2 - a_6)],\end{aligned}$$

where $a_i, i = 1, \ldots, 6$ are real constants. Using the combination angles

$$\begin{aligned}2\psi_1 &= 2\phi_1 - \phi_2 - a_2,\\ 2\psi_2 &= 2\phi_3 - \phi_2 - a_6,\end{aligned}$$

we obtain the equations of motion (with $\eta = \frac{1}{2}a_2 + \frac{1}{2}a_6 - a_4$)

$$\begin{aligned}\dot{\tau}_1 &= 2\varepsilon\sqrt{2\tau_2}[2a_1\tau_1\sin(2\psi_1) + a_3\sqrt{\tau_1\tau_3}\sin(\psi_1 + \psi_2 + \eta)],\\ \dot{\tau}_2 &= -2\varepsilon\sqrt{2\tau_2}[2a_1\tau_1\sin(2\psi_1) + a_3\sqrt{\tau_1\tau_3}\sin(\psi_1 + \psi_2 + \eta) + a_5\tau_3\sin(2\psi_2)],\\ \dot{\tau}_3 &= 2\varepsilon\sqrt{2\tau_2}[a_3\sqrt{\tau_1\tau_2}\sin(\psi_1 + \psi_2 + \eta) + 2a_5\tau_3\sin(2\psi_2)],\\ \dot{\psi}_1 &= \varepsilon\sqrt{2\tau_2}[2a_1\cos(2\psi_1) + a_3\sqrt{\frac{\tau_3}{\tau_1}}\cos(\psi_1 + \psi_2 + \eta)]\\ &-\frac{\varepsilon}{\sqrt{2\tau_2}}[a_1\tau_1\cos(2\psi_1) + a_3\sqrt{\tau_1\tau_3}\cos(\psi_1 + \psi_2 + \eta) + a_5\tau_3\cos(2\psi_2)],\\ \dot{\psi}_2 &= \varepsilon\sqrt{2\tau_2}[a_3\sqrt{\frac{\tau_1}{\tau_3}}\cos(\psi_1 + \psi_2 + \eta) + 2a_5\cos(2\psi_2)]\\ &-\frac{\varepsilon}{\sqrt{2\tau_2}}[a_1\tau_1\cos(2\psi_1) + a_3\sqrt{\tau_1\tau_3}\cos(\psi_1 + \psi_2 + \eta) + a_5\tau_3\cos(2\psi_2)].\end{aligned}$$

As predicted $\mathrm{H}^0 = \tau_1 + 2\tau_2 + \tau_3$ is an integral of the normalized system. Analyzing the critical points of the equation of motion we find in the general case 7 periodic orbits (for each value of the energy) of the following three types:

1. one unstable normal mode in the τ_2-direction;
2. two stable periodic solutions in the $\tau_2 = 0$ hyperplane;
3. two stable and two unstable periodic solutions in general position (i.e. $\tau_1\tau_2\tau_3 > 0$).

10.8.2 Integrability of the 1 : 2 : 1 Normal Form

Looking for a third integral of the normalized system. Van der Aa [259] showed that certain algebraic integrals did not exist. Duistermaat [79] (see also [78]) reconsidered the problem as follows. At first the normal form given above is simplified again by using the action of a certain linear symplectic transformation leaving H^0 invariant; this removes two parameters from the normal form, reducing the number of parameters of the original H^1 (56 parameters) to 4 instead of 6. This improved normal form has a special feature. One observes that on a special submanifold, given by

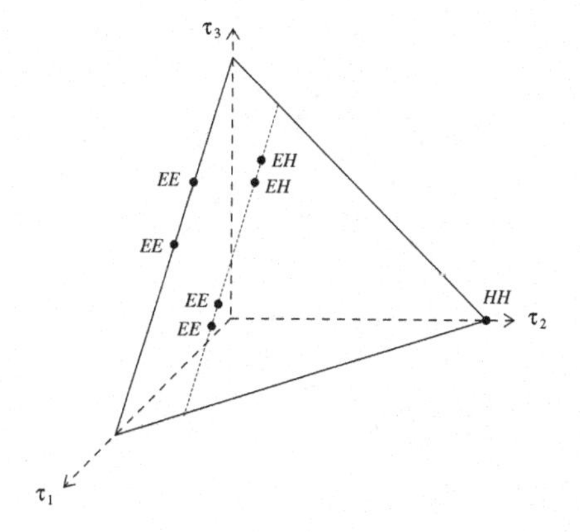

Fig. 10.9: Action simplex for the the 1 : 2 : 1-resonance.

$$\overline{\mathrm{H}}^1 = 0,$$

all solutions of the normal form Hamiltonian system are periodic. Considering complex continuations of the corresponding period function P (the period as a function of the initial conditions on the submanifold), one finds infinite branching of the manifolds P = constant. This excludes the existence of a third analytic integral on the special submanifold.

At this stage the implications for the dynamics of the Hamiltonian normal form are not clear and this is still an open question. Regarding the dynamics, it was shown in [79] that adding normal form $\overline{\mathrm{H}}^2$-terms, a corresponding Melnikov integral yields intersecting manifolds and chaotic behavior.

Symmetry assumptions

In applications, assumptions arise which induce certain symmetries in the Hamiltonian. Such symmetries cause special bifurcations and other phenomena which are of practical interest. We discuss here some of the consequences of the assumption of discrete (mirror) symmetry in the position variable q.

First we consider the case of discrete symmetry in p_1, q_1 or p_3, q_3 (or both). In the normal form this results in $a_3 = 0$, since the Hamiltonian has to be invariant under M, defined by

$$M\phi_i = \phi_i + \pi, \quad i = 1, 3.$$

Analysis of the critical points of the averaged equation shows that no periodic orbits in general position exist. There are still 7 periodic orbits, but the four in general position have moved into the $\tau_1 = 0$ and $\tau_3 = 0$ hyperplanes; see the action simplex in Figure 10.10.

Although this symmetry assumption reduces the number of terms in the normalized Hamiltonian, a third analytic integral does not exist in this case either. This can be deduced by using the analysis of [79] for this particular case.

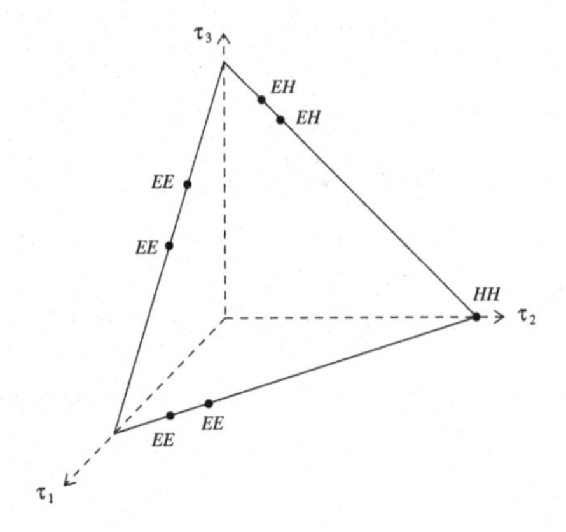

Fig. 10.10: Action simplex for the discrete symmetric 1 : 2 : 1-resonance.

It is interesting to observe that in applications the symmetry assumptions may even be stronger. An example is the three-dimensional elastic pendulum, see [180], which is a swinging spring with spring frequency 2 and swing frequencies 1. Without the spring behavior, it acts like a spherical pendulum for which an additional integral, angular momentum, is present. This integral permits a reduction with as a consequence that the normal form is integrable. Some of the physical phenomena are tied in to monodromy in [80].

We assume now discrete symmetry in p_2, q_2. The assumption turns out to have drastic consequences: the normal form to the third degree vanishes, $\overline{\mathrm{H}}^1 = 0$. This is the higher dimensional analogue of similar phenomena for the symmetric 1 : 2-resonance, described in Section 10.6.4. So in this case, higher-order averaging has to be carried out and the natural time scale of the phase flow is at least of order $1/\varepsilon^2$. The second-order normal form contains one combination angle, $\psi = 2(\phi_1 - \phi_3)$; the implication is that the resonance with this symmetry is not genuine, and that τ_2 is a third integral of the normal form.

10.8.3 The 1 : 2 : 2-Resonance

This case contains a surprise: Martinet, Magnenat and Verhulst [184] showed that the first-order normalized system, in the case that the Hamiltonian is derived from a potential, is integrable. Before the proof, this result was suggested by the consideration of numerically obtained stereoscopic projections of the flow in phase space. It is easy to generalize this result to the general Hamiltonian [259].

In action-angle coordinates, the normal form to H^1 is

$$\begin{aligned}\overline{H} &= \tau_1 + 2\tau_2 + 2\tau_3 + \\ &2\varepsilon\tau_1[a_1\sqrt{2\tau_2}\cos(2\phi_1 - \phi_2 - a_2) + a_3\sqrt{2\tau_3}\cos(2\phi_1 - \phi_3 - a_4)],\end{aligned}$$

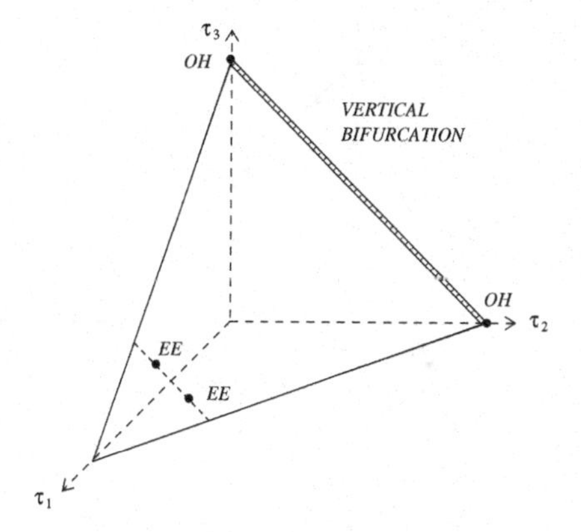

Fig. 10.11: Action simplex for the 1 : 2 : 2-resonance normalized to H^1. The vertical bifurcation at $\tau_1 = 0$ corresponds to a continuous set of periodic solutions of the normalized Hamiltonian.

where $a_i \in \mathbb{R}, i = 1, \ldots, 4$. Using the combination angles

$$2\psi_1 = 2\phi_1 - \phi_2 - a_2,$$
$$2\psi_2 = 2\phi_1 - \phi_3 - a_4,$$

we obtain the equations of motion

$$\dot{\tau}_1 = 4\varepsilon\tau_1[a_1\sqrt{2\tau_2}\sin(2\psi_1) + a_3\sqrt{2\tau_3}\sin(2\psi_2)],$$
$$\dot{\tau}_2 = -2\varepsilon a_1\tau_1\sqrt{2\tau_2}\sin(2\psi_1),$$
$$\dot{\tau}_3 = -2\varepsilon a_3\tau_1\sqrt{2\tau_3}\sin(2\psi_2),$$
$$\dot{\psi}_1 = \varepsilon\frac{a_1}{\sqrt{2\tau_2}}(4\tau_2 - \tau_1)\cos(2\psi_1) + 2\varepsilon a_3\sqrt{2\tau_3}\cos(2\psi_2),$$
$$\dot{\psi}_2 = 2\varepsilon a_1\sqrt{2\tau_2}\cos(2\psi_1) + \varepsilon\frac{a_3}{\sqrt{2\tau_3}}(4\tau_3 - \tau_1)\cos(2\psi_2).$$

Analyzing the critical points of the equations of motion we find in the energy plane $\tau_1 + 2\tau_2 + 2\tau_3 = constant$:

1. 2 normal modes (τ_2 and τ_3 direction) which are unstable;
2. 2 general position orbits which are stable;
3. 1 vertical bifurcation set in the hyperplane $\tau_1 = 0$ (all solutions with $\tau_1 = 0$ are periodic in the first-order normalized system).

Note that the phenomenon of a vertical bifurcation is nongeneric, in general it is not stable under perturbation by higher-order terms.

10.8.4 Integrability of the 1 : 2 : 2 Normal Form

Apart from H^0 and H^1, in [263] a third integral is found, a quadratic one. The existence of this integral and the existence of the vertical bifurcation set

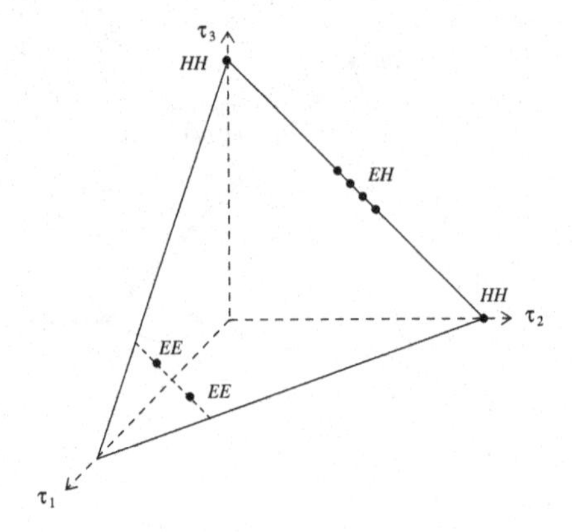

Fig. 10.12: Action simplex of the 1 : 2 : 2-resonance normalized to H^2. The vertical bifurcation at $\tau_1 = 0$ has broken up into two normal modes and four periodic solutions with $\tau_2\tau_3 \neq 0$.

are both tied in with the symmetry of the first-order normal form. According to [66] the system splits after a suitable linear transformation into a 1 : 2-resonance and a one-dimensional subsystem.

Van der Aa and Verhulst [263] considered two types of perturbation of the normal form to study the persistence of these phenomena. First, a simple deformation of the vector field is obtained by detuning the resonance. Replace H^0 by

$$\mathrm{H}^0 = \tau_1 + (2 + \Delta_1)\tau_2 + (2 + \Delta_2)\tau_3.$$

They find that in general no quadratic or cubic third integral exists in this case.

Secondly, they considered how the vertical bifurcation and the integrability break up on adding higher-order terms to the expansion of the normal form. In particular they consider the following symmetry breaking:

$$\mathrm{H}^1 = \tau_1 + 2\tau_2 + 2\tau_3 + \varepsilon(a_1 q_1^2 q_2 + a_2 q_1^2 q_3 + a_3 q_1 q_2 q_3).$$

The parameter a_3 is the deformation parameter. From the point of view of applications this is a natural approach since it reflects approximate symmetry in a problem, which seems to be quite common. The vertical bifurcation set is seen to break up into 6 periodic solutions (including the two normal modes). No third integral could be found in this case.

10.8.5 The 1 : 2 : 3-Resonance

The normal form of the general Hamiltonian was studied by Van der Aa in [259]. Kummer [161] obtained periodic solutions using the normal form while comparing these with numerical results.

As an example we present the normal form truncated at degree three: The normalized H^1 is a linear combination of the cubic generators. In action-angle coordinates the normal form to first-order is

$$\overline{\mathrm{H}} = \tau_1 + 2\tau_2 + 3\tau_3 + 2\varepsilon\sqrt{2\tau_1\tau_2}[a_1\sqrt{\tau_3}\cos(\phi_1 + \phi_2 - \phi_3 - a_2)$$
$$+a_3\sqrt{\tau_1}\cos(2\phi_1 - \phi_2 - a_4)],$$

where as usual the $a_i, i = 1, \ldots, 4$ are constants. Introducing the combination angles

$$\psi_1 = \phi_1 + \phi_2 - \phi_3 - a_2,$$
$$\psi_2 = 2\phi_1 - \phi_2 - a_4$$

we obtain the equations of motion

$$\dot{\tau}_1 = 2\varepsilon\sqrt{2\tau_1\tau_2}[a_1\sqrt{\tau_3}\sin(\psi_1) + 2a_3\sqrt{\tau_1}\sin(\psi_2)],$$
$$\dot{\tau}_2 = 2\varepsilon\sqrt{2\tau_1\tau_2}[a_1\sqrt{\tau_3}\sin(\psi_1) - a_3\sqrt{\tau_1}\sin(\psi_2)],$$
$$\dot{\tau}_3 = -2\varepsilon a_1\sqrt{2\tau_1\tau_2\tau_3}\sin(\psi_1),$$
$$\dot{\psi}_1 = \varepsilon\frac{2}{\sqrt{2\tau_1\tau_2\tau_3}}[a_1(\tau_1\tau_3 + \tau_2\tau_3 - \tau_1\tau_2)\cos(\psi_1) +$$
$$a_3\sqrt{\tau_1\tau_3}(\tau_1 + 2\tau_2)\cos(\psi_2)],$$
$$\dot{\psi}_2 = \varepsilon\frac{1}{\sqrt{\tau_1\tau_2}}[a_1\sqrt{2\tau_3}(2\tau_2 - \tau_1)\cos(\psi_1) + a_3\sqrt{2\tau_1}(4\tau_2 - \tau_1)\cos(\psi_2)].$$

Analyzing the critical points of the equation of motion we find 7 periodic solutions (see Figure 10.13):

1. 2 unstable normal modes (τ_2 and τ_3 direction);
2. 1 stable solution in the $\tau_2 = 0$ hyperplane;
3. 4 orbits in general position, two of which are stable and two unstable.

10.8.6 Integrability of the 1 : 2 : 3 Normal Form

Some aspects of the integrability of the 1 : 2 : 3-resonance were discussed in [102]; in [259] it has been shown that no quadratic or cubic third integral exists.

A new approach came from Hoveijn and Verhulst [134]. They observed that one of the seven periodic (families of) solutions is complex unstable for an open set of parameter values. It is then possible to apply Šilnikov-Devaney theory [73]. Summarized, the theory runs as follows. Suppose that one locates a complex unstable periodic solution in a Hamiltonian system with associated isolated homoclinic solution. Than, the map of the flow, transverse to the homoclinic solution, contains a horseshoe map with as a consequence, that the Hamiltonian flow is nonintegrable and chaotic.

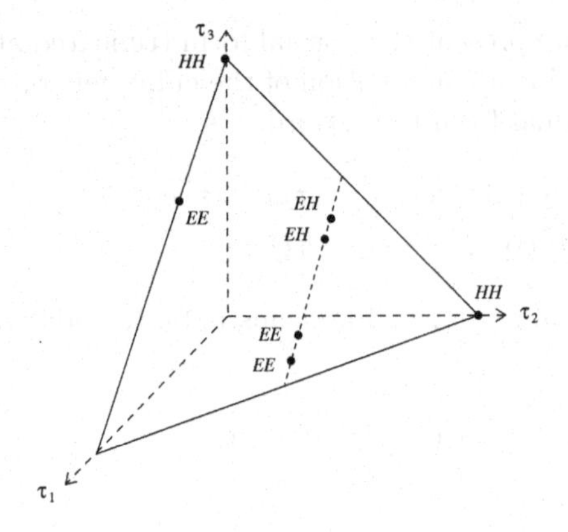

Fig. 10.13: Action simplex for the 1 : 2 : 3-resonance.

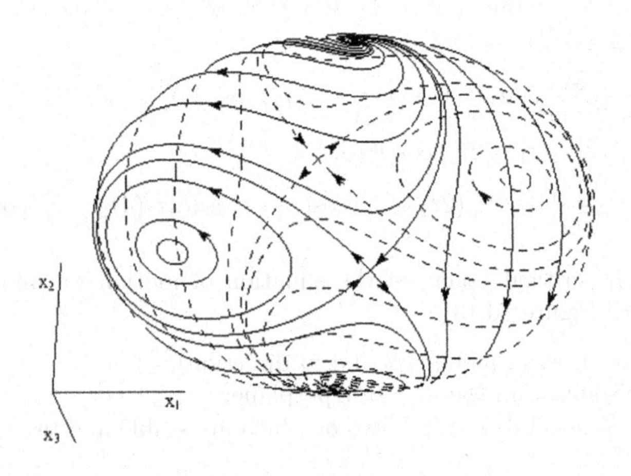

Fig. 10.14: The invariant manifold M_1 embedded in the energy manifold of the 1 : 2 : 3 normal form $\mathrm{H}^0 + \overline{\mathrm{H}}^1$. Observe that M_1 contains a one-parameter family of homoclinic solutions, a homoclinic set, and in addition two isolated heteroclinic solutions (Courtesy I. Hoveijn).

In [134] it is shown that the complex unstable periodic solution is located on an invariant manifold M_1, embedded in the energy manifold. M_1 also contains a one-parameter family of homoclinic solutions, a homoclinic set, and in addition two isolated heteroclinic solutions, see Figure 10.14. M_1 itself is embedded in an invariant manifold N, which, in its turn is embedded in the energy manifold; for a given value of the energy, N is determined by the condition $\overline{\mathrm{H}}^1 = 0$.

At this stage it is not allowed to apply Šilnikov-Devaney theory as we have a *set* of homoclinic solutions. Then $\overline{\mathrm{H}}^2$ is calculated and it is shown that the set of homoclinic solutions does not persist on adding $\overline{\mathrm{H}}^2$ to the normal form, but that one homoclinic solution survives. This permits the application of Šilnikov-Devaney theory, resulting in nonintegrability of the normal form when $\overline{\mathrm{H}}^2 = 0$ is included. The integrability of the normal form cut off at $\overline{\mathrm{H}}^1 = 0$ is still an open question, but numerical calculations in [134] suggest nonintegrability.

Note that the two heteroclinic solutions on M_1 also do not survive the $\overline{\mathrm{H}}^2$ perturbation.

In addition, in [133] a Melnikov integral is computed to prove again the existence of an isolated homoclinic solution in M_1. Moreover it becomes clear in this analysis, that the nonintegrability of the flow takes place in a set which is algebraic in the small parameter (i.e. the energy).

The integrability question for this resonance is discussed in a wider context in [282].

Symmetry assumptions

Discrete symmetry assumptions introduce drastic changes; Mirror symmetry in p_1, q_1 or p_3, q_3 (or both) produces $a_1 = 0$ in the normal form. From the equations of motion we find that τ_3 is constant, i.e. the system splits into two invariant subsystems: between the first and second degree of freedom we have a 1 : 2-resonance, in the third degree of freedom we have a nonlinear oscillator. So the system is clearly integrable with τ_3 as the third integral. One can show that these results carry through for the system normalized to H^2.

Discrete symmetry in p_2, q_2 implies $a_1 = a_3 = 0$, i.e. $\overline{\mathrm{H}}^1 = 0$ (a similar strong degeneration of the normal form has been discussed in Section 10.8.1 on the 1 : 2 : 1-resonance). Normalizing to second-order produces a system which splits into a two-dimensional and a one-dimensional subsystem which again implies integrability.

10.8.7 The 1 : 2 : 4-Resonance

In action-angle coordinates the normal form to first-order is

$$\begin{aligned}\overline{H} = \tau_1 + 2\tau_2 + 4\tau_3 + 2\varepsilon[&a_1\tau_1\sqrt{2\tau_2}\cos(2\phi_1 - \phi_2 - a_2) + \\ &a_3\tau_2\sqrt{2\tau_3}\cos(2\phi_2 - \phi_3 - a_4)],\end{aligned}$$

where $a_1, \ldots, a_4 \in \mathbb{R}$. This resonance has been studied in [287] and, more detailed, in [259]. Using combination angles

$$\begin{aligned}2\psi_1 &= 2\phi_1 - \phi_2 - a_2,\\ 2\psi_2 &= 2\phi_2 - \phi_3 - a_4,\end{aligned}$$

the equations of motion become

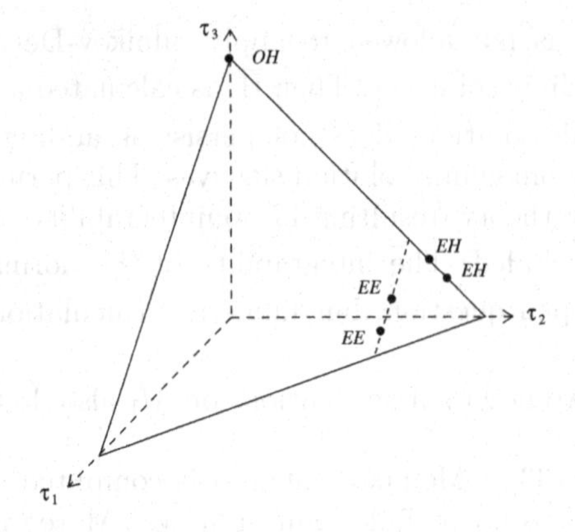

Fig. 10.15: Action simplex for the 1 : 2 : 4-resonance for $\Delta > 0$.

$$
\begin{aligned}
\dot{\tau}_1 &= 4\varepsilon a_1\tau_1\sqrt{2\tau_2}\sin(2\psi_1),\\
\dot{\tau}_2 &= -2\varepsilon\sqrt{2\tau_2}[a_1\tau_1\sin(2\psi_1) - 2a_3\sqrt{\tau_2\tau_3}\sin(2\psi_2)],\\
\dot{\tau}_3 &= -2\varepsilon a_3\tau_2\sqrt{2\tau_3}\sin(2\psi_2),\\
\dot{\psi}_1 &= \varepsilon\frac{1}{\sqrt{2\tau_2}}[a_1(4\tau_2-\tau_1)\cos(2\psi_1) - 2a_3\sqrt{\tau_2\tau_3}\cos(2\psi_2)],\\
\dot{\psi}_2 &= \varepsilon\frac{1}{\sqrt{2\tau_2\tau_3}}[2a_1\tau_1\sqrt{\tau_3}\cos(2\psi_1) + a_3(4\tau_3-\tau_2)\sqrt{\tau_2}\cos(2\psi_2)].
\end{aligned}
$$

The analysis of periodic solutions differs slightly from the treatment of the preceding first-order resonances as we have a bifurcation at the value $\Delta = 16a_1^2 - a_3^2 = 0$. From the analysis of the critical points we find:

1. 1 unstable normal mode (τ_3 direction);
2. if $\Delta < 0$ we have 2 stable periodic solutions in the $\tau_1 = 0$ hyperplane, there are no periodic orbits in general position; at $\Delta = 0$ two orbits branch off the $\tau_1 = 0$ solutions which for $\Delta > 0$ become stable orbits in general position, while the $\tau_1 = 0$ solutions are unstable.

See Figure 10.15.

10.8.8 Integrability of the 1 : 2 : 4 Normal Form

Apart from H^0 and H^1 no other independent integral of the normal form has been found, but it has been shown in [259] that no third quadratic or cubic integral exists. Discrete symmetry in p_1, q_1 does not make the system integrable.

Symmetry assumptions
Discrete symmetry in p_2, q_2 forces a_1 to be zero, and we consequently have a third integral τ_1, producing the usual splitting into a one and a two degree

of freedom system. These results carry over to second-order normal forms. Discrete symmetry in p_3, q_3 produces $a_3 = 0$ and the third integral τ_3. Again we have the usual splitting, moreover the results carry over to second order. Of course, the normal form degenerates if we assume discrete symmetry in both the second and the third degree of freedom. In this case one has to calculate higher-order normal forms.

10.8.9 Summary of Integrability of Normalized Systems

We summarize the results from the preceding sections on three degrees freedom systems with respect to integrability after normalization in Table 10.20.

Resonance		H^1	H^2	Remarks
1:2:1	general	2	2	no analytic third integral
	discr. symm. q_1	2	2	no analytic third integral
	discr. symm. q_2	3	3	$\overline{H}^1 = 0$; 2 subsystems at $\overline{H}^2$
	discr. symm. q_3	2	2	no analytic third integral
1:2:2	general	3	2	no cubic third integral at $\overline{H}^2$
	discr. symm. q_2 and q_3	3	3	$\overline{H}^1 = 0$; 2 subsystems at $\overline{H}^2$
1:2:3	general	2	2	no analytic third integral
	discr. symm. q_1	3	3	2 subsystems at $\overline{H}^1$ and $\overline{H}^2$
	discr. symm. q_2	3	3	$\overline{H}^1 = 0$
	discr. symm. q_3	3	3	2 subsystems at $\overline{H}^1$ and $\overline{H}^2$
1:2:4	general	2	2	no cubic third integral
	discr. symm. q_1	2	2	no cubic third integral
	discr. symm. q_2 or q_3	3	3	2 subsystems at $\overline{H}^1$ and $\overline{H}^2$

Table 10.20: Integrability of the normal forms of the four genuine first-order resonances.

If three independent integrals of the normalized system can be found, the normalized system is integrable; the original system is in this case called formally integrable. The integrability depends in principle on how far the normalization is carried out. The formal integrals have a precise asymptotic meaning, see Section 10.6.1. We have the following abbreviations: *no cubic integral* for no quadratic or cubic third integral; *discr. symm.* q_i for discrete symmetry in the p_i, q_i-degree of freedom; *2 subsystems at* $\overline{H}^k$ for the case that the normalized system decouples into a one and a two degrees of freedom subsystem upon normalizing to H^k. In the second and third column one finds the number of known integrals when normalizing to $\overline{H}^1$ respectively $\overline{H}^2$.

The remarks which have been added to the table reflect some of the results known on the nonexistence of third integrals. Note that the results presented here are for the general Hamiltonian and that additional assumptions may change the results.

10.8.10 Genuine Second-Order Resonances

Although the second-order resonances in three degrees of freedom are as important as the first-order resonances, not much is known about them, a few excepted.

The 1 : 1 : 1-Resonance

The symmetric $1:1:1$**-resonance** To analyze this second-order resonance in this special case we have to normalize at least to H^2. Six combination angles play a part and the technical complications are enormous. Up till now, only special systems have been considered involving symmetries that play a part in applications. For instance in studying the dynamics of elliptical galaxies, one often assumes discrete symmetry with respect to the three perpendicular galactic planes. A typical problem is then to consider the potential problem

$$\mathrm{H}^{[0]} = \mathrm{H}^0 + \varepsilon^2 \mathrm{V}^{[2]}(q_1^2, q_2^2, q_3^2), \tag{10.8.1}$$

where $\mathrm{V}^{[2]}$ has an expansion which starts with quartic terms. Even with these symmetries, no third integral of the normal form could be found.

The periodic solutions can be listed as follows. Each of the three coordinate planes contains the $1:1$-resonance as a subsystem with the corresponding periodic solutions. This produces three normal modes and six periodic solutions in the coordinate planes. In addition one can find five periodic solutions in general position. For references see [70].

The Hénon–Heiles problem, discussed in Section 10.5 has two degrees of freedom. Because of its benchmark role in Hamiltonian mechanics it was generalized to three degrees of freedom in [96] and [97]; see also [95]. The Hamiltonian is

$$\mathrm{H}^{[0]} = \mathrm{H}^0 + \varepsilon\left(a(x^2+y^2)z + bz^3\right),$$

with a, b real parameters. Choosing $a = 1, b = -\frac{1}{3}$ we have the original Hénon–Heiles problem in the $x = \dot{x} = 0$ and $y = \dot{y} = 0$ subsystems. In [96] and [97], equilibria, periodic orbits and their bifurcations are studied.

A deeper study of system (10.8.1) is [95] in which symmetry reduction and unfolding of bifurcations are used, and a geometric description is provided.

Applications in the theory of vibrating systems sometimes produces nongenuine first-order resonances. Among the second-order resonances these are $1:2:5$ and $1:2:6$. Examples are given in [285] and [125].
In action-angles coordinates the normal form to H^1 produces (with $a_1, a_2 \in \mathbb{R}$)

$$\overline{\mathrm{H}} = \tau_1 + 2\tau_2 + \omega\tau_3 + 4\varepsilon a_1 \tau_1 \sqrt{\tau_2}\cos(2\phi_1 - \phi_2 - a_2)$$

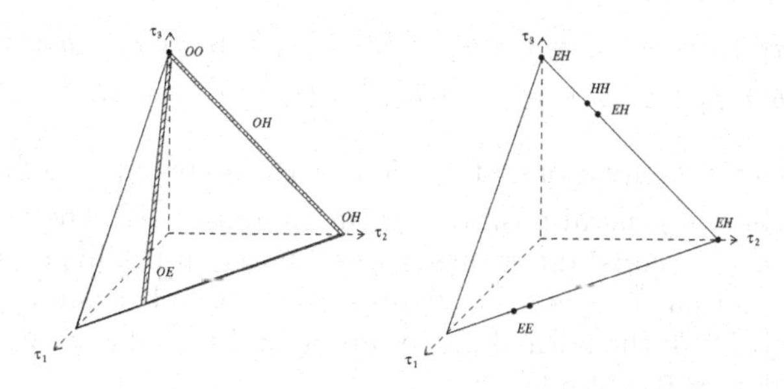

Fig. 10.16: Action simplices for the 1 : 2 : 5-resonance normalized to H^1 and to H^2. The normalization to H^2 produces a break-up of the two vertical bifurcations.

which clearly exhibits the 1 : 2-resonance between the first two modes while τ_3 is constant (integral of motion). So there are three independent integrals of the truncated normalized system, but, of course, results from such a low-order truncation are not robust. From Section 10.6.1 we have in the 1 : 2-resonance two stable periodic orbits in general position and one hyperbolic normal mode. Adding the third mode, we have for the $H^0 + \overline{H}^1$ normal form three families of periodic solutions for each value of the energy; see Figure 10.16 for this nongeneric situation. The periodic solutions are surrounded by tori, so we have families of 2-tori embedded in the 5-dimensional energy manifold.

An important analysis of the quartic normal form is given in [126]. It is shown that in the quartic normal form, there exist whiskered 2-tori and families of 3-tori. Also there is nearby chaotic dynamics in the normal form. So there are two types of tori corresponding to quasiperiodic motion on the energy manifold. The tori live in two domains, separated by hyperbolic structures which can create multi-pulse motion associated with homoclinic and heteroclinic connections between the submanifolds.

These results can also be related to diffusion effects in phase space. The diffusion is different from Arnol'd diffusion and probably also more effective as the time scales are shorter. It arises from the intersection of resonance domains when more than one resonance is present in a Hamiltonian system. For details of the analysis see [126] and for a general description of these intriguing results [125].

A normal form analysis to H^2 was carried out in [259] (the 1 : 2 : 5-resonance) and in [260] (1 : 2 : 5- and 1 : 2 : 6-resonance). We discuss the results briefly. Introducing the real constants $b_1, \ldots, b_8$ and the combination angles

$$\psi_1 = 2\phi_1 - \phi_2 - a_2,$$
$$\psi_2 = \phi_1 + 2\phi_2 - \phi_3 - b_8,$$

we have

$$\overline{\mathrm{H}} = \tau_1 + 2\tau_2 + 5\tau_3 + 4\varepsilon a_1 \tau_1 \sqrt{\tau_2} \cos(\psi_1) + 4\varepsilon^2 [b_1 \tau_1^2 + b_2 \tau_1 \tau_2 + b_3 \tau_1 \tau_3 + b_4 \tau_2^2 + b_5 \tau_2 \tau_3 + b_6 \tau_3^2 + b_7 \tau_2 \sqrt{\tau_1 \tau_3} \cos(\psi_3)].$$

The analysis of the equations of motion gives interesting results. The two families of orbits in general position vanish on adding $\overline{\mathrm{H}}^2$. The normal mode family $\tau_1 = 0$, $\tau_3 =$ *constant* breaks up as follows: in the hyperplane $\tau_1 = 0$ we have two normal modes $\tau_2 = 0$ resp. $\tau_3 = 0$ and a family of periodic orbits with $\tau_2 \tau_3 > 0$; the normal modes are hyperbolic, the family of periodic solutions near $\tau_3 = 0$ is stable.

The results are illustrated in Figure 10.16.

The 1 : 2 : 6-Resonance

A normal form analysis to H^2 was carried out in [260].

The 1 : 3 : 7-Resonance

The normal form of this resonance is characterized by a relatively large number of generators. Following [282] we note that discrete symmetry in the first degree of freedom means that we have a four-dimensional submanifold with its dynamics ruled by the 3 : 7-resonance. A study of the stability of the solutions in this submanifold implies the use of a normal form with the generators at least of degree ten.

If we have moreover discrete symmetry in the second or third degree of freedom (or both), we have to use the generators at least to degree twenty. This involves extensive computations, but we stress that it could be worse. The computational effort is facilitated by our knowledge of the finite list of generators of the normal form.

11

Classical (First–Level) Normal Form Theory

11.1 Introduction

As we have seen, one can consider averaging as the application of a near-identity transformation of the underlying space on which the differential equation is defined. In Section 3.2 this process was formalized using the Lie method. We are now going to develop a completely abstract theory of normal forms that generalizes the averaging approach. This will require of the reader a certain knowledge of algebraic concepts, which we will indicate on the way. The emphasis will be much more on the formal algebraic properties of the theory than on the analytic aspects. However, these will have to be incorporated in the theory yet to be developed.

A simple example of the procedure we are going to follow is that of matrices. Given a square matrix, one can act on it by conjugation with the group of invertible matrices of the same size. We define equivalence of two matrices A and B using similarity as follows.

Definition 11.1.1. *Let $A, B \in \mathfrak{gl}_n$. Then we say that A is* **equivalent** *to B if there exists some Q in GL_n such that $AQ = QB$.*

Choosing with each equivalence class a representantive in $\mathfrak{gl}_n$ is called a **choice of normal form**. The Jordan normal form is an example of this. The choice of normal form is what we call a **style**. Fixing the action of GL_n on $\mathfrak{gl}_n$ determines the space of equivalence classes and leaves no room for choice. It is only the representation of the equivalence classes (the choice of style) that leaves us with considerable freedom. One should keep in mind that there is also considerable freedom in the choice of spaces and the action. One could for instance replace $\mathbb{C}$ by $\mathbb{R}$, $\mathbb{Q}$, or $\mathbb{Z}$, just to name a few possibilities, each of them leading to new theories.

The choice of style is usually determined by the wish to obtain the simplest expression possible. This, however, may vary with the intended application. For instance, in the Jordan normal form one chooses 1's on the off-diagonal positions, where the choice of $1, 2, 3, \ldots$ is more natural in the representation theory of $\mathfrak{sl}_2$.

In the preceding theory we have also made some choices, for instance whether u should have zero average or zero initial value. This choice may seem like a choice of style, but it is not. By making this choice we do not use the freedom we have, as is illustrated in the discussion of hypernormal forms in Section 3.4.

In the next part we shall try to define what a normal form should look like, without using transformations explicitly, but, of course, relying on our experience with averaging.

The mathematics in this chapter is a lot more abstract looking than before. But since the object of our studies is still very concrete, the reader may use this background to better understand the abstract concepts as they are introduced. For instance, the spectral sequences that occur here are much more concrete than the spectral sequences found elsewhere in the literature and so may serve as a good introduction to them.

11.2 Leibniz Algebras and Representations

We now start our abstract formulation of normal form theory.

Definition 11.2.1. *Let R be a ring and M an additively written abelian group. M is called an* **R-module** *if there exists a map $R \times M \to M$, written as $(a, m) \mapsto am$, such that for all $m, n \in M$ and $a, b \in R$ one has*

$$\begin{aligned} a(m+n) &= am + an \\ (a+b)m &= am + bm \\ (ab)m &= a(bm) \end{aligned}$$

We already saw an example of a module in Section 7.2: the $\mathbb{Z}$-module $\omega^\perp$.

The indices that we are going to use look somewhat imposing at first sight, but they allow us to give a simple interpretation of unique normal form theory in terms of cohomology theory. The superscript is 0 or 1, depending on whether we consider an algebra or a module on which the algebra acts, respectively:

Definition 11.2.2. *A vector space or, replacing the coefficient field by a ring R, an R-module $\mathcal{Z}^0$ with a bilinear map*

$$[\cdot,\cdot] : \mathcal{Z}^0 \times \mathcal{Z}^0 \to \mathcal{Z}^0$$

is called a **Leibniz algebra** *if with $\rho_0(x)y = [x, y]$, we have*

$$\rho_0([x, y]) = \rho_0(x)\rho_0(y) - \rho_0(y)\rho_0(x).$$

If, moreover, the bilinear map (or bracket) is antisymmetric, that is, $[x, y] + [y, x] = 0$, then $\mathcal{Z}^0$ is called a **Lie algebra**. *When we use the term bilinear,* *it is with respect to the addition within the vector space or module, not necessarily with respect to the multiplication by the elements in the field or ring. If the field or ring contains $\mathbb{Q}, \mathbb{R}$, or $\mathbb{C}$, we do suppose linearity with respect to these subfields. We will call the elements of $\mathcal{Z}^0$* **generators**, *since we think of them as the generators of transformations.*

Remark 11.2.3. The terminology Leibniz algebra is introduced here to denote what is usually called a left Leibniz algebra [178]. It is introduced with the single goal to emphasize the fact that the antisymmetry of the Lie bracket hardly plays a role in the theory that we develop here. ♡

Remark 11.2.4. The defining relation for a Leibniz algebra reduces to the Jacobi identity in the case of a Lie algebra. ♡

Example 11.2.5. Let $\mathbb{A}$ be an associative algebra. Then it is also a Lie algebra by $[x, y] = xy - yx$. Taking $\mathbb{A} = \mathrm{End}(\mathcal{Z}^0)$, the linear maps from $\mathcal{Z}^0$ to itself (endomorphisms), this allows us to write the rule for ρ_0 as

$$\rho_0([x, y]_{\mathcal{Z}^0}) = [\rho_0(x), \rho_0(y)]_{\mathrm{End}(\mathcal{Z}^0)},$$

where we put a subscript on the bracket for conceptual clarity. ♢

Definition 11.2.6. *Let $\mathcal{Z}^1$ be a vector space or module, let $\mathcal{Z}^0$ be a Leibniz algebra, and let $\rho_1 : \mathcal{Z}^0 \to \mathrm{End}(\mathcal{Z}^1)$ be such that*

$$\rho_1([x, y]) = [\rho_1(x), \rho_1(y)].$$

Then we say that ρ_1 is a **representation** *of $\mathcal{Z}^0$ in $\mathcal{Z}^1$. If a representation of $\mathcal{Z}^0$ in $\mathcal{Z}^1$ exists, $\mathcal{Z}^1$ is a* **$\mathcal{Z}^0$-module**. *We think of the elements in $\mathcal{Z}^1$ as the vector fields that have to be put in normal form using the generators from $\mathcal{Z}^0$.*

Remark 11.2.7. This is somewhat simplified, for the right definition see [178], and should not be taken as the starting point of a study in representation theory of Leibniz algebras. For instance, the usual central extension construction cannot be easily generalized from Lie algebras to this simplified type of Leibniz algebra representation, so it should be seen as an ad hoc construction.

Example 11.2.8. Let $\mathcal{Z}^1 = \mathcal{Z}^0$ and $\rho_1 = \rho_0$. ♢

In doing our first averaging transformation, we compute terms modulo $\mathcal{O}(\varepsilon^2)$. In our abstract approach we will do the same by supposing that we have a filtered Leibniz module [24] and a filtered representation, that is to say, we have

$$\mathcal{Z}^0 = \mathcal{Z}^{1,0} \supset \mathcal{Z}^{2,0} \supset \cdots \supset \mathcal{Z}^{k,0} \supset \mathcal{Z}^{k+1,0} \supset \cdots$$

and

$$\mathcal{Z}^1 = \mathcal{Z}^{0,1} \supset \mathcal{Z}^{1,1} \supset \cdots \supset \mathcal{Z}^{k-1,1} \supset \mathcal{Z}^{k,1} \supset \cdots$$

(where the first superscript denotes the filtering degree, and the second is the original one indicating whether we have a Leibniz algebra or Leibniz module) such that $\rho_q(\mathcal{Z}^{k,0})\mathcal{Z}^{l,q} \subset \mathcal{Z}^{k+l,q}, q = 0, 1$. Starting with $\mathcal{Z}^{1,0}$ instead of $\mathcal{Z}^{0,0}$ is equivalent to considering only near-identity transformations.

Remark 11.2.9. One can think of the $\mathcal{Z}^{k,q}$ as open neighborhoods of $\mathcal{Z}^{\infty,q}$. It makes it easier on the topology if we require $\mathcal{Z}^{\infty,q} = \{0\}$, but if our modules are germs of smooth vector fields, this would ignore the flat vector fields, so the assumption is not always practical. The topology induced by these neighborhoods is called the **filtration topology**. So to say that f_k converges to f in the filtration topology means that $f - f_k \in \mathcal{Z}^{n(k),q}$, for some n such that $\lim_{k\to\infty} n(k) = \infty$. ♡

We can define

$$\exp(\rho_q(x)) = \sum_{k=0}^{\infty} \frac{1}{k!}\rho_q^k(x) : \mathcal{Z}^q \to \mathcal{Z}^q$$

without having to worry about convergence, since for $y \in \mathcal{Z}^{k,q}$, $\sum_{k=N}^{\infty} \frac{1}{k!}\rho_q^k(x)y \in \mathcal{Z}^{k+N,q}$ is very small in the filtration topology for large N.

Definition 11.2.10. *Let* $\mathbf{f} \in \mathcal{Z}^{1-q,q}$. *If* $\mathbf{f} - \mathbf{f}_k \in \mathcal{Z}^{k+1,q}$, *we say that* $\mathbf{f}_k$ *is a* k**-jet** *of* $\mathbf{f}$.

Remark 11.2.11. In particular, $\mathbf{f}$ is its own k-jet. Yet saying that something depends on $\mathbf{f}$ is different from saying that it depends on $\mathbf{f}_k$. Why? ♡

Example 11.2.12. Let $\mathcal{Z}^{0,1} = \mathcal{Z}^1$ be the affine space of vector fields of the form

$$\nabla_{\mathbf{f}^{[1]}} = \frac{\partial}{\partial t} + \varepsilon \mathbf{f}^{[1]} = \frac{\partial}{\partial t} + \varepsilon \sum_{i=1}^{n} \mathrm{f}_i^1(\boldsymbol{x}, t)\frac{\partial}{\partial x_i} + \mathcal{O}(\varepsilon^2),$$

where the 0 in the upper index of $\mathcal{Z}^{0,1}$ stands for the lowest ε power in the expression. Let $\mathcal{Z}^{1,0} = \mathcal{Z}^0$ be the Lie algebra of generators $\mathbf{u}^{[1]}$ of the form

$$\mathcal{D}_{\mathbf{u}^{[1]}} = \varepsilon \sum_{i=1}^{n} \mathrm{u}_i^1(\boldsymbol{x}, t)\frac{\partial}{\partial x_i} + \mathcal{O}(\varepsilon^2),$$

and take

$$\begin{aligned}\rho_1(\mathcal{D}_{\mathbf{u}^{[1]}})\nabla_{\mathbf{f}^{[1]}} &= \mathcal{L}_{\mathbf{u}^{[1]}}\nabla_{\mathbf{f}^{[1]}} = \mathcal{D}_{\mathbf{u}^{[1]}}\nabla_{\mathbf{f}^{[1]}} - \nabla_{\mathbf{f}^{[1]}}\mathcal{D}_{\mathbf{u}^{[1]}} \\ &= \mathcal{D}_{[\mathbf{u}^{[1]},\mathbf{f}^{[1]}]} - \mathcal{D}_{\mathbf{u}_t^{[1]}}.\end{aligned}$$

Notice that $[\mathbf{u}^{[1]}, \mathbf{f}^{[1]}] \in \mathcal{Z}^{2,1}$, so the important term in all our filtered calculations will be $\mathbf{u}_t^{[1]}$, as we very well now from averaging theory. In the sequel we will denote $\nabla_{\mathbf{f}^{[1]}}$ by $\hat{\mathbf{f}}^{[0]}$ and $\mathcal{D}_{\mathbf{u}^{[1]}}$ by $\mathbf{u}^{[1]}$. There is also an extended version of $\mathcal{D}_{\mathbf{u}^{[1]}}$, but it does not involve $\frac{\partial}{\partial t}$ but $\frac{\partial}{\partial \varepsilon}$. We write

$$\nabla_{\mathbf{u}^{[1]}} = \frac{\partial}{\partial \varepsilon} + \mathbf{u}^{[1]} = \frac{\partial}{\partial \varepsilon} + \sum_{i=1}^{n} \mathrm{u}_i^1(\boldsymbol{x}, t)\frac{\partial}{\partial x_i} + \mathcal{O}(\varepsilon).$$

11.3 Cohomology

Definition 11.3.1. *Let a sequence of spaces and maps*

$$\cdots \longrightarrow C^{j-1} \xrightarrow{d^{j-1}} C^j \xrightarrow{d^j} C^{j+1} \xrightarrow{d^{j+1}} C^{j+2} \quad \cdots$$

be given. We say that this defines a **complex** *if $d^{j+1}d^j = 0$ for all j. In a complex one defines* **cohomology** *spaces by*

$$H^j(C,d) = \ker\, d^j/\mathrm{im}\, d^{j-1}.$$

Definition 11.3.2. *Let $d^{0,1}_{\hat{\mathbf{f}}^{[0]}}\mathbf{u}^{[1]} = -\rho_1(\mathbf{u}^{[1]})\hat{\mathbf{f}}^{[0]}$ define a map $d^{0,1}_{\hat{\mathbf{f}}^{[0]}}$ of $\mathcal{Z}^{1,0}$ to $\mathcal{Z}^{1,1}$. Here the superscripts on the d denote the increment of the corresponding superscripts in $\mathcal{Z}^{1,0}$. Then we have a (rather short) complex*

$$0 \xrightarrow{0} \mathcal{Z}^{1,0} \xrightarrow{d^{0,1}_{\hat{\mathbf{f}}^{[0]}}} \mathcal{Z}^{1,1} \xrightarrow{0} 0.$$

Remark 11.3.3. The minus sign in the definition of $d^{0,1}_{\hat{\mathbf{f}}^{[0]}}$ adheres to the conventions in normal form theory. Of course, it does not influence the definition of cohomology, and that is all that matters here. ♡

We have solved the descriptive normal form problem if we can describe $H^0(\mathcal{Z})$ and $H^1(\mathcal{Z})$. This is, however, not so easy in general, which forces us to formulate an approximation scheme in Chapter 13 leading to a spectral sequence [5, 4]. The final result of this spectral sequence $\mathsf{E}^{\cdot,q}_{\infty}$ corresponds to $H^q(\mathcal{Z})$.

Definition 11.3.4. *The first-order* **cohomological equation** *is*

$$d^{0,1}_{\hat{\mathbf{f}}^{[0]}}\mathbf{u}^1 = \mathbf{f}^1.$$

The obvious question is, does there exist a solution to this equation?

Example 11.3.5. (Continuation of Example 11.2.12.) In the averaging case, with

$$\hat{\mathbf{f}}^{[0]} = \nabla_{\mathbf{f}^{[1]}} = \frac{\partial}{\partial t} + \varepsilon \sum_{i=1}^{n} f_i^1(\boldsymbol{x},t)\frac{\partial}{\partial x_i} + \cdots = \frac{\partial}{\partial t} + \varepsilon \mathbf{f}^1 + \cdots,$$

we have to solve

$$\mathcal{D}_{\mathbf{u}_t^1} = \mathcal{D}_{\mathbf{f}^1},$$

so the obvious answer is, not unless the f_i^1 can be written as derivatives with respect to t, a necessary and sufficient condition being that the f_i^1 have zero average (we restrict our attention here to the easier periodic case). ♢

Any obstruction to solving this equation lies in

$$\mathcal{Z}^{1,1}/(d^{0,1}_{\hat{\mathbf{f}}^{[0]}}\mathcal{Z}^{1,0}+\mathcal{Z}^{2,1}).$$

Another interesting point is that there are generators that do not do anything. We see that the kernel of $d^{0,1}_{\hat{\mathbf{f}}^{[0]}}$ consists of those u_i that have no time dependence. In other words,

$$\mathbf{u}^1 \in \ker d^{0,1}_{\hat{\mathbf{f}}^{[0]}}.$$

Now define

$$\mathsf{E}^{p,q}_0 = \mathcal{Z}^{p,q}/\mathcal{Z}^{p+1,q}$$

and (dropping $\hat{\mathbf{f}}^{[0]}$ in the notation) $\mathsf{d}^{0,1}:\mathsf{E}^{p,0}_0 \to \mathsf{E}^{p,1}_0$ by $\mathsf{d}^{0,1}[\mathbf{u}^{[p]}] = [d^{0,1}_{\hat{\mathbf{f}}^{[0]}}\mathbf{u}^{[p]}]$, where the $[\mathbf{u}^{[p]}]$ is the equivalence class of $\mathbf{u}^{[p]}$ in $\mathsf{E}^{p,0}_0$, that is, $\mathbf{u}^p$. We now have the cohomology space

$$H^{p,q}(\mathsf{E}_0,\mathsf{d}^{0,1})$$

using Definition 11.3.1:

$$H^{p,0}(\mathsf{E}_0,\mathsf{d}^{0,1}) = \ker \mathsf{d}^{0,1}$$

and

$$H^{p,1}(\mathsf{E}_0,\mathsf{d}^{0,1}) = \mathsf{E}^{p,1}_0/\mathsf{d}^{0,1}\mathsf{E}^{p,0}_0.$$

We now translate the way we actually compute normal forms in an abstract notation. We collect all the transformations in $\mathcal{Z}^{p,0}$ that end up (under $d^{0,1}_{\hat{\mathbf{f}}^{[0]}}$) in $\mathcal{Z}^{p+1,1}$ in the space $\mathcal{Z}^{p,0}$. In general,

$$\mathcal{Z}^{p,q}_1 = \{\mathbf{u}^p \in \mathcal{Z}^{p,q} | d^{0,1}_{\hat{\mathbf{f}}^{[0]}}\mathbf{u}^p \in \mathcal{Z}^{p+1,q+1}\}.$$

In the averaging case, $\mathcal{Z}^{p,0}_1$ consists of terms $\varepsilon^p\mathbf{u}^p(\boldsymbol{x})$. Then we let, with $\mathcal{Z}^{p,q}_0 = \mathcal{Z}^{p,q}$,

$$\mathsf{E}^{p,q}_1 = \mathcal{Z}^{p,q}_1/(d^{0,1}_{\hat{\mathbf{f}}^{[0]}}\mathcal{Z}^{p,q-1}_0 + \mathcal{Z}^{p+1,q}_0).$$

What are we doing here? First of all, $\mathcal{Z}^{p,1}_1 = \mathcal{Z}^{p,1}_0$. So $\mathsf{E}^{p,1}_1$ consists of the terms of degree p in normal form modulo terms of order $p+1$. This is consistent with doing the normal form calculations degree by degree, and throwing away higher-order terms till we need them.

We see that $\mathcal{Z}^{p,0}_1$ consists of those terms that carry $\mathbf{u}^{[p]}$ from degree p to $p+1$, which means that $\mathbf{u}^p$ commutes with the $\hat{\mathbf{f}}^0$ (this is equivalent to saying that $\rho_1(\mathbf{u}^{[p]})\hat{\mathbf{f}}^{[0]} \in \mathcal{Z}^{p+1,1}_0$). So $\mathsf{E}^{p,0}_1$ consists of the terms in $\mathcal{Z}^{p,0}_1$ modulo terms of order $p+1$. In other words,

$$\mathsf{E}^{p,q}_1 = H^{p,q}(\mathsf{E}^{\cdot,\cdot}_0,\mathsf{d}^{0,1}).$$

We provide a formal proof of this in general later on, in Lemma 13.2.7.

11.4 A Matter of Style

In this chapter we discuss the problem of describing and computing the normal form with respect to the linear part of the vector field (or the quadratic part of a Hamiltonian). The description problem is attacked with methods from invariant theory. In the semisimple case, the fact that the normal form has the flow of the linear equation as a symmetry group has done much to make normal form popular, since it makes the analysis of the equations in normal form much easier than the analysis of the original equations (supposedly not in normal form). When the linear part is not semisimple, the higher-order terms in normal form do have a symmetry group, but it is not the same as the flow of the linear part. The analysis therefore is not becoming much easier, although the normal form calculation at least removes all inessential parameters to higher order.

In the averaging case, we saw that the vector fields with nonzero average gave us an obstruction to solving the (co)homological equation, and we therefore considered the vector fields without explicit time dependence to be in normal form. Evidently we could add terms to this normal form with vanishing average and this would not change the obstruction, so they would stand for the same normal form. Our choice not to do so is a choice of *style*. In the case of averaging the choice is so obvious that no one gives it a second thought, but in the general context it is something to be considered with care.

The following two definitions are given for the yet to be defined spaces $\mathsf{E}_r^{p,1}$. At this stage they should be read with $r = 0$, but they keep their validity for $r > 0$, as defined in Chapter 13.

Definition 11.4.1. *Suppose, with $p > r$, $\dim \mathsf{E}_r^{p-r,0} < \infty$ and $\dim \mathsf{E}_r^{p,1} < \infty$. Then $\mathsf{d}^{r,1}$ maps $\mathsf{E}_r^{p-r,0}$ to $\mathsf{E}_r^{p,1}$. Define inner products on $\mathsf{E}_r^{p-r,0}$ and $\mathsf{E}_r^{p,1}$. Then we say that $\mathbf{f}^p \in \mathsf{E}_r^{p,1}$ is in* **inner product normal form** *(of level $r+1$) if $\mathbf{f}^p \in \ker \mathsf{d}^{-r,-1}$, where $\mathsf{d}^{-r,-1} : \mathsf{E}_r^{p,1} \to \mathsf{E}_r^{p-r,0}$ is the adjoint of $\mathsf{d}^{r,1}$ under the given inner products.*

This defines an inner product normal form style. This definition has the advantage that it is always possible under fairly weak conditions (and even those conditions are not necessary; in averaging theory the relevant spaces are usually not finite-dimensional, but we can still define an inner product and adjoint operator). A second advantage is that in applications the inner product is often already given, which makes the definition come out natural. The disadvantage of this definition is that when there is no given inner product (which is often the case in the bifurcation theory for nonlinear PDEs), it is fairly arbitrary. See also the discussion in Section 9.1.6.

Definition 11.4.2. *Suppose there exists a $\mathsf{d}^{-r,-1} : \mathsf{E}_r^{p,1} \to \mathsf{E}_r^{p-r,0}$, $\mathsf{E}_{r+1}^{p,1} = \ker \mathsf{d}^{-r,-1}$. Then we say that $\mathbf{f}^p \in \mathsf{E}_r^{p,1}$ is in* **dualistic normal form** *(of level $r+1$) if $\mathbf{f}^p \in \ker \mathsf{d}^{-r,-1}$.*

We call this style **dualistic**. It follows that inner product styles are dualistic.

Now the cohomological equation we have to solve is

$$\mathsf{d}^{r,1}\mathbf{u}^{p-r} = \mathbf{f}^p - \overline{\mathbf{f}}^p, \quad \overline{\mathbf{f}}^p \in \ker \mathsf{d}^{-r,-1},$$

or

$$\mathsf{d}^{-r,-1}\mathsf{d}^{r,1}\mathbf{u}^{p-r} = \mathsf{d}^{-r,-1}\mathbf{f}^p.$$

Suppose now that $\mathbf{E}_r^{p-r,0} = \ker \mathsf{d}^{r,1} \oplus \operatorname{im} \mathsf{d}^{-r,-1}$. Since in the homological equation $\mathbf{u}^{p-r}$ is defined up to terms in $\ker \mathsf{d}^{r,1}$, we might as well take $\mathbf{u}^{p-r} = \mathsf{d}^{-r,-1}\mathbf{g}^p$, and find the solution of

$$\mathsf{d}^{-r,-1}\mathsf{d}^{r,1}\mathsf{d}^{-r,-1}\mathbf{g}^p = \mathsf{d}^{-r,-1}\mathbf{f}^p.$$

Both $\mathbf{g}^p$ and $\mathbf{f}^p$ live in $\mathbf{E}_r^{p,1}$, so we have reduced the normal form problem to a linear algebra problem in $\mathbf{E}_r^{p,1}$.

Example 11.4.3. In the case of Example 11.3.5 this gives the complicated averaging formula (to be derived below)

$$g_i^p(\boldsymbol{x},t) = -\int^t\int^\tau\int^\sigma \frac{\partial f_i^p}{\partial t}(\boldsymbol{x},\zeta)\,\mathrm{d}\zeta\,\mathrm{d}\sigma\,\mathrm{d}\tau.$$

Here $\int^t$ stands for the right inverse of $\frac{\partial}{\partial t}$ mapping $\operatorname{im}\frac{\partial}{\partial t}$ into itself (zero mean to zero mean). Canceling the integrations against the differentations one obtains the old result, but there one has to subtract the average of f_i^p first, and in order to determine u_i^p uniquely, one has to see to it that (for instance) its average is zero. All these little details have been taken care of by this complicated formula. If we define the inner product as

$$(f,g)(\boldsymbol{x}) = \frac{1}{T}\int_0^T f(\boldsymbol{x},t)g(\boldsymbol{x},t)\,\mathrm{d}t,$$

we obtain, since $\mathsf{d}^{-r,-1} = -\frac{\partial}{\partial t}$,

$$u_i^p(\boldsymbol{x},t) = \int^t\int^\sigma \frac{\partial f_i^p}{\partial t}(\boldsymbol{x},\zeta)\,\mathrm{d}\zeta\,\mathrm{d}\sigma,$$

which is equivalent to the usual zero-mean generator in averaging (cf. Section 3.4). ◇

Going back to the abstract problem, we see that we need to have a right inverse $\overline{\mathsf{d}}^{r,1}$ of $\mathsf{d}^{-r,-1}$ and a right inverse $\overline{\mathsf{d}}^{-r,-1}$ of $\mathsf{d}^{r,1}$. This leads to

$$\mathbf{g}^p = \overline{\mathsf{d}}^{r,1}\overline{\mathsf{d}}^{-r,-1}\overline{\mathsf{d}}^{r,1}\mathsf{d}^{-r,-1}\mathbf{f}^p$$

and

$$\mathbf{u}^{p-r} = \overline{\mathsf{d}}^{-r,-1}\overline{\mathsf{d}}^{r,1}\mathsf{d}^{-r,-1}\mathbf{f}^p.$$

Observe that $\pi^r = \overline{\mathsf{d}}^{r,1}\mathsf{d}^{-r,-1}$ is a projection operator. It projects $\mathbf{f}^p$ on im $\overline{\mathsf{d}}^{-r,-1}$. In averaging language this is done by subtracting the average.

If we do not have these right inverses ready, we may try the following approach. Our problem is to solve the equation

$$\mathsf{d}^{r,1}\mathsf{d}^{-r,-1}\mathbf{g}^p = \mathbf{f}^p$$

as well as we can. We know that $\mathsf{d}^{r,1}\mathsf{d}^{-r,-1}$ is symmetric if we are in the case in which $\mathsf{d}^{-r,-1}$ is the adjoint of $\mathsf{d}^{r,1}$, and therefore semisimple, that is, its matrix A can be put in diagonal form, and its eigenvalues are nonnegative real numbers. Let p_A be the minimal polynomial of A:

$$p_A(\lambda) = \prod_{i=1}^{k}(\lambda - \lambda_i)$$

with $\lambda_i \in \mathbb{C}$ and all different. Define

$$p_A^j(\lambda) = \prod_{i \neq j}(\lambda - \lambda_i)$$

and let

$$E_i = \frac{p_A^i(A_k)}{p'(\lambda_i)},$$

where p'_A is the derivative of p_A with respect to λ. Let f be a function defined on the spectrum of A. Then we can define

$$f(A) = \sum_{i=1}^{k} f(\lambda_i)E_i.$$

This allows us to compute (with $f(\lambda) = \tau^\lambda$) $\tau^A = \sum_i \tau^{\lambda_i}E_i$. We now define $T : \mathsf{E}_r^{p,1} \to \mathsf{E}_r^{p,1}$ by

$$T\mathbf{g}^p = \left[\int_1^t \tau^A \mathbf{g}^p \frac{\mathrm{d}\tau}{\tau}\right]_{t=1}. \tag{11.4.1}$$

Observe that if A is the matrix of $\mathsf{d}^{r,1}\mathsf{d}^{-r,-1}$ then $T = (\mathsf{d}^{r,1}\mathsf{d}^{-r,-1})^{-1}$ on im $\mathsf{d}^{r,1}\mathsf{d}^{-r,-1}$ and $T = 0$ on ker $\mathsf{d}^{r,1}\mathsf{d}^{-r,-1}$. This is easily checked on an eigenvector of $\mathsf{d}^{r,1}\mathsf{d}^{-r,-1}$, and, since $\mathsf{d}^{r,1}\mathsf{d}^{-r,-1}$ is semisimple, extends to the whole space. Indeed, let $\mathbf{g}_{(s)}^p \in \mathsf{E}_r^{p,1}$ be such that $\mathsf{d}^{r,1}\mathsf{d}^{-r,-1}\mathbf{g}_{(s)}^p = \lambda_s \mathbf{g}_{(s)}^p$ for some $s \in \{1, \ldots, k\}$. Then

$$\begin{aligned}
T\mathsf{d}^{r,1}\mathsf{d}^{-r,-1}\mathbf{g}_{(s)}^p &= \lambda_s T\mathbf{g}_{(s)}^p \\
&= \lambda_s \left[\int^t \tau^A \mathbf{g}_{(s)}^p \frac{\mathrm{d}\tau}{\tau}\right]_{t=1} \\
&= \lambda_s \left[\int^t \sum_{i=1}^{k} \tau^{\lambda_i} E_i \mathbf{g}_{(s)}^p \frac{\mathrm{d}\tau}{\tau}\right]_{t=1} \\
&= \lambda_s \left[\int^t \sum_{i=1}^{k} \tau^{\lambda_i - 1}\delta_{is}\,\mathrm{d}\tau\right]_{t=1} \mathbf{v}_s^p \\
&= \begin{cases} \mathbf{v}_s^p & \text{if } \lambda_s \neq 0, \\ 0 & \text{if } \lambda_s = 0. \end{cases}
\end{aligned}$$

Thus the operator T gives us the solution to the problem

$$\mathsf{d}^{r,1}\mathsf{d}^{-r,-1}\mathbf{g}^p = \mathbf{f}^p$$

in the form $\mathbf{g}^p = T\mathbf{f}^p$. Observe that this approach is completely different from the usual, since there one identifies $\mathsf{E}_0^{0,0}$ and $\mathsf{E}_0^{0,1}$ and one is mainly interested in the spectrum of $\mathsf{d}^{0,1}$. Even if $\mathsf{d}^{0,1}$ is nilpotent, $\mathsf{d}^{0,1}\mathsf{d}^{0,-1}$ might be semisimple.

Alternatively, one can simply compute the matrices of $\mathsf{d}^{r,1}$ and $\mathsf{d}^{-r,-1}$, and compute the generalized inverse Q of $\mathsf{d}^{r,1}\mathsf{d}^{-r,-1}$, where the generalized inverse of a linear map $A : V \to V$ is a map $Q : V \to V$ such that QA is a projection on V with $AQA = A$. Then $Q\mathsf{d}^{r,1}\mathsf{d}^{-r,-1}$ is a projection on the image of $\mathsf{d}^{r,1}$. This procedure has the advantage of being rational, we do not need to compute the eigenvalues of $\mathsf{d}^{r,1}\mathsf{d}^{-r,-1}$. It has the disadvantage that there does not seem to be a smart way to do it, that is, a way induced by the lower-dimensional linear algebra on the (co)ordinates. Whether the first method can be done in a smart way remains to be seen. It would need the existence of an element in $\mathbf{g}^{p-r} \in \mathsf{E}_r^{p-r,0}$ such that $L_{\mathbf{g}^{p-r}} = \mathsf{d}^{r,1}\mathsf{d}^{-r,-1}$ on $\mathsf{E}_r^{p,1}$. A little experimentation teaches that this is not possible in general, as is illustrated by the next example.

11.4.1 Example: Nilpotent Linear Part in $\mathbb{R}^2$

We consider the formal vector fields on $\mathbb{R}^2$ with nilpotent linear part: Take

$$\begin{aligned}\mathbf{f}^{[0]} &= x_1\frac{\partial}{\partial x_2} + \left(\frac{1}{\sqrt{2}}a_1x_1^2 + a_2x_1x_2 + \frac{1}{\sqrt{2}}a_3x_2^2\right)\frac{\partial}{\partial x_1}\\ &\quad + \left(\frac{1}{\sqrt{2}}a_4x_1^2 + a_5x_1x_2 + \frac{1}{\sqrt{2}}a_6x_2^2\right)\frac{\partial}{\partial x_2} + \cdots,\end{aligned}$$

corresponding to the differential equation

$$\begin{aligned}\dot{x}_1 &= \left(\frac{1}{\sqrt{2}}a_1x_1^2 + a_2x_1x_2 + \frac{1}{\sqrt{2}}a_3x_2^2\right) + \cdots,\\ \dot{x}_2 &= x_1 + \left(\frac{1}{\sqrt{2}}a_4x_1^2 + a_5x_1x_2 + \frac{1}{\sqrt{2}}a_6^2x_2^2\right) + \cdots.\end{aligned}$$

We choose a basis for $\mathsf{E}_0^{1,1}$, the space of quadratic vector fields, as follows:

$$\begin{aligned}u_1 &= \tfrac{1}{\sqrt{2}}x_1^2\tfrac{\partial}{\partial x_1}, & u_2 &= x_1x_2\tfrac{\partial}{\partial x_1}, & u_3 &= \tfrac{1}{\sqrt{2}}x_2^2\tfrac{\partial}{\partial x_1},\\ u_4 &= \tfrac{1}{\sqrt{2}}x_1^2\tfrac{\partial}{\partial x_2}, & u_5 &= x_1x_2\tfrac{\partial}{\partial x_2}, & u_6 &= \tfrac{1}{\sqrt{2}}x_2^2\tfrac{\partial}{\partial x_2}.\end{aligned}$$

This basis is orthonormal with respect to the inner product (with $I = (i_1, \ldots, i_N)$ and $J = (j_1, \ldots, j_K)$ multi-indices, and $\boldsymbol{x}^I = x_1^{i_1} \cdots x_n^{i_n}$)

$$\left(\boldsymbol{x}^I \frac{\partial}{\partial x_{i_0}}, \boldsymbol{x}^J \frac{\partial}{\partial x_{j_0}}\right) = \delta_{i_0,j_0} \prod_{l=1}^{n} \delta_{i_l,j_l} i_l!.$$

The matrix of $\mathsf{d}^{0,1}$ with respect to a similar choice of basis for $\mathsf{E}_0^{1,0}$ is

$$N = \begin{bmatrix} 0 & \sqrt{2} & 0 & 0 & 0 & 0 \\ 0 & 0 & \sqrt{2} & 0 & 0 & 0 \\ 0 & 0 & 0 & 0 & 0 & 0 \\ -1 & 0 & 0 & 0 & \sqrt{2} & 0 \\ 0 & -1 & 0 & 0 & 0 & \sqrt{2} \\ 0 & 0 & -1 & 0 & 0 & 0 \end{bmatrix}.$$

This matrix describes how the coefficients $a_1, \ldots, a_6$ are mapped onto new coefficients. The matrix of $\mathbf{N}\mathbf{N}^\dagger$ is

$$NN^\dagger = \begin{bmatrix} 2 & 0 & 0 & 0 & -\sqrt{2} & 0 \\ 0 & 2 & 0 & 0 & 0 & -\sqrt{2} \\ 0 & 0 & 0 & 0 & 0 & 0 \\ 0 & 0 & 0 & 3 & 0 & 0 \\ -\sqrt{2} & 0 & 0 & 0 & 3 & 0 \\ 0 & -\sqrt{2} & 0 & 0 & 0 & 1 \end{bmatrix}.$$

Notice that this can never be the matrix of an action induced by a linear vector field. The generalized inverse is

$$M = \begin{bmatrix} 3/4 & 0 & 0 & 0 & 1/4\sqrt{2} & 0 \\ 0 & 2/9 & 0 & 0 & 0 & -1/9\sqrt{2} \\ 0 & 0 & 0 & 0 & 0 & 0 \\ 0 & 0 & 0 & 1/3 & 0 & 0 \\ 1/4\sqrt{2} & 0 & 0 & 0 & 1/2 & 0 \\ 0 & -1/9\sqrt{2} & 0 & 0 & 0 & 1/9 \end{bmatrix}.$$

If we multiply the generalized inverse on the left by $N^\dagger$ we obtain

$$N^\dagger M = \begin{bmatrix} 0 & 0 & 0 & -1/3 & 0 & 0 \\ 1/2\sqrt{2} & 0 & 0 & 0 & 0 & 0 \\ 0 & 1/3\sqrt{2} & 0 & 0 & 0 & -1/3 \\ 0 & 0 & 0 & 0 & 0 & 0 \\ 0 & 0 & 0 & 1/3\sqrt{2} & 0 & 0 \\ 1/2 & 0 & 0 & 0 & 1/2\sqrt{2} & 0 \end{bmatrix},$$

so that the transformation is

$$\begin{aligned} & N^\dagger M \mathbf{f}^1 = \mathbf{u}^1 \\ & = -\frac{1}{3\sqrt{2}} a_4 x_1^2 \frac{\partial}{\partial x_1} + \frac{1}{\sqrt{2}} a_1 x_1 x_2 \frac{\partial}{\partial x_1} + \frac{1}{3\sqrt{2}} (\sqrt{2} a_2 - a_6) x_2^2 \frac{\partial}{\partial x_1} \\ & + \frac{1}{3} \sqrt{2} a_4 x_1 x_2 \frac{\partial}{\partial x_2} + \frac{1}{2\sqrt{2}} (a_1 + \sqrt{2} a_5) x_2^2 \frac{\partial}{\partial x_2}. \end{aligned}$$

This leads to the following projection matrix mapping $I - NN^\dagger M$ from $\mathsf{E}_0^{1,1}$ to $\mathsf{E}_1^{1,1}$:

$$\begin{bmatrix} 0 & 0 & 0 & 0 & 0 & 0 \\ 0 & 1/3 & 0 & 0 & 0 & 1/3\sqrt{2} \\ 0 & 0 & 1 & 0 & 0 & 0 \\ 0 & 0 & 0 & 0 & 0 & 0 \\ 0 & 0 & 0 & 0 & 0 & 0 \\ 0 & 1/3\sqrt{2} & 0 & 0 & 0 & 2/3 \end{bmatrix}.$$

We see that the normal form that remains is

$$\bar{\mathbf{f}}^1 = \frac{1}{3}(a_2 + \sqrt{2} a_6) x_2 \left(x_1 \frac{\partial}{\partial x_1} + x_2 \frac{\partial}{\partial x_2} \right) + \frac{1}{\sqrt{2}} a_3 x_2^2 \frac{\partial}{\partial x_1}.$$

11.5 Induced Linear Algebra

This might be a good moment to explain what we mean by **smart methods**. Vector fields and Hamiltonians at equilibrium are sums of products of the local coordinates (and, in the case of vector fields, of the ordinates $\frac{\partial}{\partial x_i}, i = 1, \ldots, n$). As a consequence, it is enough for us to know the action of the object to be normalized (and we call this the **vector field** for short) on the (co)ordinates. The rules for the induced action are simple:

$$\mathsf{d}_{p-1}^{r,1} \sum_{i=1}^{n} \mathsf{g}_i^p \frac{\partial}{\partial x_i} = \sum_{i=1}^{n} (\mathsf{d}_p^{r,1} \mathsf{g}_i^p) \frac{\partial}{\partial x_i} + \mathsf{g}_i^p \mathsf{d}_{-1}^{r,1} \frac{\partial}{\partial x_i},$$

where $\mathsf{d}_p^{r,1}\mathsf{g}_i^p = \sum_{j=1}^n f_j^0 \frac{\partial}{\partial x_j}\mathsf{g}_i^p$ and $\mathsf{d}_{-1}^{r,1}\frac{\partial}{\partial x_i} = [\mathbf{f}^0, \frac{\partial}{\partial x_i}]$. Moreover,

$$\tau^{\mathbb{A}} x_1^{i_1}\cdots x_n^{i_n}\frac{\partial}{\partial x_{i_0}} = (\tau^{A^\star}x_1)^{i_1}\cdots(\tau^{A^\star}x_n)^{i_n}\tau^A\frac{\partial}{\partial x_{i_0}}.$$

This means that we can compute the generalized inverse of $\mathbb{A}$ from the knowledge of A, using the methods described in Section 11.4. The only thing we need here are the eigenvalues of A. If these are not known, there seems to be no smart way of inverting $\mathbb{A}$, although some claims to this effect are being made in [187].

This means that in order to compute $\tau^{\mathbb{A}}$, we need only the computation of τ^A, and this requires the computation of low-dimensional projection operators. Let us consider a well known example: the perturbed harmonic oscillator. Let $\mathbf{f}^0 = x_1\frac{\partial}{\partial x_2} - x_2\frac{\partial}{\partial x_1}$. Then let

$$u_1 = \frac{\partial}{\partial x_1}, \quad u_2 = \frac{\partial}{\partial x_2}.$$

Then $[f^0, u_1] = -u_2$ and $[f^0, u_2] = u_1$. So the matrix of $\mathsf{d}^{0,1}$ is

$$A = \begin{bmatrix} 0 & 1 \\ -1 & 0 \end{bmatrix}.$$

Then $p_A(\lambda) = \lambda^2 + 1 = (\lambda+i)(\lambda-i)$. So $p_A^1(\lambda) = (\lambda - i)$ and $p_A^2(\lambda) = (\lambda+i)$. Then

$$E_1 = \frac{p_A^1(A)}{p\prime(\lambda_1)} = -\frac{1}{2i}(A - iI) = -\frac{1}{2i}\begin{bmatrix} -i & 1 \\ -1 & -i \end{bmatrix} = \begin{bmatrix} \frac{1}{2} & \frac{i}{2} \\ -\frac{i}{2} & \frac{1}{2} \end{bmatrix}$$

and

$$E_2 = \frac{p_A^2(A)}{p\prime(\lambda_2)} = \frac{1}{2i}(A + iI) = \frac{1}{2i}\begin{bmatrix} i & 1 \\ -1 & i \end{bmatrix} = \begin{bmatrix} \frac{1}{2} & -\frac{i}{2} \\ \frac{i}{2} & \frac{1}{2} \end{bmatrix}.$$

Thus

$$\tau^A = \tau^{-i}E^1 + \tau^i E_2 = \tau^{-i}\begin{bmatrix} \frac{1}{2} & \frac{i}{2} \\ -\frac{i}{2} & \frac{1}{2} \end{bmatrix} + \tau^i\begin{bmatrix} \frac{1}{2} & -\frac{i}{2} \\ \frac{i}{2} & \frac{1}{2} \end{bmatrix} = \begin{bmatrix} \frac{1}{2}(\tau^{-i}+\tau^i) & -\frac{i}{2}(\tau^i - \tau^{-i}) \\ \frac{i}{2}(\tau^i - \tau^{-i}) & \frac{1}{2}(\tau^{-i}+\tau^i) \end{bmatrix}.$$

Replacing τ by e^t, the reader recognizes the familiar $e^{tA} = \left[\begin{smallmatrix} \cos t & \sin t \\ -\sin t & \cos t \end{smallmatrix}\right]$. Observe that $\frac{\mathsf{d}\tau}{\tau} = \mathsf{d}t$. We can now compute $\tau^{\mathbb{A}}$ on something like $x_1x_2\frac{\partial}{\partial x_2}$ by replacing x_1 by $y_1\cos t - y_2\sin t$, x_2 by $y_1\sin t + y_2\cos t$, and $\frac{\partial}{\partial x_2}$ by $-\sin t\frac{\partial}{\partial y_1} + \cos t\frac{\partial}{\partial y_2}$. This amounts to putting the equation in a comoving frame. The result is then integrated with respect to t, and then one puts $t = 0$ and $y_i = x_i, i = 1, 2$.

Remark 11.5.1. One can also stay in the comoving coordinates (instead of putting $t = 0$), since the flow is a Lie algebra homomorphism, so the normal form calculation scheme is not affected by this. Computationally it has the advantage of speed, since one does not have to carry out the transformation to comoving coordinates every time one needs to solve the cohomological equation, but the disadvantage of having a much bigger expression size. The optimal choice may well be problem and machine dependent. ♡

This is obviously a variant of the averaging method, so one can say that the general normal form method, along the lines that we have followed here, is a direct generalization of the averaging method, even if the averaging is reduced to computing residues. This is why some authors call this method of normalization with respect to the semisimple linear vector fields the averaging method, and why other authors find this confusing [203, p. 221].

11.5.1 The Nilpotent Case

We made the assumption in the decomposition of the matrix A into projection operators that the action of $\mathsf{d}^{0,1}$ is semisimple. What if it is not? Let us first consider our previous example, where it is nilpotent. In that case, we can apply the following method. We can, under certain technical conditions, embed the nilpotent element in a bigger Lie algebra, isomorphic to $\mathfrak{sl}_2$. The Lie algebra $\mathfrak{sl}_2$ is defined abstractly by the commutation relations of its three generators $\mathbf{N}_+, \mathbf{N}_-$, and $\mathbf{H}$ as follows:

$$[\mathbf{N}_+, \mathbf{N}_-] = \mathbf{H}, \quad [\mathbf{H}, \mathbf{N}_+] = 2\mathbf{N}_+, \quad [\mathbf{H}, \mathbf{N}_-] = -2\mathbf{N}_-.$$

In our example this embedding is rather trivial: let $\mathsf{d}^{0,1}$ be induced by $\mathbf{N}_- = x_1\frac{\partial}{\partial x_2}$, $\mathbf{N}_+$ by $x_2\frac{\partial}{\partial x_1}$, and $\mathbf{H}$ by $[x_2\frac{\partial}{\partial x_1}, x_1\frac{\partial}{\partial x_2}] = -x_1\frac{\partial}{\partial x_1} + x_2\frac{\partial}{\partial x_2}$. The (finite-dimensional) representation theory of $\mathfrak{sl}_2$ (cf. [136]) now tells us that every vector field can be written uniquely as the sum of a vector in the kernel of the action of $\mathbf{N}_+$ and a vector in the image of the action of $\mathbf{N}_-$. It also tells us that the space can be seen as a direct sum of irreducible spaces spanned by vectors $e_0, \ldots, e_m$ on which $\mathfrak{sl}_2$ acts as follows:

$$\begin{aligned} \mathbf{N}_- e_j &= (m-j)e_{j+1}, \\ \mathbf{N}_+ e_j &= j e_{j-1}, \\ \mathbf{H} e_j &= (m-2j)e_j. \end{aligned}$$

This implies that if we have an eigenvector e_0 of $\mathbf{H}$ in $\ker \mathbf{N}_+$, then its eigenvalue m determines the dimension of the irreducible representation $m+1$. We call the $\mathbf{H}$-eigenvalue the **weight of a vector**. One way to think of these irreducible representations is as binary forms, where X and Y are seen as coordinates in $\mathbb{R}^2$,

$$\hat{e}_0(X,Y) = \sum_{j=0}^{m} \binom{m}{j} e_j X^j Y^{m-j}.$$

Indeed, if we apply $\hat{\boldsymbol{Y}} = Y\frac{\partial}{\partial X}$ to this expression, we obtain

$$\begin{aligned} \hat{\boldsymbol{Y}}\hat{e}_0(X,Y) &= Y\frac{\partial}{\partial X}\hat{e}_m(X,Y) \\ &= \sum_{j=0}^{m} \binom{m}{j} j e_j X^{j-1} Y^{m-j+1} \end{aligned}$$

$$= \sum_{j=1}^{m} \binom{m}{j-1} (m-j+1) e_j X^{j-1} Y^{m-j+1}$$
$$= \sum_{j=0}^{m-1} \binom{m}{j} (m-j) e_{j+1} X^j Y^{m-j}$$
$$= \sum_{j=0}^{m} \binom{m}{j} \mathbf{N}_- e_j X^j Y^{m-j}$$
$$= \mathbf{N}_- \hat{e}_0(X,Y).$$

Similar expressions can be obtained for $\hat{\boldsymbol{X}} = X\frac{\partial}{\partial Y}$ and the commutator of the two. So if we have a vector field that is part of an irreducible representation, we can label it with $X^j Y^{m-j}$ to indicate where it lives, and how it behaves under the action of $\mathfrak{sl}_2$. Notice that we can rewrite $\hat{e}_m$ in terms of e_0 as

$$\hat{e}_0(X,Y) = \sum_{j=0}^{m} \frac{1}{j!} \mathbf{N}_-^j e_0 X^{m-j} Y^j.$$

The problem, of course, is that if we start with an arbitrary element in the representation space, we initially have no idea where things are. If we could somehow project a given v_0 onto the $\mathbf{H}$-eigenspaces in ker $\mathbf{N}_+$ we would be done, but this takes work. What we can do is to apply $\mathbf{N}_+$ to v_0 till the result is zero by defining $v_{j+1} = \mathbf{N}_+ v_j$. So suppose $v_{k+1} = 0$ and $v_k \neq 0$. If $k = 0$, we see that $v_0 \in \ker \mathbf{N}_+$ and we are done. So suppose $k > 0$. How do we now know the eigenvalue of v_k? We do not, because v_k may well be the sum of several vectors with different eigenvalues. But we already know how to solve this problem: we compute $X^{\mathbf{H}} v_k$, and we can do this since $\mathbf{H}$ is semisimple. We write the result as

$$X^{\mathbf{H}} v_k = \sum_{i=1}^{N} X^{\lambda_i} v_k^i.$$

The λ_i's are strictly positive, since we are in the kernel of $\mathbf{N}_+$ but also in the image of $\mathbf{N}_+$ (since $k > 0$), so that we cannot be in ker $\mathbf{N}_-$. Therefore we are not in the intersection of the kernels, and this implies that the $\mathbf{H}$-eigenvalue cannot be zero. The v_k^i have eigenvalues λ_i and so generate an irreducible representation of dimension $\lambda_i + 1$. We now want to construct v_{k-1}^i such that $\mathbf{N}_+ v_{k-1}^i = v_k^i$. As a candidate we take $\alpha_k^i \mathbf{N}_- v_k^i$. We know that $\alpha_k^i \mathbf{N}_+ \mathbf{N}_- v_k^i X^{\lambda^i} = \alpha_k^i \lambda_i \mathbf{N}_+ v_{k-1}^i X^{\lambda^i} = \alpha_k^i \lambda_i v_k^i X^{\lambda^i}$. Therefore we should take $\alpha_k^i = \frac{1}{\lambda_i}$, and we obtain

$$v_{k-1}^i = \frac{1}{\lambda_i} \mathbf{N}_- v_k^i.$$

Here v_{k-1}^i is labeled by $X^{\lambda_i - 1} Y$, and we can obtain the coefficient by applying

$$\hat{\mathcal{I}} = \frac{1}{X} \int^{Y} \int^{X} \cdot \ \frac{\mathrm{d}\xi}{\xi}$$

to X^{λ_i}. That is,

$$\hat{\mathcal{I}}X^{\lambda_i} = \frac{1}{X}\int^Y\int^X \xi^{\lambda_i}\frac{\mathrm{d}\xi}{\xi} = \frac{1}{\lambda_i}X^{\lambda_i-1}\int^Y \mathrm{d}\xi = \frac{1}{\lambda_i}X^{\lambda_i-1}Y.$$

If we apply this operation to $X^{\lambda_i-j}Y^j$ we obtain $\frac{1}{(\lambda_i-j)(j+1)}X^{\lambda_i-j-1}Y^{j+1}$, and this effectively counteracts the numerical effect of the operation $\mathbf{N}_-\mathbf{N}_+$ on the term v^i_{k-j-1}. If we now compute

$$v_{k-1} - \mathbf{N}_-\hat{\mathcal{I}}X^{\mathbf{H}}v_k|_{X=Y=1}$$

we obtain an expression that again has to be in ker $\mathbf{N}_+$, so we can scale it with $X^{\mathbf{H}}$ and add it to $\hat{\mathcal{I}}\mathbf{N}_-X^{\mathbf{H}}v_k$, replacing v_{k-1} by its scaled version. This way we can inductively describe what to do, and at the end we arrive at the situation in which, with v_1 the properly scaled version of v_1,

$$v_0 - \mathbf{N}_-\hat{\mathcal{I}}X^{\mathbf{H}}v_1|_{X=Y=1} \in \ker \mathbf{N}_+.$$

This solves the first-order cohomological equation in the nilpotent case. Obviously, the procedure is better suited for computer calculations than for hand calculations. It is rather time-consuming.

11.5.2 Nilpotent Example Revisited

We look at the example in Section 11.4.1. We list the induced action of $\mathbf{N}_-$ and $\mathbf{N}_+$ on the basis:

$$\begin{aligned}
\mathbf{N}_-u_1 &= -u_4 & \mathbf{N}_+u_1 &= \sqrt{2}u_2\\
\mathbf{N}_-u_2 &= \sqrt{2}u_1 - u_5 & \mathbf{N}_+u_2 &= \sqrt{2}u_3\\
\mathbf{N}_-u_3 &= \sqrt{2}u_2 - u_6 & \mathbf{N}_+u_3 &= 0\\
\mathbf{N}_-u_4 &= 0 & \mathbf{N}_+u_4 &= \sqrt{2}u_5 - u_1\\
\mathbf{N}_-u_5 &= \sqrt{2}u_4 & \mathbf{N}_+u_5 &= \sqrt{2}u_6 - u_2\\
\mathbf{N}_-u_6 &= \sqrt{2}u_5 & \mathbf{N}_+u_6 &= -u_3
\end{aligned}$$

The $\sqrt{2}$'s are artifacts since we want to use the same basis as before. If we would choose a basis with rational coefficients, the whole computation would be rational.

Let $v_0 = \sum_{i=1}^6 a_iu_i$. Then

$$\begin{aligned}
v_1 &= \mathbf{N}_+v_0 = \sqrt{2}a_1u_2 + \sqrt{2}a_2u_3 + a_4(\sqrt{2}u_5 - u_1) + a_5(\sqrt{2}u_6 - u_2) - a_6u_3\\
&= -a_4u_1 + (\sqrt{2}a_1 - a_5)u_2 + (\sqrt{2}a_2 - a_6)u_3 + \sqrt{2}a_4u_5 + \sqrt{2}a_5u_6,\\
v_2 &= \mathbf{N}_+v_1 = -\sqrt{2}a_4u_2 + (\sqrt{2}a_1 - a_5)\sqrt{2}u_3 + \sqrt{2}a_4(\sqrt{2}u_6 - u_2) - \sqrt{2}a_5u_3\\
&= -2\sqrt{2}a_4u_2 + (\sqrt{2}a_1 - 2a_5)\sqrt{2}u_3 + 2a_4u_6,\\
v_3 &= \mathbf{N}_+v_2 = -4a_4u_3 - 2a_4u_3 = -6a_4u_3,\\
v_4 &= \mathbf{N}_+v_3 = 0.
\end{aligned}$$

Since $X^{\mathbf{H}}u_3 = X^3 u_3$, we find that the preimage of v_3 equals

$$-2a_4(\sqrt{2}u_2 - u_6)X^2Y.$$

We find that the scaled v_2 equals

$$(\sqrt{2}a_1 - 2a_5)\sqrt{2}u_3X^3 - 2a_4(\sqrt{2}u_2 - u_6)X^2Y.$$

The preimage of this term is

$$\frac{2}{3}(a_1 - \sqrt{2}a_5)(\sqrt{2}u_2 - u_6)X^2Y - a_4(u_1 - \sqrt{2}u_5)XY^2.$$

The remaining term in the kernel is

$$(\sqrt{2}a_2 - a_6)u_3X^3 + \frac{1}{3}(\sqrt{2}a_1 + a_5)(u_2 + \sqrt{2}u_6)X.$$

Applying $\hat{\mathcal{I}}$ to the sum of these last two terms and putting $X = Y = 1$ gives us the generator of transformation to normal form:

$$-\frac{1}{3}a_4u_1 + \frac{1}{\sqrt{2}}a_1u_2 + \frac{1}{3}(\sqrt{2}a_2 - a_6)u_3 + \frac{\sqrt{2}}{3}a_4u_5 + \frac{1}{2}(a_1 + \sqrt{2}a_5)u_6$$

The normal form that remains is

$$a_3u_3 + \frac{1}{3}(a_2 + \sqrt{2}a_6)(u_2 + \sqrt{2}u_6)$$

or

$$\frac{1}{\sqrt{2}}a_3x_2^2\frac{\partial}{\partial x_1} + \frac{1}{3}(a_2 + \sqrt{2}a_6)x_2\left(x_1\frac{\partial}{\partial x_1} + x_2\frac{\partial}{\partial x_2}\right).$$

Here we can already see the module structure: the vectors $x_2\frac{\partial}{\partial x_1}$ and $x_1\frac{\partial}{\partial x_1} + x_2\frac{\partial}{\partial x_2}$ are both in ker $\mathbf{N}_+$, as well as x_2.

11.5.3 The Nonsemisimple Case

What remains to be done is the solution of the cohomological equation in the nonsemisimple case. In that case the operator can be written (under suitable technical conditions) as the commuting sum of a semisimple and a nilpotent operator; see Algorithm 11.1. Let us denote the induced matrices by $\boldsymbol{S}$ and $\mathbf{N}_-$. Then the space is the direct sum im $\boldsymbol{S} \oplus$ ker $\boldsymbol{S}$ (computed by applying $\tau^{\boldsymbol{S}}$) and ker $\boldsymbol{S}$ = im $\mathbf{N}_- \oplus$ ker $\mathbf{N}_+$. On im $\boldsymbol{S}$ the equation can be solved by inverting $\boldsymbol{S} + \mathbf{N}_-$ by

$$\sum_{i=0}^{\infty}(-1)^i(\boldsymbol{S}^{-1}\mathbf{N}_-)^i\boldsymbol{S}^{-1}.$$

The sum is finite, and can be explicitly computed.

Algorithm 11.1 MAPLE™ procedures for $S-N$ decomposition, after [172]

```
# The procedure semisimple computes the semisimple part of a matrix
with (linalg) :
semisimple := proc(A)
local xp, Q, B, x, n, lp, t, q, i, g :
xp := charpoly(eval(A), x) :
xp := numer(normal(xp)) :
g := diff(xp, x) :
q := gcdex(xp, g, x) :
lp := normal(xp/q) :
n := highest(q, x) :
g := x :
for i to n do
   q := diff(lp ∧ i, x$i) :
   gcdex(lp, q, x, 'q', 't') :
   q := diff(g, x$i) :
   g := g − lp ∧ i ∗ rem(t ∗ q, lp, x) :
od;
g := rem(g, xp, x) :
B := band([0], rowdim(A)) :
Q := band([1], rowdim(A)) :
for i from 0 to degree (g, x) do
   q := coefc(g, x, i) :
   if q ≠ 0 then
      B := evalm(B + q ∗ Q) :
   fi;
   Q := multiply(A, Q) :
od;
B := map(normal, B) :
RETURN(op(B)) :
end :
coefc := proc(f, x, n) :
RETURN(coeff(collect(f, x), x, n)) :
end :
highest := proc(f, x)
local g, n, d, sf :
sf := convert(numer(f), sqrfree, x) :
if type (sf, '∗∗') then
   d := op(2, sf)
elif degree (f, x) = 0 then
   d := 0
elif type (sf, '+') or type (sf, string) then
   d := 1
else
   n := nops(sf) :
   g := op(n, sf) :
   d := op(2, g) :
fi;
RETURN(d) :
end :
```

11.6 The Form of the Normal Form, the Description Problem

We have shown how to compute the normal form term by term, at least we have described the first step. But in many applications one is not immediately interested in the normal form of a specific vector field, but only in the general form of a vector field with a given linear part, for instance, if one wants to describe the bifurcation behavior of a system with a given linear part for all possible higher-order terms. If we look at the problem of periodic averaging,

$$\dot{\boldsymbol{x}} = \varepsilon \mathbf{f}^{[1]}(\boldsymbol{x}, t, \varepsilon),$$

the answer is simple:

$$\dot{\boldsymbol{x}} = \varepsilon \mathbf{f}_\star^{[1]}(\boldsymbol{x}, \varepsilon),$$

where $\mathbf{f}_\star^{[1]}$ is the pushforward under the formal averaging transformation. In its full generality this is a very difficult question, but there are quite a few problems that we can handle. For instance, for the anharmonic oscillator, with

$$\mathbf{f}^0 = x_1 \frac{\partial}{\partial x_2} - x_2 \frac{\partial}{\partial x_1} = \begin{bmatrix} -x_2 \\ x_1 \end{bmatrix},$$

the general normal form with respect to its linear part $\mathbf{f}^0$ is

$$F(x_1^2 + x_2^2)\left(x_1 \frac{\partial}{\partial x_2} - x_2 \frac{\partial}{\partial x_1}\right) + G(x_1^2 + x_2^2)\left(x_1 \frac{\partial}{\partial x_1} + x_2 \frac{\partial}{\partial x_2}\right).$$

This follows from the fact that the space of formal vector fields has a splitting

$$\ker \mathcal{L}_{\mathbf{f}^0} \oplus \operatorname{im} \mathcal{L}_{\mathbf{f}^0}.$$

We have only to verify that $\mathcal{D}_{\mathbf{f}^0}(x_1^2 + x_2^2) = 0$ and $\mathcal{L}_{\mathbf{f}^0}(x_1 \frac{\partial}{\partial x_1} + x_2 \frac{\partial}{\partial x_2}) = 0$.

Remark 11.6.1. This is an indication that from the theoretical point of view we are losing information if we change a system by comoving coordinates into a form that is amenable to the averaging method. While the results are correct, the description problem can no longer be solved in its most general form. If we want to do that, we have to work with systems that are in the form of comoving systems. ♡

But how do we know that we are not missing anything? Once the problems become a bit more intricate, it is easy to overlook certain terms, and it seems a worthwhile requirement that any author prove a given normal form to be complete.

In the semisimple case, the linear vector field generates a one-dimensional Lie group. The generating function for equivariant vector fields under a fixed group representation ρ is given by the Molien integral

$$P(t) = \int_G \frac{\operatorname{tr}(\rho(g^{-1}))}{\det(1 - t\rho(g))} \mathrm{d}\mu(g).$$

Here μ is the unitary Haar measure on the (compact) group, which means that $\int_G \mathrm{d}\mu(g) = 1$. For those readers who are not familiar with integration on groups, we give the details for our example, so at least the computation can be verified. In our example,

$$\rho(g) = \begin{bmatrix} \cos\theta & \sin\theta \\ -\sin\theta & \cos\theta \end{bmatrix}.$$

Thus $\operatorname{tr}(\rho(g^{-1})) = 2\cos\theta$, $\det(1 - t\rho(g)) = 1 - 2t\cos\theta + t^2$, and $\mathrm{d}\mu(g) = \frac{\mathrm{d}\theta}{2\pi}$. The result is

$$P(t) = \frac{2t}{1 - t^2}.$$

The term $2t$ stands for two linear vector fields $x_1\frac{\partial}{\partial x_2} - x_2\frac{\partial}{\partial x_1}$, $x_1\frac{\partial}{\partial x_1} + x_2\frac{\partial}{\partial x_2}$, and the term t^2 for a quadratic invariant $x_1^2 + x_2^2$, multiplying the linear vector fields. Looking back at the formula for the general normal form, we see the exact correspondence. This shows that the normal form is complete. Strictly speaking, we should also prove that there is no double counting, that is, terms that are linearly dependent in the general formula, but in this case that is easy to see. For more complicated problems the computation of the Molien integral can be quite intricate. If it cannot be computed, at least it can be computed degree by degree, expanding in t. This will give one control over the completeness of expressions up to any desired degree. An example that can still be computed is the Hamiltonian $1 : 1 : \cdots : 1$-resonance (n degrees of freedom). Its generating function is

$$P(t) = \frac{\sum_{k=0}^{n-1} \binom{n-1}{k}^2 t^{2k}}{(1 - t^2)^{2n-1}}.$$

This was the semisimple case. Can we do something similar in the nilpotent case? In principle we can, using SU(2) as the relevant group, but the computations tend to be rather complicated. In practice, it is easier to play the following guessing game. We are trying to find all terms in $\ker \mathbf{N}_+$. Every such term gives rise to an irreducible $\mathfrak{sl}_2$ representation. If m is the $\mathbf{H}$-eigenvalue of the term, then the dimension of the representation is $m + 1$. So we can characterize a term by its degree d and eigenvalue m by representing it with a term $t^d u^m$. To see in a uniform way how many terms of degree d this produces under the action of $\mathbf{N}_-$, we simply multiply by u, differentiate with respect to u, and put $u = 1$ to obtain the expected $(d+1)t^d$. If we look at our simple nilpotent example, we may guess that its normal form is of the form

$$F(x_2)\frac{\partial}{\partial x_1} + G(x_2)\left(x_1\frac{\partial}{\partial x_1} + x_2\frac{\partial}{\partial x_2}\right).$$

This has the generating function

$$\mathbb{P}^2(t,u) = \frac{t+u}{1-ut}.$$

If we now compute $\frac{\partial u Q(t,u)}{\partial u}|_{u=1}$ we obtain

$$\mathbb{P}^2(t) = \frac{2}{(1-t)^2},$$

the generating function of polynomial vector fields on $\mathbb{R}^2$. This implies that we generate everything there is to generate, which means that (unless there is linear dependency in our candidate normal form description, but this we can easily see not to be the case here) the result is correct.

Remark 11.6.2. The reader may wonder why we stress these computational details, since after all we are just solving a linear problem. However, the published (and unpublished!) literature contains errors that could easily have been caught by these simple checks. ♡

Definition 11.6.3. *A generating function $P(t,u)$ is called* (m,n)-**perfect** *if* $\frac{\partial u Q(t,u)}{\partial u}|_{u=1} = \frac{m}{(1-t)^n}$.

Remark 11.6.4. This test was found by Cushman and Sanders [68] and seems not to have been known to Sylvester, who employed other tests. It has been applied to the generating functions for the covariants that can be found in Sylvester's work [254], see also [104], and all passed the test. Regrettably, the results of Sylvester are not of direct use to us, since one cannot read off the Stanley decomposition. ♡

In the nonsemisimple case, we again play the guessing game, but now the outcome will be the generating function of all vector fields equivariant with respect to the semisimple action, which can be computed as a Molien integral.

12

Nilpotent (Classical) Normal Form

12.1 Introduction

In this chapter, we compute the normal form with respect to a nilpotent linear part for a number of cases. We employ different methods with a varying degree of sophistication. One can consider this part of normal form theory as belonging to invariant theory. Of course, this is also true for the semisimple case, but there we can usually solve the problem without any knowledge of invariant theory beyond the obvious. A good source for the background of some the mathematics we will be using in this chapter is [220].

12.2 Classical Invariant Theory

In classical invariant theory (see [129, 214] for very readable introductions to the field) one poses the following problem. Consider the so-called **ground-form**

$$\hat{\mathbf{x}}(X,Y) = \sum_{i=0}^{n} \binom{n}{i} \mathbf{x}_i X^{n-i} Y^i.$$

On this form we have the natural action of the group SL_2 acting on X, Y. After the linear transformation $g : (X,Y) \mapsto (\tilde{X}, \tilde{Y})$, $g \in \mathrm{SL}_2$, one obtains

$$\hat{\mathbf{x}}(\tilde{X},\tilde{Y}) = \sum_{i=0}^{n} \binom{n}{i} \tilde{\mathbf{x}}_i \tilde{X}^{n-i} \tilde{Y}^i.$$

We see that there is now an induced action of SL_2, given by $g : (\mathbf{x}_0, \ldots, \mathbf{x}_n) \mapsto (\tilde{\mathbf{x}}_0, \ldots, \tilde{\mathbf{x}}_n)$. If we now act at the same time with g on X, Y and with g^{-1} on $(\mathbf{x}_0, \cdots, \mathbf{x}_n)$, then $\hat{\mathbf{x}}$ is carried to

$$\hat{\mathbf{x}}(\tilde{X},\tilde{Y}) = \sum_{i=0}^{n} \binom{n}{i} \mathbf{x}_i \tilde{X}^{n-i} \tilde{Y}^i.$$

In other words, as a form $\hat{\mathbf{x}}$ remains invariant. The basic problem of classical invariant theory is now to classify all forms

$$\hat{\mathbf{g}}(X,Y) = \sum_{i=0}^{m} \binom{m}{i} \mathbf{g}_i(\mathbf{x}_0,\ldots,\mathbf{x}_n) X^{m-i} Y^i$$

such that the polynomial $\hat{\mathbf{g}}$ is invariant under the simultaneous transformation $g : (X,Y) \mapsto (\tilde{X},\tilde{Y})$ and $g^{-1} : (\mathbf{x}_0,\ldots,\mathbf{x}_n) \mapsto (\tilde{\mathbf{x}}_0,\ldots,\tilde{\mathbf{x}}_n)$. Such an invariant is called a **covariant** of degree m, and covariants of degree 0 are called **invariants**. The leading term $\mathbf{g}_0$ of a covariant is called a **seminvariant**. These correspond to elements in $\ker \mathbf{N}_+$ in our formulation in this chapter. The motivating example is well known outside invariant theory. Take

$$\hat{\mathbf{x}}(X,Y) = \mathbf{x}_0 X^2 + 2\mathbf{x}_1 XY + \mathbf{x}_2 Y^2.$$

This form has as an invariant $\hat{\mathbf{g}} = \mathbf{x}_0\mathbf{x}_2 - \mathbf{x}_1^2$. The reader may want to check, using the definitions given in Section 12.3, that $\hat{\mathbf{g}} = (\hat{\mathbf{x}},\hat{\mathbf{x}})^{(2)}$. The art in invariant theory was to find new covariants from old ones using a process called transvection (Überschiebung). The fundamental problem of classical invariant theory was to show that every covariant could be expressed as a polynomial of a finite number of transvectants, starting with the groundform. This was proved by Gordan, and the procedure was later much simplified by Hilbert, laying the foundations for commutative algebra and algebraic geometry. The computations of the transvectants had to be done by hand at the time, and this was a boring and therefore error-prone process. On the positive side it seems to have motivated Emmy Noether, who started her career under the supervision of Gordan, in her abstract formulation of algebra. Good introductions to classical invariant theory are [129] and [214].

12.3 Transvectants

Notation 12.3.1 *Let $\mathbb{V}$ and $\mathbb{W}$ be $\mathcal{I}$- and $\mathfrak{sl}_2$-modules, see Definition 11.2.1. Let $\mathcal{I}$ be a ring on which $\mathfrak{sl}_2$ acts trivially. To start with, this will be $\mathbb{R}$ or $\mathbb{C}$, but if α is an invariant in $\mathbb{V} \otimes_{\mathcal{I}} \mathbb{W}$, then we can change $\mathcal{I}$ to $\mathcal{I}[[\alpha]]$. The notation, $\otimes_{\mathcal{I}}$ is used to remind the reader that invariants can move through the tensorproduct. If $\mathbf{f}$ is an $\mathbf{H}$-eigenvector, we denote its eigenvalue by $w_{\mathbf{f}}$, the* **weight** *of $\mathbf{f}$.*

Definition 12.3.2. *Suppose $\mathbf{f} \in \mathbb{V}$ and $\mathbf{g} \in \mathbb{W}$ are $\mathbf{H}$-eigenvectors in* $\ker \mathbf{N}_+$, *with weight $w_{\mathbf{f}}$, $w_{\mathbf{g}}$, respectively. Let for any $\mathbf{f} \in \ker \mathbf{N}_+$, $\mathbf{f}_i = \mathbf{N}_-^i f$ for $i = 0,\ldots,w_{\mathbf{f}}$. For any $n \leq \min(w_{\mathbf{f}}, w_{\mathbf{g}})$, define the nth* **transvectant** [214, Chapter 5] *of $\mathbf{f}$ and $\mathbf{g}$, $\tau^n(\mathbf{f} \otimes_{\mathcal{I}} \mathbf{g}) = (\mathbf{f},\mathbf{g})^{(n)}$, by*

$$\widehat{(\mathbf{f},\mathbf{g})^{(n)}}(X,Y) = \sum_{i=0}^{n} (-1)^i \binom{n}{i} \frac{\partial^n \hat{\mathbf{f}}}{\partial X^{n-i} \partial Y^i} \otimes_{\mathcal{I}} \frac{\partial^n \hat{\mathbf{g}}}{\partial X^i \partial Y^{n-i}},$$

where

$$\hat{\mathbf{f}}(X,Y) = \sum_{j=0}^{w_{\mathbf{f}}} \frac{1}{j!}\mathbf{f}_j X^{w_{\mathbf{f}}-j}Y^j.$$

So $\tau^n : \mathbb{V} \otimes_{\mathcal{I}} \mathbb{W} \to \mathbb{V} \otimes_{\mathcal{I}} \mathbb{W}$.

Remark 12.3.3. Although the tensor definition seems to be the most natural way to define the transvectant in the light of the Clebsch–Gordan decomposition, see Remark 12.3.8, the reader can ignore the tensor symbol in the concrete calculations. We usually have a situation in which $\mathbb{W}$ is a $\mathbb{V}$-algebra, and then we contract the tensor product $\mathbf{f} \otimes_{\mathcal{I}} \mathbf{g}$ to $\mathbf{f} \cdot \mathbf{g}$, where $\cdot$ denotes the action (usually multiplication) of $\mathbb{V}$ on $\mathbb{W}$. ♡

We see that we can recover $(\mathbf{f},\mathbf{g})^{(n)}$ by taking

$$(\mathbf{f},\mathbf{g})^{(n)} = \widehat{(\mathbf{f},\mathbf{g})^{(n)}}(1,0).$$

Consider now $\hat{\mathbf{f}}$, with $\mathbf{f}$ an $\mathbf{H}$-eigenvector. Then

$$\begin{aligned}\frac{\partial^n \hat{\mathbf{f}}}{\partial X^{n-i}\partial Y^i} &= \frac{\partial^n}{\partial X^{n-i}\partial Y^i}\sum_{j=0}^{w_{\mathbf{f}}} \frac{1}{j!}\mathbf{f}_j X^{w_{\mathbf{f}}-j}Y^j \\ &= \frac{\partial^{n-i}}{\partial X^{n-i}}\sum_{j=i}^{w_{\mathbf{f}}} \frac{1}{(j-i)!}\mathbf{f}_j X^{w_{\mathbf{f}}-j}Y^{j-i} \\ &= \sum_{j=i}^{w_{\mathbf{f}}-(n-i)} \frac{(n-i)!}{(j-i)!}\binom{w_{\mathbf{f}}-j}{n-i}\mathbf{f}_j X^{w_{\mathbf{f}}-j-n+i}Y^{j-i},\end{aligned}$$

which corresponds (taking $X=1, Y=0$) to

$$(n-i)!\binom{w_{\mathbf{f}}-i}{n-i}\mathbf{f}_i.$$

Taking as the other argument for the transvectant construction $\mathbf{g}_j$, we see that

$$(\mathbf{f},\mathbf{g})^{(n)} = n! \sum_{i+j=n} (-1)^i \binom{w_{\mathbf{f}}-i}{n-i}\binom{w_{\mathbf{g}}-j}{n-j}\mathbf{f}_i \otimes_{\mathcal{I}} \mathbf{g}_j.$$

We have now derived an expression in terms of $\mathbf{f}$ and $\mathbf{g}$ in $\ker \mathbf{N}_+$.

Definition 12.3.4. *Suppose* $\mathbf{f}$ *and* $\mathbf{g}$ *are* $\mathbf{H}$*-eigenvectors, with weights* $w_{\mathbf{f}}$*,* $w_{\mathbf{g}}$*, respectively. Define the* n*th* **transvectant** [214, Chapter 5] *of* $\mathbf{f}$ *and* $\mathbf{g}$ *by*

$$(\mathbf{f},\mathbf{g})^{(n)} = n! \sum_{i+j=n} (-1)^i \binom{w_{\mathbf{f}}-i}{n-i}\mathbf{f}_i \otimes_{\mathcal{I}} \binom{w_{\mathbf{g}}-j}{n-j}\mathbf{g}_j.$$

Specific lower-order cases are

$$(\mathbf{f},\mathbf{g})^{(1)} = w_{\mathbf{f}}\mathbf{f}_0 \otimes_{\mathcal{I}} \mathbf{g}_1 - w_{\mathbf{g}}\mathbf{f}_1 \otimes_{\mathcal{I}} \mathbf{g}_0,$$
$$(\mathbf{f},\mathbf{g})^{(2)} = w_{\mathbf{f}}(w_{\mathbf{f}}-1)\mathbf{f}_0 \otimes_{\mathcal{I}} \mathbf{g}_2 - 2(w_{\mathbf{f}}-1)(w_{\mathbf{g}}-1)\mathbf{f}_1 \otimes_{\mathcal{I}} \mathbf{g}_1 + w_{\mathbf{g}}(w_{\mathbf{g}}-1)\mathbf{f}_2 \otimes_{\mathcal{I}} \mathbf{g}_0.$$

Example 12.3.5. Let $\tilde{\mathbf{N}}_- = X\frac{\partial}{\partial Y}$, $\tilde{\mathbf{N}}_+ = Y\frac{\partial}{\partial X}$, and $\tilde{\mathbf{H}} = X\frac{\partial}{\partial X} - Y\frac{\partial}{\partial Y}$. This gives a representation of $\mathfrak{sl}_2$ on the space of $\hat{\mathbf{f}}(X,Y)$ with the condition that $w_{\mathbf{f}} \in \mathbb{N}$, that is, the $\hat{\mathbf{f}}(X,Y)$ are polynomial in X and Y. Check that for such $\hat{\mathbf{f}}$ one has $\mathbf{N}_\pm\hat{f} = \tilde{\mathbf{N}}_\pm\hat{\mathbf{f}}$. One can associate $\hat{\mathbf{f}}$ with an irreducible representation of $\mathfrak{sl}_2$ of dimension $w_{\mathbf{f}}+1$. ◊

Lemma 12.3.6. *The relations*

$$[\mathbf{H},\mathbf{N}_\pm] = \pm 2\mathbf{N}_\pm, \quad [\mathbf{N}_+,\mathbf{N}_-] = \mathbf{H}$$

imply

$$[\mathbf{H},\mathbf{N}_-^k] = -2k\mathbf{N}_-^k$$

and

$$[\mathbf{N}_+,\mathbf{N}_-^k] = k\mathbf{N}_-^{k-1}(\mathbf{H}-(k-1)).$$

Proof For $k=1$ the statements follows from the above relations. We then use induction to show that

$$\mathbf{H}\mathbf{N}_-^{k+1} = \mathbf{N}_-^k(\mathbf{H}-2k)\mathbf{N}_- = \mathbf{N}_-^{k+1}(\mathbf{H}-2(k+1))$$

and

$$\begin{aligned}\mathbf{N}_+\mathbf{N}_-^{k+1} &= \mathbf{N}_-^k\mathbf{N}_+\mathbf{N}_- + \mathbf{N}_-^{k-1}k(\mathbf{H}-(k-1))\mathbf{N}_-\\ &= \mathbf{N}_-^k(\mathbf{N}_-\mathbf{N}_+ + \mathbf{H}) + k\mathbf{N}_-^{k-1}(\mathbf{N}_-\mathbf{H}-2\mathbf{N}_-) - k(k-1)\mathbf{N}_-^k\\ &= \mathbf{N}_-^{k+1}\mathbf{N}_+ + (k+1)\mathbf{N}_-^k(\mathbf{H}-k).\end{aligned}$$

In particular, for $\mathbf{f} \in \ker \mathbf{N}_+$, $\mathbf{N}_+\mathbf{f}_i = i(w_{\mathbf{f}}-(i-1))\mathbf{f}_{i-1}$. □

Theorem 12.3.7. *If* $\mathbf{f},\mathbf{g} \in \ker \mathbf{N}_+$ *are* $\mathbf{H}$*-eigenvectors, then* $(\mathbf{f},\mathbf{g})^{(n)} \in \ker \mathbf{N}_+$ *is an* $\mathbf{H}$*-eigenvector with* $w_{(\mathbf{f},\mathbf{g})^{(n)}} = w_{\mathbf{f}} + w_{\mathbf{g}} - 2n$.

Proof The proof that the transvectants is in $\ker \mathbf{N}_+$ is a straightforward computation. The eigenvalue computation is similar but easier and left for the reader:

$$\begin{aligned}\frac{1}{n!}\mathbf{N}_+(\mathbf{f},\mathbf{g})^{(n)} &= \sum_{i+j=n}(-1)^i\binom{w_{\mathbf{f}}-i}{n-i}\binom{w_{\mathbf{g}}-j}{n-j}\mathbf{N}_+\mathbf{f}_i \otimes_{\mathcal{I}} \mathbf{g}_j\\ &\quad+\sum_{i+j=n}(-1)^i\binom{w_{\mathbf{f}}-i}{n-i}\binom{w_{\mathbf{g}}-j}{n-j}\mathbf{f}_i \otimes_{\mathcal{I}} \mathbf{N}_+\mathbf{g}_j\\ &= \sum_{i+j=n}(-1)^i\binom{w_{\mathbf{f}}-i}{n-i}\binom{w_{\mathbf{g}}-j}{n-j}i(w_{\mathbf{f}}-(i-1))\mathbf{f}_{i-1} \otimes_{\mathcal{I}} \mathbf{g}_j\\ &\quad+\sum_{i+j=n}(-1)^i\binom{w_{\mathbf{f}}-i}{n-i}\binom{w_{\mathbf{g}}-j}{n-j}\mathbf{f}_i \otimes_{\mathcal{I}} j(w_{\mathbf{g}}-(j-1))\mathbf{g}_{j-1}\end{aligned}$$

$$
\begin{aligned}
&= -\sum_{i+j=n-1} (-1)^i \binom{w_\mathfrak{f}-(i+1)}{n-(i+1)} \binom{w_\mathfrak{g}-j}{n-j} (i+1)(w_\mathfrak{f}-i)\mathbf{f}_i \otimes_\mathcal{I} \mathbf{g}_j \\
&+ \sum_{i+j=n-1} (-1)^i \binom{w_\mathfrak{f}-i}{n-i} \binom{w_\mathfrak{g}-(j+1)}{n-(j+1)} \mathbf{f}_i \otimes_\mathcal{I} (j+1)(w_\mathfrak{g}-j)\mathbf{g}_j \\
&= -\sum_{i+j=n-1} (-1)^i \binom{w_\mathfrak{f}-i}{n-i} \binom{w_\mathfrak{g}-j}{n-j} (n-i)(i+1)\mathbf{f}_i \otimes_\mathcal{I} \mathbf{g}_j \\
&+ \sum_{i+j=n-1} (-1)^i \binom{w_\mathfrak{f}-i}{n-i} \binom{w_\mathfrak{g}-j}{n-j} \mathbf{f}_i \otimes_\mathcal{I} (n-j)(j+1)\mathbf{g}_j \\
&= 0.
\end{aligned}
$$

Here we have used that by definition, $\mathbf{N}_+(\mathbf{f}\otimes_\mathcal{I}\mathbf{g}) = (\mathbf{N}_+\mathbf{f})\otimes_\mathcal{I}\mathbf{g}+\mathbf{f}\otimes_\mathcal{I}\mathbf{N}_+\mathbf{g}$. □

Remark 12.3.8. The transvectant gives the explicit realization of the classical Clebsch–Gordan decomposition

$$
V_{w_\mathfrak{f}+1} \otimes_\mathcal{I} V_{w_\mathfrak{g}+1} = \bigoplus_{i=0}^{\min(w_\mathfrak{f},w_\mathfrak{g})} V_{w_\mathfrak{f}+w_\mathfrak{g}-2i+1}
$$

by identifying the irreducible representation $V_{w_\mathfrak{f}+1}$ with its leading term $\mathbf{f}$, and similarly for $\mathbf{g}$, and generating $V_{w_\mathfrak{f}+w_\mathfrak{g}-2i+1}$ by $(\mathbf{f},\mathbf{g})^{(i)}$. The proof is by simply counting the dimensions using the method described in Section 11.6: consider a $w_\mathfrak{f}+1$ by $w_\mathfrak{g}+1$ rectangle of dots, and remove the upper and rightmost layers. These layers contain $w_\mathfrak{f}+1+w_\mathfrak{g}+1-1$ elements, so this corresponds to the $i=0$ term in the direct sum. Repeat this. Each layer will now be shorter by 2. Stop when one of the sides is zero (because there are no dots left).

The symbolic representation of this formula is given by assigning to each irreducible of weight w the function u^w. The Clebsch–Gordan formula then becomes

$$
u^{w_\mathfrak{f}} \otimes_\mathcal{I} u^{w_\mathfrak{g}} = \sum_{i=0}^{\min(w_\mathfrak{f},w_\mathfrak{g})} u^{w_\mathfrak{f}+w_\mathfrak{g}-2i}. \tag{12.3.1}
$$

♡

Let $\mathcal{S}$ denote the ring of seminvariants. In the sequel we will be taking transvectants of expressions $\mathfrak{a},\mathfrak{b}\in\mathcal{S}$ and $\mathfrak{u}$, where $\mathfrak{u}$ a vector in $\ker \mathbf{N}_+$.

Lemma 12.3.9. *Let* $\mathfrak{a},\mathfrak{b}$ *and* $\mathfrak{u}\in\ker\mathbf{N}_+$ *have* $\mathbf{H}$*-eigenvalues* $w_\mathfrak{a}, w_\mathfrak{b}$ *and* $w_\mathfrak{u}$*, respectively. Then*

$$
(\mathfrak{a}^n,\mathfrak{u})^{(1)} = n\mathfrak{a}^{n-1}(\mathfrak{a},\mathfrak{u})^{(1)},
$$

and

$$
(w_\mathfrak{a}-1)^2(\mathfrak{a}^2,\mathfrak{u})^{(2)} = 2(2w_\mathfrak{a}-1)(w_\mathfrak{a}-1)\mathfrak{a}(\mathfrak{a},\mathfrak{u})^{(2)} - w_\mathfrak{u}(w_\mathfrak{u}-1)(\mathfrak{a},\mathfrak{a})^{(2)}\mathfrak{u}.
$$

Proof The proof is by straightforward application of the transvectant definition.

$$\begin{aligned}(\mathfrak{a}^n,\mathfrak{u})^{(1)} - n\mathfrak{a}^{n-1}(\mathfrak{a},\mathfrak{u})^{(1)} &= nw_\mathfrak{a}\mathfrak{a}^n\mathfrak{u}_1 - w_\mathfrak{u}n\mathfrak{a}^{n-1}\mathfrak{a}_1\mathfrak{u} - n\mathfrak{a}^{n-1}w_\mathfrak{a}\mathfrak{a}\mathfrak{u}_1\\ &\quad +n\mathfrak{a}^{n-1}w_u\mathfrak{a}_1\mathfrak{u}\\ &= 0,\end{aligned}$$

$$\begin{aligned}&(w_\mathfrak{a}-1)^2(\mathfrak{a}^2,\mathfrak{u})^{(2)} - 2(2w_\mathfrak{a}-1)(w_\mathfrak{a}-1)\mathfrak{a}(\mathfrak{a},\mathfrak{u})^{(2)} + w_\mathfrak{u}(w_\mathfrak{u}-1)(\mathfrak{a},\mathfrak{a})^{(2)}\mathfrak{u}\\ &= 2(w_\mathfrak{a}-1)^2w_\mathfrak{a}(2w_\mathfrak{a}-1)\mathfrak{a}^2\mathfrak{u}_2 - 4(w_\mathfrak{a}-1)^2(2w_\mathfrak{a}-1)(w_\mathfrak{u}-1)\mathfrak{a}\mathfrak{a}_1\mathfrak{u}_1\\ &\quad +(w_\mathfrak{a}-1)^2w_\mathfrak{u}(w_\mathfrak{u}-1)2(\mathfrak{a}\mathfrak{a}_2+\mathfrak{a}_1^2)\mathfrak{u} - 2(2w_\mathfrak{a}-1)(w_\mathfrak{a}-1)^2w_\mathfrak{a}\mathfrak{a}^2\mathfrak{u}_2\\ &\quad +4(2w_\mathfrak{a}-1)(w_\mathfrak{a}-1)^2(w_\mathfrak{u}-1))\mathfrak{a}\mathfrak{a}_1\mathfrak{u}_1 - 2(2w_\mathfrak{a}-1)(w_\mathfrak{a}-1)w_\mathfrak{u}(w_\mathfrak{u}-1)\mathfrak{a}\mathfrak{a}_2\mathfrak{u}\\ &\quad +2w_\mathfrak{u}(w_\mathfrak{u}-1)(w_\mathfrak{a}-1)(w_\mathfrak{a}\mathfrak{a}\mathfrak{a}_2 - (w_\mathfrak{a}-1)\mathfrak{a}_1^2)\mathfrak{u}\\ &= 2(w_\mathfrak{a}-1)w_\mathfrak{u}(w_\mathfrak{u}-1)\left((w_\mathfrak{a}-1)-(2w_\mathfrak{a}-1)+w_\mathfrak{a}\right)\mathfrak{a}\mathfrak{a}_2\mathfrak{u}\\ &\quad +2(w_\mathfrak{a}-1)^2w_\mathfrak{u}(w_\mathfrak{u}-1)\mathfrak{a}_1^2\mathfrak{u} - 2w_\mathfrak{u}(w_\mathfrak{u}-1)(w_\mathfrak{a}-1)^2\mathfrak{a}_1^2\mathfrak{u}\\ &= 0.\end{aligned}$$

This concludes the proof. □

Lemma 12.3.10. *With the notation as in Lemma 12.3.9, one has*

$$\begin{aligned}&\binom{w_\mathfrak{b}}{n}\mathfrak{b}(\mathfrak{a},\mathfrak{u})^{(n)} - \binom{w_\mathfrak{a}}{n}\mathfrak{a}(\mathfrak{b},\mathfrak{u})^{(n)}\\ &\quad = (w_\mathfrak{u}-n+1)\frac{\binom{w_\mathfrak{a}-1}{n-1}\binom{w_\mathfrak{b}-1}{n-1}}{\binom{w_\mathfrak{a}+w_\mathfrak{b}-2}{n-1}}((\mathfrak{a},\mathfrak{b})^{(1)},\mathfrak{u})^{(n-1)} + \cdots,\end{aligned}$$

where the $\cdots$ stands for lower-order transvectants with $\mathfrak{u}$.

Corollary 12.3.11.

$$w_\mathfrak{b}\mathfrak{b}(\mathfrak{a},\mathfrak{u})^{(1)} - w_\mathfrak{a}\mathfrak{a}(\mathfrak{b},\mathfrak{u})^{(1)} = w_\mathfrak{u}(\mathfrak{a},\mathfrak{b})^{(1)}\mathfrak{u}.$$

12.4 A Remark on Generating Functions

Classical invariant theory gives us (if the dimension is not too high) explicit expressions for the polynomials in the kernel of $\mathbf{N}_+$. What we need, however, are polynomial vector fields in ker $\mathbf{N}_+$. Let ν be a partition of the dimension n of our underlying space. Let $\mathbb{R}^{\nu_i}$ be an irreducible subspace. We can consider the polynomial vector fields of degree m as elements in the tensor product

$$\mathbb{P}_m^{\nu_1,\ldots,\overline{\nu}_i,\ldots,\nu_n} = P_m^\nu \otimes_\mathcal{I} \mathbb{R}^{\nu_i} = P_m^\nu \otimes_\mathcal{I} P_1^{\nu_i},$$

and we can obtain the elements in ker $\mathbf{N}_+$ by first finding the elements in ker $\mathbf{N}_+$ in P_m^ν and $P_1^{\nu_i}$ and then applying the transvection process. This implies that once one has the generating function for the invariants, computing the generating function of the equivariants is surprisingly easy.

Remark 12.4.1. From the point of view of representation theory, finding covariants amounts to finding $\mathfrak{sl}_2$-invariant 0-forms, where the equivariants are to be identified with $\mathfrak{sl}_2$-invariant 1-forms.

We do this by example. Let $P^2(t,u) = 1/(1-ut)$ be the generating function for $\mathbb{R}^2$. Then let $\bar{P}^2(t,u) = P^2(t,u) - T_0^0 P^2(t,u) = 1/(1-ut) - 1 = ut/(1-ut)$, where $T_0^0 P(t,u)$ stands for the Taylor expansion of $P(t,u)$ up to order 0 at $u=0$. If we look at the equivariants we basically take the tensor product of the invariants with $\mathbb{R}^{\nu_i}$. Any irreducible representation of dimension $k \geq \nu_i$ obeys (Clebsch–Gordan (12.3.1)) $V_k \otimes_{\mathcal{I}} \mathbb{R}^{\nu_i} = \oplus_{s=0}^{\nu_i - 1} V_{k+\nu_i-1-2s}$. In terms of generating functions, this defines a multiplication as follows. We subscript u by the variable it symbolizes, just to keep track of the meaning of our actions:

$$u_\alpha^{k-1} \boxtimes_{\mathcal{I}} u_{\mathfrak{u}}^{\nu_i - 1} = \sum_{s=0}^{\nu_i - 1} u_{(\alpha,\mathfrak{u})^{(s)}}^{k+\nu_i-2-2s}.$$

In other words, $\bar{P}^\nu \boxtimes_{\mathcal{I}} u^{\nu_i - 1} = \bar{P}^\nu \sum_{s=0}^{\nu_i-1} u^{\nu_i-1-2s}$. Here we use the notation $\boxtimes$ to indicate that we are not taking the tensor product of the two spaces in $\ker \mathbf{N}_+$, but the tensor product of the representation spaces induced by these two spaces. In the sequel we use generating functions with the subscripted variables u and t. To emphasize that the generating function is equivalent to a direct sum decomposition, we write $+$ as $\oplus$.

Since the zero transvectant is just multiplication, its effect on the calculation is very simple: it will just be multiplication of the original generating function for the invariants with $u_{\mathfrak{u}}^{\nu_i - 1}$. We therefore need to worry only about the higher-order transvectants. For this reason we introduce a new tensor product, defined by

$$u_\alpha^{k-1} \hat{\boxtimes}_{\mathcal{I}} u_{\mathfrak{u}}^{\nu_i - 1} = \bigoplus_{s=1}^{\nu_i - 1} u_{(\alpha,\mathfrak{u}_i)^{(s)}}^{k+\nu_i-2-2s}.$$

Since $1 \boxtimes_{\mathcal{I}} u = u$ we find that the generating function for the equivariants in the irreducible $\mathbb{R}^2$ case, with $\nu = \nu_i = 2$, is

$$\begin{aligned}
\mathbb{P}^{\overline{2}}(t,u) &= P^2(t,u) u_{\mathfrak{u}}^1 \oplus P^2(t,u) \hat{\boxtimes}_{\mathcal{I}} u_{\mathfrak{u}}^1 \\
&= P^2(t,u) u_{\mathfrak{u}}^1 \oplus 1 \hat{\boxtimes}_{\mathcal{I}} u_{\mathfrak{u}}^1 \oplus u_{\mathfrak{a}}^1 t_{\mathfrak{a}}^1 P^2(t,u) \hat{\boxtimes}_{\mathcal{I}} u_{\mathfrak{u}}^1 \\
&= P^2(t,u) u_{\mathfrak{u}}^1 \oplus u_{(\alpha,\mathfrak{u})^{(1)}}^0 t_{\mathfrak{a}}^1 P^2(t,u) \\
&= P^2(t,u) \left(u_{\mathfrak{u}}^1 \oplus u_{(\alpha,\mathfrak{u})^{(1)}}^0 t_{\mathfrak{a}}^1 \right) \\
&= \frac{u_{\mathfrak{u}}^1 \oplus u_{(\alpha,\mathfrak{u})^{(1)}}^0 t_{\mathfrak{a}}^1}{1 - u_{\mathfrak{a}}^1 t_{\mathfrak{a}}^1} = \frac{u+t}{1-ut}.
\end{aligned}$$

For practical purposes one likes to have a minimum number of terms in the numerator, cf. Definition 12.4.2.

In general, this method will give us a Stanley decomposition (cf. Definition 12.4.2), which may be far from optimal.

Definition 12.4.2. *A* **Stanley decomposition** *of* ker $\mathbf{N}_+$ *is a direct sum decomposition of the form*

$$\ker \mathbf{N}_+ = \bigoplus_\iota \mathbb{R}[[\mathfrak{a}_0, \ldots, \mathfrak{a}_{n_\iota}]] m_\iota,$$

where $\mathfrak{a}_0, \ldots, \mathfrak{a}_{n_\iota}, m_\iota \in \ker \mathbf{N}_+$. *We define the* **Stanley dimension** *of the ring (or module)* $\mathcal{S}$, $\mathrm{Sdim}\, \mathcal{S}$, *as the minimum number of direct summands.*

In the example, a Stanley decomposition would be given by

$$\ker \mathbf{N}_+ | \mathbb{P}^{\overline{2}} = \mathbb{R}[[\mathfrak{a}]]\mathfrak{u} \oplus \mathbb{R}[[\mathfrak{a}]](\mathfrak{a}, \mathfrak{u})^{(1)}$$

corresponding to

$$F(x_2) \begin{bmatrix} 0 \\ 1 \end{bmatrix} + G(x_2) \begin{bmatrix} x_1 \\ x_2 \end{bmatrix}.$$

The following lemma follows from the correctness of the tensoring procedure, but is included to illustrate the power of generating function arguments.

Lemma 12.4.3. *If* $P^\nu(t,u)$ *is* $(1, |\nu|)$*-perfect (see Definition 11.6.3), then* $\mathbb{P}^{\nu_{\overline{i}}}(t,u)$*, as obtained by the tensoring method, is* $(\nu_i, |\nu|)$*-perfect.*

Proof We write

$$P^\nu(t,u) = \sum_{i=0}^{\nu_i - 2} \frac{1}{i!} \frac{\partial^i P^\nu}{\partial u^i}(t,0) u^i + u^{\nu_i - 1} \tilde{P}^\nu(t,u),$$

which defines $\tilde{P}^\nu$. Now

$$\begin{aligned}
\mathbb{P}^{\nu_{\overline{i}}}(t,u) &= \sum_{i=0}^{\nu_i - 2} \frac{1}{i!} \frac{\partial^i P^\nu}{\partial u^i}(t,0) \sum_{j=0}^{i} u^{i+\nu_i-1-2j} + \tilde{P}^\nu(t,u) \sum_{j=0}^{\nu_i - 1} u^{2\nu_i - 2 - 2j} \\
&= \sum_{i=0}^{\nu_i - 2} \frac{1}{i!} \frac{\partial^i P^\nu}{\partial u^i}(t,0) \sum_{j=0}^{i} u^{i+\nu_i-1-2j} \\
&\quad + \left(P^\nu(t,u) - \sum_{i=0}^{\nu_i - 2} \frac{1}{i!} \frac{\partial^i P^\nu}{\partial u^i}(t,0) u^i \right) \sum_{j=0}^{\nu_i - 1} u^{\nu_i - 1 - 2j} \\
&= P^\nu(t,u) \sum_{j=0}^{\nu_i - 1} u^{\nu_i - 1 - 2j} - \sum_{i=0}^{\nu_i - 2} \frac{1}{i!} \frac{\partial^i P^\nu}{\partial u^i}(t,0) \sum_{j=i+1}^{\nu_i - 1} u^{i+\nu_i-1-2j} \\
&= P^\nu(t,u) \sum_{j=0}^{\nu_i - 1} u^{\nu_i - 1 - 2j} - \sum_{i=0}^{\nu_i - 2} \frac{1}{i!} \frac{\partial^i P^\nu}{\partial u^i}(t,0) \sum_{j=0}^{\nu_i - 2 - i} u^{\nu_i - i - 3 - 2j}.
\end{aligned}$$

We now multiply by u, differentiate with respect to u, and put $u = 1$. The first term gives

$$\nu \frac{\partial u P^{\nu}(t,u)}{\partial u}\bigg|_{u=1} + P^{\nu}(t,1)\sum_{j=0}^{\nu_i-1}(\nu_i - 1 - 2j) = \frac{\nu_i}{(1-t)^n},$$

since we know that $P^{\nu}(t,u)$ is $(1,\nu)$-perfect. The second term gives

$$\sum_{i=0}^{\nu_i-2}\frac{1}{i!}\frac{\partial^i P^{\nu}}{\partial u^i}(t,0)\sum_{j=0}^{\nu_i-2-i}(\nu_i - 2 - i - 2j) = 0.$$

This proves the lemma. □

We are now going to find all the vector fields in ker $\mathbf{N}_+$ by transvecting the polynomials in ker $\mathbf{N}_+$ with u. Murdock [202] formulates for the first time a systematic procedure to do this. Here we adapt this procedure so that it can be used together with the Clebsch–Gordan decomposition and the tensoring of generating functions. The generating function is very useful in practice to see the kind of simplification that can be done on the Stanley form that is obtained (so that it will have fewer direct summands).

Definition 12.4.4. *Let $\nu = \nu_1, \ldots, \nu_m$ be a partition of n. Suppose we have a nilpotent $\mathbf{N}_-$ with irreducible blocks of dimension $\nu_1, \ldots, \nu_m$. This will be called the N_ν case. Let $\nu_{\bar{i}} = (\nu_1, \ldots, \overline{\nu}_i, \ldots, \nu_n)$. Then we define the defect $\Delta_S^{(\nu_i,n)}$, $i = 1, \ldots, m$, by*

$$\Delta_S^{\nu_{\bar{i}}} = \operatorname{Sdim}\ker \mathbf{N}_+|\mathbb{P}^{\nu_{\bar{i}}} - \nu_i \operatorname{Sdim}\ker \mathbf{N}_+|P^{\nu}.$$

Conjecture 12.4.5. We conjecture that

$$\Delta_S^{\nu_{\bar{i}}} \geq 0,$$

based on the fact the relations among the invariants induce syzygies among the vector fields.

12.5 The Jacobson–Morozov Lemma

Given $\mathbf{N}_-$, finding the corresponding $\mathbf{N}_+$ and $\mathbf{H}$ is not a completely trivial problem. For reductive Lie algebras the existence of an $\mathfrak{sl}_2$ extending $\mathbf{N}_-$ is guaranteed by the Jacobson–Morozov lemma [151, Section X.2]; but since computing the extension is sufficient proof in concrete cases, we sketch the computation now. A simplified presentation can be found in [203, algorithm 2.7.2, page 64].

Let N be the square matrix of $\mathbf{N}_-$ acting on the coordinates. Take M_1 to be an arbitrary matrix of the same size. Then solve the linear equation

$$2N + [[M_1, N], N] = 0.$$

Then put $H = [M_1, N]$ and remove superfluous elements. Let Y be the general solution of $[[Y, N], N] = 0$. Solve, with M_2 arbitrary, the equation

$$[H, (M_1 - M_2)] = 2(M_1 - M_2)$$

and remove superfluous elements from $[N, M_2]$. Put $M = M_1 - M_2$. The matrices H and M immediately lead to operators $\mathbf{H}$ and $\mathbf{N}_+$.

The solution of this problem is not unique, so we can even make nice choices. MAPLE code, implementing the above algorithm, is given in Algorithms 12.1–12.2.

12.6 A GL_n-Invariant Description of the First Level Normal Forms for $n < 6$

In the following sections we solve the description problem for the first level normal form of equations with nilpotent linear part in $\mathbb{R}^n, n < 6$. In each case we start with a specific nilpotent operator $\mathbf{N}_-$, but we remark here that the whole procedure is equivariant with respect to GL_n conjugation, since the transvectants are GL_n-homomorphisms. This implies that the final description that is given is a general answer that depends only on the irreducible blocks of the nilpotent. We denote each case by an N subscripted with the dimensions of the irreducible blocks. We mention that beyond these results the description problem is solved in the case $N_{2,2,\ldots,2}$ [65, 182].

12.6.1 The N_2 Case

A simple example of the construction of a solution to the description problem using transvectants is the following. Take on $\mathbb{R}^2$ the linear vector field $\mathbf{N}_- = x_1\frac{\partial}{\partial x_2}$, $\mathbf{N}_+ = x_2\frac{\partial}{\partial x_1}$, and $\mathbf{H} = [\mathbf{N}_+, \mathbf{N}_-] = -x_1\frac{\partial}{\partial x_1} + x_2\frac{\partial}{\partial x_2}$. Then let $\mathfrak{a} = \mathfrak{a}_0 = x_2 \in \ker \mathbf{N}_+$ with $w(\mathfrak{a}) = 1$ and $\mathfrak{u} = \mathfrak{u}_0 = \frac{\partial}{\partial x_1} \in \ker \mathbf{N}_+$ with weight $w(\mathfrak{u}) = 1$. Define $\mathfrak{a}_1 = \mathbf{N}_-\mathfrak{a} = x_1, \mathfrak{u}_1 = \mathbf{N}_-\mathfrak{u} = -\frac{\partial}{\partial x_2}$, and we obtain

$$\mathfrak{v} = (\mathfrak{a}, \mathfrak{u})^{(1)} = \mathfrak{a}_1\mathfrak{u}_0 - \mathfrak{a}_0\mathfrak{u}_1 = x_1\frac{\partial}{\partial x_1} + x_2\frac{\partial}{\partial x_2}, \quad w_{\mathfrak{v}} = 0.$$

(In this kind of calculation we can ignore all multiplicative constants, since they play no role in the description problem.) This exhausts all possibilities of taking i-transvectants with $i > 0$. So this indicates that we are done. We conjecture that a vector field in $\ker \mathbf{N}_+$ can always be written as

$$F_0(\mathfrak{a})\mathfrak{u} + F_1(\mathfrak{a})(\mathfrak{a}, \mathfrak{u})^{(1)}.$$

Let us now try to do this more systematically, using the generating function. We start with the basic element in $\ker \mathbf{N}_+|P^2$, $\mathfrak{a}$, and write down the generating function

$$P^2(t, u) = \frac{1}{1 - u_{\mathfrak{a}}t_{\mathfrak{a}}},$$

Algorithm 12.1 MAPLE code: Jacobson–Morozov, part 1

```
#This is the gl_n implementation of Jacobson-Morozov.
#If there is no matrix given, one first has to compute
# ad(n) and form its matrix N on the generators of the Poisson algebra.
# Given is N, computed are H and M, forming sl_2
with (linalg);
N := array([[0,0,0],[aa,0,0],[2,1,0]]);
n := rowdim(N) :
M1 := array(1..n,1..n);
M2 := array(1..n,1..n);
X := evalm(2*N+(M1&*N-N&*M1)&*N-N&*(M1&*N-N&*M1)) :
eqs := {};
vars := {};
for i to n do
  for j to n do
    if X[i,j] ≠ 0 then
      eqs := eqs union {X[i,j]};
      vars := vars union indets(X[i,j]);
    fi;
  od;
od;
ps := solve(eqs, vars) :
for l in ps do
  if op(1,l) = op(2,l) then
    ps := ps minus {l};
  fi;
od;
assign(ps);
H := evalm(M1&*N - N&*M1);
eqs := {};
vars := {};
for i to n do
  for j to n do
    H[i,j] := eval(H[i,j]) :
    ps := solve({H[i,j]}, indets(H[i,j])) :
    for l in ps do
      if op(1,l) = op(2,l) then
        ps := ps minus {l};
      fi;
    od;
    assign(ps) :
  od;
od;
```

Algorithm 12.2 MAPLE code: Jacobson–Morozov, part 2

```
X := evalm(N&*M2 - M2&*N) :
Y := evalm(H&*(M1 - M2) - (M1 - M2)&*H - 2*(M1 - M2)) :
for i to n do
  for j to n do
    X[i,j] := eval(X[i,j]) :
    ps := solve({X[i,j]}, indets(X[i,j])) :
    for l in ps do
      if op(1,l) = op(2,l) then
        ps := ps minus {l};
      fi;
    od;
    assign(ps) :
    X[i,j] := eval(X[i,j]) :
    Y[i,j] := expand(eval(Y[i,j])) :
    ps := solve({Y[i,j]}, indets(Y[i,j])) :
    for l in ps do
      if op(1,l) = op(2,l) then
        ps := ps minus {l};
      fi;
    od;
    assign(ps) :
    Y[i,j] := eval(Y[i,j]) :
  od;
od;
vars := {};
for i to n do
  for j to n do
    M1[i,j] := eval(M1[i,j]) :
    M2[i,j] := eval(M2[i,j]) :
    H[i,j] := eval(H[i,j]) :
    vars := vars union indets(M1[i,j]) union indets(M2[i,j]) union
            indets(H[i,j]) :
  od;
od;
for i to n do
  for j to n do
    if evaln(M2[i,j]) in vars then
      M2[i,j] := 0
    fi;
  od;
od;
for i to n do
  for j to n do
    M1[i,j] := eval(M1[i,j]) :
    M2[i,j] := eval(M2[i,j]) :
    H[i,j] := eval(H[i,j]) :
  od;
od;
M := evalm(M1 - M2) :
```

corresponding to the fact that $\ker \mathbf{N}_+|P^2 = \mathbb{R}[[\mathfrak{a}]]$. As we already saw in Section 12.4, the corresponding generating function for the vector fields is

$$\mathbb{P}^{\overline{2}}(t,u) = \frac{u^1_{\mathfrak{u}} \oplus u^0_{(\mathfrak{a},\mathfrak{u})^{(1)}} t^1_{\mathfrak{a}}}{1 - u^1_{\mathfrak{a}} t^1_{\mathfrak{a}}} = \frac{u+t}{1-ut},$$

corresponding to the fact that $\ker \mathbf{N}_+|\mathbb{P}^{\overline{2}} = \mathbb{R}[[\mathfrak{a}]]\mathfrak{u} \oplus \mathbb{R}[[\mathfrak{a}]](\mathfrak{a},\mathfrak{u})^{(1)}$ (and using the fact that $t^1_{(\mathfrak{a},\mathfrak{u})^{(1)}} = t^1_{\mathfrak{a}}$). We leave it to the reader to show that

$$F_0(\mathfrak{a})\mathfrak{u} + F_1(\mathfrak{a})(\mathfrak{a},\mathfrak{u})^{(1)} = 0$$

implies that $F_0 = F_1 = 0$. We see that $\Delta^{\overline{2}}_S = 0$, in accordance with Conjecture 12.4.5.

12.6.2 The N_3 Case

Let us now illustrate the techniques on a less-trivial example. We take

$$\mathbf{N}_- = x_1 \frac{\partial}{\partial x_2} + 2x_1 \frac{\partial}{\partial x_3} + x_2 \frac{\partial}{\partial x_3}.$$

We then obtain (using the methods described in Section 12.5)

$$\begin{aligned}\mathbf{N}_+ &= 2x_2 \frac{\partial}{\partial x_1} - 4x_2 \frac{\partial}{\partial x_2} - 8x_2 \frac{\partial}{\partial x_3} + 2x_3 \frac{\partial}{\partial x_2} + 4x_3 \frac{\partial}{\partial x_3},\\ \mathbf{H} &= -2x_1 \frac{\partial}{\partial x_1} - 4x_2 \frac{\partial}{\partial x_3} + 2x_3 \frac{\partial}{\partial x_3}.\end{aligned}$$

We see that $\mathfrak{a} = 2x_2 - x_3$ and $\mathfrak{u} = \frac{\partial}{\partial x_1}$ are both in $\ker \mathbf{N}_+$, with $w(\mathfrak{a}) = 2$ and $w(\mathfrak{u}) = 2$. The invariants, that is, $\ker \mathbf{N}_+|\mathbb{R}[[x_1,x_2,x_3]]$, can be found by computing the second transvectant $\mathfrak{b} = (\mathfrak{a},\mathfrak{a})^{(2)}$ of $\mathfrak{a}$ with itself (the first is automatically zero). The result is

$$\mathfrak{b} = (\mathfrak{a},\mathfrak{a})^{(2)} = -2x_1(2x_2 - x_3) - x_2^2.$$

This gives us a generating function

$$P^3(t,u) = \frac{1}{(1-u^2_{\mathfrak{a}} t^1_{\mathfrak{a}})(1-t^2_{\mathfrak{b}})},$$

which is easily shown to be $(1,3)$-perfect. Furthermore, $\mathfrak{a}$ and $\mathfrak{b}$ are clearly algebraically independent (look at the $\mathbf{H}$-eigenvalues). We let $\mathcal{I} = \mathbb{R}[[\mathfrak{b}]]$ and we find that $\ker \mathbf{N}_+|\mathbb{R}[[x_1,x_2,x_3]] = \mathcal{I}[[\mathfrak{a}]]$.

We now start the Clebsch–Gordan calculation, leading to $\mathbb{P}^{\overline{3}}$. First of all,

$$P^3(t,u) = \frac{1}{(1-u^2_{\mathfrak{a}} t^1_{\mathfrak{a}})(1-t^2_{\mathfrak{b}})} = \frac{1}{1-t^2_{\mathfrak{b}}} \oplus \frac{u^2_{\mathfrak{a}} t^1_{\mathfrak{a}}}{(1-u^1_{\mathfrak{a}} t^1_{\mathfrak{a}})(1-t^2_{\mathfrak{b}})}.$$

Tensoring with $u_{\mathfrak{u}}^2$, we obtain

$$\begin{aligned}\mathbb{P}^{\overline{3}}(t,u) &= P^3(t,u)u_{\mathfrak{u}}^2 \oplus \frac{1}{1-t_{\mathfrak{b}}^2}\hat{\boxtimes}_{\mathcal{I}} u_{\mathfrak{u}}^2 \oplus u_{\mathfrak{a}}^2 t_{\mathfrak{a}}^1 P^3(t,u)\hat{\boxtimes}_{\mathcal{I}} u_{\mathfrak{u}}^2 \\ &= P^3(t,u)\left(u_{\mathfrak{u}}^2 \oplus u_{(\mathfrak{a},\mathfrak{u})^{(1)}}^2 t_{\mathfrak{a}}^1 \oplus u_{(\mathfrak{a},\mathfrak{u})^{(2)}}^0 t_{\mathfrak{a}}^1\right) \\ &= \frac{u_{\mathfrak{u}}^2 \oplus u_{(\mathfrak{a},\mathfrak{u})^{(1)}}^2 t_{\mathfrak{a}}^1 \oplus u_{(\mathfrak{a},\mathfrak{u})^{(2)}}^0 t_{\mathfrak{a}}^1}{(1-u_{\mathfrak{a}}^2 t_{\mathfrak{a}}^1)(1-t_{\mathfrak{b}}^2)}.\end{aligned}$$

We compute $(\mathfrak{a},\mathfrak{u})^{(1)}$ and $(\mathfrak{a},\mathfrak{u})^{(2)}$, with weights 2 and 0, respectively. We obtain

$$\begin{aligned}\mathfrak{v} = (\mathfrak{a},\mathfrak{u})^{(1)} &= 2(2x_2 - x_3)\left(\frac{\partial}{\partial x_2} + 2\frac{\partial}{\partial x_3}\right) - 2x_2\frac{\partial}{\partial x_1}, \\ \mathfrak{w} = (\mathfrak{a},\mathfrak{u})^{(2)} &= -x_1\frac{\partial}{\partial x_1} - x_2\frac{\partial}{\partial x_2} - x_3\frac{\partial}{\partial x_3},\end{aligned}$$

so the result is

$$\ker \mathbf{N}_+|\mathbb{R}^3[[x_1,x_2,x_3]] = \mathcal{I}[[\mathfrak{a}]]\mathfrak{u} + \mathcal{I}[[\mathfrak{a}]]\mathfrak{v} + \mathcal{I}[[\mathfrak{a}]]\mathfrak{w}.$$

To be complete, one should now verify algebraic independence, proving

$$\ker \mathbf{N}_+|\mathbb{R}^3[[x_1,x_2,x_3]] = \mathcal{I}[[\mathfrak{a}]]\mathfrak{u} \oplus \mathcal{I}[[\mathfrak{a}]]\mathfrak{v} \oplus \mathcal{I}[[\mathfrak{a}]]\mathfrak{w}.$$

We see that $\Delta_S^{\overline{3}} \leq 0$, in accordance with Conjecture 12.4.5.

12.6.3 The N_4 Case

We now turn to the problem in $\mathbb{R}^4$. We take our linear field in the standard Jordan normal form

$$\mathbf{N}_- = x_1\frac{\partial}{\partial x_2} + x_2\frac{\partial}{\partial x_3} + x_3\frac{\partial}{\partial x_4}.$$

We then obtain

$$\begin{aligned}\mathbf{N}_+ &= 3x_2\frac{\partial}{\partial x_1} + 4x_3\frac{\partial}{\partial x_2} + 3x_4\frac{\partial}{\partial x_3}, \\ \mathbf{H} &= -3x_1\frac{\partial}{\partial x_1} - x_2\frac{\partial}{\partial x_2} + x_3\frac{\partial}{\partial x_3} + 3x_4\frac{\partial}{\partial x_4}.\end{aligned}$$

We see that $\mathfrak{a} = x_4$ and $\mathfrak{u} = \frac{\partial}{\partial x_1} \in \ker \mathbf{N}_+$, with $w(\mathfrak{a}) = 3$ and $w(\mathfrak{u}) = 3$. Observe that this is basically where we can forget about the newly constructed operators. They play a computationally very small role by providing the starting point (the groundform, in classical invariant theory language) from which we can compute everything using the transvection construction. As a theoretical statement this is a result of Gordan, the proof of which was later simplified by Hilbert.

In the sequel we give the result of the transvectant computation modulo integer multiplication, just to keep the integers from growing too much. We compute

$$\mathfrak{b} = (\mathfrak{a}, \mathfrak{a})^{(2)} = 4(3x_2x_4 - 2x_3^2), \quad w(\mathfrak{b}) = 2.$$

The corresponding generating function is

$$\frac{1}{(1 - u_{\mathfrak{a}}^3 t_{\mathfrak{a}}^1)(1 - u_{\mathfrak{b}}^2 t_{\mathfrak{b}}^2)}.$$

We now multiply by u, differentiate with respect to u, and put $u = 1$. The difference with $\frac{1}{(1-t)^4}$ is

$$\frac{4t^3}{(1-t)^2(1-t^2)^2} = 4t^3 + \mathcal{O}(t^4),$$

and so we should start looking for a function of degree three, if possible with weight three. The obvious candidate is

$$\mathfrak{c} = (\mathfrak{a}, \mathfrak{b})^{(1)} = 4(9x_1x_4^2 - 9x_2x_3x_4 + 4x_3^3), \quad w(\mathfrak{c}) = 3.$$

The corresponding generating function is

$$\frac{1}{(1 - u_{\mathfrak{a}}^3 t_{\mathfrak{a}}^1)(1 - u_{\mathfrak{b}}^2 t_{\mathfrak{b}}^2)(1 - u_{\mathfrak{c}}^3 t_{\mathfrak{c}}^3)}.$$

With three functions we might be done. However, we obtain a difference of

$$\frac{(t^2 + 4t + 1)\, t^4}{(1-t^2)^2\,(1-t^3)^2} = t^4 + \mathcal{O}(t^5),$$

which indicates a function of degree 4 and weight 0. Candidates are $(\mathfrak{a}, \mathfrak{c})^{(3)}$ and $(\mathfrak{b}, \mathfrak{b})^{(2)}$. Since there is only one such function, these are either equal or zero. We obtain

$$\mathfrak{d} = (\mathfrak{b}, \mathfrak{b})^{(2)} = 16(18x_1x_2x_3x_4 + 3x_2^2x_3^2 - 6x_3^3x_4 - 8x_1x_3^3 - 9x_1^2x_4^2), \quad w(\mathfrak{d}) = 0.$$

We now have one function too many, so we expect there to be a relation among the four. Looking at the generating function we obtain the difference $7t^6$, which makes us look at terms u^6t^6:

$$\mathfrak{c}^2, \quad \mathfrak{b}^3, \quad \mathfrak{a}^2\mathfrak{d}.$$

Indeed, there is a relation: $2\mathfrak{c}^2 + \mathfrak{b}^3 + 18\mathfrak{a}^2\mathfrak{d} = 0$ (the reader is encouraged to check this!), and then the generating function becomes

$$P^4(t,u) = \frac{1 \oplus u_{\mathfrak{c}}^3 t_{\mathfrak{c}}^3}{(1-u_{\mathfrak{a}}^3 t_{\mathfrak{a}}^1)(1-u_{\mathfrak{b}}^2 t_{\mathfrak{b}}^2)(1-t_{\mathfrak{d}}^4)},$$

and this is indeed $(1,4)$-perfect. So we let $\mathcal{I} = \mathbb{R}[[\mathfrak{d}]]$. This implies that $\ker \mathbf{N}_+$ can be written in the form

$$\ker \mathbf{N}_+ | \mathbb{R}[[x_1,\ldots,x_4]] = \mathcal{I}[[\mathfrak{a},\mathfrak{b}]] \oplus \mathcal{I}[[\mathfrak{a},\mathfrak{b}]]\mathfrak{c}.$$

Thus $\operatorname{Sdim} \ker \mathbf{N}_+ | \mathbb{R}[[x_1,\ldots,x_4]] = 2$. We are now going to find all the vector fields in $\ker \mathbf{N}_+$ by transvection with $\mathfrak{u}$. We write

$$\begin{aligned}
P^4(t,u) &= \frac{1 \oplus u_{\mathfrak{c}}^3 t_{\mathfrak{c}}^3}{(1-u_{\mathfrak{a}}^3 t_{\mathfrak{a}}^1)(1-u_{\mathfrak{b}}^2 t_{\mathfrak{b}}^2)(1-t_{\mathfrak{d}}^4)} \\
&= \frac{1}{(1-u_{\mathfrak{a}}^3 t_{\mathfrak{a}}^1)(1-u_{\mathfrak{b}}^2 t_{\mathfrak{b}}^2)(1-t_{\mathfrak{d}}^4)} \oplus \frac{u_{\mathfrak{c}}^3 t_{\mathfrak{c}}^3}{(1-u_{\mathfrak{a}}^3 t_{\mathfrak{a}}^1)(1-u_{\mathfrak{b}}^2 t_{\mathfrak{b}}^2)(1-t_{\mathfrak{d}}^4)} \\
&= \frac{1}{(1-u_{\mathfrak{b}}^2 t_{\mathfrak{b}}^2)(1-t_{\mathfrak{d}}^4)} \oplus \frac{u_{\mathfrak{a}}^3 t_{\mathfrak{a}}^1 + u_{\mathfrak{c}}^3 t_{\mathfrak{c}}^3}{(1-u_{\mathfrak{a}}^3 t_{\mathfrak{a}}^1)(1-u_{\mathfrak{b}}^2 t_{\mathfrak{b}}^2)(1-t_{\mathfrak{d}}^4)} \\
&= \frac{1+u_{\mathfrak{b}}^2 t_{\mathfrak{b}}^2}{1-t_{\mathfrak{d}}^4} \oplus \frac{u_{\mathfrak{b}}^4 t_{\mathfrak{b}}^4}{(1-u_{\mathfrak{b}}^2 t_{\mathfrak{b}}^2)(1-t_{\mathfrak{d}}^4)} \oplus \frac{u_{\mathfrak{a}}^3 t_{\mathfrak{a}}^1 + u_{\mathfrak{c}}^3 t_{\mathfrak{c}}^3}{(1-u_{\mathfrak{a}}^3 t_{\mathfrak{a}}^1)(1-u_{\mathfrak{b}}^2 t_{\mathfrak{b}}^2)(1-t_{\mathfrak{d}}^4)}.
\end{aligned}$$

This is in the right form to apply the Clebsch–Gordan procedure:

$$\begin{aligned}
\mathbb{P}^{\overline{4}}(t,u) &= P^4(t,u)u_{\mathfrak{u}}^3 + P^4(t,u)\hat{\boxtimes}_{\mathcal{I}} u_{\mathfrak{u}}^3 = P^4(t,u)u_{\mathfrak{u}}^3 \\
&\oplus \frac{(u^3_{(\mathfrak{b},\mathfrak{u})^{(1)}} \oplus u^1_{(\mathfrak{b},\mathfrak{u})^{(2)}})t_{\mathfrak{b}}^2}{(1-t_{\mathfrak{d}}^4)} \oplus \frac{(u^5_{(\mathfrak{b}^2,\mathfrak{u})^{(1)}} \oplus u^3_{(\mathfrak{b}^2,\mathfrak{u})^{(2)}} \oplus u^1_{(\mathfrak{b}^2,\mathfrak{u})^{(3)}})t_{\mathfrak{b}}^4}{(1-u_{\mathfrak{b}}^2 t_{\mathfrak{b}}^2)(1-t_{\mathfrak{d}}^4)} \\
&\oplus \frac{(u^4_{(\mathfrak{a},\mathfrak{u})^{(1)}} \oplus u^2_{(\mathfrak{a},\mathfrak{u})^{(2)}} \oplus u^0_{(\mathfrak{a},\mathfrak{u})^{(3)}})t_{\mathfrak{a}}^1}{(1-u_{\mathfrak{a}}^3 t_{\mathfrak{a}}^1)(1-u_{\mathfrak{b}}^2 t_{\mathfrak{b}}^2)(1-t_{\mathfrak{d}}^4)} \\
&\oplus \frac{(u^4_{(\mathfrak{c},\mathfrak{u})^{(1)}} \oplus u^2_{(\mathfrak{c},\mathfrak{u})^{(2)}} \oplus u^0_{(\mathfrak{c},\mathfrak{u})^{(3)}})t_{\mathfrak{c}}^3}{(1-u_{\mathfrak{a}}^3 t_{\mathfrak{a}}^1)(1-u_{\mathfrak{b}}^2 t_{\mathfrak{b}}^2)(1-t_{\mathfrak{d}}^4)}.
\end{aligned}$$

This corresponds to

$$\begin{aligned}
\ker \mathbf{N}_+ | \mathbb{R}^4[[x_1,\ldots,x_4]] &= \mathcal{I}[[\mathfrak{a},\mathfrak{b}]]\mathfrak{u} \oplus \mathcal{I}[[\mathfrak{a},\mathfrak{b}]]\mathfrak{c}\mathfrak{u} \\
&\oplus \mathcal{I}(\mathfrak{b},\mathfrak{u})^{(1)} \oplus \mathcal{I}(\mathfrak{b},\mathfrak{u})^{(2)} \\
&\oplus \mathcal{I}[[\mathfrak{b}]](\mathfrak{b}^2,\mathfrak{u})^{(1)} \oplus \mathcal{I}[[\mathfrak{b}]](\mathfrak{b}^2,\mathfrak{u})^{(2)} \oplus \mathcal{I}[[\mathfrak{b}]](\mathfrak{b}^2,\mathfrak{u})^{(3)} \\
&\oplus \mathcal{I}[[\mathfrak{a},\mathfrak{b}]](\mathfrak{a},\mathfrak{u})^{(1)} \oplus \mathcal{I}[[\mathfrak{a},\mathfrak{b}]](\mathfrak{a},\mathfrak{u})^{(2)} \oplus \mathcal{I}[[\mathfrak{a},\mathfrak{b}]](\mathfrak{a},\mathfrak{u})^{(3)} \\
&\oplus \mathcal{I}[[\mathfrak{a},\mathfrak{b}]](\mathfrak{c},\mathfrak{u})^{(1)} \oplus \mathcal{I}[[\mathfrak{a},\mathfrak{b}]](\mathfrak{c},\mathfrak{u})^{(2)} \oplus \mathcal{I}[[\mathfrak{a},\mathfrak{b}]](\mathfrak{c},\mathfrak{u})^{(3)}.
\end{aligned}$$

Remark 12.6.1. In this section and the following we carry out a number of simplifications to the decomposition obtained by the tensoring. These simplifications are based on the fact that one can simplify the generating function (without subscripts to u and t). For these simplifications to hold in a non-symbolic way, we need to prove certain relations. Those are the relations mentioned in the text. ♡

To further simplify this expression, we note that it follows from Lemma 12.3.9 that

$$(\mathfrak{b}^2, \mathfrak{u})^{(1)} = 2\mathfrak{b}(\mathfrak{b}, \mathfrak{u})^{(1)},$$

implying that

$$\mathcal{I}[[\mathfrak{b}]](\mathfrak{b}^2, \mathfrak{u})^{(1)} \oplus \mathcal{I}(\mathfrak{b}, \mathfrak{u})^{(1)} = \mathcal{I}[[\mathfrak{b}]]\mathfrak{b}(\mathfrak{b}, \mathfrak{u})^{(1)} \oplus \mathcal{I}(\mathfrak{b}, \mathfrak{u})^{(1)} = \mathcal{I}[[\mathfrak{b}]](\mathfrak{b}, \mathfrak{u})^{(1)},$$

and

$$(\mathfrak{b}^2, \mathfrak{u})^{(2)} = 6\mathfrak{b}(\mathfrak{b}, \mathfrak{u})^{(2)} - 6\mathfrak{d}\mathfrak{u},$$

implying that

$$\begin{aligned} &\mathcal{I}\ [[\mathfrak{b}]](\mathfrak{b}^2, \mathfrak{u})^{(2)} \oplus \mathcal{I}(\mathfrak{b}, \mathfrak{u})^{(2)} \oplus \mathcal{I}[[\mathfrak{a}, \mathfrak{b}]]\mathfrak{u} \\ &= \mathcal{I}[[\mathfrak{b}]]\mathfrak{b}(\mathfrak{b}, \mathfrak{u})^{(2)} \oplus \mathcal{I}(\mathfrak{b}, \mathfrak{u})^{(2)} \oplus \mathcal{I}[[\mathfrak{a}, \mathfrak{b}]]\mathfrak{u} \\ &= \mathcal{I}[[\mathfrak{b}]](\mathfrak{b}, \mathfrak{u})^{(2)} \oplus \mathcal{I}[[\mathfrak{a}, \mathfrak{b}]]\mathfrak{u}. \end{aligned}$$

At this point the reader might well ask how one is to find all these relations. The answer is very simple and similar to the ideas in Lemma 12.3.10. For example, the expression $(\mathfrak{b}^2, \mathfrak{u})^{(2)}$ is equivalent (mod im $\mathbf{N}_-$) to its leading term $\mathfrak{b}^2\mathfrak{u}_2$. The same can be said for $\mathfrak{b}(\mathfrak{b}, \mathfrak{u})^{(2)}$. Using the formula for the transvectants we can compute the constant (6 in this case) between the two. Then $(\mathfrak{b}^2, \mathfrak{u})^{(2)} - 6\mathfrak{b}(\mathfrak{b}, \mathfrak{u})^{(2)}$ has to lie in ker $\mathbf{N}_+$, and only terms $\mathfrak{u}_i, i = 0, 1$, may appear in it. This is an algorithmic way to find all the syzygies. We now have

$$\begin{aligned} &\ker \mathbf{N}_+ \ |\mathbb{R}^4[[x_1, \ldots, x_4]] \\ &= \ \mathcal{I}[[\mathfrak{a}, \mathfrak{b}]]\mathfrak{u} \oplus \mathcal{I}[[\mathfrak{a}, \mathfrak{b}]](\mathfrak{a}, \mathfrak{u})^{(1)} \oplus \mathcal{I}[[\mathfrak{a}, \mathfrak{b}]](\mathfrak{a}, \mathfrak{u})^{(2)} \oplus \mathcal{I}[[\mathfrak{a}, \mathfrak{b}]](\mathfrak{a}, \mathfrak{u})^{(3)} \\ &\quad \oplus \mathcal{I}[[\mathfrak{a}, \mathfrak{b}]]\mathfrak{c}\mathfrak{u} \oplus \mathcal{I}[[\mathfrak{a}, \mathfrak{b}]](\mathfrak{c}, \mathfrak{u})^{(1)} \oplus \mathcal{I}[[\mathfrak{a}, \mathfrak{b}]](\mathfrak{c}, \mathfrak{u})^{(2)} \oplus \mathcal{I}[[\mathfrak{a}, \mathfrak{b}]](\mathfrak{c}, \mathfrak{u})^{(3)} \\ &\quad \oplus \mathcal{I}[[\mathfrak{b}]](\mathfrak{b}, \mathfrak{u})^{(1)} \oplus \mathcal{I}[[\mathfrak{b}]](\mathfrak{b}, \mathfrak{u})^{(2)} \oplus \mathcal{I}[[\mathfrak{b}]](\mathfrak{b}^2, \mathfrak{u})^{(3)}. \end{aligned}$$

This result is equivalent to the decomposition found in [202] and it is of the form Expected plus Correction terms, see the formula for $\delta^{\overline{4}}(t, u)$ in Section 12.6.4. But there is still room for improvement. Proceeding as before we obtain

$$4(\mathfrak{c}, \mathfrak{u})^{(1)} = 3\mathfrak{b}(\mathfrak{a}, \mathfrak{u})^{(2)} - 9\mathfrak{a}(\mathfrak{b}, \mathfrak{u})^{(2)},$$

which implies

$$\begin{aligned} &\mathcal{I}\ [[\mathfrak{a}, \mathfrak{b}]](\mathfrak{c}, \mathfrak{u})^{(1)} \oplus \mathcal{I}[[\mathfrak{a}, \mathfrak{b}]](\mathfrak{a}, \mathfrak{u})^{(2)} \oplus \mathcal{I}[[\mathfrak{b}]](\mathfrak{b}, \mathfrak{u})^{(2)} \\ &= \mathcal{I}[[\mathfrak{a}, \mathfrak{b}]]\mathfrak{a}(\mathfrak{b}, \mathfrak{u})^{(2)} \oplus \mathcal{I}[[\mathfrak{a}, \mathfrak{b}]](\mathfrak{a}, \mathfrak{u})^{(2)} \oplus \mathcal{I}[[\mathfrak{b}]](\mathfrak{b}, \mathfrak{u})^{(2)} \\ &= \mathcal{I}[[\mathfrak{a}, \mathfrak{b}]](\mathfrak{b}, \mathfrak{u})^{(2)} \oplus \mathcal{I}[[\mathfrak{a}, \mathfrak{b}]](\mathfrak{a}, \mathfrak{u})^{(2)}. \end{aligned}$$

One also has, as follows from Corollary 12.3.11,

$$3\mathfrak{c}\mathfrak{u} = 2\mathfrak{b}(\mathfrak{a}, \mathfrak{u})^{(1)} - 3\mathfrak{a}(\mathfrak{b}, \mathfrak{u})^{(1)},$$

implying

$$\begin{aligned}
&\mathcal{I}[[\mathfrak{a}, \mathfrak{b}]]\mathfrak{c}\mathfrak{u} \oplus \mathcal{I}[[\mathfrak{b}]](\mathfrak{b}, \mathfrak{u})^{(1)} \oplus \mathcal{I}[[\mathfrak{a}, \mathfrak{b}]](\mathfrak{a}, \mathfrak{u})^{(1)} \\
&= \mathcal{I}[[\mathfrak{a}, \mathfrak{b}]]\mathfrak{a}(\mathfrak{b}, \mathfrak{u})^{(1)} \oplus \mathcal{I}[[\mathfrak{b}]](\mathfrak{b}, \mathfrak{u})^{(1)} \oplus \mathcal{I}[[\mathfrak{a}, \mathfrak{b}]](\mathfrak{a}, \mathfrak{u})^{(1)} \\
&= \mathcal{I}[[\mathfrak{a}, \mathfrak{b}]](\mathfrak{b}, \mathfrak{u})^{(1)} \oplus \mathcal{I}[[\mathfrak{a}, \mathfrak{b}]](\mathfrak{a}, \mathfrak{u})^{(1)}
\end{aligned}$$

This results in the following decomposition for $\ker \mathsf{N}_+|\mathbb{R}^4[[x_1, \ldots, x_4]]$:

$$\begin{aligned}
&\oplus \mathcal{I}[[\mathfrak{a}, \mathfrak{b}]](\mathfrak{a}, \mathfrak{u})^{(3)} \oplus \mathcal{I}[[\mathfrak{a}, \mathfrak{b}]](\mathfrak{a}, \mathfrak{u})^{(2)} \oplus \mathcal{I}[[\mathfrak{a}, \mathfrak{b}]](\mathfrak{a}, \mathfrak{u})^{(1)} \oplus \mathcal{I}[[\mathfrak{a}, \mathfrak{b}]]\mathfrak{u} \\
&\oplus \mathcal{I}[[\mathfrak{b}]](\mathfrak{b}^2, \mathfrak{u})^{(3)} \oplus \mathcal{I}[[\mathfrak{a}, \mathfrak{b}]](\mathfrak{b}, \mathfrak{u})^{(2)} \oplus \mathcal{I}[[\mathfrak{a}, \mathfrak{b}]](\mathfrak{b}, \mathfrak{u})^{(1)} \\
&\oplus \mathcal{I}[[\mathfrak{a}, \mathfrak{b}]](\mathfrak{c}, \mathfrak{u})^{(3)} \oplus \mathcal{I}[[\mathfrak{a}, \mathfrak{b}]](\mathfrak{c}, \mathfrak{u})^{(2)}.
\end{aligned}$$

The final result is the same as the one obtained in [68]. We see that $\Delta_S^{\overline{4}} \leq 1$, which is not in contradiction to Conjecture 12.4.5.

12.6.4 Intermezzo: How Free?

When one first starts to compute seminvariants and the corresponding equivariant modules, one at first gets the impression that the equivariant vector field can simply be described by n seminvariant functions, where n is the dimension of the vector space. This idea works fine for $n = 2, 3$, but for $n = 4$ it fails. Can we measure the degree to which it fails? We propose the following test here. Suppose that the generating function is $P^n(t,u) + u^{n-1}Q^n(t,u)$. Then we would expect the generating function of the vector fields of dimension n to look like

$$\mathbb{E}^I(t,u) = P^n(t,u)\left(u^{n-1} + \sum_{i=1}^{n-1} u^{2(n-1-i)}t\right) + \sum_{i=0}^{n-1} u^{2(n-1-i)}Q^n(t,u).$$

By subtracting this expected generating function from the real one, we measure the difference between reality and simplicity. Let us call this function $\delta^I(t,u)$, where I indicates the case we are in. The decomposition $P^n(t,u) + u^{n-1}Q^n(t,u)$ is motivated by the cancellation of factors $1 - u^{n-1}t$ in all computed examples.

Observe that the generating functions are independent of the choice of Stanley decomposition, so that δ^I is, apart from the splitting in $P^n(t,u) + u^{n-1}Q^n(t,u)$, an invariant of the block decomposition of the linear part of the equation. On the basis of subsequent results we make the following conjecture.

Conjecture 12.6.2. The coefficients in the Taylor expansion of $\delta^I(t,u)$ are natural numbers, that is, integers ≥ 0.

One has

$$\mathbb{E}^{\overline{2}}(t,u) = \frac{u+t}{1-ut} = \mathbb{P}^{\overline{2}}(t,u),$$

so that $\delta^{\overline{2}} = 0$, as expected. Also

$$\mathbb{E}^{\overline{3}}(t,u) = \frac{u^2 + (1+u^2)t}{(1-u^2t)(1-t^2)} = \mathbb{P}^{\overline{3}}(t,u),$$

so that $\delta^{\overline{3}} = 0$, as expected. The next one is more interesting:

$$\mathbb{E}^{\overline{4}}(t,u) = \frac{u^3 + (1+u^2+u^4)t + (1+u^2+u^4+u^6)t^3}{(1-u^3t)(1-u^2t^2)(1-t^4)},$$

while

$$\mathbb{P}^{\overline{4}}(t,u) = \frac{u^3 + (1+u^2+u^4)t + (u+u^3)t^2 + (1+u^2)t^3}{(1-u^3t)(1-u^2t^2)(1-t^4)} + \frac{ut^4}{(1-u^2t^2)(1-t^4)}.$$

Thus

$$\delta^{\overline{4}}(t,u) = \frac{ut^2(1+u^2+t^2)}{(1-u^2t^2)(1-t^4)}.$$

12.6.5 The $N_{2,2}$ Case

In this section we introduce a method to compute a Stanley decomposition in the reducible case from Stanley decompositions of the components. This gives us *addition formulas* for normal form problems. Another presentation of this method, written in a different style and including all the necessary proofs, may be found in [207]. The question of computing the generating function for the invariants is addressed in [42].

As we have seen in Section 12.6.1, the polynomials in ker $\mathbf{N}_+$ are generated by one element $\mathfrak{a}$ in the case of one irreducible 2-dimensional block, so in the case of two irreducible 2-dimensional blocks we can at least expect to need two generators $\mathfrak{a}$ and $\tilde{\mathfrak{a}}$. Identifying the space of polynomials in two variables of degree n with $\mathbb{R}^{2\otimes n} = \mathbb{R}^2 \otimes_{\mathcal{I}} \cdots \otimes_{\mathcal{I}} \mathbb{R}^2$, we can compute the generating function as follows. In the following computation we drop the tensor product whenever we take a zero transvectant, since the zero transvectant is bilinear over the ring. We identify $\mathbb{R}[[x_1,x_2]]$ and $\mathbb{R}[[\mathfrak{a}]]$ with the generating function $\frac{1}{1-u_{\mathfrak{a}}t_{\mathfrak{a}}}$. If we expand this generating function, the kth power of $u_{\mathfrak{a}}t_{\mathfrak{a}}$ is associated with a k-fold tensor product of $\mathbb{R}^2$ with itself. This way the irreducible $\mathfrak{sl}_2$-representations are associated with $\mathfrak{a}^k$ or $u_{\mathfrak{a}}^k t_{\mathfrak{a}}^k$. We use the fact that $\mathbb{R}[[x_1,x_2,x_3,x_4]] = \mathbb{R}[[x_1,x_2]] \otimes_{\mathcal{I}} \mathbb{R}[[x_3,x_4]]$, that is, it has generating function $\frac{1}{1-u_{\mathfrak{a}}t_{\mathfrak{a}}} \boxtimes_{\mathcal{I}} \frac{1}{1-u_{\tilde{\mathfrak{a}}}t_{\tilde{\mathfrak{a}}}}$. When we encounter the expression

$$\frac{u_{\mathfrak{a}}^p t_{\mathfrak{a}}^\mu}{1-u_{\mathfrak{a}}^p t_{\mathfrak{a}}^\mu} \boxtimes_{\mathcal{I}} \frac{u_{\tilde{\mathfrak{a}}}^q t_{\tilde{\mathfrak{a}}}^\nu}{1-u_{\tilde{\mathfrak{a}}}^q t_{\tilde{\mathfrak{a}}}^\nu}$$

we can use Clebsch–Gordan as follows.

Lemma 12.6.3. *Let $p,q \in \mathbb{N}$, with $\hat{p} = p/\gcd(p,q), \hat{q} = q/\gcd(p,q)$. We denote an irreducible $\mathfrak{sl}_2$-representation of dimension n by V_n. Then, with $r = p\hat{q} = \hat{p}q$,*

$$V_{k+pn} \otimes_{\mathcal{I}} V_{l+qm} = \oplus_{i=0}^{r} V_{k+l+pn+qm-2i} \oplus \left(V_{k+p(n-\hat{q})} \otimes_{\mathcal{I}} V_{l+q(m-\hat{p})}\right).$$

Proof The proof is a direct consequence of the Clebsch–Gordan decomposition:

$$\begin{aligned}
V_{k+pn} \otimes_{\mathcal{I}} V_{l+qm} &= \bigoplus_{i=0}^{\min(k+pn,l+qm)} V_{k+l+pn+qm-2i} \\
&= \bigoplus_{i=0}^{r-1} V_{k+l+pn+qm-2i} \oplus \bigoplus_{i=r}^{\min(k+pn,l+qm)} V_{k+l+pn+qm-2i} \\
&= \bigoplus_{i=0}^{r-1} V_{k+l+pn+qm-2i} \oplus \bigoplus_{i=0}^{\min(k+pn,l+qm)-r} V_{k+l+pn+qm-2i-2r} \\
&= \bigoplus_{i=0}^{r-1} V_{k+l+pn+qm-2i} \oplus \bigoplus_{i=0}^{\min(k+p(n-\hat{q}),l+q(m-\hat{p})} V_{k+l+p(n-\hat{q})+q(m-\hat{p})-2i} \\
&= \bigoplus_{i=0}^{r-1} V_{k+l+pn+qm-2i} \oplus (V_{k+p(n-\hat{q})} \otimes_{\mathcal{I}} V_{l+q(m-\hat{p})}),
\end{aligned}$$

which gives us the desired recursive formula. □

In generating function language this implies (ignoring the t-dependence)

$$u^{k+r} \boxtimes_{\mathcal{I}} u^{l+r} = \sum_{i=0}^{r-1} u^{k+l+2(r-i)} + u^k \boxtimes_{\mathcal{I}} u^l,$$

which we can then index as (with $w_{\mathfrak{a}} = k$, $w_{\mathfrak{b}} = p$, $w_{\mathfrak{c}} = l$ and $w_{\mathfrak{d}} = q$)

$$u^{k+r}_{\mathfrak{a}\mathfrak{b}^{\hat{q}}} \boxtimes_{\mathcal{I}} u^{l+r}_{\mathfrak{c}\mathfrak{d}^{\hat{p}}} = \bigoplus_{i=0}^{r-1} u^{k+l+2(r-i)}_{\mathfrak{a}\mathfrak{c}(\mathfrak{b}^{\hat{q}},\mathfrak{d}^{\hat{p}})^{(i)}} \oplus u^0_{(\mathfrak{b}^{\hat{q}},\mathfrak{d}^{\hat{p}})^{(r)}} (u^k_{\mathfrak{a}} \boxtimes_{\mathcal{I}} u^l_{\mathfrak{c}}).$$

Here we identify the representation starting with $(\mathfrak{a}\mathfrak{b}^n, \mathfrak{c}\mathfrak{d}^m)^{(r)}$ with the one starting with $\mathfrak{a}\mathfrak{c}(\mathfrak{b}^n, \mathfrak{d}^m)^{(r)}$. Another way of writing this is as

$$\begin{aligned}
\mathcal{I}[[\mathfrak{a}, \mathfrak{b}, \ldots]]\mathfrak{b}^{\hat{q}} \boxtimes_{\mathcal{I}} \mathcal{I}[[\mathfrak{c}, \mathfrak{d}, \ldots]]\mathfrak{d}^{\hat{p}} &= \bigoplus_{i=0}^{r-1} \mathcal{I}[[\mathfrak{a}, \mathfrak{b}, \ldots, \mathfrak{c}, \mathfrak{d}, \ldots]](\mathfrak{b}^n, \mathfrak{d}^m)^{(i)} \\
&\oplus (\mathfrak{b}^n, \mathfrak{d}^m)^{(r)} (\mathcal{I}[[\mathfrak{a}, \mathfrak{b}, \ldots]] \boxtimes_{\mathcal{I}} \mathcal{I}[[\mathfrak{c}, \mathfrak{d}, \ldots]]).
\end{aligned}$$

The idea is now that in order to compute the Stanley decomposition of $\mathcal{I}[[\mathfrak{a}, \mathfrak{b}, \ldots]] \boxtimes_{\mathcal{I}} \mathcal{I}[[\mathfrak{c}, \mathfrak{d}, \ldots]]$ we write

$$\mathcal{I}[[\mathfrak{a}, \mathfrak{b}, \ldots]] = \mathcal{I}[[\mathfrak{a}, \ldots]] \oplus \cdots \oplus \mathcal{I}[[\mathfrak{a}, \ldots]]\beta^{n-1} \oplus \mathcal{I}[[\mathfrak{a}, \mathfrak{b}, \ldots]]\beta^n.$$

By this expansion we either simplify the modules in the Stanley decomposition or we can use the relation we just derived so that we obtain a recursive formula. The following computation illustrates this approach in a simple case of $P[[x_1, x_2]] \otimes_{\mathcal{I}} P[[x_3, x_4]]$. We use our result with $p = q = 1$.

The generating function is given by

$$\frac{1}{1-u_{\mathfrak{a}}t_{\mathfrak{a}}} \boxtimes_{\mathcal{I}} \frac{1}{1-u_{\tilde{\mathfrak{a}}}t_{\tilde{\mathfrak{a}}}} = (1 \oplus \frac{u_{\mathfrak{a}}t_{\mathfrak{a}}}{1-u_{\mathfrak{a}}t_{\mathfrak{a}}}) \boxtimes_{\mathcal{I}} (1 \oplus \frac{u_{\tilde{\mathfrak{a}}}t_{\tilde{\mathfrak{a}}}}{1-u_{\tilde{\mathfrak{a}}}t_{\tilde{\mathfrak{a}}}})$$
$$= 1 \oplus \frac{u_{\mathfrak{a}}t_{\mathfrak{a}}}{1-u_{\mathfrak{a}}t_{\mathfrak{a}}} \oplus \frac{u_{\tilde{\mathfrak{a}}}t_{\tilde{\mathfrak{a}}}}{1-u_{\tilde{\mathfrak{a}}}t_{\tilde{\mathfrak{a}}}} \oplus \frac{u_{\mathfrak{a}}u_{\tilde{\mathfrak{a}}}t_{\mathfrak{a}}t_{\tilde{\mathfrak{a}}}}{(1-u_{\mathfrak{a}}t_{\mathfrak{a}})(1-u_{\tilde{\mathfrak{a}}}t_{\tilde{\mathfrak{a}}})} \oplus \frac{t^2_{(\mathfrak{a},\tilde{\mathfrak{a}})^{(1)}}}{1-u_{\mathfrak{a}}t_{\mathfrak{a}}} \boxtimes_{\mathcal{I}} \frac{1}{1-u_{\tilde{\mathfrak{a}}}t_{\tilde{\mathfrak{a}}}}$$
$$= \frac{1}{(1-u_{\mathfrak{a}}t_{\mathfrak{a}})(1-u_{\tilde{\mathfrak{a}}}t_{\tilde{\mathfrak{a}}})} \oplus t^2_{(\mathfrak{a},\tilde{\mathfrak{a}})^{(1)}} \frac{1}{1-u_{\mathfrak{a}}t_{\mathfrak{a}}} \boxtimes_{\mathcal{I}} \frac{1}{1-u_{\tilde{\mathfrak{a}}}t_{\tilde{\mathfrak{a}}}},$$

and we see that the generating function is

$$\frac{1}{1-u_{\mathfrak{a}}t_{\mathfrak{a}}} \boxtimes_{\mathcal{I}} \frac{1}{1-u_{\tilde{\mathfrak{a}}}t_{\tilde{\mathfrak{a}}}} = \frac{1}{(1-t^2_{(\mathfrak{a},\tilde{\mathfrak{a}})^{(1)}})(1-u_{\mathfrak{a}}t_{\mathfrak{a}})(1-u_{\tilde{\mathfrak{a}}}t_{\tilde{\mathfrak{a}}})},$$

and this is seen to be $(1,4)$-perfect. Let $\mathfrak{b} = (\mathfrak{a},\tilde{\mathfrak{a}})^{(1)}$. Thus we can write any function in $\ker \mathbf{N}_+$ as

$$f(\mathfrak{a},\tilde{\mathfrak{a}},\mathfrak{b}).$$

We let as usual $\mathcal{I} = \mathbb{R}[[\mathfrak{b}]]$.

We now give the same computation in a different notation. We compute

$$\mathbb{R}[[\mathfrak{a}]] \boxtimes_{\mathcal{I}} \mathbb{R}[[\tilde{\mathfrak{a}}]] = (\mathbb{R} \oplus \mathbb{R}[[\mathfrak{a}]]\mathfrak{a}) \boxtimes_{\mathcal{I}} (\mathbb{R} \oplus \mathbb{R}[[\tilde{\mathfrak{a}}]]\tilde{\mathfrak{a}})$$
$$= \mathbb{R} \oplus \mathbb{R}[[\mathfrak{a}]]\mathfrak{a} \oplus \mathbb{R}[[\tilde{\mathfrak{a}}]]\tilde{\mathfrak{a}} \oplus \mathbb{R}[[\mathfrak{a}]]\mathfrak{a} \boxtimes_{\mathcal{I}} \mathbb{R}[[\tilde{\mathfrak{a}}]]\tilde{\mathfrak{a}}$$
$$= \mathbb{R} \oplus \mathbb{R}[[\mathfrak{a}]]\mathfrak{a} \oplus \mathbb{R}[[\tilde{\mathfrak{a}}]]\tilde{\mathfrak{a}} \oplus \mathbb{R}[[\mathfrak{a},\tilde{\mathfrak{a}}]]\mathfrak{a}\tilde{\mathfrak{a}} \oplus (\mathfrak{a},\tilde{\mathfrak{a}})^{(1)}\mathbb{R}[[\mathfrak{a}]] \boxtimes_{\mathcal{I}} \mathbb{R}[[\tilde{\mathfrak{a}}]]$$
$$= \mathbb{R} \oplus \mathbb{R}[[\mathfrak{a},\tilde{\mathfrak{a}}]]\mathfrak{a} \oplus \mathbb{R}[[\mathfrak{a},\tilde{\mathfrak{a}}]]\tilde{\mathfrak{a}} \oplus (\mathfrak{a},\tilde{\mathfrak{a}})^{(1)}\mathbb{R}[[\mathfrak{a}]] \boxtimes_{\mathcal{I}} \mathbb{R}[[\tilde{\mathfrak{a}}]]$$
$$= \mathbb{R}[[\mathfrak{a},\tilde{\mathfrak{a}}]] \oplus (\mathfrak{a},\tilde{\mathfrak{a}})^{(1)})\mathbb{R}[[\mathfrak{a}]] \boxtimes_{\mathcal{I}} \mathbb{R}[[\tilde{\mathfrak{a}}]]$$
$$= \mathbb{R}[[(\mathfrak{a},\tilde{\mathfrak{a}})^{(1)}]][[\mathfrak{a},\tilde{\mathfrak{a}}]] = \mathcal{I}[[\mathfrak{a},\tilde{\mathfrak{a}}]].$$

So the ring of seminvariants can be seen as $\mathcal{I}[[\mathfrak{a},\tilde{\mathfrak{a}}]]$. If we now compute the generating function for the vector field by tensoring with $u_{\mathfrak{u}}$ (cf. Section 12.4) or reading it off from the transvectant list (obtained by transvecting with $\mathfrak{u}$), we obtain

$$\frac{1}{(1-u_{\mathfrak{a}}t_{\mathfrak{a}})(1-u_{\tilde{\mathfrak{a}}}t_{\tilde{\mathfrak{a}}})(1-t^2_{\mathfrak{b}})} \hat{\boxtimes}_{\mathcal{I}} u_{\mathfrak{u}}$$
$$= \frac{1}{(1-u_{\tilde{\mathfrak{a}}}t_{\tilde{\mathfrak{a}}})(1-t^2_{\mathfrak{b}})} \hat{\boxtimes}_{\mathcal{I}} u_{\mathfrak{u}} \oplus \frac{u_{\mathfrak{a}}t_{\mathfrak{a}}}{(1-u_{\mathfrak{a}}t_{\mathfrak{a}})(1-u_{\tilde{\mathfrak{a}}}t_{\tilde{\mathfrak{a}}})(1-t^2_{\mathfrak{b}})} \hat{\boxtimes}_{\mathcal{I}} u_{\mathfrak{u}}$$
$$= \frac{u_{\tilde{\mathfrak{a}}}t_{\tilde{\mathfrak{a}}}}{(1-u_{\tilde{\mathfrak{a}}}t_{\tilde{\mathfrak{a}}})(1-t^2_{\mathfrak{b}})} \hat{\boxtimes}_{\mathcal{I}} u_{\mathfrak{u}} \oplus \frac{u_{\mathfrak{a}}t_{\mathfrak{a}}}{(1-u_{\mathfrak{a}}t_{\mathfrak{a}})(1-u_{\tilde{\mathfrak{a}}}t_{\tilde{\mathfrak{a}}})(1-t^2_{\mathfrak{b}})} \hat{\boxtimes}_{\mathcal{I}} u_{\mathfrak{u}}$$
$$= \frac{u^0_{(\tilde{\mathfrak{a}},\mathfrak{u})^{(1)}} t_{\tilde{\mathfrak{a}}}}{(1-u_{\tilde{\mathfrak{a}}}t_{\tilde{\mathfrak{a}}})(1-t^2_{\mathfrak{b}})} \oplus \frac{u^0_{(\mathfrak{a},\mathfrak{u})^{(1)}} t_{\mathfrak{a}}}{(1-u_{\mathfrak{a}}t_{\mathfrak{a}})(1-u_{\tilde{\mathfrak{a}}}t_{\tilde{\mathfrak{a}}})(1-t^2_{\mathfrak{b}})}.$$

Thus we find that

$$\mathbb{P}^{\bar{2},2}(t,u) = \frac{u^0_{(\tilde{\mathfrak{a}},\mathfrak{u})^{(1)}} t_{\tilde{\mathfrak{a}}}}{(1-u_{\tilde{\mathfrak{a}}}t_{\tilde{\mathfrak{a}}})(1-t^2_{\mathfrak{b}})} \oplus \frac{u_{\mathfrak{u}} + u^0_{(\mathfrak{a},\mathfrak{u})^{(1)}} t_{\mathfrak{a}}}{(1-u_{\mathfrak{a}}t_{\mathfrak{a}})(1-u_{\tilde{\mathfrak{a}}}t_{\tilde{\mathfrak{a}}})(1-t^2_{\mathfrak{b}})}.$$

We can draw the conclusion that

$$\ker \mathbf{N}_+|\mathbb{R}^4[[x_1,\ldots,x_4]] = \mathcal{I}[[\tilde{\mathfrak{a}}]](\tilde{\mathfrak{a}},\mathfrak{u})^{(1)} \oplus \mathcal{I}[[\mathfrak{a},\tilde{\mathfrak{a}}]]\mathfrak{u} \oplus \mathcal{I}[[\mathfrak{a},\tilde{\mathfrak{a}}]](\mathfrak{a},\mathfrak{u})^{(1)}$$
$$\oplus \mathcal{I}[[\mathfrak{a}]](\mathfrak{a},\tilde{\mathfrak{u}})^{(1)} \oplus \mathcal{I}[[\mathfrak{a},\tilde{\mathfrak{a}}]]\tilde{\mathfrak{u}} \oplus \mathcal{I}[[\mathfrak{a},\tilde{\mathfrak{a}}]](\tilde{\mathfrak{a}},\tilde{\mathfrak{u}})^{(1)}.$$

With

$$\mathbb{E}^{\overline{2},2} = \frac{u+t}{(1-ut)^2(1-t^2)},$$

one can conclude with the observation that

$$\delta^{\overline{2},2} = \frac{t}{(1-ut)(1-t^2)}.$$

12.6.6 The N_5 Case

We treat here only the problem of going from the polynomial case (which is basically classical invariant theory) to the generating function of the vector fields in $\ker \mathbf{N}_+$. We use here the classical notation (with $\mathfrak{a} = f$) for the invariants $f, H = (f,f)^{(2)}, i = (f,f)^{(4)}, \tau = (f,H)^{(1)}$, and $j = (f,H)^{(4)}$ and refer to [111, Section 89, Irreducible system for the quartic] for the detailed analysis (we use τ instead of t in order to avoid confusion in the generating function). The generating function for the covariants is

$$P^5_{\mathbb{R}}(t,u) = \frac{1 \oplus u_\tau^6 t_\tau^3}{(1-u_f^4 t_f)(1-u_H^4 t_H^2)(1-t_i^2)(1-t_j^3)}.$$

Ignoring for the moment the factors $1-t_i^2$ and $1-t_j^3$, we obtain

$$P^5_{\mathbb{R}(t^2,t^3)}(t,u) = \frac{1 \oplus u_\tau^6 t_\tau^3}{(1-u_f^4 t_f)(1-u_H^4 t_H^2)}.$$

We expand this to

$$P^5_{\mathbb{R}(t^2,t^3)}(t,u) = \frac{1 \oplus u_\tau^6 t_\tau^3}{(1-u_f^4 t_f)(1-u_H^4 t_H^2)}$$
$$= \frac{1}{(1-u_f^4 t_f)(1-u_H^4 t_H^2)} \oplus \frac{u_\tau^6 t_\tau^3}{(1-u_f^4 t_f)(1-u_H^4 t_H^2)}$$
$$= \frac{1}{(1-u_H^4 t_H^2)} \oplus \frac{u_f^4 t_f}{(1-u_f^4 t_f)(1-u_H^4 t_H^2)} \oplus \frac{u_\tau^6 t_\tau^3}{(1-u_f^4 t_f)(1-u_H^4 t_H^2)}$$
$$= 1 \oplus \frac{u_H^4 t_H^2}{(1-u_H^4 t_H^2)} \oplus \frac{u_f^4 t_f \oplus u_\tau^6 t_\tau^3}{(1-u_f^4 t_f)(1-u_H^4 t_H^2)}.$$

By tensoring with $\mathbb{R}^5$ this leads to a decomposition of $\ker \mathbf{N}_+$ of the following type (where as usual, $\mathcal{I} = \mathbb{R}[[i,j]]$):

$$\begin{aligned}&\mathcal{I}[[f,H]]\mathfrak{u}\oplus\mathcal{I}[[f,H]]\tau\mathfrak{u}\\&\oplus\mathcal{I}[[f,H]](f,\mathfrak{u})^{(1)}\oplus\mathcal{I}[[f,H]](f,\mathfrak{u})^{(2)}\oplus\mathcal{I}[[f,H]](f,\mathfrak{u})^{(3)}\oplus\mathcal{I}[[f,H]](f,\mathfrak{u})^{(4)}\\&\oplus\mathcal{I}[[f,H]](\tau,\mathfrak{u})^{(1)}\oplus\mathcal{I}[[f,H]](\tau,\mathfrak{u})^{(2)}\oplus\mathcal{I}[[f,H]](\tau,\mathfrak{u})^{(3)}\oplus\mathcal{I}[[f,H]](\tau,\mathfrak{u})^{(4)}\\&\oplus\mathcal{I}[[H]](H,\mathfrak{u})^{(1)}\oplus\mathcal{I}[[H]](H,\mathfrak{u})^{(2)}\oplus\mathcal{I}[[H]](H,\mathfrak{u})^{(3)}\oplus\mathcal{I}[[H]](H,\mathfrak{u})^{(4)}\end{aligned}$$

or (cf. [64])

$$\begin{aligned}\mathbb{P}^{\bar{5}}_{\mathbb{R}(t^2,t^3)}(t,u)=&\frac{u^4+(1+u^2+u^4+u^6)t+(1+u^2+u^4+u^6+u^8)u^2t^3}{(1-u^4t)(1-u^4t^2)}\\&+\frac{(1+u^2+u^4+u^6)t^2}{(1-u^4t^2)}.\end{aligned}$$

It follows that

$$\mathbb{P}^{\bar{5}}_{\mathbb{R}}(t,u)=\frac{u^4+(1+u^2+u^4+u^6)(t+t^2)+u^2t^3}{(1-u^4t)(1-u^4t^2)(1-t^2)(1-t^3)}$$

and

$$\delta^{\bar{5}}(t,u)=\frac{(1+u^2+u^4+u^6)t^2}{(1-u^4t^2)(1-t^2)(1-t^3)}.$$

12.6.7 The $N_{2,3}$ Case

Suppose we have a nilpotent with two irreducible invariant spaces of dimensions 2 and 3, with generators $\mathfrak{a}$ and $\mathfrak{b}$ on the coordinate side, and $\mathfrak{u}$ and $\mathfrak{v}$ on the ordinate side, with weights 1 and 2, respectively. The generating function is

$$\frac{1}{1-u_{\mathfrak{a}}t_{\mathfrak{a}}}\boxtimes_{\mathcal{I}}\frac{1}{(1-u_{\mathfrak{b}}^2t_{\mathfrak{b}})(1-t_{\mathfrak{c}}^2)},$$

where $\mathfrak{c}=(\mathfrak{b},\mathfrak{b})^{(2)}$. Notice that $\mathfrak{b}^k$ generates a space V_{2k}. It follows from Lemma 12.6.3 that (with $\mathfrak{d}=(\mathfrak{a},\mathfrak{b})^{(1)}$ and using $(\mathfrak{a}^2,\mathfrak{b})^{(1)}=2\mathfrak{a}\mathfrak{d}$)

$$\begin{aligned}&\frac{u_{\mathfrak{a}}^2t_{\mathfrak{a}}^2}{1-u_{\mathfrak{a}}t_{\mathfrak{a}}}\boxtimes_{\mathcal{I}}\frac{u_{\mathfrak{b}}^2t_{\mathfrak{b}}}{1-u_{\mathfrak{b}}^2t_{\mathfrak{b}}}\\&=\frac{u_{\mathfrak{a}}^2t_{\mathfrak{a}}^2}{1-u_{\mathfrak{a}}t_{\mathfrak{a}}}\frac{u_{\mathfrak{b}}^2t_{\mathfrak{b}}}{1-u_{\mathfrak{b}}^2t_{\mathfrak{b}}}\oplus\frac{u_{\mathfrak{a}}t_{\mathfrak{a}}}{1-u_{\mathfrak{a}}t_{\mathfrak{a}}}\frac{u_{\mathfrak{d}}t_{\mathfrak{d}}^2}{1-u_{\mathfrak{b}}^2t_{\mathfrak{b}}}\oplus t^3_{(\mathfrak{a},\mathfrak{d})^{(1)}}\frac{1}{1-u_{\mathfrak{a}}t_{\mathfrak{a}}}\boxtimes_{\mathcal{I}}\frac{1}{1-u_{\mathfrak{b}}^2t_{\mathfrak{b}}}.\end{aligned}$$

We can now compute

$$\begin{aligned}
&\frac{1}{1-u_{\mathfrak{a}}t_{\mathfrak{a}}} \boxtimes_{\mathcal{I}} \frac{1}{(1-u_{\mathfrak{b}}^2t_{\mathfrak{b}})(1-t_{\mathfrak{c}}^2)} \\
&= (1 \oplus u_{\mathfrak{a}}t_{\mathfrak{a}} \oplus \frac{u_{\mathfrak{a}}^2t_{\mathfrak{a}}^2}{1-u_{\mathfrak{a}}t_{\mathfrak{a}}}) \boxtimes_{\mathcal{I}} (1 \oplus \frac{u_{\mathfrak{b}}^2t_{\mathfrak{b}}}{1-u_{\mathfrak{b}}^2t_{\mathfrak{b}}}) \frac{1}{(1-t_{\mathfrak{c}}^2)} \\
&= \frac{1}{(1-t_{\mathfrak{c}}^2)} \left(1 \oplus u_{\mathfrak{a}}t_{\mathfrak{a}} \oplus \frac{u_{\mathfrak{a}}^2t_{\mathfrak{a}}^2}{1-u_{\mathfrak{a}}t_{\mathfrak{a}}} \oplus \frac{u_{\mathfrak{b}}^2t_{\mathfrak{b}}}{1-u_{\mathfrak{b}}^2t_{\mathfrak{b}}} \oplus u_{\mathfrak{a}}t_{\mathfrak{a}} \boxtimes_{\mathcal{I}} \frac{u_{\mathfrak{b}}^2t_{\mathfrak{b}}}{1-u_{\mathfrak{b}}^2t_{\mathfrak{b}}} \right. \\
&\quad \left. \oplus \frac{u_{\mathfrak{a}}^2t_{\mathfrak{a}}^2}{1-u_{\mathfrak{a}}t_{\mathfrak{a}}} \boxtimes_{\mathcal{I}} \frac{u_{\mathfrak{b}}^2t_{\mathfrak{b}}}{1-u_{\mathfrak{b}}^2t_{\mathfrak{b}}} \right) \\
&= \frac{1}{(1-t_{\mathfrak{c}}^2)} \left(\frac{1}{1-u_{\mathfrak{a}}t_{\mathfrak{a}}} \oplus \frac{u_{\mathfrak{b}}^2t_{\mathfrak{b}}}{1-u_{\mathfrak{b}}^2t_{\mathfrak{b}}} \oplus u_{\mathfrak{a}}t_{\mathfrak{a}} \frac{u_{\mathfrak{b}}^2t_{\mathfrak{b}}}{1-u_{\mathfrak{b}}^2t_{\mathfrak{b}}} \oplus \frac{u_{\mathfrak{d}}t_{\mathfrak{d}}^2}{1-u_{\mathfrak{b}}^2t_{\mathfrak{b}}} \right. \\
&\quad \left. \oplus \frac{u_{\mathfrak{a}}^2t_{\mathfrak{a}}^2}{1-u_{\mathfrak{a}}t_{\mathfrak{a}}} \frac{u_{\mathfrak{b}}^2t_{\mathfrak{b}}}{1-u_{\mathfrak{b}}^2t_{\mathfrak{b}}} \oplus \frac{u_{\mathfrak{a}}t_{\mathfrak{a}}}{1-u_{\mathfrak{a}}t_{\mathfrak{a}}} \frac{u_{\mathfrak{d}}t_{\mathfrak{d}}^2}{1-u_{\mathfrak{b}}^2t_{\mathfrak{b}}} \oplus \frac{t^3_{(\mathfrak{a},\mathfrak{d})^{(1)}}}{1-u_{\mathfrak{a}}t_{\mathfrak{a}}} \boxtimes_{\mathcal{I}} \frac{1}{1-u_{\mathfrak{b}}^2t_{\mathfrak{b}}} \right) \\
&= \frac{1 \oplus u_{\mathfrak{d}}t_{\mathfrak{d}}^2}{(1-u_{\mathfrak{a}}t_{\mathfrak{a}})(1-u_{\mathfrak{b}}^2t_{\mathfrak{b}})(1-t_{\mathfrak{c}}^2)} \oplus t^3_{(\mathfrak{a},\mathfrak{d})^{(1)}} \frac{1}{1-u_{\mathfrak{a}}t_{\mathfrak{a}}} \boxtimes_{\mathcal{I}} \frac{1}{1-u_{\mathfrak{b}}^2t_{\mathfrak{b}}} \frac{1}{1-t_{\mathfrak{c}}^2}.
\end{aligned}$$

The generating function is (with $\mathfrak{e} = (\mathfrak{a}, \mathfrak{d})^{(1)}$)

$$\frac{1 \oplus u_{\mathfrak{d}}t_{\mathfrak{d}}^2}{(1-u_{\mathfrak{a}}t_{\mathfrak{a}})(1-u_{\mathfrak{b}}^2t_{\mathfrak{b}})(1-t_{\mathfrak{c}}^2)(1-t_{\mathfrak{e}}^3)},$$

and one can easily check that it is indeed $(1,5)$-perfect.

Let $\mathcal{I} = \mathbb{R}[[\mathfrak{c}, \mathfrak{e}]]$. We can now calculate the tensoring of

$$\mathcal{I}[[\mathfrak{a}, \mathfrak{b}]] \oplus \mathcal{I}[[\mathfrak{a}, \mathfrak{b}]]\mathfrak{d}$$

with respect to $\mathfrak{u}$ and $\mathfrak{v}$, respectively. We compute

$$\begin{aligned}
&\frac{1 \oplus u_{\mathfrak{d}}t_{\mathfrak{d}}^2}{(1-u_{\mathfrak{a}}t_{\mathfrak{a}})(1-u_{\mathfrak{b}}^2t_{\mathfrak{b}})(1-t_{\mathfrak{c}}^2)(1-t_{\mathfrak{e}}^3)} \hat{\boxtimes}_{\mathcal{I}} u_{\mathfrak{u}} \\
&= \frac{u_{\mathfrak{b}}^2t_{\mathfrak{b}}}{(1-u_{\mathfrak{b}}^2t_{\mathfrak{b}})(1-t_{\mathfrak{c}}^2)(1-t_{\mathfrak{e}}^3)} \hat{\boxtimes}_{\mathcal{I}} u_{\mathfrak{u}} \oplus \frac{u_{\mathfrak{a}}t_{\mathfrak{a}} \oplus u_{\mathfrak{d}}t_{\mathfrak{d}}^2}{(1-u_{\mathfrak{a}}t_{\mathfrak{a}})(1-u_{\mathfrak{b}}^2t_{\mathfrak{b}})(1-t_{\mathfrak{c}}^2)(1-t_{\mathfrak{e}}^3)} \hat{\boxtimes}_{\mathcal{I}} u_{\mathfrak{u}} \\
&= \frac{u^1_{(\mathfrak{b},\mathfrak{u})^{(1)}} t_{\mathfrak{b}}}{(1-u_{\mathfrak{b}}^2t_{\mathfrak{b}})(1-t_{\mathfrak{c}}^2)(1-t_{\mathfrak{e}}^3)} \oplus \frac{u^0_{(\mathfrak{a},\mathfrak{u})^{(1)}} t_{\mathfrak{a}} \oplus u^0_{(\mathfrak{d},\mathfrak{u})^{(1)}} t_{\mathfrak{d}}^2}{(1-u_{\mathfrak{a}}t_{\mathfrak{a}})(1-u_{\mathfrak{b}}^2t_{\mathfrak{b}})(1-t_{\mathfrak{c}}^2)(1-t_{\mathfrak{e}}^3)}.
\end{aligned}$$

This implies that the generating function for the 2-dimensional vectors is given by

$$\mathbb{P}^{\overline{2},3}(t,u) = \frac{u^1_{(\mathfrak{b},\mathfrak{u})^{(1)}} t_{\mathfrak{b}}}{(1-u_{\mathfrak{b}}^2t_{\mathfrak{b}})(1-t_{\mathfrak{c}}^2)(1-t_{\mathfrak{e}}^3)} \oplus \frac{u_{\mathfrak{u}} \oplus u^0_{(\mathfrak{a},\mathfrak{u})^{(1)}} t_{\mathfrak{a}} \oplus u^2_{\mathfrak{d}\mathfrak{u}} t_{\mathfrak{d}}^2 \oplus u^0_{(\mathfrak{d},\mathfrak{u})^{(1)}} t_{\mathfrak{d}}^2}{(1-u_{\mathfrak{a}}t_{\mathfrak{a}})(1-u_{\mathfrak{b}}^2t_{\mathfrak{b}})(1-t_{\mathfrak{c}}^2)(1-t_{\mathfrak{e}}^3)},$$

and this is $(2,5)$-perfect. It follows immediately that

$$\delta^{\overline{2},3}(t,u) = \frac{ut}{(1-u^2t)(1-t^2)(1-t^3)}.$$

So this part of the space looks like

$$\mathcal{I}[[\mathfrak{a},\mathfrak{b}]]\mathfrak{u} \oplus \mathcal{I}[[\mathfrak{a},\mathfrak{b}]](\mathfrak{a},\mathfrak{u})^{(1)} \oplus \mathcal{I}[[\mathfrak{a},\mathfrak{b}]]\mathfrak{d}\mathfrak{u} \oplus \mathcal{I}[[\mathfrak{a},\mathfrak{b}]](\mathfrak{d},\mathfrak{u})^{(1)} \oplus \mathcal{I}[[\mathfrak{b}]](\mathfrak{b},\mathfrak{u})^{(1)}.$$

Since $\mathfrak{d} = (\mathfrak{a},\mathfrak{b})^{(1)}$,

$$\mathcal{I}[[\mathfrak{a},\mathfrak{b}]]\mathfrak{d}\mathfrak{u} \oplus \mathcal{I}[[\mathfrak{a},\mathfrak{b}]](\mathfrak{a},\mathfrak{u})^{(1)} \oplus \mathcal{I}[[\mathfrak{b}]](\mathfrak{b},\mathfrak{u})^{(1)} = \mathcal{I}[[\mathfrak{a},\mathfrak{b}]](\mathfrak{b},\mathfrak{u})^{(1)} \oplus \mathcal{I}[[\mathfrak{a},\mathfrak{b}]](\mathfrak{a},\mathfrak{u})^{(1)},$$

the Stanley decomposition simplifies to

$$\mathcal{I}[[\mathfrak{a},\mathfrak{b}]]\mathfrak{u} \oplus \mathcal{I}[[\mathfrak{a},\mathfrak{b}]](\mathfrak{a},\mathfrak{u})^{(1)} \oplus \mathcal{I}[[\mathfrak{a},\mathfrak{b}]](\mathfrak{d},\mathfrak{u})^{(1)} \oplus \mathcal{I}[[\mathfrak{a},\mathfrak{b}]](\mathfrak{b},\mathfrak{u})^{(1)}.$$

We see that $\Delta_S^{\overline{2},3} \leq 0$. We now turn to those vector fields generated by $\mathfrak{v}$. We ignore the zero-transvectant part for the moment, but include it in the decomposition:

$$\begin{aligned}
\hat{\mathbb{P}}^{2,\overline{3}}(t,u) &= \frac{1 \oplus u_{\mathfrak{d}} t_{\mathfrak{d}}^2}{(1-u_{\mathfrak{a}}t_{\mathfrak{a}})(1-u_{\mathfrak{b}}^2 t_{\mathfrak{b}})(1-t_{\mathfrak{c}}^2)(1-t_{\mathfrak{e}}^3)} \hat{\boxtimes}_{\mathcal{I}} u_{\mathfrak{v}}^2 \\
&= \frac{u_{\mathfrak{b}}^2 t_{\mathfrak{b}}(1 \oplus u_{\mathfrak{d}} t_{\mathfrak{d}}^2)}{(1-u_{\mathfrak{a}}t_{\mathfrak{a}})(1-u_{\mathfrak{b}}^2 t_{\mathfrak{b}})(1-t_{\mathfrak{c}}^2)(1-t_{\mathfrak{e}}^3)} \hat{\boxtimes}_{\mathcal{I}} u_{\mathfrak{v}}^2 \\
&\oplus \frac{1 \oplus u_{\mathfrak{d}} t_{\mathfrak{d}}^2}{(1-u_{\mathfrak{a}}t_{\mathfrak{a}})(1-t_{\mathfrak{c}}^2)(1-t_{\mathfrak{e}}^3)} \hat{\boxtimes}_{\mathcal{I}} u_{\mathfrak{v}}^2 \\
&= \frac{u_{\mathfrak{b}}^2 t_{\mathfrak{b}}(1 \oplus u_{\mathfrak{d}} t_{\mathfrak{d}}^2)}{(1-u_{\mathfrak{a}}t_{\mathfrak{a}})(1-u_{\mathfrak{b}}^2 t_{\mathfrak{b}})(1-t_{\mathfrak{c}}^2)(1-t_{\mathfrak{e}}^3)} \hat{\boxtimes}_{\mathcal{I}} u_{\mathfrak{v}}^2 \oplus \frac{u_{\mathfrak{d}} t_{\mathfrak{d}}^2 \oplus u_{\mathfrak{a}} t_{\mathfrak{a}}}{(1-t_{\mathfrak{c}}^2)(1-t_{\mathfrak{e}}^3)} \hat{\boxtimes}_{\mathcal{I}} u_{\mathfrak{v}}^2 \\
&\oplus \frac{u_{\mathfrak{a}\mathfrak{d}}^2 t_{\mathfrak{a}\mathfrak{d}}^3}{(1-u_{\mathfrak{a}}t_{\mathfrak{a}})(1-t_{\mathfrak{c}}^2)(1-t_{\mathfrak{e}}^3)} \hat{\boxtimes}_{\mathcal{I}} u_{\mathfrak{v}}^2 \oplus \frac{u_{\mathfrak{a}^2}^2 t_{\mathfrak{a}^2}^2}{(1-u_{\mathfrak{a}}t_{\mathfrak{a}})(1-t_{\mathfrak{c}}^2)(1-t_{\mathfrak{e}}^3)} \hat{\boxtimes}_{\mathcal{I}} u_{\mathfrak{v}}^2 \\
&= \frac{(u_{(\mathfrak{b},\mathfrak{v})^{(1)}}^2 t_{(\mathfrak{b},\mathfrak{v})^{(1)}} \oplus t_{(\mathfrak{b},\mathfrak{v})^{(2)}})(1 + u_{\mathfrak{d}} t_{\mathfrak{d}}^2)}{(1-u_{\mathfrak{a}}t_{\mathfrak{a}})(1-u_{\mathfrak{b}}^2 t_{\mathfrak{b}})(1-t_{\mathfrak{c}}^2)(1-t_{\mathfrak{e}}^3)} \\
&\oplus \frac{u_{(\mathfrak{d},\mathfrak{v})^{(1)}} t_{(\mathfrak{d},\mathfrak{v})^{(1)}}^2 \oplus u_{(\mathfrak{a},\mathfrak{v})^{(1)}} t_{(\mathfrak{a},\mathfrak{v})^{(1)}}}{(1-t_{\mathfrak{c}}^2)(1-t_{\mathfrak{e}}^3)} \\
&\oplus \frac{u_{(\mathfrak{a}\mathfrak{d},\mathfrak{v})^{(1)}}^2 t_{(\mathfrak{a}\mathfrak{d},\mathfrak{v})^{(1)}}^3 \oplus t_{(\mathfrak{a}\mathfrak{d},\mathfrak{v})^{(2)}}^3 \oplus u_{(\mathfrak{a}^2,\mathfrak{v})^{(1)}}^2 t_{(\mathfrak{a}^2,\mathfrak{v})^{(1)}}^2 \oplus t_{(\mathfrak{a}^2,\mathfrak{v})^{(2)}}^2}{(1-u_{\mathfrak{a}}t_{\mathfrak{a}})(1-t_{\mathfrak{c}}^2)(1-t_{\mathfrak{e}}^3)}.
\end{aligned}$$

This implies that

$$\delta^{2,\overline{3}}(t,u) = \frac{ut^2 + ut + t^3 + t^2}{(1-ut)(1-t^2)(1-t^3)}.$$

The decomposition is

$$\begin{aligned}
&\mathcal{I}[[\mathfrak{a},\mathfrak{b}]]\mathfrak{v} \oplus \mathcal{I}[[\mathfrak{a},\mathfrak{b}]](\mathfrak{b},\mathfrak{v})^{(1)} \oplus \mathcal{I}[[\mathfrak{a},\mathfrak{b}]](\mathfrak{b},\mathfrak{v})^{(2)} \\
&\oplus \mathcal{I}[[\mathfrak{a},\mathfrak{b}]]\mathfrak{d}\mathfrak{v} \oplus \mathcal{I}[[\mathfrak{a},\mathfrak{b}]]\mathfrak{d}(\mathfrak{b},\mathfrak{v})^{(1)} \oplus \mathcal{I}[[\mathfrak{a},\mathfrak{b}]]\mathfrak{d}(\mathfrak{b},\mathfrak{v})^{(2)} \oplus \mathcal{I}(\mathfrak{d},\mathfrak{v})^{(1)} \oplus \mathcal{I}(\mathfrak{a},\mathfrak{v})^{(1)} \\
&\oplus \mathcal{I}[[\mathfrak{a}]](\mathfrak{a}\mathfrak{d},\mathfrak{v})^{(1)} \oplus \mathcal{I}[[\mathfrak{a}]](\mathfrak{a}\mathfrak{d},\mathfrak{v})^{(2)} \oplus \mathcal{I}[[\mathfrak{a}]](\mathfrak{a}^2,\mathfrak{v})^{(1)} \oplus \mathcal{I}[[\mathfrak{a}]](\mathfrak{a}^2,\mathfrak{v})^{(2)}.
\end{aligned}$$

The first simplification is, since $(\mathfrak{a}^2, \mathfrak{v})^{(1)} = 2\mathfrak{a}(\mathfrak{a}, \mathfrak{v})^{(1)}$,

$$\mathcal{I}[[\mathfrak{a}]](\mathfrak{a}^2, \mathfrak{v})^{(1)} \oplus \mathcal{I}(\mathfrak{a}, \mathfrak{v})^{(1)} = \mathcal{I}[[\mathfrak{a}]](\mathfrak{a}, \mathfrak{v})^{(1)}.$$

The second simplification is, since $(\mathfrak{a}\mathfrak{d}, \mathfrak{v})^{(1)} = \mathfrak{a}(\mathfrak{d}, \mathfrak{v})^{(1)} + \mathfrak{d}(\mathfrak{a}, \mathfrak{v})^{(1)} = 2\mathfrak{a}(\mathfrak{d}, \mathfrak{v})^{(1)} + 2\mathfrak{e}\mathfrak{v}$,

$$\mathcal{I}[[\mathfrak{a}]](\mathfrak{a}\mathfrak{d}, \mathfrak{v})^{(1)} \oplus \mathcal{I}[[\mathfrak{a}]](\mathfrak{a}, \mathfrak{v})^{(1)} \oplus \mathcal{I}[[\mathfrak{a}]](\mathfrak{a}\mathfrak{d}, \mathfrak{v})^{(2)} \oplus \mathcal{I}[[\mathfrak{a}]](\mathfrak{a}^2, \mathfrak{v})^{(2)}$$

The third simplification is obtained by noticing that $\mathfrak{d}\mathfrak{v} = \mathfrak{b}(\mathfrak{a}, \mathfrak{v})^{(1)} - \frac{1}{2}\mathfrak{a}(\mathfrak{b}, \mathfrak{v})^{(1)}$. Thus one has

$$\mathcal{I}[[\mathfrak{a}, \mathfrak{b}]]\mathfrak{d}\mathfrak{v} \oplus \mathcal{I}[[\mathfrak{a}]](\mathfrak{a}, \mathfrak{v})^{(1)} \oplus \mathcal{I}[[\mathfrak{a}, \mathfrak{b}]](\mathfrak{b}, \mathfrak{v})^{(1)} = \mathcal{I}[[\mathfrak{a}, \mathfrak{b}]](\mathfrak{a}, \mathfrak{v})^{(1)} \oplus \mathcal{I}[[\mathfrak{a}, \mathfrak{b}]](\mathfrak{b}, \mathfrak{v})^{(1)}.$$

We obtain the decomposition

$$\begin{aligned} &\mathcal{I}[[\mathfrak{a}, \mathfrak{b}]]\mathfrak{v} \oplus \mathcal{I}[[\mathfrak{a}, \mathfrak{b}]](\mathfrak{b}, \mathfrak{v})^{(1)} \oplus \mathcal{I}[[\mathfrak{a}, \mathfrak{b}]](\mathfrak{b}, \mathfrak{v})^{(2)} \\ &\oplus \mathcal{I}[[\mathfrak{a}, \mathfrak{b}]]\mathfrak{d}(\mathfrak{b}, \mathfrak{v})^{(1)} \oplus \mathcal{I}[[\mathfrak{a}, \mathfrak{b}]]\mathfrak{d}(\mathfrak{b}, \mathfrak{v})^{(2)} \oplus \mathcal{I}[[\mathfrak{a}, \mathfrak{b}]](\mathfrak{a}, \mathfrak{v})^{(1)} \\ &\oplus \mathcal{I}[[\mathfrak{a}]](\mathfrak{d}, \mathfrak{v})^{(1)} \oplus \mathcal{I}[[\mathfrak{a}]](\mathfrak{a}\mathfrak{d}, \mathfrak{v})^{(2)} \oplus \mathcal{I}[[\mathfrak{a}]](\mathfrak{a}^2, \mathfrak{v})^{(2)}. \end{aligned}$$

The fourth simplification is obtained by noticing that $\mathfrak{d}(\mathfrak{b}, \mathfrak{v})^{(1)} = 2\mathfrak{b}(\mathfrak{d}, \mathfrak{v})^{(1)} + 2(\mathfrak{b}, \mathfrak{d})^{(1)}\mathfrak{v} = 2\mathfrak{b}(\mathfrak{d}, \mathfrak{v})^{(1)} + \mathfrak{a}\mathfrak{c}\mathfrak{v}$. This implies that

$$\mathcal{I}[[\mathfrak{a}, \mathfrak{b}]]\mathfrak{d}(\mathfrak{b}, \mathfrak{v})^{(1)} \oplus \mathcal{I}[[\mathfrak{a}]](\mathfrak{d}, \mathfrak{v})^{(1)} \oplus \mathcal{I}[[\mathfrak{a}, \mathfrak{b}]]\mathfrak{v} = \mathcal{I}[[\mathfrak{a}, \mathfrak{b}]](\mathfrak{d}, \mathfrak{v})^{(1)} \oplus \mathcal{I}[[\mathfrak{a}, \mathfrak{b}]]\mathfrak{v}.$$

We obtain the decomposition, first obtained in [85],

$$\begin{aligned} &\ker \mathsf{N}_+|\mathbb{R}^5[[x_1, \ldots, x_5]] \\ &= \mathcal{I}[[\mathfrak{a}, \mathfrak{b}]]\mathfrak{u} \oplus \mathcal{I}[[\mathfrak{a}, \mathfrak{b}]](\mathfrak{a}, \mathfrak{u})^{(1)} \oplus \mathcal{I}[[\mathfrak{a}, \mathfrak{b}]](\mathfrak{d}, \mathfrak{u})^{(1)} \oplus \mathcal{I}[[\mathfrak{b}]](\mathfrak{b}, \mathfrak{u})^{(1)} \\ &\oplus \mathcal{I}[[\mathfrak{a}, \mathfrak{b}]]\mathfrak{v} \oplus \mathcal{I}[[\mathfrak{a}, \mathfrak{b}]](\mathfrak{b}, \mathfrak{v})^{(1)} \oplus \mathcal{I}[[\mathfrak{a}, \mathfrak{b}]](\mathfrak{b}, \mathfrak{v})^{(2)} \oplus \mathcal{I}[[\mathfrak{a}, \mathfrak{b}]](\mathfrak{d}, \mathfrak{v})^{(1)} \\ &\oplus \mathcal{I}[[\mathfrak{a}, \mathfrak{b}]]\mathfrak{d}(\mathfrak{b}, \mathfrak{v})^{(2)} \oplus \mathcal{I}[[\mathfrak{a}, \mathfrak{b}]](\mathfrak{a}, \mathfrak{v})^{(1)} \oplus \mathcal{I}[[\mathfrak{a}]](\mathfrak{a}\mathfrak{d}, \mathfrak{v})^{(2)} \oplus \mathcal{I}[[\mathfrak{a}]](\mathfrak{a}^2, \mathfrak{v})^{(2)}. \end{aligned}$$

We see that $\Delta_S^{2,\overline{3}} \leq 2$.

12.7 A GL_n-Invariant Description of the Ring of Seminvariants for $n \geq 6$

12.7.1 The $N_{2,2,2}$ Case

We denote the generators by $\mathfrak{a}^i, i = 1, 2, 3$, and their first transvectants by $\mathfrak{d}^{ij} = (\mathfrak{a}^i, \mathfrak{a}^j)^{(1)}$. Then we have to compute

$$\frac{1}{1 - u_{\mathfrak{a}^1} t_{\mathfrak{a}^1}} \boxtimes_{\mathcal{I}} \frac{1}{(1 - t_{\mathfrak{d}^{23}}^2)(1 - u_{\mathfrak{a}^2} t_{\mathfrak{a}^2})(1 - u_{\mathfrak{a}^3} t_{\mathfrak{a}^3})}.$$

We rewrite this as

$$\frac{1}{1-u_{\mathfrak{a}^1}t_{\mathfrak{a}^1}} \boxtimes_{\mathcal{I}} \frac{1}{(1-t_{\mathfrak{d}^{23}}^2)(1-u_{\mathfrak{a}^2}t_{\mathfrak{a}^2})}\left(1 \oplus \frac{u_{\mathfrak{a}^3}t_{\mathfrak{a}^3}}{(1-u_{\mathfrak{a}^3}t_{\mathfrak{a}^3})}\right),$$

and this equals, using again Lemma 12.6.3,

$$\begin{aligned}&\frac{1}{1-u_{\mathfrak{a}^1}t_{\mathfrak{a}^1}} \boxtimes_{\mathcal{I}} \frac{1}{(1-t_{\mathfrak{d}^{23}}^2)(1-u_{\mathfrak{a}^2}t_{\mathfrak{a}^2})}\\ \oplus &\frac{1}{(1-t_{\mathfrak{d}^{23}}^2)(1-u_{\mathfrak{a}^2}t_{\mathfrak{a}^2})}\frac{1}{(1-u_{\mathfrak{a}^1}t_{\mathfrak{a}^1})}\frac{u_{\mathfrak{a}^3}t_{\mathfrak{a}^3}}{(1-u_{\mathfrak{a}^3}t_{\mathfrak{a}^3})}\\ \oplus &\frac{t_{\mathfrak{d}^{13}}^2}{1-u_{\mathfrak{a}^1}t_{\mathfrak{a}^1}} \boxtimes_{\mathcal{I}} \frac{1}{(1-t_{\mathfrak{d}^{23}}^2)(1-u_{\mathfrak{a}^2}t_{\mathfrak{a}^2})}\frac{1}{(1-u_{\mathfrak{a}^3}t_{\mathfrak{a}^3})}.\end{aligned}$$

Using the results from Section 12.6.5 we see that this equals

$$\begin{aligned}&\frac{1}{(1-t_{\mathfrak{d}^{12}}^2)}\frac{1}{(1-t_{\mathfrak{d}^{23}}^2)}\frac{1}{(1-u_{\mathfrak{a}^1}t_{\mathfrak{a}^1})}\frac{1}{(1-u_{\mathfrak{a}^2}t_{\mathfrak{a}^2})}\\ \oplus &\frac{1}{(1-t_{\mathfrak{d}^{23}}^2)(1-u_{\mathfrak{a}^2}t_{\mathfrak{a}^2})}\frac{1}{(1-u_{\mathfrak{a}^1}t_{\mathfrak{a}^1})}\frac{u_{\mathfrak{a}^3}t_{\mathfrak{a}^3}}{(1-u_{\mathfrak{a}^3}t_{\mathfrak{a}^3})}\\ \oplus &\frac{t_{\mathfrak{d}^{13}}^2}{(1-u_{\mathfrak{a}^1}t_{\mathfrak{a}^1})} \boxtimes_{\mathcal{I}} \frac{1}{(1-t_{\mathfrak{d}^{23}}^2)(1-u_{\mathfrak{a}^2}t_{\mathfrak{a}^2})}\frac{1}{(1-u_{\mathfrak{a}^3}t_{\mathfrak{a}^3})}.\end{aligned}$$

This implies that the generating function is

$$\begin{aligned}P^{2,2,2}(t,u) = &\frac{1}{(1-t_{\mathfrak{d}^{12}}^2)}\frac{1}{(1-t_{\mathfrak{d}^{13}}^2)}\frac{1}{(1-t_{\mathfrak{d}^{23}}^2)}\frac{1}{(1-u_{\mathfrak{a}^1}t_{\mathfrak{a}^1})}\frac{1}{(1-u_{\mathfrak{a}^2}t_{\mathfrak{a}^2})}\\ \oplus &\frac{1}{(1-t_{\mathfrak{d}^{13}}^2)}\frac{1}{(1-t_{\mathfrak{d}^{23}}^2)(1-u_{\mathfrak{a}^2}t_{\mathfrak{a}^2})}\frac{1}{(1-u_{\mathfrak{a}^1}t_{\mathfrak{a}^1})}\frac{u_{\mathfrak{a}^3}t_{\mathfrak{a}^3}}{(1-u_{\mathfrak{a}^3}t_{\mathfrak{a}^3})}.\end{aligned}$$

12.7.2 The $N_{3,3}$ Case

This case was first addressed in [64].

We denote the generators by $\mathfrak{a}^i, i = 1, 2$, and their transvectants by $\mathfrak{d}_k^{ij} = (\mathfrak{a}^i, \mathfrak{a}^j)^{(k)}$. Then we have to compute

$$\frac{1}{(1-u_{\mathfrak{a}^1}^2t_{\mathfrak{a}^1})(1-t_{\mathfrak{d}_2^{11}}^2)} \boxtimes_{\mathcal{I}} \frac{1}{(1-u_{\mathfrak{a}^2}^2t_{\mathfrak{a}^2})(1-t_{\mathfrak{d}_2^{22}}^2)}.$$

We rewrite this as

$$\begin{aligned}&\frac{1}{(1-u_{\mathfrak{a}^2}^2t_{\mathfrak{a}^2})(1-t_{\mathfrak{d}_2^{11}}^2)(1-t_{\mathfrak{d}_2^{22}}^2)} \oplus \frac{u_{\mathfrak{a}^1}^2t_{\mathfrak{a}^1}}{(1-u_{\mathfrak{a}^1}^2t_{\mathfrak{a}^1})(1-t_{\mathfrak{d}_2^{11}}^2)(1-t_{\mathfrak{d}_2^{22}}^2)}\\ \oplus &\frac{u_{\mathfrak{a}^1}^2t_{\mathfrak{a}^1}}{(1-u_{\mathfrak{a}^1}^2t_{\mathfrak{a}^1})(1-t_{\mathfrak{d}_2^{11}}^2)} \boxtimes_{\mathcal{I}} \frac{u_{\mathfrak{a}^2}^2t_{\mathfrak{a}^2}}{(1-u_{\mathfrak{a}^2}^2t_{\mathfrak{a}^2})(1-t_{\mathfrak{d}_2^{22}}^2)}.\end{aligned}$$

Using again Lemma 12.6.3, we see that

$$\frac{1}{(1-u^2_{\mathfrak{a}^2}t_{\mathfrak{a}^2})(1-t^2_{\mathfrak{d}^{11}_2})(1-t^2_{\mathfrak{d}^{22}_2})} \oplus \frac{u^2_{\mathfrak{a}^1}t_{\mathfrak{a}^1}}{(1-u^2_{\mathfrak{a}^1}t_{\mathfrak{a}^1})(1-t^2_{\mathfrak{d}^{11}_2})(1-t^2_{\mathfrak{d}^{22}_2})}$$
$$\oplus \frac{u^2_{\mathfrak{a}^1}t_{\mathfrak{a}^1}}{(1-u^2_{\mathfrak{a}^1}t_{\mathfrak{a}^1})(1-t^2_{\mathfrak{d}^{11}_2})} \frac{u^2_{\mathfrak{a}^2}t_{\mathfrak{a}^2}}{(1-u^2_{\mathfrak{a}^2}t_{\mathfrak{a}^2})(1-t^2_{\mathfrak{d}^{22}_2})}$$
$$\oplus \frac{u^2_{\mathfrak{d}^{12}_1}t^2_{\mathfrak{d}^{12}_1}}{(1-u^2_{\mathfrak{a}^1}t_{\mathfrak{a}^1})(1-t^2_{\mathfrak{d}^{11}_2})(1-u^2_{\mathfrak{a}^2}t_{\mathfrak{a}^2})(1-t^2_{\mathfrak{d}^{22}_2})}$$
$$\oplus t^2_{\mathfrak{d}^{12}_2} \frac{1}{(1-u^2_{\mathfrak{a}^1}t_{\mathfrak{a}^1})(1-t^2_{\mathfrak{d}^{11}_2})} \boxtimes_{\mathcal{I}} \frac{1}{(1-u^2_{\mathfrak{a}^2}t_{\mathfrak{a}^2})(1-t^2_{\mathfrak{d}^{22}_2})}$$
$$= \frac{1+u^2_{\mathfrak{d}^{12}_1}t^2_{\mathfrak{d}^{12}_1}}{(1-u^2_{\mathfrak{a}^1}t_{\mathfrak{a}^1})(1-t^2_{\mathfrak{d}^{11}_2})(1-u^2_{\mathfrak{a}^2}t_{\mathfrak{a}^2})(1-t^2_{\mathfrak{d}^{22}_2})}$$
$$\oplus t^2_{\mathfrak{d}^{12}_2} \frac{1}{(1-u^2_{\mathfrak{a}^1}t_{\mathfrak{a}^1})(1-t^2_{\mathfrak{d}^{11}_2})} \boxtimes_{\mathcal{I}} \frac{1}{(1-u^2_{\mathfrak{a}^2}t_{\mathfrak{a}^2})(1-t^2_{\mathfrak{d}^{22}_2})}.$$

It follows that the generating function is

$$\frac{1\oplus u^2_{\mathfrak{d}^{12}_1}t^2_{\mathfrak{d}^{12}_1}}{(1-t^2_{\mathfrak{d}^{12}_2})(1-u^2_{\mathfrak{a}^1}t_{\mathfrak{a}^1})(1-t^2_{\mathfrak{d}^{11}_2})(1-u^2_{\mathfrak{a}^2}t_{\mathfrak{a}^2})(1-t^2_{\mathfrak{d}^{22}_2})}.$$

12.7.3 The $N_{3,4}$ Case

It follows from the results in Sections 12.6.2 and 12.6.3 that the generating function is given by

$$\frac{1}{(1-u^2_{\mathfrak{a}}t^1_{\mathfrak{a}})(1-t^2_{(\mathfrak{a},\mathfrak{a})^{(2)}})} \boxtimes_{\mathcal{I}} \frac{1\oplus u^3_{\mathfrak{d}}t^3_{\mathfrak{d}}}{(1-u^3_{\mathfrak{b}}t^1_{\mathfrak{b}})(1-u^2_{\mathfrak{c}}t^2_{\mathfrak{c}})(1-t^4_{(\mathfrak{c},\mathfrak{c})^{(2)}})},$$

with $\mathfrak{c}=(\mathfrak{b},\mathfrak{b})^{(2)}$ and $\mathfrak{d}=(\mathfrak{b},\mathfrak{c})^{(1)}$. Ignoring the invariants for the moment, we now concentrate on computing

$$P(t,u) = \frac{1}{(1-u^2_{\mathfrak{a}}t^1_{\mathfrak{a}})} \boxtimes_{\mathcal{I}} \frac{1\oplus u^3_{\mathfrak{d}}t^3_{\mathfrak{d}}}{(1-u^3_{\mathfrak{b}}t^1_{\mathfrak{b}})(1-u^2_{\mathfrak{c}}t^2_{\mathfrak{c}})}.$$

This equals

$$P(t,u) = \frac{1}{(1-u^2_{\mathfrak{a}}t^1_{\mathfrak{a}})} \boxtimes_{\mathcal{I}} \frac{1\oplus u^3_{\mathfrak{d}}t^3_{\mathfrak{d}}}{(1-u^3_{\mathfrak{b}}t^1_{\mathfrak{b}})}$$
$$\oplus \frac{1}{(1-u^2_{\mathfrak{a}}t^1_{\mathfrak{a}})} \boxtimes_{\mathcal{I}} \frac{u^2_{\mathfrak{c}}t^2_{\mathfrak{c}}(1\oplus u^3_{\mathfrak{d}}t^3_{\mathfrak{d}})}{(1-u^3_{\mathfrak{b}}t^1_{\mathfrak{b}})(1-u^2_{\mathfrak{c}}t^2_{\mathfrak{c}})}$$

Let $G(t,u) = \frac{1}{(1-u^2_{\mathfrak{a}}t^1_{\mathfrak{a}})} \boxtimes_{\mathcal{I}} \frac{1\oplus u^3_{\mathfrak{d}}t^3_{\mathfrak{d}}}{(1-u^3_{\mathfrak{b}}t^1_{\mathfrak{b}})}$ and $H(t,u) = \frac{1}{(1-u^2_{\mathfrak{a}}t^1_{\mathfrak{a}})} \boxtimes_{\mathcal{I}} \frac{u^2_{\mathfrak{c}}t^2_{\mathfrak{c}}(1\oplus u^3_{\mathfrak{d}}t^3_{\mathfrak{d}})}{(1-u^3_{\mathfrak{b}}t^1_{\mathfrak{b}})(1-u^2_{\mathfrak{c}}t^2_{\mathfrak{c}})}$. Then

$$
\begin{aligned}
H(t,u) &= \frac{1}{(1-u_{\mathfrak{a}}^2 t_{\mathfrak{a}}^1)} \otimes_{\mathcal{I}} \frac{u_{\mathfrak{c}}^2 t_{\mathfrak{c}}^2 (1 \oplus u_{\mathfrak{d}}^3 t_{\mathfrak{d}}^3)}{(1-u_{\mathfrak{b}}^3 t_{\mathfrak{b}}^1)(1-u_{\mathfrak{c}}^2 t_{\mathfrak{c}}^2)} \\
&= \frac{u_{\mathfrak{c}}^2 t_{\mathfrak{c}}^2 (1 \oplus u_{\mathfrak{d}}^3 t_{\mathfrak{d}}^3)}{(1-u_{\mathfrak{b}}^3 t_{\mathfrak{b}}^1)(1-u_{\mathfrak{c}}^2 t_{\mathfrak{c}}^2)} \oplus \frac{u_{\mathfrak{a}}^2 t_{\mathfrak{a}}^1}{(1-u_{\mathfrak{a}}^2 t_{\mathfrak{a}}^1)} \otimes_{\mathcal{I}} \frac{u_{\mathfrak{c}}^2 t_{\mathfrak{c}}^2 (1 \oplus u_{\mathfrak{d}}^3 t_{\mathfrak{d}}^3)}{(1-u_{\mathfrak{b}}^3 t_{\mathfrak{b}}^1)(1-u_{\mathfrak{c}}^2 t_{\mathfrak{c}}^2)} \\
&= \frac{(u_{\mathfrak{c}}^2 t_{\mathfrak{c}}^2 \oplus u^2_{(\mathfrak{a},\mathfrak{c})^{(1)}} t^3_{(\mathfrak{a},\mathfrak{c})^{(1)}})(1 \oplus u_{\mathfrak{d}}^3 t_{\mathfrak{d}}^3)}{(1-u_{\mathfrak{a}}^2 t_{\mathfrak{a}}^1)(1-u_{\mathfrak{b}}^3 t_{\mathfrak{b}}^1)(1-u_{\mathfrak{c}}^2 t_{\mathfrak{c}}^2)} \oplus t^3_{(\mathfrak{a},\mathfrak{c})^{(2)}} P(t,u)
\end{aligned}
$$

and

$$
\begin{aligned}
G(t,u) &= \frac{1}{(1-u_{\mathfrak{a}}^2 t_{\mathfrak{a}}^1)} \otimes_{\mathcal{I}} (1 \oplus u_{\mathfrak{b}}^3 t_{\mathfrak{b}}^1)(1 \oplus u_{\mathfrak{d}}^3 t_{\mathfrak{d}}^3) \\
&\oplus (1 \oplus u_{\mathfrak{a}}^2 t_{\mathfrak{a}}^1 \oplus u_{\mathfrak{a}^2}^4 t_{\mathfrak{a}^2}^2 \oplus \frac{u_{\mathfrak{a}^3}^6 t_{\mathfrak{a}^3}^3}{(1-u_{\mathfrak{a}}^2 t_{\mathfrak{a}}^1)}) \otimes_{\mathcal{I}} \frac{u_{\mathfrak{b}^2}^6 t_{\mathfrak{b}^2}^2 (1 \oplus u_{\mathfrak{d}}^3 t_{\mathfrak{d}}^3)}{(1-u_{\mathfrak{b}}^3 t_{\mathfrak{b}}^1)} \\
= \frac{1}{1-u_{\mathfrak{a}}^2 t_{\mathfrak{a}}^1} &\left(\sum_{i=1}^{2} u^{5-2i}_{(\mathfrak{a},\mathfrak{d})^{(i)}} t^4_{(\mathfrak{a},\mathfrak{d})^{(i)}} \oplus u_{(\mathfrak{a}^2,\mathfrak{d})^{(3)}} t^5_{(\mathfrak{a}^2,\mathfrak{d})^{(3)}} \oplus u^2_{(\mathfrak{a}^2,\mathfrak{b}\mathfrak{d})^{(4)}} t^6_{(\mathfrak{a}^2,\mathfrak{b}\mathfrak{d})^{(4)}} \right. \\
&\oplus \sum_{i=5}^{6} u^{12-2i}_{(\mathfrak{a}^3,\mathfrak{b}\mathfrak{d})^{(i)}} t^7_{(\mathfrak{a}^3,\mathfrak{b}\mathfrak{d})^{(i)}} \frac{1 \oplus u_{\mathfrak{d}}^3 t_{\mathfrak{d}}^3}{1-u_{\mathfrak{b}}^3 t_{\mathfrak{b}}^1} \left(1 \oplus \sum_{i=1}^{2} u^{5-2i}_{(\mathfrak{a},\mathfrak{b})^{(i)}} t^2_{(\mathfrak{a},\mathfrak{b})^{(i)}} u_{(\mathfrak{a}^2,\mathfrak{b})^{(3)}} t^3_{(\mathfrak{a}^2,\mathfrak{b})^{(3)}} \right. \\
&\left.\left. \oplus \oplus u^2_{(\mathfrak{a}^2,\mathfrak{b}^2)^{(4)}} t^4_{(\mathfrak{a}^2,\mathfrak{b}^2)^{(4)}} \oplus u^2_{(\mathfrak{a}^3,\mathfrak{b}^2)^{(5)}} t^5_{(\mathfrak{a}^3,\mathfrak{b}^2)^{(5)}} \right) \right) \oplus t^5_{(\mathfrak{a}^3,\mathfrak{b}^2)^{(6)}} G(t,u).
\end{aligned}
$$

This gives us $G(t,u)$. With $H(t,u)$ already expressed in terms of $P(t,u)$, this is enough to compute $P(t,u)$ explicitly.

Therefore the generating function is

$$
\begin{aligned}
P^{3,4}(t,u) = &\frac{1}{(1-t^2_{(\mathfrak{a},\mathfrak{a})^{(2)}})(1-t^4_{(\mathfrak{c},\mathfrak{c})^{(2)}})(1-t^3_{(\mathfrak{a},\mathfrak{c})^{(2)}})(1-u_{\mathfrak{a}}^2 t_{\mathfrak{a}}^1)} \Bigg(\\
&\frac{1 \oplus u_{\mathfrak{d}}^3 t_{\mathfrak{d}}^3}{1-u_{\mathfrak{b}}^3 t_{\mathfrak{b}}^1} \frac{(u_{\mathfrak{c}}^2 t_{\mathfrak{c}}^2 \oplus u^2_{(\mathfrak{a},\mathfrak{c})^{(1)}} t^3_{(\mathfrak{a},\mathfrak{c})^{(1)}})}{1-u_{\mathfrak{c}}^2 t_{\mathfrak{c}}^2} \\
&\oplus \frac{1}{1-t^5_{(\mathfrak{a}^3,\mathfrak{b}^2)^{(6)}}} \left(\sum_{i=1}^{2} u^{5-2i}_{(\mathfrak{a},\mathfrak{d})^{(i)}} t^4_{(\mathfrak{a},\mathfrak{d})^{(i)}} \oplus u_{(\mathfrak{a}^2,\mathfrak{d})^{(3)}} t^5_{(\mathfrak{a}^2,\mathfrak{d})^{(3)}} \right. \\
&\oplus u^2_{(\mathfrak{a}^2,\mathfrak{b}\mathfrak{d})^{(4)}} t^6_{(\mathfrak{a}^2,\mathfrak{b}\mathfrak{d})^{(4)}} \oplus \sum_{i=5}^{6} u^{12-2i}_{(\mathfrak{a}^3,\mathfrak{b}\mathfrak{d})^{(i)}} t^7_{(\mathfrak{a}^3,\mathfrak{b}\mathfrak{d})^{(i)}} \\
&\oplus \frac{1 \oplus u_{\mathfrak{d}}^3 t_{\mathfrak{d}}^3}{1-u_{\mathfrak{b}}^3 t_{\mathfrak{b}}^1} \left(1 \oplus \sum_{i=1}^{2} u^{5-2i}_{(\mathfrak{a},\mathfrak{b})^{(i)}} t^2_{(\mathfrak{a},\mathfrak{b})^{(i)}} \oplus u_{(\mathfrak{a}^2,\mathfrak{b})^{(3)}} t^3_{(\mathfrak{a}^2,\mathfrak{b})^{(3)}} \right. \\
&\left.\left. \oplus\ u^2_{(\mathfrak{a}^2,\mathfrak{b}^2)^{(4)}} t^4_{(\mathfrak{a}^2,\mathfrak{b}^2)^{(4)}} \oplus u^2_{(\mathfrak{a}^3,\mathfrak{b}^2)^{(5)}} t^5_{(\mathfrak{a}^3,\mathfrak{b}^2)^{(5)}} \right) \right) \Bigg)
\end{aligned}
$$

and this can be easily checked to be $(1,7)$-perfect (just in case one would suspect calculational errors).

This corresponds, with $\mathcal{I} = \mathbb{R}[[(\mathfrak{a},\mathfrak{a})^{(2)}, (\mathfrak{a},\mathfrak{c})^{(2)}, (\mathfrak{c},\mathfrak{c})^{(2)}, (\mathfrak{a}^3,\mathfrak{b}^2)^{(6)}]]$, to the following Stanley decomposition of the ring of seminvariants:

$$
\begin{aligned}
&\mathcal{I}[[\mathfrak{a}, \mathfrak{b}, \mathfrak{c}]]\mathfrak{c} \oplus \mathcal{I}[[\mathfrak{a}, \mathfrak{b}, \mathfrak{c}]](\mathfrak{a}, \mathfrak{c})^{(1)} \oplus \mathcal{I}[[\mathfrak{a}, \mathfrak{b}, \mathfrak{c}]]\mathfrak{d} \oplus \mathcal{I}[[\mathfrak{a}, \mathfrak{b}, \mathfrak{c}]]\mathfrak{d}(\mathfrak{a}, \mathfrak{c})^{(1)}) \\
&\oplus \mathcal{I}[[\mathfrak{a}, (\mathfrak{a}^3, \mathfrak{b}^2)^{(6)}]](\mathfrak{a}, \mathfrak{d})^{(1)} \oplus \mathcal{I}[[\mathfrak{a}, (\mathfrak{a}^3, \mathfrak{b}^2)^{(6)}]](\mathfrak{a}, \mathfrak{d})^{(2)} \\
&\oplus \mathcal{I}[[\mathfrak{a}, (\mathfrak{a}^3, \mathfrak{b}^2)^{(6)}]](\mathfrak{a}^2, \mathfrak{d})^{(3)} \oplus \mathcal{I}[[\mathfrak{a}, (\mathfrak{a}^3, \mathfrak{b}^2)^{(6)}]](\mathfrak{a}^2, \mathfrak{b}\mathfrak{d})^{(4)} \\
&\oplus \mathcal{I}[[\mathfrak{a}, (\mathfrak{a}^3, \mathfrak{b}^2)^{(6)}]](\mathfrak{a}^3, \mathfrak{b}\mathfrak{d})^{(5)} \oplus \mathcal{I}[[\mathfrak{a}, (\mathfrak{a}^3, \mathfrak{b}^2)^{(6)}]](\mathfrak{a}^3, \mathfrak{b}\mathfrak{d})^{(6)} \\
&\oplus \mathcal{I}[[\mathfrak{a}, \mathfrak{b}, (\mathfrak{a}^3, \mathfrak{b}^2)^{(6)}]] \oplus \mathcal{I}[[\mathfrak{a}, \mathfrak{b}, (\mathfrak{a}^3, \mathfrak{b}^2)^{(6)}]]\mathfrak{d} \\
&\oplus \mathcal{I}[[\mathfrak{a}, \mathfrak{b}, (\mathfrak{a}^3, \mathfrak{b}^2)^{(6)}]](\mathfrak{a}, \mathfrak{b})^{(1)} \oplus \mathcal{I}[[\mathfrak{a}, \mathfrak{b}, (\mathfrak{a}^3, \mathfrak{b}^2)^{(6)}]]\mathfrak{d}(\mathfrak{a}, \mathfrak{b})^{(1)} \\
&\oplus \mathcal{I}[[\mathfrak{a}, \mathfrak{b}, (\mathfrak{a}^3, \mathfrak{b}^2)^{(6)}]](\mathfrak{a}, \mathfrak{b})^{(2)} \oplus \mathcal{I}[[\mathfrak{a}, \mathfrak{b}, (\mathfrak{a}^3, \mathfrak{b}^2)^{(6)}]]\mathfrak{d}(\mathfrak{a}, \mathfrak{b})^{(2)} \\
&\oplus \mathcal{I}[[\mathfrak{a}, \mathfrak{b}, (\mathfrak{a}^3, \mathfrak{b}^2)^{(6)}]](\mathfrak{a}^2, \mathfrak{b})^{(3)} \oplus \mathcal{I}[[\mathfrak{a}, \mathfrak{b}, (\mathfrak{a}^3, \mathfrak{b}^2)^{(6)}]]\mathfrak{d}(\mathfrak{a}^2, \mathfrak{b})^{(3)} \\
&\oplus \mathcal{I}[[\mathfrak{a}, \mathfrak{b}, (\mathfrak{a}^3, \mathfrak{b}^2)^{(6)}]](\mathfrak{a}^2, \mathfrak{b}^2)^{(4)} \oplus \mathcal{I}[[\mathfrak{a}, \mathfrak{b}, (\mathfrak{a}^3, \mathfrak{b}^2)^{(6)}]]\mathfrak{d}(\mathfrak{a}^2, \mathfrak{b}^2)^{(4)} \\
&\oplus \mathcal{I}[[\mathfrak{a}, \mathfrak{b}, (\mathfrak{a}^3, \mathfrak{b}^2)^{(6)}]](\mathfrak{a}^3, \mathfrak{b}^2)^{(5)} \oplus \mathcal{I}[[\mathfrak{a}, \mathfrak{b}, (\mathfrak{a}^3, \mathfrak{b}^2)^{(6)}]]\mathfrak{d}(\mathfrak{a}^3, \mathfrak{b}^2)^{(5)}.
\end{aligned}
$$

12.7.4 Concluding Remark

Having computed the Stanley decomposition of the seminvariants in these reducible cases, there remains the problem of computing the Stanley decomposition of the normal form. The technique should be clear by now, but it will be a rather long computation.

13

Higher–Level Normal Form Theory

13.1 Introduction

In this chapter we continue the abstract treatment of normal form theory that we started in Chapter 11 with the introduction of the first two terms of the spectral sequence $\mathsf{E}_r^{\cdot,\cdot}$.

A sequence in which

$$\mathsf{E}_{r+1}^{p,q} = H^{p,q}(\mathsf{E}_r^{\cdot,\cdot}, \mathsf{d}^{\cdot,1})$$

is called a spectral sequence, and we compute the first term of such a spectral sequence when we compute the normal form at first level. Can we generalize the construction to get a full sequence in a meaningful way? The generalization seems straightforward enough:

$$\mathcal{Z}_r^{p,q} = \{\mathbf{u}^{[p]} \in \mathcal{Z}_0^{p,q} | d_{\hat{\mathbf{f}}^{[0]}}^{r,1} \mathbf{u}^{[p]} \in \mathcal{Z}_0^{p+r,q+1}\}$$

and

$$\mathsf{E}_{r+1}^{p,q} = \mathcal{Z}_{r+1}^{p,q} / (d_{\hat{\mathbf{f}}^{[0]}}^{r,1} \mathcal{Z}_r^{p-r,q-1} + \mathcal{Z}_r^{p+1,q}).$$

This describes in formulas that to compute the normal form at **level** $r+1$, we use the generators that so far were useless. In the process we ignore the ones that are of higher degree (in $\mathcal{Z}_r^{p+1,q}$) and so trivially end up in $\mathcal{Z}_{r+1}^{p,q}$. The generators of degree $p-r$ in $\mathcal{Z}_r^{p-r,q-1}$ are mapped by $d_{\hat{\mathbf{f}}^{[0]}}^{r,1}$ to terms of degree p.

Before we continue, one should remark that this formulation does not solve any problems. Whole books have been written on the computation of $\mathsf{E}_1^{p,1}$, and there are not many examples in which the higher terms of this sequence have been computed. Nevertheless, by giving the formulation like this and showing that we have indeed a spectral sequence we pose the problem that needs to be solved in a very concrete way, so that one can now give a name to some higher-level calculation, which has a well-defined mathematical meaning.

Another issue that we need to discuss is the following. We have constructed the spectral sequence using $\mathbf{f}^{[0]}$. But in the computational practice one first computes the normal form of level 1, and then one uses this normal form to continue to level 2. Since the normal form computation depends only on the $(r-1)$-jet of $\mathbf{f}^{[0]}$, this approach makes a lot of sense in the description problem, where one is interested in what the normal form looks like given a certain jet of the (normal form of the) vector field. If one is interested only in computing the unique normal form to a certain order, this is less important. We can adapt to this situation by introducing $\mathbf{f}_j^{[0]}$ such that $\mathbf{f}_0^{[0]} = \mathbf{f}^{[0]}$ and the r-jet of $\mathbf{f}_r^{[0]}$ is in normal form with respect to the $(r-1)$-jet of $\mathbf{f}_{r-1}^{[0]}$ for $r > 0$. It turns out that the usual proof that our sequence is indeed a spectral sequence can be easily adapted to this new situation. We also will prove that the spectral sequence is independent of the coordinate transformations, so one can choose any definition one likes.

13.1.1 Some Standard Results

Here we formulate some well-known results, which we need in the next section.

Proposition 13.1.1. *Let* $\mathbf{u}_0^{[0]}, \mathbf{v}_0^{[0]} \in \mathcal{Z}_0^{0,0}$. *Then for any representation* $\rho_q, q = 0, 1,$ *one has*

$$\rho_q^k(\mathbf{v}_0^{[0]})\rho_q(\mathbf{u}_0^{[0]}) = \sum_{i=0}^{k} \binom{k}{i} \rho_q(\rho_0^i(\mathbf{v}_0^{[0]})\mathbf{u}_0^{[0]})\rho_q^{k-i}(\mathbf{v}_0^{[0]}).$$

Proof We prove this by induction on k. For $k = 1$ the statement reads

$$\rho_q(\mathbf{v}_0^{[0]})\rho_q(\mathbf{u}_0^{[0]}) = \rho_q(\mathbf{u}_0^{[0]})\rho_q(\mathbf{v}_0^{[0]}) + \rho_q(\rho_0(\mathbf{v}_0^{[0]})\mathbf{u}_0^{[0]}),$$

and this follows immediately from the fact that ρ_q is a representation. Then it follows from the induction assumption for $k-1$ that

$$\begin{aligned}
&\rho_q^k(\mathbf{v}_0^{[0]})\rho_q(\mathbf{u}_0^{[0]}) = \rho_q(\mathbf{v}_0^{[0]})\rho_q^{k-1}(\mathbf{v}_0^{[0]})\rho_q(\mathbf{u}_0^{[0]}) \\
&= \sum_{i=0}^{k-1} \binom{k-1}{i} \rho_q(\mathbf{v}_0^{[0]})\rho_q(\rho_0^i(\mathbf{v}_0^{[0]})\mathbf{u}_0^{[0]})\rho_q^{k-1-i}(\mathbf{v}_0^{[0]}) \\
&= \sum_{i=0}^{k-1} \binom{k-1}{i} (\rho_q(\rho_0^i(\mathbf{v}_0^{[0]})\mathbf{u}_0^{[0]})\rho(\mathbf{v}_0^{[0]}) + [\rho_q(\mathbf{v}_0^{[0]}), \rho_q(\rho_0^i(\mathbf{v}_0^{[0]})\mathbf{u}_0^{[0]})])\rho_q^{k-1-i}(\mathbf{v}_0^{[0]}) \\
&= \sum_{i=0}^{k-1} \binom{k-1}{i} \rho_q(\rho_0^i(\mathbf{v}_0^{[0]})\mathbf{u}_0^{[0]})\rho_q^{k-i}(\mathbf{v}_0^{[0]}) + \rho_q(\rho_0^{i+1}(\mathbf{v}_0^{[0]})\mathbf{u}_0^{[0]})\rho_q^{k-1-i}(\mathbf{v}_0^{[0]}) \\
&= \sum_{i=0}^{k-1} \binom{k-1}{i} \rho_q(\rho_0^i(\mathbf{v}_0^{[0]})\mathbf{u}_0^{[0]})\rho_q^{k-i}(\mathbf{v}_0^{[0]}) + \sum_{i=1}^{k} \binom{k-1}{i-1} \rho_q(\rho_0^i(\mathbf{v}_0^{[0]})\mathbf{u}_0^{[0]})\rho_q^{k-i}(\mathbf{v}_0^{[0]}) \\
&= \sum_{i=0}^{k} \binom{k}{i} \rho_q(\rho_0^i(\mathbf{v}_0^{[0]})\mathbf{u}_0^{[0]})\rho_q^{k-i}(\mathbf{v}_0^{[0]}),
\end{aligned}$$

and this proves the proposition. □

The little **ad** in the following lemma refers to ρ_0, the big **Ad** is defined in the statement as conjugation.

Lemma 13.1.2 (big Ad–little ad). *Let* $\mathbf{u}_0^{[0]} \in \mathcal{Z}_0^{0,0}$ *and* $\mathbf{v}_0^{[1]} \in \mathcal{Z}_0^{1,0}$. *Then for any filtered representation* ρ_1 *one has*

$$\rho_1(e^{\rho_0(\mathbf{v}_0^{[1]})}\mathbf{u}_0^{[0]}) = e^{\rho_1(\mathbf{v}_0^{[1]})}\rho_1(\mathbf{u}_0^{[0]})e^{-\rho_1(\mathbf{v}_0^{[1]})} = Ad(e^{\rho_1(\mathbf{v}_0^{[1]})})\rho_1(\mathbf{u}_0^{[0]}).$$

Proof We simply compute

$$\begin{aligned}
e^{\rho_1(\mathbf{v}_0^{[1]})}\rho_1(\mathbf{u}_0^{[0]}) &= \sum_{k=0}^{\infty} \frac{1}{k!}\rho_1^k(\mathbf{v}_0^{[1]})\rho_1(\mathbf{u}_0^{[0]}) \\
&= \sum_{k=0}^{\infty} \frac{1}{k!} \sum_{i=0}^{k} \binom{k}{i} \rho_1(\rho_0^i(\mathbf{v}_0^{[1]})\mathbf{u}_0^{[0]})\rho_1^{k-i}(\mathbf{v}_0^{[1]}) \\
&= \sum_{i=0}^{\infty}\sum_{k=i}^{\infty} \frac{1}{i!(k-i)!}\rho_1(\rho_0^i(\mathbf{v}_0^{[1]})\mathbf{u}_0^{[0]})\rho_1^{k-i}(\mathbf{v}_0^{[1]}) \\
&= \sum_{i=0}^{\infty}\sum_{k=0}^{\infty} \frac{1}{i!k!}\rho_1(\rho_0^i(\mathbf{v}_0^{[1]})\mathbf{u}_0^{[0]})\rho_1^{k}(\mathbf{v}_0^{[1]}) \\
&= \sum_{i=0}^{\infty} \frac{1}{i!}\rho_1(\rho_0^i(\mathbf{v}_0^{[1]})\mathbf{u}_0^{[0]})e^{\rho_1(\mathbf{v}_0^{[1]})} \\
&= \rho_1(e^{\rho_0(\mathbf{v}_0^{[1]})}\mathbf{u}_0^{[0]})e^{\rho_1(\mathbf{v}_0^{[1]})},
\end{aligned}$$

and this proves the lemma. □

Corollary 13.1.3. *Let* $\mathbf{u}_0^{[0]} \in \mathcal{Z}_0^{0,0}, \mathbf{v}_0^{[1]} \in \mathcal{Z}_0^{1,0}$, *where* $\mathcal{Z}_0^{0,0}$ *is a filtered Leibniz algebra. Then*

$$\rho_1(e^{\rho_0(\mathbf{v}_0^{[1]})}\mathbf{u}_0^{[0]}) = e^{\rho_1(\mathbf{v}_0^{[1]})}\rho_1(\mathbf{u}_0^{[0]})e^{-\rho_1(\mathbf{v}_0^{[1]})} = Ad(e^{\rho_1(\mathbf{v}_0^{[1]})})\rho_1(\mathbf{u}_0^{[0]}).$$

Corollary 13.1.4. *Let* $\mathbf{u}_0^{[0]} \in \mathcal{Z}_0^{0,0}, \mathbf{v}_0^{[1]} \in \mathcal{Z}_0^{1,0}, \mathbf{f}_0^{[0]} \in \mathcal{Z}_0^{0,1}$, *where* $\mathcal{Z}_0^{0,0}$ *is a filtered Leibniz algebra. Then*

$$d_{\exp(\rho_1(\mathbf{v}_0^{[1]}))\mathbf{f}_0^{[0]}} \exp(\rho_0(\mathbf{v}_0^{[1]}))\mathbf{u}_0^{[0]} = \exp(\rho_1(\mathbf{v}_0^{[1]}))d_{\mathbf{f}_0^{[0]}}\mathbf{u}_0^{[0]}.$$

Proof Follows immediately from Corollary 13.1.3. □

13.2 Abstract Formulation of Normal Form Theory

Let $\mathbf{f}_r^{[0]}$ be a sequence such that each $\mathbf{f}_r^p$ is in $\mathsf{E}_r^{p,1}$ for $p \leq r$ and $\mathbf{f}_r^{[0]}$ is an r-jet of $\mathbf{f}_{r+1}^{[0]}$. Define for $\mathbf{u}_r^{[p]} \in \mathcal{Z}_r^{p,0}$, $d_{\hat{\mathbf{f}}^{[0]}}^{r,1}\mathbf{u}_r^{[p]} = -\rho_1(\mathbf{u}_r^{[p]})\mathbf{f}_r^{[0]} \in \mathcal{Z}_r^{p,1}$, and for $\mathbf{u}_r^{[p]} \in \mathcal{Z}_r^{p,q}, q \neq 0$, $d_{\hat{\mathbf{f}}^{[0]}}^{r,1}\mathbf{u}_r^{[p]} = 0$.

Remark 13.2.1. The following proposition touches on the main difference with the classical formulation in [107] of spectral sequences, where the coboundary operators do not depend on r. For a formulation of normal form theory with an r-independent coboundary operator, see [29].

Proposition 13.2.2. *The* $d^{r,1}_{\hat{\mathbf{f}}^{[0]}}$ *are* **stable**, *that is, if* $\mathbf{u}^{[p]}_r \in \mathcal{Z}^{p,q}_r$ *then* $d^{r,1}_{\hat{\mathbf{f}}^{[0]}}\mathbf{u}^{[p]}_r - d^{r-1,1}_{\hat{\mathbf{f}}^{[0]}}\mathbf{u}^{[p]}_r \in \mathcal{Z}^{p+r,q+1}_0$.

Proof We need to prove this only for $q = 0$. The expression $d^{r,1}_{\hat{\mathbf{f}}^{[0]}}\mathbf{u}^{[p]}_r - d^{r-1,1}_{\hat{\mathbf{f}}^{[0]}}\mathbf{u}^{[p]}_r$ equals $-\rho_1(\mathbf{u}^{[p]}_r)(\mathbf{f}^{[0]}_r - \mathbf{f}^{[0]}_{r-1})$, and we know that $\mathbf{f}^{[0]}_r - \mathbf{f}^{[0]}_{r-1} \in \mathcal{Z}^{r,1}_0$. It follows that $\rho_1(\mathbf{u}^{[p]}_r)(\mathbf{f}^{[0]}_r - \mathbf{f}^{[0]}_{r-1}) \in \mathcal{Z}^{r+p,1}_0$. □

We now formulate two lemmas to show that the spectral sequence $\mathsf{E}^{p,q}_r$ is well defined.

Lemma 13.2.3. $\mathcal{Z}^{p+1,q}_{r-1} \subset \mathcal{Z}^{p,q}_r$.

Proof Let $\mathbf{u}^{[p+1]}_{r-1} \in \mathcal{Z}^{p+1,q}_{r-1}$. Then $\mathbf{u}^{[p+1]}_{r-1} \in \mathcal{Z}^{p+1,q}_0 \subset \mathcal{Z}^{p,q}_0$ and $d^{r-2,1}_{\hat{\mathbf{f}}^{[0]}}\mathbf{u}^{[p+1]}_{r-1} = d^{r-2,1}_{\hat{\mathbf{f}}^{[0]}}\mathbf{u}^{[p+1]}_{r-1} \in \mathcal{Z}^{p+r,q+1}_0$. Therefore $d^{r-1,1}_{\hat{\mathbf{f}}^{[0]}}\mathbf{u}^{[p+1]}_{r-1} = d^{r-2,1}_{\hat{\mathbf{f}}^{[0]}}\mathbf{u}^{[p+1]}_{r-1} + \mathcal{Z}^{p+r,q+1}_p$ and this implies that $d^{r-1,1}_{\hat{\mathbf{f}}^{[0]}}\mathbf{u}^{[p+1]}_{r-1} \in \mathcal{Z}^{p+r,q+1}_0$, that is, $\mathbf{u}^{[p+1]}_{r-1} \in \mathcal{Z}^{p,q}_r$. □

Lemma 13.2.4. $d^{r-1,1}_{\hat{\mathbf{f}}^{[0]}}\mathcal{Z}^{p-r+1,q}_{r-1} \subset \mathcal{Z}^{p,q+1}_r$.

Proof If $\mathbf{f} \in d^{r-1,1}_{\hat{\mathbf{f}}^{[0]}}\mathcal{Z}^{p-r+1,q}_{r-1}$ then there exists a $\mathbf{u}^{[p-r+1]}_{r-1} \in \mathcal{Z}^{p-r+1,q}_{r-1}$ such that $\mathbf{f} = d^{r-1,1}_{\hat{\mathbf{f}}^{[0]}}\mathbf{u}^{[p-r+1]}_{r-1}$. This implies $\mathbf{u}^{[p-r+1]}_{r-1} \in \mathcal{Z}^{p-r+1,q}_0$ and $d^{r-2,1}_{\hat{\mathbf{f}}^{[0]}}\mathbf{u}^{[p-r+1]}_{r-1} \in \mathcal{Z}^{p,q+1}_0$. It follows that $\mathbf{f} = d^{r-2,1}_{\hat{\mathbf{f}}^{[0]}}\mathbf{u}^{[p-r+1]}_{r-1} + (d^{r-1,1}_{\hat{\mathbf{f}}^{[0]}} - d^{r-2,1}_{\hat{\mathbf{f}}^{[0]}})\mathbf{u}^{[p-r+1]}_{r-1} \in \mathcal{Z}^{p,q+1}_0$, and, since $d^{r-1,1}_{\hat{\mathbf{f}}^{[0]}}\mathbf{f} = 0 \in \mathcal{Z}^{p+r,q+1}_0$, $\mathbf{f} \in \mathcal{Z}^{p,q+1}_r$. □

Definition 13.2.5. *We define for* $r \geq 0$,

$$\mathsf{E}^{p,q}_{r+1} = \mathcal{Z}^{p,q}_{r+1}/(d^{r,1}_{\hat{\mathbf{f}}^{[0]}}\mathcal{Z}^{p-r,q-1}_r + \mathcal{Z}^{p+1,q}_r).$$

Lemma 13.2.6. *The coboundary operator* $d^{r,1}_{\hat{\mathbf{f}}^{[0]}}$ *induces a unique (up to coordinate transformations) coboundary operator* $\mathsf{d}^{r,1} : \mathsf{E}^{p,q}_r \to \mathsf{E}^{p+r,q+1}_r$.

Proof It is clear that $d^{r,1}_{\hat{\mathbf{f}}^{[0]}}$ maps $\mathcal{Z}^{p,q}_r$ into $\mathcal{Z}^{p+r,q+1}_r$, by the definition of $\mathcal{Z}^{p,q}_r$ and the fact that $d^{r,1}_{\hat{\mathbf{f}}^{[0]}}$ is a coboundary operator. Furthermore, it maps $d^{p-r+1,1}_{\hat{\mathbf{f}}^{[0]}}\mathcal{Z}^{p-r+1,q-1}_{r-1} + \mathcal{Z}^{p+1,q}_{r-1}$ into $d^{r,1}_{\hat{\mathbf{f}}^{[0]}}\mathcal{Z}^{p+1,q}_{r-1}$. Let $\mathbf{f} \in d^{r,1}_{\hat{\mathbf{f}}^{[0]}}\mathcal{Z}^{p+1,q}_{r-1}$. Then there is a $\mathbf{u}^{[p+1]}_{r-1} \in \mathcal{Z}^{p+1,q}_{r-1}$ such that $\mathbf{f} = d^{r,1}_{\hat{\mathbf{f}}^{[0]}}\mathbf{u}^{[p+1]}_{r-1}$. It follows from Proposition 13.2.2 that $d^{r,1}_{\hat{\mathbf{f}}^{[0]}}\mathbf{u}^{[p+1]}_{r-1} \in d^{r-1,1}_{\hat{\mathbf{f}}^{[0]}}\mathcal{Z}^{p+1,q}_{r-1} + \mathcal{Z}^{p+r+1,q+1}_{r-1}$. Since

$$\mathsf{E}_r^{p+r,q+1} = \mathcal{Z}_r^{p+r,q+1}/(d_{\hat{\mathbf{f}}^{[0]}}^{r-1,1}\mathcal{Z}_{r-1}^{p+1,q} + \mathcal{Z}_{r-1}^{p+r+1,q+1}),$$

it follows that $d_{\hat{\mathbf{f}}^{[0]}}^{r,1}$ induces a coboundary operator $\mathsf{d}^{r,1} : \mathsf{E}_r^{p,q} \to \mathsf{E}_r^{p+r,q+1}$.

Now for the uniqueness: Let $\mathbf{f}_r^{[0]} = \exp(\rho_1(\mathbf{t}_r^{[1]}))\mathbf{f}^{[0]}$, with $\mathbf{t}_r^{[1]} \in \mathcal{Z}_0^{1,0}$. Then, with $\mathbf{u}_r^{[p,q]} \in \mathcal{Z}_r^{p,q}$, and using Corollary 13.1.4,we find that

$$\begin{aligned}
d_{\mathbf{f}^{[0]}}\mathbf{u}_r^{[p,q]} &= -\rho_1(\mathbf{u}_r^{[p,q]})\mathbf{f}^{[0]} \\
&= -\rho_1(\mathbf{u}_r^{[p,q]})\exp(-\rho_1(\mathbf{t}_r^{[1]}))\mathbf{f}_r^{[0]} \\
&= -\exp(-\rho_1(\mathbf{t}_r^{[1]}))\rho_1(\exp(\rho_0(\mathbf{t}_r^{[1]})\mathbf{u}_r^{[p,q]})\mathbf{f}_r^{[0]} \\
&= \exp(-\rho_1(\mathbf{t}_r^{[1]}))d_{\mathbf{f}_r^{[0]}}\exp(\rho_0(\mathbf{t}_r^{[1]}))\mathbf{u}_r^{[p,q]}.
\end{aligned}$$

This shows that

$$d_{\mathbf{f}^{[0]}}\exp(\rho_0(\mathbf{t}_r^{[1]}))\mathcal{Z}_r^{p-r,q-1} = \exp(\rho_1(\mathbf{t}_r^{[1]}))d_{\hat{\mathbf{f}}^{[0]}}^{r,1}\mathcal{Z}_r^{p-r,q-1}.$$

This means that in the definition of $\mathsf{E}_{r+1}^{p,1}$ all terms are transformed with $\exp(\rho_1(\mathbf{t}_r^{[1]}))$, but since higher-order terms are divided out, this acts as the identity. This is illustrated by the following commutative diagram:

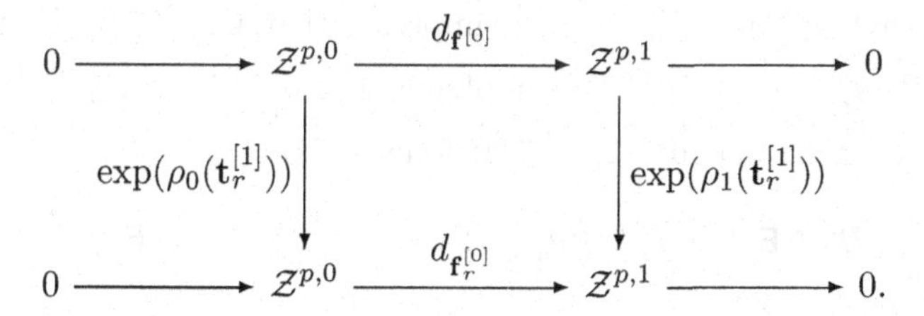

$0 \to \mathcal{Z}^{p,0} \xrightarrow{d_{\mathbf{f}^{[0]}}} \mathcal{Z}^{p,1} \to 0$
$\exp(\rho_0(\mathbf{t}_r^{[1]}))$
$\exp(\rho_1(\mathbf{t}_r^{[1]}))$
$0 \to \mathcal{Z}^{p,0} \xrightarrow{d_{\mathbf{f}_r^{[0]}}} \mathcal{Z}^{p,1} \to 0.$

This shows that $\exp(\rho_\cdot(\mathbf{t}_r^{[1]}))$ is an isomorphism of the $d_{\mathbf{f}^{[0]}}$ complex to the $d_{\mathbf{f}_r^{[0]}}$ complex. We remark, for later use, that this induces an isomorphism of the respective cohomology spaces. □

Lemma 13.2.7. *There exists on the bigraded module $\mathsf{E}_r^{p,q}$ a differential $\mathsf{d}^{r,1}$ such that $H^{p,q}(\mathsf{E}_r)$ is canonically isomorphic to $\mathsf{E}_{r+1}^{p,q}$, $r \geq 0$.*

Proof We follow [107] with modifications to allow for the stable boundary operators. For $\mathbf{u}_r^{[p]} \in \mathcal{Z}_r^{p,q}$ to define a cocycle of degree p on $\mathsf{E}_r^{p,q}$ it is necessary and sufficient that $d_{\hat{\mathbf{f}}^{[0]}}^{r,1}\mathbf{u}_r^{[p]} \in d_{\hat{\mathbf{f}}^{[0]}}^{r-1,1}\mathcal{Z}_{r-1}^{p+1,q} + \mathcal{Z}_{r-1}^{p+r+1,q+1}$, i.e., $d_{\hat{\mathbf{f}}^{[0]}}^{r,1}\mathbf{u}_r^{[p]} = d_{\hat{\mathbf{f}}^{[0]}}^{r-1,1}\mathbf{v}_{r-1}^{[p+1]} + \mathbf{f}_{r-1}^{[p+r+1]}$ with $\mathbf{v}_{r-1}^{[p+1]} \in \mathcal{Z}_{r-1}^{p+1,q}$ and $\mathbf{f}_{r-1}^{[p+r+1]} \in \mathcal{Z}_{r-1}^{p+r+1,q+1}$. Putting $\mathbf{w}^{[p]} = \mathbf{u}_r^{[p]} - \mathbf{v}_{r-1}^{[p+1]} \in \mathcal{Z}_r^{p,q} + \mathcal{Z}_{r-1}^{p+1,q} \subset \mathcal{Z}_0^{p,q}$, with $d_{\hat{\mathbf{f}}^{[0]}}^{r,1}\mathbf{w}^{[p]} = d_{\hat{\mathbf{f}}^{[0]}}^{r-1,1}\mathbf{v}_{r-1}^{[p+1]} - d_{\hat{\mathbf{f}}^{[0]}}^{r,1}\mathbf{v}_{r-1}^{[p+1]} + \mathbf{f}_{r-1}^{[p+r+1]} \in \mathcal{Z}_0^{p+r+1,q+1}$, one has $\mathbf{w}^{[p]} \in \mathcal{Z}_{r+1}^{p,q}$. In other words, $\mathbf{u}_r^{[p]} = \mathbf{v}_{r-1}^{[p+1]} + \mathbf{w}_{r+1}^{[p]} \in \mathcal{Z}_{r-1}^{p+1,q} + \mathcal{Z}_{r+1}^{p,q}$. It follows that the p-cocycles are given by

$$\mathcal{Z}^{p,q}(\mathbf{E}_r) = (\mathcal{Z}_{r+1}^{p,q} + \mathcal{Z}_{r-1}^{p+1,q})/(d_{\hat{\mathbf{f}}^{[0]}}^{r-1,1}\mathcal{Z}_{r-1}^{p-r+1,q-1} + \mathcal{Z}_{r-1}^{p+1,q}). \quad (13.2.1)$$

The space of p-coboundaries $B^{p,q}(\mathbf{E}_r)$ consists of elements of $d_{\hat{\mathbf{f}}^{[0]}}^{r,1}\mathcal{Z}_r^{p-r,q-1}$ and one has

$$B^{p,q}(\mathbf{E}_r) = (d_{\hat{\mathbf{f}}^{[0]}}^{r,1}\mathcal{Z}_r^{p-r,q-1} + \mathcal{Z}_{r-1}^{p+1,q})/(d_{\hat{\mathbf{f}}^{[0]}}^{r-1,1}\mathcal{Z}_{r-1}^{p-r+1,q-1} + \mathcal{Z}_{r-1}^{p+1,q}).$$

It follows, using the isomorphisms $U/(W + U \cap V) \simeq (U + V)/(W + V)$ and $(M/V)/(U/V) \simeq M/U$ for submodules $W \subset U$ and V, that

$$\begin{aligned} H^{p,q}(\mathbf{E}_r) &= (\mathcal{Z}_{r+1}^{p,q} + \mathcal{Z}_{r-1}^{p+1,q})/(d_{\hat{\mathbf{f}}^{[0]}}^{r,1}\mathcal{Z}_r^{p-r,q-1} + \mathcal{Z}_{r-1}^{p+1,q}) \\ &= \mathcal{Z}_{r+1}^{p,q}/(d_{\hat{\mathbf{f}}^{[0]}}^{r,1}\mathcal{Z}_r^{p-r,q-1} + \mathcal{Z}_{r+1}^{p,q} \cap \mathcal{Z}_{r-1}^{p+1,q}), \end{aligned} \quad (13.2.2)$$

since $d_{\hat{\mathbf{f}}^{[0]}}^{r,1}\mathcal{Z}_r^{p-r,q-1} \subset \mathcal{Z}_{r+1}^{p,q}$. We now first prove that $\mathcal{Z}_{r+1}^{p,q} \cap \mathcal{Z}_{r-1}^{p+1,q} = \mathcal{Z}_r^{p+1,q}$. Let $\mathbf{u} \in \mathcal{Z}_{r+1}^{p,q} \cap \mathcal{Z}_{r-1}^{p+1,q}$. Then $\mathbf{u} \in \mathcal{Z}_0^{p+1,q}$ and $d_{\hat{\mathbf{f}}^{[0]}}^{r,1}\mathbf{u} \in \mathcal{Z}_0^{p+r+1}$. This implies $\mathbf{u} \in \mathcal{Z}_r^{p+1,q}$.

On the other hand, if $\mathbf{u} \in \mathcal{Z}_r^{p+1,q}$ we have $\mathbf{u} \in \mathcal{Z}_0^{p+1,q} \subset \mathcal{Z}_0^{p,q}$ and $d_{\hat{\mathbf{f}}^{[0]}}^{r-1,1}\mathbf{u} \in \mathcal{Z}_0^{p+r+1,q+1} \subset \mathcal{Z}_0^{p+r,q+1}$. Thus $\mathbf{u} \in \mathcal{Z}_0^{p,q}$ and $d_{\hat{\mathbf{f}}^{[0]}}^{r-1,1}\mathbf{u} \in \mathcal{Z}_0^{p+r+1,q+1}$. Again it follows that $d_{\hat{\mathbf{f}}^{[0]}}^{r,1}\mathbf{u} \in \mathcal{Z}_0^{p+r+1,q+1}$, implying that $\mathbf{u} \in \mathcal{Z}_{r+1}^{p,q}$. Furthermore, $\mathbf{u} \in \mathcal{Z}_0^{p+1,q}$, $d_{\hat{\mathbf{f}}^{[0]}}^{r-1,1}\mathbf{u} \in \mathcal{Z}_0^{p+r,q+1}$, implying that $d_{\hat{\mathbf{f}}^{[0]}}^{r-1,1}\mathbf{u} \in \mathcal{Z}_0^{p+r+1,q+1}$, from which we conclude that $\mathbf{u} \in \mathcal{Z}_{r-1}^{p+1,q}$. It follows that

$$H^{p,q}(\mathbf{E}_r) = \mathcal{Z}_{r+1}^{p,q}/(d_{\hat{\mathbf{f}}^{[0]}}^{r,1}\mathcal{Z}_r^{p-r,q-1} + \mathcal{Z}_r^{p+1,q}) = \mathbf{E}_{r+1}^{p,q}. \quad (13.2.3)$$

In this way we translate normal form problems into cohomology. □

13.3 The Hilbert–Poincaré Series of a Spectral Sequence

Definition 13.3.1. *Let the* **Euler characteristic** χ_r^p *be defined by*

$$\chi_r^p = \sum_{i=0}^{1}(-1)^{i+1}\dim_{R/\mathfrak{m}}\mathbf{E}_r^{p,i}.$$

Then we define the **Hilbert–Poincaré series** *of* $\mathbf{E}_r^{\cdot}$ *as*

$$P[\mathbf{E}_r^{\cdot}](t) = \sum_{p=0}^{\infty}\chi_r^p t^p.$$

If it exists, we call $I(\mathbf{E}_r^{\cdot}) = P[\mathbf{E}_r^{\cdot}](1)$ the **index of the spectral sequence** at r.

13.4 The Anharmonic Oscillator

Let us, just to get used to the notation, treat the simplest[1] normal form problem we can think of, the anharmonic oscillator. The results we obtain were obtained first in [15, 13].

We will take our coefficients from a local ring R containing $\mathbb{Q}$.[2] Then the noninvertible elements are in the maximal ideal, say $\mathfrak{m}$, and although subsequent computations are going to affect terms that we already consider as fixed in the normal form calculation, they will not affect their equivalence class in the residue field $R/\mathfrak{m}$. So the convergence of the spectral sequence is with respect to the residue field. The actual normal form will contain formal power series that converge in the $\mathfrak{m}$-adic topology. In [13] it is assumed that there is no maximal ideal, and $R = \mathbb{R}$. We denote the units, that is, the invertible elements, in R by $R^\star$. Saying that $x \in R^\star$ amounts to saying that $[x] \neq 0 \in R/\mathfrak{m}$. When in the sequel we say that something is in the kernel of a coboundary operator, this means that the result has its coefficients in $\mathfrak{m}$. When we compute the image then this is done first in the residue field to check invertibility, and then extended to the whole of R. This gives us more accurate information than simply listing the normal form with coefficients in a field, since it allows for terms that have nonzero coefficients, by which we do not want to divide, either because they are very small or because they contain a deformation parameter in such a way that the coefficient is zero for one or more values of this parameter. In the anharmonic oscillator problem, $P[\mathsf{E}_0^{\cdot}](t) = 0$. Let, with $k \geq -1, l \geq 0, q \in \mathbb{Z}/4$, $A^{k+l}_{q,k-l} = i^q(x^{k+1}y^l\frac{\partial}{\partial x} + i^{2q}x^l y^{k+1}\frac{\partial}{\partial y})$. Since $A^k_{q+2,l} = -A^k_{q,l}$, a basis is given by $\langle A^k_{q,l}\rangle_{k=-1,\ldots,l=0,\ldots,q=0,1}$, but we have to compute in $\mathbb{Z}/4$. The commutation relation is

$$[A^{k+l}_{p,k-l}, A^{m+n}_{q,m-n}] = (m-k)A^{k+m+l+n}_{p+q,k-l+m-n} + nA^{k+m+l+n}_{q-p,m-n-(k-l)} \\ -lA^{k+m+l+n}_{p-q,k-l-(m-n)}.$$

Then the anharmonic oscillator is of the form

$$v = A^0_{1,0} + \sum_{q=0}^{1}\sum_{k+l=1}^{\infty} \alpha^{k-l,q}_{k+l} A^{k+l}_{q,k-l}, \alpha^l_k \in R.$$

Since

$$[A^0_{1,0}, A^{k+l}_{q,k-l}] = (k-l)A^{k+l}_{q+1,k-l}$$

we see that the kernel of $\mathsf{d}^{0,1}$ consists of those $A^{k+l}_{q,k-l}$ with $k = l$, and the image of $\mathsf{d}^{0,1}$of those with $k \neq l$. We are now in a position to compute $\mathsf{E}_1^{p,\cdot}$:

---

[1]The analysis of the general one-dimensional vector field is even simpler and makes a good exercise for the reader.

[2]One can think, for instance, of formal power series in a deformation parameter λ, which is the typical situation in bifurcation problems. Then a term $\lambda x^2\frac{\partial}{\partial x}$ has coefficient λ that is neither zero nor invertible, since $\frac{1}{\lambda}$ is not a formal power series.

$$\mathsf{E}_1^{p,0} = \{\mathbf{u}_0^p \in \mathsf{E}_0^{p,0} | [\mathbf{u}_0^p, A_{1,0}^0] = 0\},$$
$$\mathsf{E}_1^{p,1} = \mathsf{E}_0^{p,1} / \mathrm{im}\ \mathsf{d}^{r,1} | \mathsf{E}_0^{p,0} \equiv \{\mathbf{f}_0^p \in \mathsf{E}_0^{p,1} | [\mathbf{f}_0^p, A_{1,0}^0] = 0\},$$

since $\mathfrak{F}^p = \mathrm{im}\ \mathrm{ad}(v_0^0)|_{\mathfrak{F}^p} \oplus \ker\ \mathrm{ad}(v_0^0)|_{\mathfrak{F}^p}$, due to the semisimplicity of $\mathrm{ad}(A_{1,0}^0)$. Here $\mathfrak{F}^p$ denotes both the transformation generators and vector fields of total degree $p+1$ in x and y. It follows that $P[\mathsf{E}_1^{\cdot}](t) = 0$. In general we have

$$\begin{aligned} \mathsf{E}_1^{2p,0} &= \mathsf{E}_1^{2p,1} = \langle A_{0,0}^{2p}, A_{1,0}^{2p} \rangle_{R/\mathfrak{m}}, \\ \mathsf{E}_1^{2p+1,0} &= \mathsf{E}_1^{2p+1,1} = 0. \end{aligned} \tag{13.4.1}$$

Since from now on every $A_{m,l}^k$ has $l = 0$, we write $A_{m,l}^k$ as A_m^k. One has the following commutation relations:

$$[A_p^{2k}, A_q^{2m}] = (m-k)A_{p+q}^{2k+2m} + mA_{q-p}^{2k+2m} - kA_{p-q}^{2k+2m}.$$

For later use we write out the three different cases:

$$\begin{aligned} [A_0^{2k}, A_0^{2m}] &= 2(m-k)A_0^{2k+2m}, \\ [A_0^{2k}, A_1^{2m}] &= 2mA_1^{2k+2m}, \\ [A_1^{2k}, A_1^{2m}] &= 0. \end{aligned}$$

Let $\mathcal{A}_q = \prod_{m \in \mathbb{N}} \langle A_q^{2m} \rangle_{R/\mathfrak{m}}$. It follows that $\mathcal{A}_1 \subset \prod_{m \in \mathbb{N}} \mathsf{E}_1^{2m,1}$ is an invariant module under the action of $\mathsf{E}_1^{\cdot,0}$, which itself is an $\mathbb{N} \times \mathbb{Z}/2$-graded Lie algebra. We can consider $\mathsf{E}_1^{\cdot,0}$ as a central extension of $\mathcal{A}_0 \subset \mathsf{E}_1^{\cdot,0}$ with $\mathcal{A}_1 \subset \mathsf{E}_1^{\cdot,1}$.

We now continue our normal form calculations until we hit a term $v_{2r}^{2r} = \beta_{2r}^0 A_0^{2r} + \beta_{2r}^1 A_1^{2r}$ with either β_{2r}^0 or β_{2r}^1 invertible. We have $\mathsf{E}_{2r}^{\cdot,q} = \mathsf{E}_1^{\cdot,q}$. A general element in $\mathsf{E}_{2r}^{2p,0}$ is given by

$$\mathbf{u}_{2r}^{2p} = \sum_{q=0}^{1} \gamma_{2p}^q A_q^{2p}.$$

We have, with $p > r$,

$$\begin{aligned} \mathsf{d}^{2r,1}\mathbf{u}_{2r}^{2p} &= \beta_{2r}^0 \gamma_{2p}^0 [A_0^{2r}, A_0^{2p}] + \beta_{2r}^0 \gamma_{2p}^1 [A_0^{2r}, A_1^{2p}] + \beta_{2r}^1 \gamma_{2p}^0 [A_1^{2r}, A_0^{2p}] \\ &= 2(p-r)\beta_{2r}^0 \gamma_{2p}^0 A_0^{2p+2r} + 2p\beta_{2r}^0 \gamma_{2p}^1 A_1^{2p+2r} - 2r\beta_{2r}^1 \gamma_{2p}^0 A_1^{2p+2r}. \end{aligned}$$

We view this as a map from the coefficients at $\mathsf{E}_{2r}^{2p,0}$ to those at $\mathsf{E}_{2r}^{2p+2r,1}$ with matrix representation

$$\begin{bmatrix} 2(p-r)\beta_{2r}^0 & 0 \\ -2r\beta_{2r}^1 & 2p\beta_{2r}^0 \end{bmatrix} \begin{bmatrix} \gamma_{2p}^0 \\ \gamma_{2p}^1 \end{bmatrix},$$

and we see that for $0 < p \neq r$ the map is surjective if β_{2r}^0 is invertible; if it is not, it has a one-dimensional image since we assume that in this case β_{2r}^1 is invertible.

13.4.1 Case $\mathcal{A}^r$: β^0_{2r} Is Invertible.

In this subsection we assume that β^0_{2r} is invertible. The following analysis is equivalent to the one in [13, Theorem 4.11, case (3)], $j = r$, if $\beta^1_{2r} = 0$. For $\beta^1_{2r} \neq 0$, see Section 13.4.2. Since $\ker\ \mathsf{d}^{2r,1}|\mathbf{E}^{2r,0}_{2r} = \langle A_0^{2r}\rangle_{R/\mathfrak{m}}$ and $\ker\ \mathsf{d}^{2r,1}|\mathbf{E}^{2r,0}_{2r} = 0$ for $0 < p \neq r$, then $\mathbf{E}^{2r,0}_{2r+2} = H^{2r}(\mathbf{E}^{\cdot,0}_{2r}) = \ker \mathsf{d}^{2r,1} = \langle A_0^{2r}\rangle_{R/\mathfrak{m}}$ and $\mathbf{E}^{2p,0}_{2r+2} = H^{2r}(\mathbf{E}^{\cdot,0}_{2r}) = 0$ for $p > r$. We have already shown that $\mathrm{im}\, \mathsf{d}^{2r,1}|\mathbf{E}^{2p,0}_{2r} = \langle A_0^{2p+2r}, A_1^{2p+2r}\rangle_{R/\mathfrak{m}} = \mathbf{E}^{2p+2r,1}_{2r}$ for $0 < p \neq r$ and $\mathrm{im}\, \mathsf{d}^{2r,1}|\mathbf{E}^{2r,0}_{2r} = \langle A_1^{4r}\rangle_{R/\mathfrak{m}}$. For $0 < p \neq 2r$ we obtain $\mathbf{E}^{2p+2r,1}_{2r+2} = H^{2p+2r}(\mathbf{E}^{\cdot,1}_{2r}) = 0$, while for $p = 2r$ we have $\mathbf{E}^{4r,1}_{2r+2} = H^{4r}(\mathbf{E}^{\cdot,1}_{2r}) = \langle A_0^{2r}\rangle_{R/\mathfrak{m}}$. One has

$$\mathbf{E}^{2p}_{\infty} = \mathbf{E}^{2p}_{2r+2} = \left\{ \begin{array}{llllll} \multicolumn{2}{l}{\mathbf{E}^{2p,1}_{2r+2}} & \multicolumn{2}{l}{\mathbf{E}^{2p,0}_{2r+2}} & \\ & & & & \\ A_0^{2p} & A_1^{2p} & A_0^{2p} & A_1^{2p} & p \\ & & & & \\ \mathfrak{m} & R^\star & R & R & 0 \\ \mathfrak{m} & \mathfrak{m} & 0 & 0 & 1,\ldots,r-1 \\ R^\star & R & R & 0 & r \\ 0 & 0 & 0 & 0 & r+1,\ldots,2r-1 \\ R & 0 & 0 & 0 & 2r \\ 0 & 0 & 0 & 0 & 2r+1,\ldots \end{array} \right. .$$

We see that

$$P^r[\mathbf{E}^{\cdot}_{\infty}](t) = \sum_{i=1}^{r-1} 2t^{2i} + t^{2r} + t^{4r}$$

and $I(\mathbf{E}^{\cdot}_{\infty}) = 2r$. The codimension of the sequence, which we obtain by looking at the dimension of the space with coefficients in $\mathfrak{m}$, is $2r - 1$. We can reconstruct the normal form out of this result. Here $c_{2p} \in \mathfrak{m}$ at position $A^{2p}_{\cdot}$ means that the coefficient of $A^{2p}_{\cdot}$ in c_{2p} cannot be invertible. And $c_{2p} \in R^\star$ means that it should be invertible, while $c_{2p} \in R$ indicates that the coefficient could be anything in R. By ignoring the $\mathfrak{m}A^{2p}_{\cdot}$ terms we obtain the results in [13]. The $R^\star$-terms indicate the organizing center of the corresponding bifurcation problem.

Since we have no more effective transformations at our disposal, all cohomology after this will be trivial, and we have reached the end of our spectral sequence calculation.

13.4.2 Case $\mathcal{A}_r$: β^0_{2r} Is Not Invertible, but β^1_{2r} Is

The following analysis is equivalent to the one in [13, Theorem 4.11, case (4)], $k = r, l = q$. Since

$$\mathsf{d}^{2r,1}\mathbf{u}^{2p}_{2r} = 2(p-r)\beta^0_{2r}\gamma^0_{2p}A_0^{2p+2r} + 2p\beta^0_{2r}\gamma^1_{2p}A_1^{2p+2r} - 2r\beta^1_{2r}\gamma^0_{2p}A_1^{2p+2r}$$

we can remove, using A_0^{2p}, all terms A_1^{2p+2r} for $p > 0$ by taking $\gamma_{2p}^1 = 0$. This contributes only terms in $\mathfrak{m}A_0^{2p+2r}$ since $\beta_{2r}^0 \in \mathfrak{m}$. We obtain

$$\mathsf{E}_{2r+2}^{2p} = \begin{cases} \mathsf{E}_{2r+2}^{2p,1} & \mathsf{E}_{2r+2}^{2p,0} & \\ A_0^{2p}\ A_1^{2p} & A_0^{2p}\ A_1^{2p} & p \\ \mathfrak{m}\quad R^\star & R\quad R & 0 \\ \mathfrak{m}\quad \mathfrak{m} & 0\quad R & 1,\dots,r-1 \\ \mathfrak{m}\quad R^\star & 0\quad R & r \\ R\quad 0 & 0\quad R & r+1,\dots \end{cases}.$$

We see that

$$P_r[\mathsf{E}^{\cdot}_{2r+2}](t) = \sum_{i=1}^{r} t^{2i}$$

and $I(\mathsf{E}^{\cdot}_{2r+2}) = r$. The codimension is $2r$.

Case $\mathcal{A}_r^q$: β_{2q}^0 Is Invertible

We now continue our normal form calculation until at some point we hit on a term

$$\beta_{2q}^0 A_0^{2q}$$

with β_{2q}^0 invertible and $q > r$. The following argument is basically the tic-tic-toe lemma [38, Proposition 12.1], and this was a strong motivation to consider spectral sequences as a framework for normal form theory. The idea is to add the $\mathbb{Z}/2$-grading to our considerations. We view $\mathrm{ad}(A_1^0 + \beta_{2r}^1 A_1^{2r})$ as one coboundary operator $\mathsf{d}^{1,1}$ and $\mathrm{ad}(\beta_{2q}^0 A_0^{2q})$ as another, $\mathsf{d}^{0,1}$. Both operators act completely homogeneously with respect to the gradings induced by the filtering and allow us to consider the bicomplex spanned by $\mathsf{E}^{\cdot}_{1,0}$ and $\mathsf{E}^{\cdot}_{1,1}$, where $\mathsf{E}_{1,0}^{2p,0} = \langle A_0^{2p}\rangle_{R/\mathfrak{m}}$, $\mathsf{E}_{1,0}^{2p,1} = \langle A_0^{2p}\rangle_{R/\mathfrak{m}}$ and $\mathsf{E}_{1,1}^{2p,0} = \langle A_1^{2p}\rangle_{R/\mathfrak{m}}$, $\mathsf{E}_{1,1}^{2p,1} = \langle A_1^{2p}\rangle_{R/\mathfrak{m}}$.

To compute the image of $\mathsf{d}^{1,1} + \mathsf{d}^{0,1}$ we start with the $\mathsf{E}_{2r+2,1}^{2s,0}$-term. Take $\mathbf{u}_1^{2s} \in A_1^{2s}$. Then $\mathsf{d}^{0,1}A_1^{2s} = \beta_{2q}^0[A_0^{2q}, A_1^{2s}] = 2s\beta_{2q}^0 A_1^{2q+2s} \in \mathsf{E}_{2r+2,1}^{2q+2s,1}$. But $\mathsf{E}_{2r+2,1}^{2q+2s,1}$ is trivial, so we can write $\mathsf{d}^{0,1}A_1^{2s} = -\mathsf{d}^{1,1}\mathbf{u}_0^{2s+2q-2r}$, with $\mathbf{u}_0^{2s+2q-2r} \in \mathsf{E}_{2r,0}^{2s+2q-2r,0}$. The reason that this trivial element enters the computation is that it is of higher order than $\mathbf{u}_1^{2s}$, and the cohomology considerations apply only to the lowest-order terms. If we now compute $(\mathsf{d}^{1,1} + \mathsf{d}^{0,1})(\mathbf{u}_1^{2s} + \mathbf{u}_0^{2s+2q-2r})$, we obtain

$$(\mathsf{d}^{1,1} + \mathsf{d}^{0,1})(\mathbf{u}_1^{2s} + \mathbf{u}_0^{2s+2q-2r}) = \mathsf{d}^{0,1}\mathbf{u}_0^{2s+2q-2r}.$$

We see that this gives us a nonzero result under the condition $0 < s \neq r$, since $[A_0^{2q}, A_0^{2s+2q-2r}] = 4(s-r)A_0^{2(2q+s-r)}$. So we find that $\mathsf{E}_{2q,0}^{2s,1} = 0$ for $2q - r < s \neq 2q$, and $\mathsf{E}_{2q,0}^{4q,1} = \langle A_0^{4q}\rangle_{R/\mathfrak{m}}$. And on the other hand, $\mathsf{E}_{2q,1}^{2s,0} = 0$ for

$0 < s \neq r$, and $\mathsf{E}_{2q,1}^{2r,0} = \langle A_0^{2r} \rangle_{R/\mathfrak{m}}$. The term A_0^{2r} stands for the equation itself, which is a symmetry of itself, so it can never be used as an effective normal form transformation. This means we are done. Thus

$$\mathsf{E}_\infty^{2p} = \mathsf{E}_{2q}^{2p} = \left\{ \begin{array}{llllc} \mathsf{E}_{2r+2}^{2p,1} & & \mathsf{E}_{2r+2}^{2p,0} & & \\ A_0^{2p} & A_1^{2p} & A_0^{2p} & A_1^{2p} & p \\ \mathfrak{m} & R^\star & R & R & 0 \\ \mathfrak{m} & \mathfrak{m} & 0 & 0 & 1,\ldots,r-1 \\ \mathfrak{m} & R^\star & 0 & R & r \\ \mathfrak{m} & 0 & 0 & 0 & r+1,\ldots,q-1 \\ R^\star & 0 & 0 & 0 & q \\ R & 0 & 0 & 0 & q+1,\ldots,2q-r \\ 0 & 0 & 0 & 0 & 2q-r+1,\ldots,2q-1 \\ R & 0 & 0 & 0 & 2q \\ 0 & 0 & 0 & 0 & 2q+1,\ldots \end{array} \right. .$$

We see that

$$P_r^q[\mathsf{E}_\infty^{\cdot}](t) = \sum_{i=1}^{r-1} 2t^{2i} + t^{2r} + \sum_{i=r+1}^{q-1} t^{2i} + \sum_{i=q}^{2q-r} t^{2i} + t^{4q}$$

and $I(\mathsf{E}_\infty^{\cdot}) = 2q$. The codimension is $r+q-1$. The A_0^0-term may be used to scale one of the coefficients in $R^\star$ to unity.

Case $\mathcal{A}_r^\infty$: No β_{2q}^0 Is Invertible

The following analysis is equivalent to the one in [13, Theorem 4.11, case (2)], $k = r$.

Since we can eliminate all terms of type A_1^{2p}, and we find no terms of type A_0^{2p} with invertible coefficients, we can draw the conclusion that the cohomology is spanned by the A_0^{2p}, but does not show up in the normal form.

$$\mathsf{E}_\infty^{2p} = \mathsf{E}_{2r}^{2p} = \left\{ \begin{array}{llllc} \mathsf{E}_{2r+2}^{2p,1} & & \mathsf{E}_{2r+2}^{2p,0} & & \\ A_0^{2p} & A_1^{2p} & A_0^{2p} & A_1^{2p} & p \\ \mathfrak{m} & R^\star & R & R & 0 \\ \mathfrak{m} & \mathfrak{m} & 0 & R & 1,\ldots,r-1 \\ \mathfrak{m} & R^\star & 0 & R & r \\ \mathfrak{m} & 0 & 0 & R & r+1,\ldots \end{array} \right. .$$

We see that

$$P_r^\infty[\mathsf{E}_\infty^{\cdot}](t) = \sum_{i=1}^{r} t^{2i}$$

and $I(\mathsf{E}^{\cdot}_{\infty}) = r$. The codimension is infinite. Scaling the coefficient of A_1^{2r} to unity uses up the action of A_0^0. Although we still have some freedom in our choice of transformation, this freedom cannot effectively be used, so it remains in the final result. We summarize the index results as follows.

Corollary 13.4.1. *The index of $\mathcal{A}_r^q$ is $2q$ if $q \in \mathbb{N}$ and r otherwise.*

13.4.3 The $\mathfrak{m}$-adic Approach

So far we have done all computations modulo $\mathfrak{m}$. One can now continue doing the same thing, but now on level $\mathfrak{m}$, and so on. The result will be a finite sequence of $\mathfrak{m}^p \mathcal{A}_{r_p}^{q_p}$ describing exactly what remains. Here the lower index can be either empty, a natural number, or infinity, and the upper index can be a (bigger) natural number or infinity. The generating function will be

$$P[\mathsf{E}^{\cdot}_{\infty}](t) = \sum_p u^p P_{r_p}^{q_p}[\mathsf{E}^{\cdot}_{\infty}](t),$$

with u^p standing for an element in $\mathfrak{m}^p \setminus \mathfrak{m}^{p+1}$.

13.5 The Hamiltonian 1 : 2-Resonance

In this section we analyze the spectral sequence of the Hamiltonian 1 : 2 resonance. This problem was considered in [243], but this paper contains numerous typographical errors, which we hope to repair here. We work in $T^*\mathbb{R}^2$, with coordinates x_1, x_2, y_1, y_2. A Poisson structure is given, with basic bracket relations

$$\{x_i, y_i\} = 1, \{x_i, x_j\} = \{y_i, y_j\} = 0, \quad i, j = 1, 2.$$

Hamiltonians are linear combinations of terms $x_1^k x_2^l y_1^m y_2^n$, and we put a grading deg on these terms by

$$\deg(x_1^k x_2^l y_1^m y_2^n) = k + l + m + n - 2.$$

One verifies that $\deg(\{f, g\}) = \deg(f) + \deg(g)$. The grading induces a filtering, and the *linear fields* consist of quadratic Hamiltonians. In our case, the quadratic Hamiltonian to be considered is

$$\mathrm{H}_{\pm}^0 = \frac{1}{2}(x_1^2 + y_1^2) \pm (x_2^2 + y_2^2).$$

We restrict our attention now to H_+^0 for the sake of simplicity. The computation of $\mathsf{E}^{\cdot}_1$ is standard. We have to determine $\ker \operatorname{ad}(\mathrm{H}_+^0)$, and we find that it is equal to the direct sum of two copies of

$$\mathbb{R}[[B_1, B_2, R_1]] \oplus R_2\mathbb{R}[[B_1, B_2, R_1]],$$

where $B_1 = \mathrm{H}^0_+$, $B_2 = \mathrm{H}^0_-$, and

$$R_1 = x_2(x_1^2 - y_1^2) + 2x_1y_1y_2,$$
$$R_2 = 2x_1x_2y_1 - y_2(x_1^2 - y_1^2),$$

and we have the relation

$$R_1^2 + R_2^2 = \frac{1}{2}(B_1 + B_2)^2(B_1 - B_2).$$

The Poisson brackets are (ignoring B_1, since it commutes with everything)

$$\{B_2, R_1\} = -4R_2,$$
$$\{B_2, R_2\} = 4R_1,$$
$$\{R_1, R_2\} = 3B_2^2 + 2B_1B_2 - B_1^2.$$

We now suppose that our first-level normal form Hamiltonian is $\mathsf{J}^1\overline{\mathrm{H}}^{[0]} = \mathrm{H}^0_+ + \varepsilon_1 R_1 + \varepsilon_2 R_2$, with $\varepsilon = \sqrt{\varepsilon_1^2 + \varepsilon_2^2} \neq 0$. For a complete analysis of this problem, one should also consider the remaining cases, but this has never been attempted, it seems. We now do something that is formally outside the scope of our theory, namely we use a linear transformation in the R_1, R_2-plane, generated by B_2, to transform the Hamiltonian to $\mathsf{J}^1\overline{\mathrm{H}}_1^{[0]} = \mathrm{H}^0_+ + \varepsilon R_1$. One should realize here that the formalism is by no means as general as could be, but since it is already intimidating enough, we have tried to keep it simple. The reader may want to go through the whole theory again to expand it to include this simple linear transformation properly. One should remark that it involves a change of topology, since convergence in the filtration topology will no longer suffice.

Having done this, we now have to determine the image of $\mathrm{ad}(R_1)$. One finds that

$$\mathrm{ad}(\mathsf{J}^1\overline{\mathrm{H}}_1^{[0]})B_2^n R_1^k = 4nB_2^{n-1}R_1^k R_2,$$
$$\mathrm{ad}(\mathsf{J}^1\overline{\mathrm{H}}_1^{[0]})B_2^n R_1^k R_2 = B_2^{n-1}R_1^k(-4nR_1^2 + 2nB_1^3)$$
$$+B_2^n R_1^k((3-2n)B_2^2 + 2(1-n)B_1B_2 + (2n-1)B_1^2))$$

The first relation allows one to remove all terms in $R_2\mathbb{R}[[B_1, B_2, R_1]]$, while the second allows one to remove all terms in $B_2^2\mathbb{R}[[B_1, B_2, R_1]]$, since $2n-3$ is never zero. We now have

$$\mathsf{E}_2^{\cdot;1} = \mathbb{R}[[B_1, R_1]] + B_2\mathbb{R}[[B_1, R_1]] \oplus \mathbb{R}[[B_1, R_1]]$$

(the last statement follows from the first relation with $n = 0$). A moment of consideration shows that this is also the final result, that is, $\mathsf{E}_\infty^{\cdot;1} = \mathsf{E}_2^{\cdot;1}$. It says that the unique normal form is of the form

$$\mathsf{J}^\infty\overline{\mathrm{H}}_\infty^{[0]} = \mathrm{H}^0_+ + F_1(B_1, R_1) + B_2F_2(B_1, R_1),$$

with $\frac{\partial F_1}{\partial R_1}(0,0) = \varepsilon \neq 0$ and $F_2(0,0) = 0$. Furthermore,

$$\mathsf{E}_\infty^{\cdot,0} = \mathbb{R}[[B_1, \mathsf{J}^\infty \overline{\mathrm{H}}^{[0]}]].$$

We have now computed the normal form of the 1 : 2-resonance Hamiltonian under the formal group of symplectic transformations. The reader may want to expand the transformation group to include all formal transformations to see what happens, and to compare the result with the normal form given in [43, page 55].

13.6 Averaging over Angles

For a theoretical method to be the right method, it needs to work in situations that arise in practice. Let us have a look at equations of the form

$$\dot{\boldsymbol{x}} = \sum_{i=1}^{\infty} \varepsilon^i \mathbf{X}^i(\boldsymbol{x}, \theta), \quad \boldsymbol{x} \in \mathbb{R}^n,$$

$$\dot{\theta} = \Omega^0(\boldsymbol{x}) + \sum_{i=1}^{\infty} \varepsilon^i \Omega^i(\boldsymbol{x}, \theta), \quad \theta \in S^1.$$

This equation has a given filtering in powers of ε, and the zeroth-level normal form is

$$\begin{aligned} \dot{\boldsymbol{x}} &= 0, \quad \boldsymbol{x} \in \mathbb{R}^n, \\ \dot{\theta} &= \Omega^0(\boldsymbol{x}), \quad \theta \in S^1. \end{aligned}$$

This means that in our calculations on the spectral sequence level we can consider $\boldsymbol{x}$ as an element of the ring, that is, the ring will be $C^\infty(\mathbb{R}^n, \mathbb{R})$ and the Lie algebra of periodic vector fields on $\mathbb{R}^n \times S^1$ acts on it, but in such a way that the filtering degree increases if we act with the original vector field or one of its normal forms, so that we can effectively assume that the $\boldsymbol{x}$ is a constant in the first-level normal form calculations. The only thing we need to worry about is that we may have to divide through $\Omega^0(\boldsymbol{x})$ in the course of our calculations, thereby introducing small divisors; see Chapter 7 and [236, 238]. This leads to interesting problems, but the formal computation of the normal form is not affected as long as we stay outside the resonance domain. The first-order normal form homological equation is

$$\Omega^0(\boldsymbol{x}) \frac{\partial}{\partial \theta} \begin{bmatrix} \mathbf{Y}^1 \\ \Phi^1 \end{bmatrix} - \begin{bmatrix} 0 \\ \mathbf{Y}^1 \cdot \mathsf{D}\Omega^0 \end{bmatrix} = \begin{bmatrix} \mathbf{X}^1 - \overline{\mathbf{X}}^1 \\ \Omega^1 - \overline{\Omega}^1 \end{bmatrix},$$

and we can solve this equation by taking (with $\mathrm{d}\varphi$ the Haar measure on S^1)

$$\overline{\mathbf{X}}^1(\boldsymbol{x}) = \frac{1}{2\pi} \int_0^{2\pi} \mathbf{X}^1(\boldsymbol{x}, \varphi)\, \mathrm{d}\varphi,$$

and

$$\mathbf{Y}^1(\boldsymbol{x},\theta) = \frac{1}{\Omega^0(\boldsymbol{x})}\int^{\theta}\left(\mathbf{X}^1(\boldsymbol{x},\varphi) - \overline{\mathbf{X}}^1\right)\mathrm{d}\varphi.$$

We let $\overline{\mathbf{Y}}^1(\boldsymbol{x}) = \frac{1}{2\pi}\int_0^{2\pi}\mathbf{Y}^1(\boldsymbol{x},\varphi)\,\mathrm{d}\varphi$ and observe that it is not fixed yet by the previous calculations. We now put

$$\begin{aligned}\overline{\Omega}^1(\boldsymbol{x}) &= \frac{1}{2\pi}\int_0^{2\pi}\Omega^1(\boldsymbol{x},\varphi) + \mathbf{Y}^1(\boldsymbol{x},\varphi)\cdot \mathsf{D}\Omega^0(\boldsymbol{x})\,\mathrm{d}\varphi \\ &= \frac{1}{2\pi}\int_0^{2\pi}\Omega^1(\boldsymbol{x},\varphi)\,\mathrm{d}\varphi + \overline{\mathbf{Y}}^1(\boldsymbol{x})\cdot \mathsf{D}\Omega^0(\boldsymbol{x}),\end{aligned}$$

and we observe that if $\mathsf{D}\Omega^0(\boldsymbol{x}) \neq 0$ we can take $\overline{\mathbf{Y}}^1(\boldsymbol{x})$ in such a way as to make $\overline{\Omega}^1(\boldsymbol{x}) = 0$. All this indicates that the second-level normal form computation will be messy, since there is still a lot of freedom in the choice of $\overline{\mathbf{Y}}^1(\boldsymbol{x})$, and this will have to be carefully used. There do not seem to be any results in the literature on this problem apart from [245, Section 6.3]. We have the following theorem.

Theorem 13.6.1. *Assuming that $\Omega^0(\boldsymbol{x}) \neq 0$ and $\mathsf{D}\Omega^0(\boldsymbol{x}) \neq 0$, one has that $\mathsf{E}_2^{1,1}$ is the space generated by vector fields of the form*

$$\begin{aligned}&\dot{\boldsymbol{x}} = \varepsilon\overline{\mathbf{X}}^1(\boldsymbol{x}), \quad \boldsymbol{x} \in \mathbb{R}^n,\\ &\dot{\theta} = 0, \quad \theta \in S^1,\end{aligned}$$

and $\mathsf{E}_2^{1,0}$ is the space generated by transformations $\overline{\mathbf{Y}}^1(\boldsymbol{x})$ such that

$$\overline{\mathbf{Y}}^1(\boldsymbol{x})\cdot \mathsf{D}\Omega^0(\boldsymbol{x}) = 0.$$

This illustrates that the computation of the spectral sequence is not going to be easy, but also that it mimics the usual normal form analysis exactly.

13.7 Definition of Normal Form

The definition of normal form will now be given as follows.

Definition 13.7.1. *We say that $\mathsf{d}^{\cdot,1} : \mathsf{E}_r^{p+r,1} \to \mathsf{E}_r^{p,0}$ defines an* **operator style** *if $\mathsf{E}_{r+1}^{p,1} = \ker \mathsf{d}^{\cdot,1}|\mathsf{E}_r^{p,1}$.*

Definition 13.7.2. *We say that $\mathbf{f}_m^{[0]}$ is in mth-level* **normal form** *(in a style conforming to Definition 13.7.1) to order q if $\mathbf{f}_m^i \in \ker \mathsf{d}^{\cdot,1}|\mathsf{E}_{m-1}^{i,1}$ for $i = 1,\dots,q$. We say that $\mathbf{u}_m^{[0]}$ is in mth-level* **conormal form** *to order q if $\mathbf{u}_m^i \in \ker \mathsf{d}^{m-1,1}|\mathsf{E}_{m-1}^{i,0}$ for $i = 0,\dots,q$.*

A first-level normal form then is such that $\mathbf{f}_1^i \in \ker \mathsf{d}^{\cdot,1}|\mathsf{E}_0^{i,1}$ for $i > 0$. We drop the notation $\overline{\mathbf{f}}^i$ at this point, since it will not enable us to write down the higher-level normal forms.

13.8 Linear Convergence, Using the Newton Method

We now show how to actually go about computing the normal form, once we can do the linear algebra correctly. We show how convergence in the filtration topology can be obtained using Newton's method once the normal form stabilizes.

Proposition 13.8.1. *Let* $\mathbf{v}^{[1]} \in \mathcal{Z}^{1,0}$ *and* $\mathbf{u}_p^{[1]} \in \mathcal{Z}_p^{1,0}, p \geq 1$. *Then we have the following equality modulo terms containing* $\rho_1^2(\mathbf{u}_p^{[1]})$*:*

$$e^{\rho_1(\mathbf{v}^{[1]}+\mathbf{u}_p^{[1]})} - e^{\rho_1(\mathbf{v}^{[1]})} \simeq \rho_1\left(\frac{1-e^{-\rho_0(\mathbf{v}^{[1]})}}{\rho_0(\mathbf{v}^{[1]})}\mathbf{u}_p^{[1]}\right)e^{\rho_1(\mathbf{v}^{[1]})}.$$

Proof We compute the derivative of exp at $\mathbf{v}^{[1]}$:

$$\begin{aligned}
&e^{\rho_1(\mathbf{v}^{[1]}+\mathbf{u}_p^{[1]})} - e^{\rho_1(\mathbf{v}^{[1]})}\\
&= \sum_{i=0}^{\infty}\frac{1}{i!}\rho_1^i(\mathbf{v}^{[1]}+\mathbf{u}_p^{[1]}) - \sum_{i=0}^{\infty}\frac{1}{i!}\rho_1^i(\mathbf{v}^{[1]})\\
&\simeq \sum_{i=0}^{\infty}\sum_{j=0}^{i-1}\frac{1}{i!}\rho_1^j(\mathbf{v}^{[1]})\rho_1(\mathbf{u}_p^{[1]})\rho_1^{i-ju-1}(\mathbf{v}^{[1]})\\
&= \sum_{i=0}^{\infty}\sum_{j=0}^{i-1}\sum_{k=0}^{j}\frac{1}{i!}\binom{j}{k}\rho_1(\rho_0^k(\mathbf{v}^{[1]})\mathbf{u}_p^{[1]})\rho_1^{i-k-1}(\mathbf{v}^{[1]})\\
&= \sum_{i=0}^{\infty}\sum_{k=0}^{i-1}\frac{(i-k)}{k+1}\binom{i}{k}\frac{1}{i!}\rho_1(\rho_0^k(\mathbf{v}^{[1]})\mathbf{u}_p^{[1]})\rho_1^{i-k-1}(\mathbf{v}^{[1]})\\
&= \sum_{k=0}^{\infty}\sum_{i=k+1}^{\infty}\frac{1}{(k+1)!(i-k-1)!}\rho_1(\rho_0^k(\mathbf{v}^{[1]})\mathbf{u}_p^{[1]})\rho_1^{i-k-1}(\mathbf{v}^{[1]})\\
&= \sum_{k=0}^{\infty}\frac{1}{(k+1)!}\rho_1(\rho_0^k(\mathbf{v}^{[1]})\mathbf{u}_p^{[1]})\exp(\rho_1(\mathbf{v}^{[1]}))\\
&= \rho_1\left(\frac{e^{\rho_0(\mathbf{v}^{[1]})}-1}{\rho_0(\mathbf{v}^{[1]})}\mathbf{u}_p^{[1]}\right)e^{\rho_1(\mathbf{v}^{[1]})},
\end{aligned}$$

and this proves the lemma. □

In the sequel we construct a sequence μ_m, starting with $\mu_0 = 0$. These μ_m indicate the accuracy to which we have a stable normal form $\sum_{i=0}^{\mu_m}\mathbf{f}_{\mu_m}^i$.

In this section we want to consider the linear problem, that is, we want to consider equations of the form

$$\mathsf{d}^{\mu_m,1}\mathbf{u}_{\mu_m}^j = \mathbf{f}_{\mu_m}^{\mu_m+j} - \mathbf{f}_{\mu_m+1}^{\mu_m+j}, \quad j = 1,\ldots,\mu_m+1. \tag{13.8.1}$$

Observe that if these can be solved we obtain quadratic convergence, since the error term was μ_m+1 and it will now be $2(\mu_m+1)$. Notice, however, that a term $\mathbf{u}^j_{\mu_m}$ generates an uncontrolled term $\rho_1^2(\mathbf{u}^j_{\mu_m})\mathbf{f}^{[0]} \in \mathcal{Z}^{\mu_m+2j,1}$, and so would interfere (if we were to recompute the exponential, but now with the correction term $\mathbf{u}^j_{\mu_m}$) with the present system of equations if $2j \leq \mu_m + 1$. The obstruction to quadratic convergence at stage m lies in

$$\mathcal{O}^j_{\mu_m} = \mathsf{E}^{j,0}_{\mu_m} \otimes \mathsf{E}^{\mu_m+j,1}_{\mu_m} \text{ for } 2(j-1) < \mu_m,$$

since we need both an element $\mathbf{f}^{\mu_m+j}_{\mu_m} \in \mathsf{E}^{\mu_m+j,1}_{\mu_m}$ that is to be normalized and an element $\mathbf{u}^j_{\mu_m} \in \mathsf{E}^{j,0}_{\mu_m}$ that can do the job. Choose the minimal j for which $\mathcal{O}^j_{\mu_m} \neq 0$ and put $\mu_{m+1} = \mu_m + 2j - 1 \leq 2\mu_m$. (Since the first uncontrolled term is in $\mathcal{Z}^{\mu_m+2j,1}$, this implies that we can safely continue to solve our homological equations (13.8.1) to order $\mu_m + 2j - 1$.) If no such j exists we put $\mu_{m+1} = 2\mu_m + 1$, thereby guaranteeing quadratic convergence. Since $j \geq 1$, the μ_m-sequence is strictly increasing. We have

$$\mu_m + 1 < \mu_{m+1} + 1 \leq 2(\mu_m + 1).$$

This implies that $\mu_1 = 1$ and μ_2 equals 2 (if $\mathcal{O}^1_1 \neq 0$) or 3 (if $\mathcal{O}^1_1 = 0$, as is always the case in Section 13.4 by equation (13.4.1) since there $\mathsf{E}^{1,0}_1 = 0$).

Exercise 13.8.2. For which cases $\mathcal{A}^q_r$ in Section 13.4 can the normal form be computed with quadratic convergence?

If $\mu_{m+1} = \mu_m + 1$ we speak of linear convergence, when $\mu_{m+1} + 1 = 2(\mu_m + 1)$ of quadratic convergence at step m. The choice of this sequence is determined by our wish to make the computation of the normal form a completely linear problem, where the number of (computationally expensive) exponentiations is minimized. In a concrete application one need not bother with the spectral sequence itself, it is enough to consider the relevant term $\mathbf{f}^{\mu_m+j}_{\mu_m}$ and the corresponding transformation $\mathbf{u}^j_{\mu_m}$. If the transformation is nonzero, we have an obstruction and we put $\mu_{m+1} = \mu_m + 2j - 1$.

Remark 13.8.3. In practice, the difficulty with this approach is that it changes the number of equations to be solved, that is, the order of accuracy to which we compute. To redo the exponential computation every time we increase μ_{m+1} would be self-defeating, since we want to minimize the exponential calculations. One way of handling this would be simply to assume we can double the accuracy and compute enough terms. This is obviously rather wasteful if we have subquadratic convergence. Another (untested) approach is to calculate the spectral sequence from the original vector field. Since the result does not have to be exact, one could even do this over a finite field, to speed things up. This would then give an indication of the optimal choice of accuracy at each exponentiation. ♡

Let $\mathbf{v}^{[1]}_{(0)} = 0$ and suppose $\mathbf{v}^{[1]}_{(m)}$ to be the transformation that brings $\mathbf{f}^{[0]}$ into μ_mth-level normal form to order μ_m, that is, with $m > 0$,

$$\exp\left(\rho_1(\mathbf{v}_{(m)}^{[1]})\right)\mathbf{f}^{[0]} = \sum_{i=0}^{\mu_m} \mathbf{f}_{\mu_m}^i + \mathbf{f}_\star^{[\mu_m+1]}.$$

We now construct $\mathbf{u}_{\mu_m}^{[1]}$ such that

$$\exp\left(\rho_1(\mathbf{v}_{(m)}^{[1]} + \mathbf{u}_{\mu_m}^{[1]})\right)\mathbf{f}^{[0]} = \sum_{i=0}^{\mu_{m+1}} \mathbf{f}_{\mu_{m+1}}^i + \mathbf{f}_\star^{[\mu_{m+1}+1]}.$$

We compute (modulo $\mathcal{Z}_0^{\mu_{m+1}+1,1}$, which we indicate by $\equiv$):

$$\begin{aligned}
&\exp(\rho_1(\mathbf{v}_{(m)}^{[1]} + \mathbf{u}_{\mu_m}^{[1]}))\mathbf{f}^{[0]}\\
&\equiv \exp(\rho_1(\mathbf{v}_{(m)}^{[1]}))\mathbf{f}^{[0]} + \rho_1\left(\frac{e^{\rho_0(\mathbf{v}_{(m)}^{[1]})}-1}{\rho_0(\mathbf{v}_{(m)}^{[1]})}\mathbf{u}_{\mu_m}^{[1]}\right)\exp\rho_1(\mathbf{v}_{(m)}^{[1]})\mathbf{f}^{[0]}\\
&\equiv \sum_{i=0}^{\mu_m}\mathbf{f}_{\mu_m}^i + \sum_{i=\mu_m+1}^{\mu_{m+1}}\mathbf{f}_\star^i + \rho_1\left(\frac{e^{\rho_0(\mathbf{v}_{(m)}^{[1]})}-1}{\rho_0(\mathbf{v}_{(m)}^{[1]})}\mathbf{u}_{\mu_m}^{[1]}\right)\sum_{i=0}^{\mu_m}\mathbf{f}_{\mu_m}^i\\
&\equiv \sum_{i=0}^{\mu_m}\mathbf{f}_{\mu_m}^i + \sum_{i=\mu_m+1}^{\mu_{m+1}}\mathbf{f}_\star^i - \mathsf{d}^{\mu_m,1}\frac{e^{\rho_0(\mathbf{v}_{(m)}^{[1]})}-1}{\rho_0(\mathbf{v}_{(m)}^{[1]})}\mathbf{u}_{\mu_m}^{[1]}.
\end{aligned}$$

Now define $\mathbf{w}_{\mu_m}^{[1]}$ by $\mathsf{d}^{\mu_m,1}\mathbf{w}_{\mu_m}^{[1]} = \sum_{i=\mu_m+1}^{\mu_{m+1}}(\mathbf{f}_\star^i - \mathbf{f}_{\mu_{m+1}}^i)$, and let

$$\mathbf{u}_{\mu_m}^{[1]} = \frac{\rho_0(\mathbf{v}_{(m)}^{[1]})}{e^{\rho_0(\mathbf{v}_{(m)}^{[1]})}-1}\mathbf{w}_{\mu_m}^{[1]}.$$

Now put $\mathbf{v}_{(m+1)}^{[1]} = \mathbf{v}_{(m)}^{[1]} + \mathbf{u}_{\mu_m}^{[1]}$.

After exponentiation we can repeat the whole procedure with m increased by 1. It follows from the relation $\mu_{m-1} < \mu_m$ that we make progress this way, but it may be only one order of accuracy at each step, with $\mu_m = \mu_{m-1} + 1$.

Remark 13.8.4. So far we have not included the filtering of our local ring R in our considerations. There seem to be two ways of doing that.

The first way to look at this is the following: we build a sieve, which filters out those terms that can be removed by normal form computations computing modulo $\mathfrak{m}^i\mathcal{Z}_k^{l,1}$ starting with $i = 1$. We then increase i by one, and repeat the procedure on what is left. Observe that our transformations have their coefficients in $\mathfrak{m}R$, not in $\mathfrak{m}^iR$, in the same spirit of higher-level normal form as we have seen in general. This way, taking the limit for $i \to \infty$, we compute the truly unique normal form description of a certain class of vector fields. Of course, in the description of this process one has to make i an index for the spectral sequence that is being constructed. There seems to be no problem in writing this all out explicitly, but we have not done so in order to avoid unnecessary complications in the main text, but to do so might make a good exercise for the reader.

The second way is to localize with respect to certain divisors. For instance, if δ is some small parameter (maybe a detuning parameter), that is to say, $\delta \in \mathfrak{m}$, then one can encounter terms like $1-\delta$ in the computation (we are not computing modulo $\mathcal{Z}_k^{l,1}$ here!). This may force one to divide through by $1-\delta$, and in doing so repeatedly, one may run into convergence problems, since the zeros of the divisors may approach zero when the order of the computation goes to infinity. Since this is very difficult to realize in practice, this small divisor problem is a theoretical problem for the time being, which may ruin, however, the asymptotic validity of the intermediate results if we want to think of them as approximations of reality. ♡

In general, at each step we can define the rate of progress as the number $\alpha_m \in \mathbb{Q}, m \geq 2$, satisfying $\mu_m = \alpha_m \mu_{m-1} + 1$. One has $1 \leq \alpha_m \leq 2$.

Ideally, one can double the accuracy at each step in the normalization process which consists in solving a linear problem and computing an exponential at each step. Thus we can (ideally) normalize the $2\mu_{m-1}+1$-jet $\sum_{i=\mu_{m-1}+1}^{2\mu_{m-1}+1} \mathbf{f}_{\mu_{m-1}}^i$. We proved that we could normalize the μ_m-jet $\sum_{i=\mu_{m-1}}^{\mu_m} \mathbf{f}_{\mu_{m-1}}^i$. We therefore call $\Delta_m = 2\mu_{m-1} - \mu_m + 1 \overset{m\geq 2}{=} (2-\alpha_m)\mu_{m-1}$ the m-**defect**. If $\Delta_m \leq 0$, the obvious strategy is normalize up to $2\mu_{m-1}+1$. Sooner or later either we will have a positive defect, or we are done normalizing, because we reached our intended order of accuracy. In the next section we discuss what to do in case of positive defect if one still wants quadratic convergence.

Theorem 13.8.5. *The transformation connecting $\mathbf{f}^{[0]} \in \mathcal{Z}^{0,1}$ with its normal form with coefficients in the residue field can be computed at a linear rate at least and at a quadratic rate at theoretical optimum.*

Remark 13.8.6. If $\mathbf{f}^{[0]}$ has an **infinitesimal symmetry**, that is, a $\mathbf{s}^{[0]} \in \mathcal{Z}^{0,0}$ (extending the transformation space to allow for linear group actions) then one can restrict one's attention to $\ker \rho_i(\mathbf{s}^{[0]}), i = 0, 1$, to set the whole thing up, so that the normal form will preserve the symmetry, since $\rho_1(\mathbf{s}^{[0]})\rho_1(\mathbf{t}^{[1]})\mathbf{f}^{[0]} = \rho_1(\mathbf{t}^{[1]})\rho_1(\mathbf{s}^{[0]})\mathbf{f}^{[0]} + \rho_1(\rho_0(\mathbf{s}^{[0]})\mathbf{t}^{[1]})\mathbf{f}^{[0]} = 0$. If one has two of these symmetries $\mathbf{s}_0^{[0]}, \mathbf{q}_0^{[0]}$, then $\rho_0(\mathbf{s}^{[0]})\mathbf{q}^{[0]}$ is again a symmetry, that is, $\rho_1(\rho_0(\mathbf{s}^{[0]})\mathbf{q}^{[0]})\mathbf{f}^{[0]} = [\rho_1(\mathbf{s}^{[0]}), \rho_1(\mathbf{q}^{[0]})]\mathbf{f}^{[0]} = 0$, so the set of all symmetries forms again a Leibniz algebra. By the way, it is not a good idea to do this for every symmetry of the original vector field (why not?).

If G is a group acting on $\mathcal{Z}^{0,i}, i = 0, 1$, then similar remarks apply if the group action respects the Leibniz algebra structure, i.e.,

$$g\rho_i(x)y = \rho_i(gx)gy, \quad x \in \mathcal{Z}^{0,0}, \quad y \in \mathcal{Z}^{0,i} \quad i = 0, 1, \quad \forall g \in G.$$

Indeed, if x and y are G-invariant, so is $\rho_i(x)y$. This can be extended to the case in which the elements in $\mathcal{Z}^{0,1}$ are not invariant, but transform according to the rule $gy = \chi^g y$, where χ is a character taking its values in the ring R. Assuming the representation ρ_1 to be R-linear, that is, $\rho_1(x)ry = r\rho_1(x)y$, it follows that $g\rho_i(x)y = \rho_i(gx)gy = \chi^g\rho_i(x)y$. A familiar example of this situation is that of time-reversible vector fields. ♡

Remark 13.8.7. While we allow for the existence of a nonzero linear part of the vector field $\mathbf{f}^{[0]}$, we do not require it: the whole theory covers the computation of normal forms of vector fields with zero linear part. ♡

Corollary 13.8.8. *If for some m the representation $\boldsymbol{\rho}_1^m$ as induced on the graded Leibniz algebra $\mathsf{E}_m^{\cdot}$ becomes trivial (either for lack of transformations or because the possible transformations can no longer change the normal form), then $\sum_{i=0}^m \mathbf{f}_i^i$ is the* **unique normal form**, *unique in the sense that if it is the normal form of some $\mathbf{g}^{[0]}$, then $\mathbf{f}^{[0]} \equiv \mathbf{g}^{[0]}$.*

13.9 Quadratic Convergence, Using the Dynkin Formula

As we have seen in Section 13.8, one can in the worst case scenario get convergence at only a linear rate using the Newton method. In order to obtain quadratic convergence, we now allow for extra exponential computations within the linear step, hoping that these are less expensive since they are done with *small* transformations. To this end we now introduce the Dynkin formula, which generalizes the results from Proposition 13.8.1.

Lemma 13.9.1. *Let $\mathbf{g}^{[0]} = \exp(\rho_1(\mathbf{u}^{[1]}))\mathbf{f}^{[0]}$ and $\mathbf{h}^{[0]} = \exp(\rho_1(\mathbf{v}^{[k]}))\mathbf{g}^{[0]}$, with $\mathbf{f}^{[0]}, \mathbf{g}^{[0]}, \mathbf{h}^{[0]} \in \mathcal{Z}^{0,1}, \mathbf{u}^{[1]} \in \mathcal{Z}^{1,0}$, and $\mathbf{v}^{[k]} \in \mathcal{Z}^{k,0}, k \geq 1$. Then $\mathbf{h}^{[0]} = \exp(\rho_1(\mathbf{w}^{[1]})\mathbf{f}^{[0]}$, where $\mathbf{w}^{[1]}$ is given by*

$$\mathbf{w}^{[1]} = \mathbf{u}^{[1]} + \int_0^1 \psi[\exp(\rho_0(\varepsilon\mathbf{v}^{[k]}))\exp(\rho_0(\mathbf{u}^{[1]}))]\mathbf{v}^{[k]}\, d\varepsilon,$$

where $\psi(z) = \log(z)/(z-1)$.

Proof This is the right-invariant formulation, which is more convenient in our context, where we think of $\mathbf{v}^{[k]}$ as a perturbation of $\mathbf{u}^{[1]}$. A proof of the left-invariant formulation can be found in [123]. Observe that in the filtration topology all the convergence issues become trivial, so one is left with checking the formal part of the proof, which is fairly easy. The idea is to consider $Z(\varepsilon) = \exp(\rho_0(\varepsilon\mathbf{v}^{[k]}))\exp(\rho_0(\mathbf{u}^{[1]}))$. Then $\frac{dZ}{d\varepsilon}Z^{-1}(\varepsilon) = \rho_0(\mathbf{v}^{[k]})$, and the left hand side is invariant under right-multiplication of $Z(\varepsilon)$ by some ε-independent invertible operator. One then proceeds to solve this differential equation. □

Since the first powers of two are the consecutive numbers $2^0, 2^1$, we can always start our calculation with quadratic convergence. Suppose now for some m, with $\mu_{m-1} = 2^p - 1$, we find $\Delta_m > 0$. So we have $\mu_m = 2^{p+1} - 1 - \Delta_m$ and

$$\mathbf{h}^{[0]} = \exp(\rho_1(\mathbf{u}^{[1]}))\mathbf{f}^{[0]}.$$

Consider now $\mathbf{h}^{[0]}$ as the vector field to be normalized up to level and order $2\mu_{m-1} + 1$. In the next step, until we apply the Dynkin formula, we compute mod $\mathcal{Z}^{2(\mu_{m-1}+1),1}$.

We use the method from Section 13.8 to put $\mathbf{h}^{[0]}$ into μ_mth-level and -order normal form and compute the induced vector field. Then we compute Δ_{m+1} and repeat the procedure until we can put $\mu_m = 2^{p+1} - 1$ and the transformation $\mathbf{v}^{[k]}$ connecting $\mathbf{h}^{[0]}$ with the vector field in $(2^{p+1} - 1)$-level and -order normal form $\mathbf{k}^{[0]}$ by

$$\mathbf{k}^{[0]} = \exp(\rho_1(\mathbf{v}^{[k]}))\mathbf{h}^{[0]}.$$

Then we apply the Dynkin formula and continue our procedure with increased m, until we are done.

With all the intermediate exponentiations, one can not really call this quadratic convergence. Maybe one should use the term **pseudoquadratic convergence** for this procedure. It remains to be seen in practice which method functions best. One may guess that the advantages of the method sketched here will show only at high-order calculations. This has to be weighted against the cost of implementing the Dynkin formula. The Newton method is easy to implement, since it just involves a variation of exponentiation, and: certainly better than just doing things term by term and exponentiating until everything is right. One should also keep in mind that the Newton method keeps on trying to double its accuracy: one may be better off with a sequence $1, 2, 3, 6$ than with $1, 2, 3, 4, 6$. The optimum may depend on the desired accuracy. In principle one could try to develop measures to decide these issues, but that does not seem to be a very attractive course. Computer algebra computations depend on many factors, and it will not be easy to get a realistic cost estimate. If one can just assign some costs to the several actions, this will at best lead to upper estimates, but how is one to show that the best estimated method indeed gives the best actual performance? A more realistic approach is just to experiment with the alternatives until one gets a good feel for their properties.

A

The History of the Theory of Averaging

A.1 Early Calculations and Ideas

Perturbation methods for differential equations became important when scientists in the 18th century were trying to relate Newton's theory of gravitation to the observations of the motion of planets and satellites. Right from the beginning it became clear that a dynamical theory of the solar system based on a superposition of only two-body motions, one body being always the Sun and the other body being formed by the respective planets, produces a reasonable but not very accurate fit to the observations. To explain the deviations one considered effects as the influence of satellites such as the Moon in the case of the Earth, the interaction of large planets such as Jupiter and Saturn, the resistance of the ether and other effects. These considerations led to the formulation of perturbed two-body motion and, as exact solutions were clearly not available, the development of perturbation theory.
The first attempts took place in the first half of the 18th century and involve a numerical calculation of the increments of position and velocity variables from the differentials during successive small intervals of time. The actual calculations involve various ingenious expansions of the perturbation terms to make the process tractable in practice. It soon became clear that this process leads to the construction of astronomical tables but not necessarily to general insight into the dynamics of the problem. Moreover, the tables were not very accurate as to obtain high accuracy one has to take very small intervals of time. An extensive study of early perturbation theory and the construction of astronomical tables has been presented by Wilson [289] and the reader is referred to this work for details and references.

New ideas emerged in the second half of the 18th century by the work of Clairaut, Lagrange and Laplace. It is difficult to settle priority claims as the scientific gentlemen of that time did not bother very much with the acknowledgment of ideas or references. It is clear however that Clairaut had some elegant ideas about particular problems at an early stage and that Lagrange was able to extend and generalize this considerably, while presenting the theory in a clear and to the general public understandable way. Clairaut [58] wrote the solution of the (unperturbed) two-body problem in the form

$$\frac{p}{r} = 1 - c\ \cos(v),$$

where r is the distance of the two bodies, v the longitude measured from aphelion, p is a parameter, c the eccentricity of the conic section. Admitting a perturbation Ω, Clairaut derives the integral equation by a variation of constants procedure; he finds

$$\frac{p}{r} = 1 - c\ \cos(v) + \sin(v)\int \Omega\cos(u)\,du - \cos(v)\int \Omega\sin(u)\,du.$$

The perturbation Ω depends on r, v and maybe other quantities; in the expression for Ω we replace r by the solution of the unperturbed problem and we assume that we may expand in cosines of multiples of v

$$\Omega = A\cos(av) + B\cos(bv) + \cdots .$$

The perturbation part of Clairaut's integral equation contains upon integration terms such as

$$-\frac{A}{a^2-1}\cos(av) - \frac{B}{b^2-1}\cos(bv), \quad a, b \neq 1.$$

If the series for the perturbation term Ω contains a term of the form $\cos(v)$, integration yields terms such as $v\sin(v)$ which represent secular (unbounded) behavior of the orbit. In this case Clairaut adjusts the expansion to eliminate this effect. Although this process of calculating perturbation effects is not what we call averaging now, it has some of its elements. First there is the technique of integrating while keeping slowly varying quantities such as c fixed; secondly there is a procedure to avoid secular terms which is related to the modern approach (see Section 3.3.1). This technique of obtaining approximate solutions is developed and is used extensively by Lagrange and Laplace. The treatment in Laplace's *Traité de Mécanique Céleste* [69] is however very technical and the underlying ideas are not presented to the reader in a comprehensive way. One can find the ingredients of the method of averaging and also higher-order perturbation procedures in Laplace's study of the Sun–Jupiter–Saturn configuration; see for instance Book 2, Chapter 5-8 and Book 6. We shall turn now to the expositions of Lagrange who describes the perturbation method employed in his work in a transparent way. Instead of referring to various papers by Lagrange we cite from the Mécanique Analytique, published in 1788 [165]. After discussing the formulation of motion in dynamics Lagrange argues that to analyze the influence of perturbations one has to use a method which we now call 'variation of parameters'. The start of the 2nd Part, 5th Section, Art. 1 reads in translation:

> 1. All approximations suppose (that we know) the exact solution of the proposed equation in the case that one has neglected some elements or quantities which one considers very small. This solution forms the first-order approximation and one improves this by taking successively into account the neglected quantities.
> In the problems of mechanics which we can only solve by approximation one usually finds the first solution by taking into account only the main forces which act on the body; to extend this solution to other forces which one can call perturbations, the simplest course is to conserve the form of the first solution while making variable the arbitrary constants which it contains; the reason for this is that if the quantities which we have neglected and which we want to take into account are very small, the new variables will be almost constant and we can apply to them the usual methods of approximation. So we have reduced the problem to finding the equations between these variables.

Lagrange then continues to derive the equations for the new variables, which we now call the perturbation equations in the standard form. In Art. 16 of the 2nd Part, 5th Section the decomposition is discussed of the perturbing forces in periodic functions which leads to averaging. In art. 20-24 of the same section a perturbation formulation is given which describes the variation of quantities as the energy. To illustrate the relation with averaging we give Lagrange's discussion of secular perturbations in planetary systems. Lagrange introduces a perturbation term Ω in the discussion of [165, 2nd Part, 7th Section, Art. 76]. This reads in translation:

> To determine the secular variations one has only to substitute for Ω the nonperiodic part of this function, i.e. the first term of the expansion of Ω in the sine and cosine series which depend on the motion of the perturbed planet and the perturbing planets. Ω is only a function of the elliptical coordinates of these planets and provided that the eccentricities and the inclinations are of no importance, we can always reduce these coordinates to a sine and cosine series in angles which are related to anomalies and average longitudes; so we can also expand the function Ω in a series of the same type and the first term which contains no sine or cosine will be the only one which can produce secular equations.

Comparing the method of Lagrange with our introduction of the averaging method in Section 2.8, we note that Lagrange starts by transforming the problem to the standard form

$$\dot{\boldsymbol{x}} = \varepsilon \mathbf{f}^1(\boldsymbol{x}, t) + \mathcal{O}(\varepsilon^2), \ \mathbf{x}(0) = \boldsymbol{x}_0$$

by *variations des constantes.* Then the function $\mathbf{f}^1$ is expanded in what we now call a Fourier series with respect to t, involving coefficients depending on x only

$$\dot{\boldsymbol{x}} = \varepsilon \overline{\mathbf{f}}^1(\boldsymbol{x}) + \varepsilon \sum_{n=1}^{\infty} [a_n(\boldsymbol{x}) \cos(nt) + b_n(\boldsymbol{x}) \sin(nt)] > + \mathcal{O}(\varepsilon^2).$$

Keeping the first, time-independent term yields the secular equation

$$\dot{\boldsymbol{y}} = \varepsilon \overline{\mathbf{f}}^1(\boldsymbol{y}) \ , \ \mathrm{y}(0) = \boldsymbol{x}_0.$$

This equation produces the secular changes of the solutions according to Lagrange [165, 2nd Part, Section 45, §3, Art. 19]. It is precisely the equation obtained by first-order averaging as described in Section 2.8. At the same time no unique meaning is attributed to what we call a first correction to the unperturbed problem. If $\overline{\mathbf{f}}^1 = 0$ it sometimes means replacing in the equation $\boldsymbol{x}$ by $\boldsymbol{x}_0$ so that we have a first-order correction like

$$\boldsymbol{x}(t) = \boldsymbol{x}_0 + \varepsilon \int_0^t \mathbf{f}^1(\boldsymbol{x}, s)\, \mathrm{d}s.$$

Sometimes the first-order correction involves more complicated expressions. This confusion of terminology will last until in the 20th century definitions and proofs have been formulated.

A.2 Formal Perturbation Theory and Averaging

Perturbation theory as developed by Clairaut, Laplace and Lagrange has been used from 1800 onwards as a collection of formal techniques. The theory can be traced in many 19th and 20th century books on celestial mechanics and dynamics; we shall discuss some of its aspects in the work of Jacobi, Poincaré and Van der Pol. See also the book by Born [37].

A.2.1 Jacobi

The lectures of Jacobi [137] on dynamics show a remarkable development of the theoretical foundations of mechanics: the discussion of Hamilton equations of motion, the partial differential equations called after Hamilton-Jacobi and many other aspects. In Jacobi's 36th lecture on dynamics perturbation theory is discussed. The main effort of this lecture is directed towards the use of Lagrange's *variation des constantes* in a canonical way. After presenting the unperturbed problem by Hamilton's equations of motion, Jacobi assumes that the perturbed problem is characterized by a Hamiltonian function. If certain transformations are introduced, the perturbation equations in the standard form are shown to have again the same Hamiltonian structure. This formulation of what we now call canonical perturbation theory has many advantages and it has become the standard formulation in perturbation theory of Hamiltonian mechanics. Note however that this treatment concerns only the way in which the standard perturbation form

$$\dot{\boldsymbol{x}} = \varepsilon \mathbf{f}^1(\boldsymbol{x}, t) + \mathcal{O}(\varepsilon^2)$$

is derived. It represents an extension of the first part of the perturbation theory of Lagrange. The second part, i.e. how to treat these perturbation equations, is discussed by Jacobi in a few lines in which the achievements of Lagrange are more or less ignored. About the introduction of the standard form involving the perturbation Ω Jacobi states (in translation):

> This system of differential equations has the advantage that the first correction of the elements is obtained by simple quadrature. This is obtained on considering the elements as constant in Ω while giving them the values which they had in the unperturbed problem. Then Ω becomes simply a function of time t and the corrected elements are obtained by simple quadrature. The determination of higher corrections is a difficult problem which we do not go into here.

Jacobi does not discuss why Lagrange's secular equation is omitted in this Hamiltonian framework; in fact, his procedure is incorrect as we *do* need the secular equation for a correct description.

A.2.2 Poincaré

We shall base ourselves in this discussion on the two series of books written by Poincaré on celestial mechanics: *Les méthodes nouvelles de la Mécanique Céleste* [218, 219] and the *Leçons de Mécanique Céleste.* The first one, which we shall indicate by *Méthodes*, is concerned with the mathematical foundations of celestial mechanics and dynamical systems; the second one, which we shall indicate by Leçons, aims at the practical use of mathematical methods in celestial mechanics. The *Méthodes* is still a rich source of ideas and methods in mathematical analysis; we only consider here the relation with perturbation theory. In [218, Chapter 3], Poincaré considers the determination of periodic solutions by series expansion with respect to a small parameter. Consider for instance the equation

$$\ddot{x} + x = \varepsilon f(x, \dot{x})$$

and suppose that an isolated periodic solution exists for $0 < \varepsilon << 1$; if $\varepsilon = 0$ all solutions are periodic. Note that this example has some similarity with the case of perturbed Kepler motion. Under certain conditions Poincaré proves that we can describe the periodic solution by a *convergent* series in entire powers of ε, where the coefficients are bounded functions of time. In volume II of the *Méthodes*, Poincaré demonstrates the application of the method and, if the conditions have not been satisfied, its failures to produce convergent series. In the actual calculations Poincaré employs Lagrange's and Jacobi's perturbation formulation supplemented by a secularity condition which is justified for periodic solutions. The conditions which we do not discuss here, are connected with the possibility of continuation or branching of solutions.

It is interesting to note that in the *Méthodes*, Poincaré has also justified the use of divergent series by the introduction of the concept of asymptotic series. It is this concept which nowadays enables us to give a precise meaning to series expansion by averaging methods.
The *Leçons* is concerned with the actual application of the *Méthodes* in celestial mechanics. The first volume deals with the theory of planetary perturbations and contains a very complete discussion of Lagrange's secular perturbation theory (the theory of averaging); moreover, the theory is added to by the study of many details and special cases. The approximations remain formal except in the case of periodic solutions.

A.2.3 Van der Pol

In the theory of nonlinear oscillations the method of Van der Pol is concerned with obtaining approximate solutions for equations of the type

$$\ddot{x} + x = \varepsilon f(x, \dot{x}).$$

In particular for the Van der Pol equation we have

$$f(x, \dot{x}) = (1 - x^2)\dot{x}$$

which arises in studying triode oscillations [269]. Van der Pol introduces the transformation $(x, \dot{x}) \mapsto (a, \phi)$ by

$$x = a\sin(t + \phi),$$
$$\dot{x} = -a\cos(t + \phi).$$

The equation for a can be written as

$$\frac{da^2}{dt} = \varepsilon a^2(1 - \frac{1}{4}a^2) + \cdots ,$$

where the dots stand for higher-order harmonics. Omitting the terms represented by the dots, as they have zero average, Van der Pol obtains an equation which can be integrated to produce an approximation of the amplitude a.
Note that the transformation $x = a\sin(t + \phi)$ is an example of Lagrange's *variation des constantes*. The equation for the approximation of a is the secular equation of Lagrange for the amplitude. Altogether Van der Pol's method is an interesting special example of the perturbation method described by Lagrange in [165].
One might wonder whether Van der Pol realized that the technique which he employed is an example of classical perturbation techniques. The answer is very probably affirmative. Van der Pol graduated in 1916 at the University of Utrecht with main subjects physics, he defended his doctorate thesis in 1920 at the same university. In that period and for many years thereafter the

study of mathematics and physics at the Dutch universities involved celestial mechanics which often contained some perturbation theory. A more explicit answer can be found in [268, pg. 704] on the amplitude of triode vibrations; here Van der Pol states that the equation under consideration

> is closely related to some problems which arise in the analytical treatment of the perturbations of planets by other planets.

This seems to establish the relation of Van der Pol's analysis for triodes with celestial mechanics.

A.3 Proofs of Asymptotic Validity

The first proof of the asymptotic validity of the averaging method was given by Fatou [89]. Assuming periodicity with respect to t and continuous differentiability of the vector field, Fatou uses the Picard-Lindelöf iteration procedure to obtain $\mathcal{O}(\varepsilon)$ estimates on the time scale $1/\varepsilon$. The proof is based essentially on the iteration (contraction) results developed at the end of the 19th century. In the Soviet Union similar results were obtained by Mandelstam and Papalexi [183]. An important step forward is the development and proof of the averaging method in the case of almost-periodic vector fields by Krylov and Bogoliubov in [158]. This is followed by Bogoliubov's averaging results in the general case where for the equation

$$\dot{\boldsymbol{x}} = \varepsilon \mathbf{f}^1(\boldsymbol{x}, t)$$

Bogoliubov requires that the general average exists:

$$\lim_{T\to\infty} \frac{1}{T} \int_0^T \mathbf{f}^1(\boldsymbol{x}, s)\, \mathrm{d}s.$$

An important part has been played by the monograph on nonlinear oscillations by Bogoliubov and Mitropolsky [35]. The book has been very influential because of its presentation of both many examples and an elaborate discussion of the theory. An account of Mitropolsky's theory for systems with coefficients slowly varying with time can also be found in this book.
The theory of averaging has been developed after this for many branches of nonlinear analysis. A transparent proof using the Gronwall inequality for the case of periodic differential equations has been provided by Roseau [228]. Some notes on the literature of new developments in the theory of averaging have already been given in Section 4.1 of the present monograph.

B

A 4-Dimensional Example of Hopf Bifurcation

B.1 Introduction

We present here the essentials of the Hopf bifurcation theory, as far as they might be of use to the actual user, and, on the other hand, we boil down the amount of computations needed, to the point where they will not present the reason for not computing anything at all.
There are many computational techniques and it is difficult to make a choice, not only for the engineer who wants to apply all these ideas to some real life problem, but also for the mathematician, who wants to know what "theorems" can in fact be proven about some asymptotic or numerical approximation. It has been one of our goals to make life a bit easier for both kind of people; we do not believe that practical computability and provability are contradictory requirements on a theory, and certainly not here. On the contrary, it often proves easier to prove something when the computations involved are easy and systematic, than to do the same thing in a method requiring a lot of experience and understanding of the problem, like the method of multiple time scales.

The actual problem to be treated here as the model problem for the application of our techniques, was partly solved in [246]. For two values of the parameter the asymptotic computation was carried out and (successfully) compared to the numerical result obtained separately. Since the asymptotic computations were only done for numerical values of the parameters, no general formula for the bifurcation behavior was obtained by these authors. In this appendix we shall derive such a formula, and we shall also be able to give the approximating solutions, derived by the method of averaging. One of the nice things of the method of averaging is, that we have at our disposal a rather strong result on the validity of the approximations obtained. This has been described in Chapter 5, and we shall not give any details here. It should be noted, however, that one needs in fact a slight generalization of this result, since here we are dealing with two time scales on which attraction occurs.

This is easy to do, when one is familiar with the theory of extension of time scales, but rather complicated by the sheer mass of detail, when one is not.

We shall obtain the following asymptotic results: suppose that a pair of eigenvalues is very nearly purely imaginary, then we use the method of averaging to obtain $\mathcal{O}(\varepsilon)$-approximations with validity on the time scale $0 \leq \varepsilon t \leq L$ for all components of the solution, which is the usual result, but also with validity on $[0, \infty)$ for all components but the angular. The term angular refers to the change to polar coordinates in the stable manifold, that is the plane to which all orbits are attracted.

B.2 The Model Problem

We take our model problem describing a follower-force system from [246] where we refer the reader to for details and explanation; the equations are

$$\begin{aligned}&(m_1+m_2)l^2\ddot{\phi}_1 + m_2 l^2\ddot{\phi}_2\cos(\phi_2-\phi_1) - m_2 l^2\dot{\phi}_2^2\sin(\phi_2-\phi_1)\\ &+2d\dot{\phi}_1 - d\dot{\phi}_2 + 2c\phi_1 - c\phi_2 = -Pl\sin(\phi_2-\phi_1),\\ &m_2 l^2\ddot{\phi}_2 + m_2 l^2\ddot{\phi}_1\cos(\phi_2-\phi_1) + m_2 l^2\dot{\phi}_1^2\sin(\phi_2-\phi_1) - d\dot{\phi}_1\\ &+d\dot{\phi}_2 - c\phi_1 + c\phi_2 = 0.\end{aligned} \tag{B.2.1}$$

Let

$$\tau = \left(\frac{c}{m_2}\right)^{\frac{1}{2}}\frac{t}{l}, \quad B = \frac{d}{l}(cm_2)^{-\frac{1}{2}},$$
$$\theta = \frac{Pl}{c}, \quad \mu = \frac{m_1}{m_2}$$

and scale

$$\phi_i = \varepsilon^{\frac{1}{2}} q_i, \quad i = 1, 2,$$

where ε is a small, positive parameter. Then the system (B.2.1) can be written as a vector field as follows:

$$\mathbb{A}\boldsymbol{q}'' + \mathbb{B}\boldsymbol{q}' + \mathbb{C}\boldsymbol{q} = \varepsilon\mathbf{g}^1(\boldsymbol{q}, \boldsymbol{q}') + \mathcal{O}(\varepsilon^2),$$

where

$$\boldsymbol{q} = \begin{bmatrix} q_1 \\ q_2 \end{bmatrix}, \quad {}' = \frac{d}{d\tau}$$

and, if we take $\mu = 2$,

$$\mathbb{A} = \begin{bmatrix} 3 & 1 \\ 1 & 1 \end{bmatrix}, \quad \mathbb{B} = B\begin{bmatrix} 2 & -1 \\ -1 & 1 \end{bmatrix}, \quad \mathbb{C} = \begin{bmatrix} 2-\theta & \theta-1 \\ -1 & 1 \end{bmatrix}, \quad \mathbf{g}^1 = \begin{bmatrix} g_1 \\ g_2 \end{bmatrix},$$

with

$$\begin{aligned} g_1 &= \frac{1}{4}(q_2 - q_1)^2[5Bp_1 - 4Bp_2 - (\theta - 5)q_1 - (4 - \theta)q_2] + (q_2 - q_1)p_2^2 \\ &\quad + \frac{1}{6}\theta(q_2 - q_1)^3, \\ g_2 &= \frac{1}{4}(q_2 - q_1)^2[-3Bp_1 + 2Bp_2 + (\theta - 3)q_1 + (2 - \theta)q_2] - (q_2 - q_1)p_1^2 \end{aligned}$$

([246, formula 57] does contain two printing errors: in g_1 the cube was written as a square, and in g_2 the factor $\frac{1}{4}$ has been omitted). In the next section we will write the equation as a first order system, and after some simplifying transformations, compute the eigenvalues and -spaces of its linear part.

B.3 The Linear Equation

Consider the equation

$$\mathbb{A}\boldsymbol{q}'' + \mathbb{B}\boldsymbol{q}' + \mathbb{C}\boldsymbol{q} = \boldsymbol{0}.$$

Let $\boldsymbol{p} = \boldsymbol{q}'$, then

$$\frac{d}{d\tau}\begin{bmatrix}\boldsymbol{q}\\ \boldsymbol{p}\end{bmatrix} = \begin{bmatrix}0 & 1\\ -\mathbb{A}^{-1}\mathbb{C} & -\mathbb{A}^{-1}\mathbb{B}\end{bmatrix}\begin{bmatrix}\boldsymbol{q}\\ \boldsymbol{p}\end{bmatrix},$$

provided, of course, that $\mathbb{A}$ is invertible, as is the case in our problem, Let

$$\overline{\boldsymbol{q}} = \mathbb{S}\boldsymbol{q}, \quad \overline{\boldsymbol{p}} = \mathbb{S}\boldsymbol{p},$$

where

$$\mathbb{S} = \begin{bmatrix}1 & 1\\ -1 & 1\end{bmatrix}.$$

Then

$$\frac{d}{d\tau}\begin{bmatrix}\overline{\boldsymbol{q}}\\ \overline{\boldsymbol{p}}\end{bmatrix} = \begin{bmatrix}0 & 1\\ -\mathbb{S}\mathbb{A}^{-1}\mathbb{C}\mathbb{S}^{-1} & -\mathbb{S}\mathbb{A}^{-1}\mathbb{B}\mathbb{S}^{-1}\end{bmatrix}\begin{bmatrix}\overline{\boldsymbol{q}}\\ \overline{\boldsymbol{p}}\end{bmatrix}$$

or

$$\frac{d}{d\tau}\begin{bmatrix}\overline{q}_1\\ \overline{q}_2\\ \overline{p}_1\\ \overline{p}_2\end{bmatrix} = \begin{bmatrix}0 & 0 & 1 & 0\\ 0 & 0 & 0 & 1\\ 0 & -1 & 0 & -B\\ \frac{1}{2} & \theta - \frac{7}{2} & \frac{1}{2}B & -\frac{7}{2}B\end{bmatrix}\begin{bmatrix}\overline{q}_1\\ \overline{q}_2\\ \overline{p}_1\\ \overline{p}_2\end{bmatrix} =: A\begin{bmatrix}\overline{\boldsymbol{q}}\\ \overline{\boldsymbol{p}}\end{bmatrix}.$$

The characteristic equation of A is:

$$\lambda^4 + \frac{7}{2}B\lambda^3 - (\theta - \frac{7}{2} - \frac{1}{2}\,B^2)\lambda^2 + B\lambda + \frac{1}{2} = 0.$$

The Routh–Hurwitz criteria (for stability) are:

$$0 < D_1 = \frac{7}{2}B,$$
$$0 < D_2 = (\frac{7}{4} - (\theta - \theta_{cr}))\frac{7}{2}B, \text{ with } \theta_{cr} = \frac{1}{2}B^2 + \frac{41}{28},$$
$$0 < D_3 = -\frac{7}{2}B^2(\theta - \theta_{cr}).$$

If $B > 0$ and $\theta < \theta_{cr}$, 0 is asymptotically stable. We are interested in the situation where $\delta = \theta - \theta_{cr}$ is small, say $\mathcal{O}(\varepsilon)$. At $\theta = \theta_{cr}$, we find that the equation splits as follows:

$$(\lambda^2 + 2/7)(\lambda^2 + 7B/2\ \lambda + 7/4) = 0$$

and we see that two conjugate eigenvalues are crossing the imaginary axis, while the other two still have strictly negative real parts.

We will show that it suffices to analyze the eigenspaces at $\theta = \theta_{cr}$ in order to compute the eigenvalues and -spaces with $\mathcal{O}(\eta)$-error.

B.4 Linear Perturbation Theory

We split A as follows:

$$A = A_0 + \delta A_p.$$

Suppose we found a transformation T such that

$$T^{-1}A_0T = \begin{bmatrix} \Lambda_1 & 0 \\ 0 & \Lambda_2 \end{bmatrix}$$

and such that the eigenvalues of Λ_1 are on the imaginary axis and the eigenvalues of Λ_2 on the left. Define $A_{i,j}, i, j = 1, 2$ by

$$\begin{bmatrix} A_{11} & A_{12} \\ A_{21} & A_{22} \end{bmatrix} = T^{-1}A_pT, \quad A_{ij} \in M(2, \mathbb{R}).$$

We define a near-identity transform U:

$$U = I + \delta \begin{bmatrix} 0 & X \\ Y & 0 \end{bmatrix}.$$

Then

$$U^{-1} = I - \delta \begin{bmatrix} 0 & X \\ Y & 0 \end{bmatrix} + \mathcal{O}(\delta^2)$$

and

$$U^{-1}T^{-1}ATU = \begin{bmatrix} \Lambda_1 & 0 \\ 0 & \Lambda_2 \end{bmatrix} + \delta \begin{bmatrix} A_{11} & A_{12} + \Lambda_1 X - X\Lambda_2 \\ A_{21} + \Lambda_2 Y - Y\Lambda_1 & A_{22} \end{bmatrix} + \mathcal{O}(\delta^2).$$

Since Λ_1 and Λ_2 have no common eigenvalues, it is possible to solve the equations

$$A_{12} = X\Lambda_2 - \Lambda_1 X,$$
$$A_{21} = Y\Lambda_1 - \Lambda_2 Y$$

(cf, e.g. [26]) and we obtain

$$U^{-1}T^{-1}ATU = \begin{bmatrix} \Lambda_1 + \delta A_{11} & 0 \\ 0 & \Lambda_2 + \delta A_{22} \end{bmatrix} + \mathcal{O}(\delta^2).$$

It is not necessary to compute T^{-1}: it suffices to know only one block; this is due to the simple form of A_p, in the computation of A_{11}. The reader will find no difficulties in following this remark, but since we did compute T^{-1} anyway, we shall follow the straightforward route without thinking. We can take T as follows

$$T = \begin{bmatrix} 1 & B & \frac{4}{7} & -B \\ \frac{2}{7} & 0 & 1 & 0 \\ -\frac{2}{7}B & 1 & -B & 1 \\ 0 & \frac{2}{7} & -\frac{7}{2}B & \frac{7}{4} \end{bmatrix}$$

and then

$$T^{-1} = \frac{-1}{\Delta} \begin{bmatrix} -\frac{41}{28} & \frac{41}{49} - \frac{139}{28}B^2 & \frac{57}{28}B & -2B \\ -B & B(\frac{57}{28} - B^2) & -(\frac{41}{28} + B^2) & \frac{41}{49} \\ \frac{41}{98} & -\frac{1}{14}(\frac{41}{2} + \frac{57}{7}B^2) & -\frac{57}{98}B & \frac{4}{7}B \\ B & B(\frac{82}{7^3} - \frac{7}{2} - B^2) & \frac{82}{7^3} - B^2 & -\frac{41}{49} \end{bmatrix}$$

where

$$\Delta = 2B^2 + \frac{41^2}{4.\ 7^3}.$$

It follows that

$$T^{-1}A^oT = \begin{bmatrix} 0 & 1 & 0 & 0 \\ -\frac{2}{7} & 0 & 0 & 0 \\ 0 & 0 & -\frac{7}{2}B & \frac{7}{4} \\ 0 & 0 & -1 & 0 \end{bmatrix},$$

i.e.

$$\Lambda_1 = \begin{bmatrix} 0 & 1 \\ -\frac{2}{7} & 0 \end{bmatrix};\ \Lambda_2 = \begin{bmatrix} -\frac{7}{2}B & \frac{7}{4} \\ -1 & 0 \end{bmatrix}$$

and

$$A_{11} = \frac{2}{\Delta} \begin{bmatrix} \frac{2}{7}B & 0 \\ -\frac{41}{7^3} & 0 \end{bmatrix}.$$

B.5 The Nonlinear Problem and the Averaged Equations

After the transformations in the linear system, we obtain nonlinear equations of the form

$$\frac{d}{d\tau}\begin{bmatrix} \boldsymbol{x} \\ \boldsymbol{y} \end{bmatrix} = \begin{bmatrix} \Lambda_1 & 0 \\ 0 & \Lambda_2 \end{bmatrix}\begin{bmatrix} \boldsymbol{x} \\ \boldsymbol{y} \end{bmatrix} + \delta\begin{bmatrix} A_{11} & 0 \\ 0 & A_{22} \end{bmatrix}\begin{bmatrix} \boldsymbol{x} \\ \boldsymbol{y} \end{bmatrix} + \varepsilon \mathbf{g}^1_\star(\boldsymbol{x},\boldsymbol{y}) + O((\varepsilon+\delta)^2),$$

with

$$\mathbf{g}^1_\star(\boldsymbol{x},\boldsymbol{y}) = T^{-1}\begin{bmatrix} 0 \\ \mathbb{S}\mathbb{A}^{-1}\mathbf{g}^1(\mathbb{S}^{-1}T_{11}\boldsymbol{x} + \mathbb{S}^{-1}T_{12}\boldsymbol{y}, \mathbb{S}^{-1}T_{21}\boldsymbol{x} + \mathbb{S}^{-1}T_{22}\boldsymbol{y}) \end{bmatrix},$$

where T_{ij} are the 2×2-blocks of T.

The idea is now to average over the action induced by $\exp \Lambda_1 t$ on the vector field and to get rid of the $\boldsymbol{y}$-coordinate, since it is exponentially decreasing and does not influence the system in the first-order approximation.

We refer to [56] for details.

The easiest way to see what is going on, is to transform $\boldsymbol{x}$ to polar coordinates:

$$x_1 = r\sin\omega\phi,$$
$$x_2 = \omega r\cos\omega\phi,$$

where $\omega^2 = \frac{2}{7}$.
The unperturbed equation ($\varepsilon = \delta = 0$) transforms to

$$\dot{r} = 0,$$
$$\dot{\phi} = 1,$$
$$\dot{\boldsymbol{y}} = \Lambda_2\boldsymbol{y},$$

The perturbed equations are a special case of the following type:

$$\dot{r} = \delta\sum_j Y^j_\delta(r,\phi) + \varepsilon\sum_{\alpha,j} Y^j_\alpha(r,\phi)\boldsymbol{y}^\alpha + O((\varepsilon+\delta)^2),$$

$$\dot{\phi} = 1 + \delta\sum\nolimits_j X^j_\delta(r,\phi) + \varepsilon\sum\nolimits_{\alpha,j} X_\alpha{}^j(r,\phi)\boldsymbol{y}^\alpha + O((\varepsilon+\delta)^2),$$

$$\dot{\boldsymbol{y}} = \Lambda_2\boldsymbol{y} + \delta A_{22}\boldsymbol{y} + \varepsilon\sum\nolimits_{\alpha,j} Z_\alpha{}^j(r,\phi)\boldsymbol{y}^\alpha + O((\varepsilon+\delta)^2).$$

Let

$$\mathbb{X}^j_\alpha = \begin{bmatrix} X^j_\alpha \\ Y^j_\alpha \\ Z^j_\alpha \end{bmatrix} \text{ and } \mathbb{X}^j_\delta = \begin{bmatrix} X\delta^j \\ Y^j_\delta \end{bmatrix}.$$

$\mathbb{X}$ and $\mathbb{X}_\delta$ are defined by

$$\frac{\partial^2}{\partial \phi^2}\mathbb{X}^j_{\boldsymbol{\alpha}} + \omega^2 j^2 \mathbb{X}^j_{\boldsymbol{\alpha}} = 0, \quad \frac{\partial^2}{\partial \phi^2}\mathbb{X}^j_{\delta} + \omega^2 j^2 \mathbb{X}^j_{\delta} = 0.$$

$\mathbb{X}^o_{\boldsymbol{\alpha}}$ and $\mathbb{X}^o$ do not depend on ϕ. The notation $\boldsymbol{y}^{\alpha}$ stands for

$$y_1^{\alpha_1} y_2^{\alpha_2}, \alpha_1, \alpha_2 \in \mathbb{N}.$$

It follows from the usual averaging theory (see again [56] for details) that the solutions of these equations can be approximated by the solutions of the averaged system:

$$\begin{aligned}
\dot{r} &= \delta \overline{Y}_\delta(r) + \varepsilon \overline{Y}_0(r), \\
\dot{\phi} &= 1 + \delta \overline{X}_\delta(r) + \varepsilon \overline{X}_0(r), \\
\dot{\boldsymbol{y}} &= \Lambda_2 \boldsymbol{y}.
\end{aligned}$$

These approximations have $\mathcal{O}(\varepsilon + \delta)$-error on the time-interval

$$0 \leq (\varepsilon + \delta)t \leq L$$

(for the $\boldsymbol{y}$-component the interval is $[0, \infty)$).
Clearly, this estimate is sharpest if ε and δ are of the same order of magnitude.

If the averaged equation has an attracting limit-cycle as a solution, then in the domain of attraction the time-interval of validity of the $\mathcal{O}(\varepsilon + \delta)$ -estimate is $[0, \infty)$ for the r and $\boldsymbol{y}$ component.

This makes it possible, in principle, to obtain estimates for the ϕ -component on arbitrary long time scales (in powers of ε, that is) by simply computing higher-order averaged vector fields.
We shall not follow this line of thought here, due to the considerable amount of work and the fact that the results can never be spectacular, since it can only be a regular perturbation of the approximations which we are going to find (This follows from the first-order averaged equations and represents the generic case; in practice one may meet exceptions).

After some calculations, we find the following averaged equations for our problem:

$$\begin{aligned}
\dot{r} &= \frac{B}{\Delta}\left(\frac{2}{7}\delta r - \frac{1}{8 \cdot 7^3}\varepsilon r^3 (\frac{10441}{4 \cdot\ 7^2} + \frac{277}{14}B^2)\right), \\
\dot{\phi} &= 1 + \frac{1}{\Delta}\left(\frac{41}{2 \cdot 7^2}\delta + \frac{3}{8 \cdot 7^2}\varepsilon r^2 (-\frac{41 \cdot\ 109}{8 \cdot 7^3} + \frac{517}{4 \cdot 7^2}B^2 + B^4)\right), \\
\dot{\boldsymbol{y}} &= \Lambda_2 \boldsymbol{y}.
\end{aligned}$$

It is, of course, easy to solve this equation directly, but it is more fun to obtain an asymptotic approximation for large t, without actually solving it:
Consider, with new coefficients $\alpha, \beta, \gamma, \delta \in \mathbb{R}$

$$\begin{aligned}
\dot{\phi} &= 1 + \gamma + \delta r^2, \quad \phi(0) = \phi_0, \\
\dot{r} &= \alpha r - \beta r^3, \quad r(0) = r_0.
\end{aligned}$$

Let

$$r_\infty^2 = \frac{\alpha}{\beta},$$

then

$$\frac{d}{d\tau} \log r = \beta(r_\infty^2 - r^2)$$

and

$$\log\left(\frac{r_\infty}{r_0}\right) = \beta \int_0^\infty (r_\infty^2 - \mathrm{r}^2(s))\, \mathrm{ds}.$$

Clearly

$$\dot{\phi} = 1 + \gamma + \delta r^2 = 1 + \gamma + \delta r_\infty^2 + \delta(r^2 - r_\infty^2)$$

or

$$\begin{aligned}
\phi(t) &= \phi_0 + (1+\gamma+\delta r_\infty^2)t + \delta \int_0^t (\mathrm{r}^2(s) - r_\infty^2)\, \mathrm{ds} \\
&= \phi_0 + (1+\gamma+\delta r_\infty^2)t + \delta \int_0^\infty (\mathrm{r}^2(s) - r_\infty^2)\, \mathrm{ds} - \delta \int_t^\infty (\mathrm{r}^2(s) - r_\infty^2)\, \mathrm{ds} \\
&= \phi_0 + (1+\gamma+\delta r_\infty^2)t + \frac{\delta}{\beta} \log\left(\frac{r_0}{r_\infty}\right) - \delta \int_t^\infty (\mathrm{r}^2(s) - r_\infty^2)\, \mathrm{ds}.
\end{aligned}$$

Now

$$\mathrm{r}^2(t) = r_\infty^2 + \mathcal{O}(e^{-2\alpha t}) \text{ for } t \to \infty,$$

which is clear from the equation for r^2 and the Lyapunov stability estimate, so

$$\phi(t) = \phi_0 + \frac{\delta}{\beta} \log\left(\frac{r_0}{r_\infty}\right) + (1+\gamma+\delta r_\infty^2)t + \mathcal{O}(e^{-2\alpha t}).$$

The phase-shift $\frac{\delta}{\beta}\log\left(\frac{r_0}{r_\infty}\right)$ and especially the frequency-shift $\gamma + \delta r_\infty^2$, can be used to check the asymptotic computational results numerically, and to check the numerical results, by extrapolation in ε, asymptotically.

C

Invariant Manifolds by Averaging

C.1 Introduction

In studying dynamical systems, either generated by maps, ordinary differential equations, partial differential equations, or other deterministic systems, a basic approach is to locate and to characterize the classical ingredients of such systems. These ingredients are critical points (equilibrium solutions), periodic solutions, invariant manifolds (in particular quasiperiodic tori), homoclinics, heteroclinics, and in general stable and unstable manifolds of special solutions.

Here we will discuss invariant manifolds such as slow manifolds, tori, cylinders, with emphasis on the dissipative case. Consider a system such as

$$\dot{\boldsymbol{x}} = \mathbf{f}(\boldsymbol{x}) + \varepsilon \mathbf{f}^{[1]}(\boldsymbol{x}, t, \varepsilon),$$

where ε will indicate a small positive parameter and $\mathbf{f}^{[1]}$ represents a smooth perturbation. Suppose, for instance, that we have found an isolated torus $\mathbb{T}_a$ by first-order averaging or another normalizing technique. Does this manifold persist, slightly deformed as a torus $\mathbb{T}$, when one considers the original equation? Note that the original equation can be seen as a perturbation of an averaged or normalized equation, and the question can then be rephrased as the question of persistence of the torus $\mathbb{T}_a$ under perturbation.

If the invariant manifold in the averaged equation is *normally hyperbolic*, the answer is affirmative (**normally hyperbolic** means loosely speaking that the strength of the flow along the manifold is weaker than the rate of attraction or repulsion to the manifold). We will discuss such cases. In many applications, however, the normal hyperbolicity is not easy to establish. In the Hamiltonian case, the tori arise in families and they will not even be hyperbolic.

We will look at different scenarios for the emergence of tori in some examples. A torus is generated by various independent rotational motions—at least two—and we shall find different time scales characterizing these rotations.

Our emphasis on the analysis of invariant manifolds should be supplemented by appropriate numerical schemes. In [154] and [153], continuation of quasiperiodic invariant tori is studied with a discussion of an algorithm, examples, and extensive references. Another important aspect, which we shall not discuss, is the breakup, or more generally the bifurcations, of tori. Bifurcations of invariant manifolds invoke much more complicated dynamics than bifurcations of equilibria or periodic solutions, and there are still many problems to study; for more details and references see [281].

C.2 Deforming a Normally Hyperbolic Manifold

Consider the dynamical system in $\mathbb{R}^n$, described by the equation

$$\dot{\boldsymbol{x}} = \mathbf{f}(\boldsymbol{x}),$$

and assume that the system contains a smooth (C^r) invariant manifold M. The smoothness enables us to define a *tangent* bundle $\mathsf{T}(\mathsf{M})$ and a *normal* bundle $\mathsf{N}(\mathsf{M})$ of M. A typical situation in mechanics involves N coupled two-dimensional oscillators containing an m-dimensional torus, where $2 \leq m \leq N$. In this case, $n = 2N$, the tangent bundle is m-dimensional, the normal bundle $(2N - m)$-dimensional.

Hyperbolicity is introduced as follows. Assume that we can split the corresponding normal bundle of M with respect to the flow generated by the dynamical system in an exponentially stable one N^s and an exponentially unstable one N^u, with no other components. In differential-geometric terms the flow near the invariant manifold M takes place on

$$\mathsf{N}^s \oplus \mathsf{T}(\mathsf{M}) \oplus \mathsf{N}^u.$$

In this case the manifold M is called hyperbolic. If this hyperbolic splitting does not contain an unstable manifold N^u, M is stable. For a more detailed discussion of these classical matters see, for instance, [130].

Note that the smoothness of M is needed in this description. In many cases the manifolds under consideration lose smoothness at certain bifurcation points when parameters are varied. In such cases, Lyapunov exponents can still be used to characterize the stability.

Moreover, the manifold M is *normally hyperbolic* if, measured in the matrix and vector norms in $\mathbb{R}^n$, N^u expands more sharply than the flow associated with $\mathsf{T}(\mathsf{M})$, and N^s contracts more sharply than $\mathsf{T}(\mathsf{M})$ under the flow.

A number of details and refinements of the concept can be found in [130]; see also [248], [48].

Interestingly, the concept of normal hyperbolicity is used often without explicit definition or even mentioning the term, but is implicitly present in the conditions. Normal hyperbolicity in the case of a smooth manifold can be checked in a relatively simple way; in the case of nonsmoothness we have to adapt the definition.

In many applications, the situation is simpler because a small parameter is present that induces slow and fast dynamics in the dynamical system. Consider the system

$$\dot{\boldsymbol{x}} = \varepsilon \mathbf{f}^1(\boldsymbol{x}, \boldsymbol{y}), \quad \boldsymbol{x} \in D \subset \mathbb{R}^n, t \geq 0,$$
$$\dot{\boldsymbol{y}} = \mathbf{g}^0(\boldsymbol{x}, \boldsymbol{y}), \quad \boldsymbol{y} \in G \subset \mathbb{R}^m,$$

with $\mathbf{f}^1$ and $\mathbf{g}^0$ sufficiently smooth vector functions in $\boldsymbol{x}, \boldsymbol{y}$. Putting $\varepsilon = 0$ we have $\mathbf{x}(t) = \mathbf{x}(0) = \boldsymbol{x}_0$ and from the second equation $\dot{\boldsymbol{y}} = \mathbf{g}^0(\boldsymbol{x}_0, \boldsymbol{y})$, for which we assume $\overline{\boldsymbol{y}} = \phi(\boldsymbol{x}_0)$ to be an isolated root corresponding to a compact manifold ($\phi(\boldsymbol{x})$ is supposed to be a continuous function near $\boldsymbol{x} = \boldsymbol{x}_0$). Fenichel has shown in [92], [93], [91], and [94] that if this root is hyperbolic, it corresponds to a nearby hyperbolic invariant manifold of the full system, a so-called *slow manifold.* In the analysis, the fact that if this root is hyperbolic, the corresponding manifold is also normally hyperbolic, is inherent in the problem formulation. For the fibers of the slow manifold are ruled by the fast time variable t, while the dynamics of the drift along the manifold is ruled by the time variable εt.

A simple example of a normally hyperbolic torus with small perturbations is the following system:

Example C.2.1.

$$\ddot{x} + x = \mu(1 - x^2)\dot{x} + \varepsilon f(x, y),$$
$$\ddot{y} + \omega^2 y = \mu(1 - y^2)\dot{y} + \varepsilon g(x, y),$$

with ε-independent positive constants ω and μ (fixed positive numbers, $\mathcal{O}(1)$ with respect to ε) and smooth perturbations f, g. Omitting the perturbations f, g we have two uncoupled normally hyperbolic oscillations. In general, if ω is irrational, the combined oscillations attract to a torus in 4-space, the product of the two periodic attractors, filled with quasiperiodic motion. Adding the perturbations f, g cannot destroy this torus but only deforms it. In this example the torus is two-dimensional but the time scales of rotation, if μ is large enough, are in both directions determined by the time scales of relaxation oscillation; see [112]. ◊

There are natural extensions to nonautonomous systems by introducing the so-called stroboscopic map. We demonstrate this by an example derived from [48]. See also the monograph [46].

Example C.2.2. Consider the forced Van der Pol oscillator

$$\ddot{x} + x = \mu(1 - x^2)\dot{x} + \varepsilon \cos \omega t,$$

which we write as the system

$$\dot{x} = y,$$
$$\dot{y} = -y + \mu(1 - x^2)y + \varepsilon \cos \tau,$$
$$\dot{\tau} = \omega.$$

The 2π-periodic forcing term $\varepsilon \cos \tau$ produces a stroboscopic map of the x, y-plane into itself. For $\varepsilon = 0$ this is just the map of the periodic solution of the Van der Pol equation, an invariant circle, into itself, and the closed orbit is normally hyperbolic. In the extended phase space $\mathbb{R}^2 \times \mathbb{R}/2\pi\mathbb{Z}$ this invariant circle for $\varepsilon = 0$ corresponds to a normally hyperbolic torus that is persistent for small positive values of ε.

Actually, the authors, choosing $\mu = 0.4, \omega = 0.9$, consider what happens if ε increases. At $\varepsilon = 0.3634$ the normal hyperbolicity is destroyed by a saddle-node bifurcation. ◊

C.3 Tori by Bogoliubov-Mitropolsky-Hale Continuation

The branching off of tori is more complicated than the emergence of periodic solutions in dynamical system theory. The emergence of tori was considered extensively in [35], using basically continuation of quasiperiodic motion under perturbations; for a summary and other references see also [34]. Another survey together with new results can be found in [121]; see the references there. A modern formulation in the more general context of bifurcation theory can be found in [55].

We present several theorems from [121] in an adapted form; see also [120].

Theorem C.3.1. *Consider the system S,*

$$\begin{aligned} \dot{\boldsymbol{x}} &= \mathbf{A}^0(\boldsymbol{\theta})\boldsymbol{x} + \varepsilon \mathbf{A}^1(\boldsymbol{x}, \boldsymbol{y}, \boldsymbol{\theta}, t) + \varepsilon^2 \cdots , \\ \dot{\boldsymbol{y}} &= \mathbf{B}^0(\boldsymbol{\theta})\boldsymbol{y} + \varepsilon \mathbf{B}^1(\boldsymbol{x}, \boldsymbol{y}, \boldsymbol{\theta}, t) + \varepsilon^2 \cdots , \\ \dot{\boldsymbol{\theta}} &= \omega(\boldsymbol{\theta}, t) + \varepsilon \omega^1(\boldsymbol{x}, \boldsymbol{y}, \boldsymbol{\theta}, t) + \varepsilon^2 \cdots , \end{aligned}$$

with $\boldsymbol{x} \in \mathbb{R}^n, \boldsymbol{y} \in \mathbb{R}^m, \boldsymbol{\theta} \in \mathbb{T}^k$; all vector functions on the right-hand side are periodic in $\boldsymbol{\theta}$ and t.

Such a system arises naturally from local perturbations of differential equations in a neighborhood of an invariant manifold where the "unperturbed" system

$$\dot{\boldsymbol{x}} = \mathbf{A}^0(\boldsymbol{\theta})\boldsymbol{x}, \quad \dot{\boldsymbol{y}} = \mathbf{B}^0(\boldsymbol{\theta})\boldsymbol{y}, \quad \dot{\boldsymbol{\theta}} = \boldsymbol{\omega}^0(\boldsymbol{\theta}, t),$$

is assumed to have an invariant manifold M_0 given by

$$\mathsf{M}_0 = \{(\boldsymbol{x}, \boldsymbol{y}, \boldsymbol{\theta}, t) : \boldsymbol{x} = \boldsymbol{y} = 0\}.$$

We also assume for system S that

1. *All vector functions on the right-hand side are continuous and bounded; the $\mathcal{O}(\varepsilon^2)$ terms represent vector functions that are smooth on the domain and that can be estimated $\mathcal{O}(\varepsilon^2)$.*
2. *The functions on the right-hand side are Lipschitz continuous with respect to θ, the function $\boldsymbol{\omega}^0(\boldsymbol{\theta}, t)$ with Lipschitz constant λ_{ω^0}.*
3. *The functions $\mathbf{A}^1, \mathbf{B}^1, \boldsymbol{\omega}^1$ are Lipschitz continuous with respect to $\boldsymbol{x}, \boldsymbol{y}$.*

4. *There exist positive constants* K *and* α *such that for any continuous* $\boldsymbol{\theta}(t)$ *the fundamental matrices of* $\dot{\boldsymbol{x}} = \mathbf{A}^0(\boldsymbol{\theta})\boldsymbol{x}, \dot{\boldsymbol{y}} = \mathbf{B}^0(\boldsymbol{\theta})\boldsymbol{y}$ *can be estimated by* $Ke^{-\alpha t}, Ke^{\alpha t}$ *respectively.*
5. $\alpha > \lambda_{\boldsymbol{\omega}^0}$ *(normal hyperbolicity).*

Then there exists an invariant manifold M *of system* $\mathcal{S}$ *near* M_0 *with Lipschitz continuous parametrization that is periodic in* θ.

Note that although α and $\lambda_{\boldsymbol{\omega}^0}$ are independent of ε, the difference may be small. In the applications one should take care that $\varepsilon = o(\alpha - \lambda_{\boldsymbol{\omega}^0})$.

Another remark is that Hale's results are much more general than Theorem C.3.1. For instance, the vector functions need not be periodic in $\boldsymbol{\theta}$, but only bounded. If the vector functions are almost periodic, the parametrization of M inherits almost periodicity.

Even more importantly, the perturbations $\varepsilon\mathbf{A}^1, \varepsilon\mathbf{B}^1$ in the equations for $\boldsymbol{x}$ and $\boldsymbol{y}$ can be replaced by $\mathcal{O}(1)$ vector functions. However, this complicates the conditions of the corresponding theorem. Also, to check the conditions in these more general cases is not so easy.

We turn now to a case arising often in applications.

C.4 The Case of Parallel Flow

In a number of important applications the frequency vector $\boldsymbol{\omega}^0(\boldsymbol{\theta}, t)$ of system $\mathcal{S}$ is constant; this will cause the flow on M to be parallel. In this case $\lambda_{\boldsymbol{\omega}^0} = 0$, and the fifth condition of Theorem C.3.1 is automatically satisfied.

In addition, the case of parallel flow makes it easier to consider cases in which the attraction or expansion is weak:

Theorem C.4.1. *Consider the system* $\mathcal{S}_w$,

$$\begin{aligned} \dot{\boldsymbol{x}} &= \varepsilon\mathbf{A}^0(\boldsymbol{\theta})\boldsymbol{x} + \varepsilon\mathbf{A}^1(\boldsymbol{\theta}, \boldsymbol{x}, \boldsymbol{y}, t) + \mathcal{O}(\varepsilon^2), \\ \dot{\boldsymbol{y}} &= \varepsilon\mathbf{B}^0(\boldsymbol{\theta})\boldsymbol{y} + \varepsilon\mathbf{B}^1(\boldsymbol{\theta}, \boldsymbol{x}, \boldsymbol{y}, t) + \mathcal{O}(\varepsilon^2), \\ \dot{\boldsymbol{\theta}} &= \boldsymbol{\omega}^0 + \varepsilon\boldsymbol{\omega}^1(\boldsymbol{\theta}, \boldsymbol{x}, \boldsymbol{y}, t) + \mathcal{O}(\varepsilon^2), \end{aligned}$$

with constant frequency vector $\boldsymbol{\omega}^0$. *As before, this* t- *and* $\boldsymbol{\theta}$-*periodic system is obtained by local perturbation of an invariant manifold* M_0 *in the system*

$$\dot{\boldsymbol{x}} = \varepsilon\mathbf{A}^0(\boldsymbol{\theta})\boldsymbol{x}, \quad \dot{\boldsymbol{y}} = \varepsilon\mathbf{B}^0(\boldsymbol{\theta})\boldsymbol{y}, \quad \dot{\boldsymbol{\theta}} = \boldsymbol{\omega}^0,$$

for $\varepsilon = 0$. *In the equations for* $\boldsymbol{x}$ *and* $\boldsymbol{y}$, $\mathbf{A}^0(\boldsymbol{\theta})\boldsymbol{x}$ *and* $\mathbf{B}^0(\boldsymbol{\theta})\boldsymbol{y}$ *represent the linearizations near* $(\boldsymbol{x}, \boldsymbol{y}) = (\mathbf{0}, \mathbf{0})$, *so* $\mathbf{A}^1, \mathbf{B}^1$ *are* $o(\|\boldsymbol{x}\|, \|\boldsymbol{y}\|)$. *Assume that*

1. *All vector functions on the right-hand side are continuous and bounded; the* $\mathcal{O}(\varepsilon^2)$ *terms represent vector functions that are smooth on the domain and that can be estimated* $\mathcal{O}(\varepsilon^2)$.
2. *The functions on the right-hand side are Lipschitz continuous with respect to* $\boldsymbol{\theta}$, *the function* ω_1 *with Lipschitz constant* $\lambda^{\boldsymbol{\theta}}_{\boldsymbol{\omega}^1}$.

3. *The functions* $\boldsymbol{\omega}^1, \mathbf{A}^1, \mathbf{B}^1$ *are Lipschitz continuous with respect to* $\boldsymbol{x}, \boldsymbol{y}$.
4. *There exist positive constants* K *and* α *such that for any continuous* $\boldsymbol{\theta}(t)$ *the fundamental matrices of* $\dot{\boldsymbol{x}} = \varepsilon \mathbf{A}^0(\boldsymbol{\theta})\boldsymbol{x}, \dot{\boldsymbol{y}} = \varepsilon \mathbf{B}^0(\boldsymbol{\theta})\boldsymbol{y}$ *can be estimated by* $Ke^{-\varepsilon\alpha t}, Ke^{\varepsilon\alpha t}$ *respectively.*
5. $\alpha > \lambda^{\boldsymbol{\theta}}_{\boldsymbol{\omega}^1}$ *(normal hyperbolicity at higher order).*

Then there exists an invariant manifold M *of system* $\mathcal{S}_w$ *near* M_0 *with Lipschitz continuous parametrization that is periodic in* $\boldsymbol{\theta}$.

The frequency vector being constant in system $\mathcal{S}_w$ enables us to introduce slowly varying phases by putting

$$\boldsymbol{\theta}(t) = \boldsymbol{\omega} t + \boldsymbol{\psi}(t).$$

The resulting system $\mathcal{S}_w$ is of the form

$$\dot{\boldsymbol{X}} = \varepsilon \mathbf{F}^1(\boldsymbol{X}, t) + \mathcal{O}(\varepsilon^2),$$

where we have replaced $(\boldsymbol{\psi}, \boldsymbol{x}, \boldsymbol{y})$ by $\boldsymbol{X}$. The system is quasiperiodic in t. The near-identity transformation

$$\boldsymbol{X} = \boldsymbol{z} + \varepsilon \mathbf{u}^1(\boldsymbol{z}, t), \ \mathbf{u}^1(\boldsymbol{z}, t) = \int_0^t (\mathbf{F}^1(\boldsymbol{z}, \tau) - \overline{\mathbf{F}}^1(\boldsymbol{z}))\, \mathrm{d}\tau$$

(with $\overline{\mathbf{F}}^1$ the average over the periods of $\mathbf{F}^1$ in t) leads to the equation

$$\dot{\boldsymbol{z}} = \varepsilon \overline{\mathbf{F}}^1(\boldsymbol{z}) + \mathcal{O}(\varepsilon^2).$$

Note that as yet we have not introduced any approximation. Usually we can relate Theorem C.4.1 to the equation for $\boldsymbol{z}$ which will in general - at least to $\mathcal{O}(\varepsilon)$ - be much simpler than the system $\mathcal{S}_w$.

We will present a few illustrative examples.

Example C.4.2. Consider the system

$$\ddot{x} + x = \varepsilon(2x + 2\dot{x} - \frac{8}{3}\dot{x}^3 + y^2x^2 + \dot{y}^2x^2) + \varepsilon^2 f(x, y),$$
$$\ddot{y} + \omega^2 y = \varepsilon(\dot{y} - \dot{y}^3 + x^2y^2 + \dot{x}^2y^2) + \varepsilon^2 g(x, y),$$

where f and g are smooth, bounded functions. This looks like a bad case: if $\varepsilon = 0$ we have a family of (nonhyperbolic) 2-tori in 4-space. We introduce amplitude-angle coordinates by

$$x = r_1 \cos\theta_1, \quad \dot{x} = -r_1 \sin\theta_1, \quad y = r_2 \cos\omega\theta_2, \quad \dot{y} = -\omega r_2 \sin\omega\theta_2.$$

The system transforms to

$$
\begin{aligned}
\dot r_1 &= \varepsilon\Big(-r_1\sin 2\theta_1 + 2r_1\sin^2\theta_1 - \frac{8}{3}r_1^3\sin^4\theta_1 \\
&\quad - r_1^2 r_2^2\sin\theta_1\cos^2\theta_1(\cos^2\omega\theta_2 + \omega^2\sin^2\omega\theta_2)\Big) + \mathcal{O}(\varepsilon^2),\\
\dot r_2 &= \varepsilon\left(r_2\sin^2\omega\theta_2 + \omega^2 r_2^3\sin^4\omega\theta_2 - \frac{r_1^2 r_2^2}{\omega}\sin\omega\theta_2\cos^2\omega\theta_2\right) + \mathcal{O}(\varepsilon^2),\\
\dot\theta_1 &= 1 - \varepsilon\Big(2\cos^2\theta_1 - \sin 2\theta_1 + \frac{8}{3}r_1^2\sin^3\theta_1\cos\theta_1 \\
&\quad + r_1 r_2^2\cos^3\theta_1(\cos^2\omega\theta_2 + \omega^2\sin^2\omega\theta_2)\Big) + \mathcal{O}(\varepsilon^2),\\
\dot\theta_2 &= 1 + \varepsilon\left(\frac{1}{2\omega}\sin(2\omega\theta_2) + \omega r_2^2\sin^3\omega\theta_2\cos\omega\theta_2 - \frac{r_1^2 r_2}{\omega^2}\cos^3\omega\theta_2\right) + \mathcal{O}(\varepsilon^2).
\end{aligned}
$$

Putting $\theta_1 = t + \psi_1, \theta_2 = t + \psi_2$ and using the near-identity transformation introduced above but keeping—with some abuse of notation—the same symbols, we obtain the much simpler system

$$
\begin{aligned}
\dot r_1 &= \varepsilon r_1(1 - r_1^2) + \mathcal{O}(\varepsilon^2), \quad \dot\psi_1 = -\varepsilon + \mathcal{O}(\varepsilon^2),\\
\dot r_2 &= \varepsilon\frac{r_2}{2}(1 - \frac{3}{4}r_2^2) + \mathcal{O}(\varepsilon^2), \quad \dot\psi_2 = \mathcal{O}(\varepsilon^2).
\end{aligned}
$$

The part of $(\boldsymbol{x}, \boldsymbol{y}) = (\mathbf{0}, \mathbf{0})$ is played by $(r_1, r_2) = (1, \frac{2}{\sqrt{3}})$. The averaged (normalized) equations contain a torus in phase space approximated by the parametrization

$$
\begin{aligned}
x_a(t) &= \cos(t - \varepsilon t + \psi_1(0)), \quad \dot x_a(t) = -\sin(t - \varepsilon t + \psi_1(0)),\\
y_a(t) &= \frac{2}{3}\sqrt{3}\cos(\omega t + \psi_2(0)), \quad \dot y_a(t) = -\frac{2\omega}{3}\sqrt{3}\sin(\omega t + \psi_2(0)).
\end{aligned}
$$

From linearization of the averaged equations, it is clear that the torus is attracting: it is normally hyperbolic with attraction rate $\mathcal{O}(\varepsilon)$. If the ratio of $1 - \varepsilon$ and ω is rational, the torus is filled up with periodic solutions. If the ratio is irrational we have a quasiperiodic (two-frequency) flow over the torus. Theorem C.4.1 tells us that in the original equations a torus exists in an $\mathcal{O}(\varepsilon)$ neighborhood of the torus found by normalization. It has the same stability properties. The torus is two-dimensional and the time scales of rotation are in both directions $\mathcal{O}(1)$. ◇

In the next example we return to the forced Van der Pol equation (C.2.2).

Example C.4.3. Consider the equation

$$\ddot x + x = \varepsilon(1 - x^2)\dot x + a\cos\omega t$$

with a and ω constants. The difference with (C.2.2) is that the nonlinearity is small and the forcing can be $\mathcal{O}(1)$ as $\varepsilon \to 0$.

1. Case $a = \mathcal{O}(\varepsilon)$.
   If ω is ε-close to 1, standard averaging leads to the existence of periodic solutions only. If ω takes different values, first-order averaging is not conclusive, but see the remark below.

2. Case $a = \mathcal{O}(1)$, ω is not ε-close to 1 (if ω is near to 1, the solutions move away from an $\mathcal{O}(1)$ neighborhood of the origin because of linear resonance). We introduce the transformation $x, \dot{x} \rightarrow r, \psi$,

$$x = r\cos(t+\psi) + \frac{a}{1-\omega^2}\cos\omega t, \quad \dot{x} = -r\sin(t+\psi) - \frac{a\omega}{1-\omega^2}\sin\omega t.$$

The resulting slowly varying system can be averaged, producing periodic solutions in which various values of ω play a part. Returning to the corresponding expressions for x and $\dot{x}$, we infer the presence of tori in the extended phase space.

Remark C.4.4. In some of the cases near-identity transformation leads to a slowly varying system of the form

$$\dot{r} = \varepsilon\frac{1}{2}r\left(1 - \frac{1}{4}r^2\right) + \mathcal{O}(\varepsilon^2),$$
$$\dot{\psi} = \mathcal{O}(\varepsilon^2).$$

Instead of computing higher-order normal forms to establish the behavior of ψ, we can apply slow manifold theory, see [140], [143], or [144], to conclude the existence of a slow manifold ε-close to $r = 2$. In the case of $a = \mathcal{O}(1)$ the corresponding solutions will be ε-close to the torus described by

$$x = 2\cos(t+\psi_0) + \frac{a}{1-\omega^2}\cos\omega t, \quad \dot{x} = -2\sin(t+\psi_0) - \frac{a\omega}{1-\omega^2}\sin\omega t.$$

C.5 Tori Created by Neimark–Sacker Bifurcation

Another important scenario for creating a torus arises from the Neimark–Sacker bifurcation. For an instructive and detailed introduction see [163]. Suppose that we have obtained an averaged equation $\dot{\boldsymbol{x}} = \varepsilon\mathbf{f}^1(\boldsymbol{x}, \boldsymbol{a})$, with dimension 3 or higher, by variation of constants and subsequent averaging; $\boldsymbol{a}$ is a parameter or a set of parameters. It is well known that if this equation contains a hyperbolic critical point, the original equation contains a periodic solution. The first-order approximation of this periodic solution is characterized by the time variables t and εt.

Suppose now that by varying the parameter $\boldsymbol{a}$ a pair of eigenvalues of the critical point becomes purely imaginary. For this value of $\boldsymbol{a}$ the averaged equation undergoes a Hopf bifurcation producing a periodic solution of the averaged equation; the typical time variable of this periodic solution is εt, and so the period will be $\mathcal{O}(1/\varepsilon)$. As it branches off an existing periodic solution in the original equation, it will produce a torus; it is associated with a Hopf bifurcation of the corresponding Poincaré map, and the bifurcation has a different name: Neimark–Sacker bifurcation. The result will be a two-dimensional torus that contains two-frequency oscillations, one on a time scale of order 1 and the other with time scale $\mathcal{O}(1/\varepsilon)$.

A typical example runs as follows.

Example C.5.1. A special case of a system studied by [17] is

$$\ddot{x} + \varepsilon\kappa\dot{x} + (1 + \varepsilon\cos 2t)x + \varepsilon xy = 0,$$
$$\ddot{y} + \varepsilon\dot{y} + 4(1+\varepsilon)y - \varepsilon x^2 = 0.$$

This is a system with parametric excitation and nonlinear coupling; κ is a positive damping coefficient that is independent of ε. Away from the coordinate planes we may use amplitude-phase variables by

$$x = r_1\cos(t+\psi_1), \quad \dot{x} = -r_1\sin(t+\psi_1),$$
$$y = r_2\cos(2t+\psi_2), \quad \dot{y} = -2r_2\sin(2t+\psi_1);$$

after first-order averaging we obtain

$$\dot{r}_1 = \varepsilon r_1\left(\frac{r_2}{4}\sin(2\psi_1-\psi_2) + \frac{1}{4}\sin 2\psi_1 - \frac{1}{2}\kappa\right),$$
$$\dot{\psi}_1 = \varepsilon\left(\frac{r_2}{4}\cos(2\psi_1-\psi_2) + \frac{1}{4}\cos 2\psi_1\right),$$
$$\dot{r}_2 = \varepsilon\frac{r_2}{2}\left(\frac{r_1^2}{4r_2}\sin(2\psi_1-\psi_2) - 1\right),$$
$$\dot{\psi}_2 = \frac{\varepsilon}{2}\left(-\frac{r_1^2}{4r_2}\cos(2\psi_1-\psi_2) + 2\right).$$

Putting the right-hand sides equal to zero produces a nontrivial critical point corresponding to a periodic solution of the system for the amplitudes and phases and so a quasiperiodic solution of the original coupled system in x and y. We find for this critical point the relations

$$r_1^2 = 4\sqrt{5}r_2, \quad \cos(2\psi_1-\psi_2) = \frac{2}{\sqrt{5}},$$
$$r_1 = 2\sqrt{2\kappa + \sqrt{5-16\kappa^2}}, \quad \sin(2\psi_1-\psi_2) = \frac{1}{\sqrt{5}}.$$

This periodic solution exists if the damping coefficient is not too large: $0 \leq \kappa < \frac{\sqrt{5}}{4}$. Linearization of the averaged equations at the critical point while using these relations produces the matrix

$$A = \begin{bmatrix} 0 & 0 & \frac{r_1}{4\sqrt{5}} & -\frac{r_1^3}{40} \\ 0 & -\kappa & \frac{1}{2\sqrt{5}} & \frac{r_1^2}{80} \\ \frac{r_1}{4\sqrt{5}} & \frac{r_1^2}{2\sqrt{5}} & -\frac{1}{2} & -\frac{r_1^2}{4\sqrt{5}} \\ -\frac{2}{r_1} & 1 & \frac{4\sqrt{5}}{r_1^2} & -\frac{1}{2} \end{bmatrix}.$$

Another condition for the existence of the periodic solution is that the critical point be hyperbolic, i.e., the eigenvalues of the matrix A have no real part zero. It is possible to express the eigenvalues explicitly in terms of κ by using a software package like MATHEMATICA. However, the expressions are cumbersome. Hyperbolicity is the case if we start with values of κ just below $\frac{1}{4}\sqrt{5} = 0.559$. Diminishing κ we find that, when $\kappa = 0.546$, the real part of two eigenvalues vanishes. This value corresponds to a Hopf bifurcation producing a nonconstant periodic solution of the averaged equations. This in turn corresponds to a torus in the original equations (in x and y) by a Neimark–Sacker bifurcation. As stated before, the result will be a two-dimensional torus that contains two-frequency oscillations, one on a time scale of order 1 and the other with time scale $\mathcal{O}(1/\varepsilon)$. ◇

D

Some Elementary Exercises in Celestial Mechanics

D.1 Introduction

For centuries celestial mechanics has been an exceptional rich source of problems and results in mathematics. To some extent this is still the case. Today one can discern, rather artificially, three problem fields. The first one is the study of classical problems like perturbed Kepler motion, orbits in the three-body problem, the theory of asteroids and comets, etc. The second one is a small but relatively important field in which the astrophysicists are interested; we are referring to systems with evolution like for instance changes caused by tidal effects or by exchange of mass. The third field is what one could call 'mathematical celestial mechanics', a subject which is part of the theory of dynamical systems. The distinction between the fields is artificial. There is some interplay between the fields and hopefully, this will increase in the future. An interesting example of a study combining the first and the third field is the paper by Brjuno [41]. A typical example of an important mathematical paper which has found already some use in classical celestial mechanics is Moser's study on the geometrical interpretation of the Kepler problem [193]. Surveys of mathematical aspects of celestial mechanics have been given in [194] and [3].

Here we shall be concerned with simple examples of the use of averaging theory. Apart from being an exercise it may serve as an introduction to the more complicated general literature. One of the difficulties of the literature is the use of many different coordinate systems. We have chosen here for the perturbed harmonic oscillator formulation which eases averaging and admits a simple geometric interpretation. For reasons of comparison we shall demonstrate the use of another coordinate system in a particular problem.

In celestial mechanics thousands of papers have been published and a large number of elementary results are being rediscovered again and again. Our reference list will therefore do no justice to all scientists whose efforts were directed towards the problems mentioned here. For a survey of theory and results see [253] and [118]. However, also there the reference lists are far from complete and the mathematical discussion is sometimes confusing.

D.2 The Unperturbed Kepler Problem

Consider the motion of two point masses acting upon each other by Newtonian force fields. In relative coordinates $\boldsymbol{r}$, with norm $r = \| \boldsymbol{r} \|$ we have for the gravitational potential

$$V_0(\boldsymbol{r}) = -\frac{\mu}{r}, \tag{D.2.1}$$

with μ the gravitational constant; the equations of motion are

$$\ddot{\boldsymbol{r}} = -\frac{\mu}{r^3}\boldsymbol{r}. \tag{D.2.2}$$

The angular momentum vector

$$\boldsymbol{h} = \boldsymbol{r} \times \dot{\boldsymbol{r}} \tag{D.2.3}$$

is an (vector valued) integral of motion; this follows from

$$\frac{d\boldsymbol{h}}{dt} = \dot{\boldsymbol{r}} \times \dot{\boldsymbol{r}} + \boldsymbol{r} \times \ddot{\boldsymbol{r}} = \boldsymbol{r} \times (-\frac{\mu}{r^3}\boldsymbol{r}) = 0.$$

The energy of the system

$$E = \frac{1}{2} \| \dot{\boldsymbol{r}} \|^2 + V_0(r) \tag{D.2.4}$$

is also an integral of motion. Note that the equations of motion represent a three degree of freedom system, derived from a Hamiltonian. Three independent integrals suffice to make the system integrable. The integrals (D.2.3) and (D.2.4) however represent already four independent integrals which implies that the integrability of the unperturbed Kepler problem is characterized by an unusual degeneration. We recognize this also by concluding from the constancy of the angular momentum vector $\boldsymbol{h}$ that the orbits are planar. Choosing for instance $\mathrm{z}(0) = \dot{\mathrm{z}}(0) = 0$ implies that $\mathrm{z}(t), \dot{\mathrm{z}}(t)$ are zero for all time. It is then natural to choose such initial conditions and to introduce polar coordinates $x = r\cos(\phi)$, $y = r\sin(\phi)$ in the plane. Equation (D.2.2) yields

$$\ddot{r} - r\dot{\phi}^2 = -\frac{\mu}{r^2}, \tag{D.2.5a}$$

$$\frac{d}{dt}(r^2\dot{\phi}) = 0. \tag{D.2.5b}$$

The last equation corresponds to the component of the angular momentum vector which is unequal to zero and we have

$$r^2\dot{\phi} = h \tag{D.2.6}$$

with $h = \| \boldsymbol{h} \|$. We could solve equation (D.2.5a) in various ways which have all some special advantages. If $E < 0$, the orbits are periodic and they describe conic sections. A direct description of the orbits is possible by using geometric variables like the *eccentricity* e, *semimajor axis* a and dynamical variables like the period P, the time of *peri-astron passage* T, etc. Keeping an eye on perturbation theory a useful presentation of the solution is the harmonic oscillator formulation. Introduce ϕ as a time-like variable and put

$$u = \frac{1}{r}. \tag{D.2.7}$$

Transforming $r, t \mapsto u, \phi$ in equation (D.2.5a) produces

$$\frac{d^2u}{d\phi^2} + u = \frac{\mu}{h^2}. \tag{D.2.8}$$

The solution can be written as

$$u = \frac{\mu}{h^2} + \alpha \cos(\phi + \beta) \tag{D.2.9}$$

or equivalently

$$u = \frac{\mu}{h^2} + A\cos(\phi) + B\sin(\phi),$$

with $\alpha, \beta, A, B \in \mathbb{R}$.

D.3 Perturbations

In the sequel we shall consider various perturbations of the Kepler problem. One of those is to admit variation of μ by changes of the gravitational field with time or change of the total mass. These problems will be formulated later on. For an examination of various perturbing forces, see [106]. In general we can write the equation of the perturbed Kepler problem

$$\ddot{\boldsymbol{r}} = -\frac{\mu}{r^3}\boldsymbol{r} + \boldsymbol{F} \tag{D.3.1}$$

in which $\boldsymbol{r}$ is again the relative position vector, μ the gravitational constant; $\boldsymbol{F}$ stands for the as yet unspecified perturbation. The angular momentum vector (D.2.3) will in general only be constant if $\boldsymbol{F}$ lies along $\boldsymbol{r}$ as we have

$$\frac{d\boldsymbol{h}}{dt} = \boldsymbol{r} \times \boldsymbol{F}. \tag{D.3.2}$$

It will be useful to introduce spherical coordinates $x = r\cos(\phi)\sin(\theta)$, $y = r\sin(\phi)\sin(\theta)$ and $z = r\cos(\theta)$ in which θ is the *colatitude*, ϕ the *azimuthal angle* in the equatorial plane. Specifying $\boldsymbol{F} = (F_x, F_y, F_z)$ we find from equation (D.3.1)

$$\begin{aligned}&\ddot{r} - r\dot{\phi}^2\sin^2\theta - r\dot{\theta}^2\\&= -\frac{\mu}{r^2} + (F_x\cos(\phi) + F_y\sin(\phi))\sin(\theta) + F_z\cos(\theta).\end{aligned} \quad \text{(D.3.3)}$$

It is useful to write this equation in a different form using the angular momentum. From (D.2.3) we calculate

$$h^2 = \| \boldsymbol{h}\|^2 = r^4\dot{\theta}^2 + r^4\dot{\phi}^2\sin^2(\theta).$$

Equation (D.3.3) can then be written as

$$\ddot{r} - \frac{h^2}{r^3} = -\frac{\mu}{r^2} + (F_x\cos(\phi) + F_y\sin(\phi))\sin(\theta) + F_z\cos(\theta).$$

Equation (D.3.1) is of dimension 6 so we need two more second order equations. A combination of the first two components of angular momentum produces

$$\begin{aligned}&\frac{d}{dt}(r^2\dot{\theta}) - r^2(\dot{\phi})^2\sin(\theta)\cos(\theta)\\&= -rF_z\sin(\theta) + r\cos(\theta)(F_y\sin(\phi) + F_x\cos(\phi)).\end{aligned} \quad \text{(D.3.4)}$$

The third (z) component of angular momentum is described by

$$\frac{d}{dt}(r^2\dot{\phi}\sin^2\theta) = -r\sin(\theta)(F_x\sin(\phi) - F_y\cos(\phi)). \quad \text{(D.3.5)}$$

Note that the components of $\boldsymbol{F}$ still have to be rewritten in spherical coordinates.

D.4 Motion Around an 'Oblate Planet'

To illustrate this formulation of the perturbed Kepler problem we consider the case that one of the bodies can be considered a point mass, the other body is axisymmetric and flattened at the poles. The description of the motion of a point mass around such an *oblate planet* has some relevance for satellite mechanics. Suppose that the polar axis is taken as the z-axis and the x and y axes are taken in the equatorial plane. The gravitational potential can be represented by a convergent series

$$V = -\frac{\mu}{r}\left[1 - \sum\nolimits_{n=2}^{\infty}\frac{1}{r^n}J_nP_n(\frac{z}{r})\right] \quad \text{(D.4.1)}$$

where the units are such that the equatorial radius corresponds to $r = 1$. The P_n are the standard Legendre polynomials of degree n, the J_n are constants determined by the axisymmetric distribution of mass (they have nothing to do with Bessel functions). In the case of the planet Earth we have

$$J_2 = 1 \cdot 1 \times 10^{-3},$$
$$J_3 = -2 \cdot 3 \times 10^{-6},$$
$$J_4 = -1 \cdot 7 \times 10^{-6}.$$

The constants J_5, J_6 etc. do not exceed the order of magnitude 10^{-6}. A first-order study of satellite orbits around the Earth involves the truncation of the series after J_2; we put

$$V_1 = -\frac{\mu}{r} + \frac{\mu}{r^3} J_2 P_2(\frac{z}{r}) = -\frac{\mu}{r} + \frac{1}{2} J_2 \frac{\mu}{r^3}(1 - 3\frac{z^2}{r^2})$$

or

$$V_1 = -\frac{\mu}{r} + \varepsilon \frac{\mu}{r^3}(1 - 3\cos^2(\theta)), \tag{D.4.2}$$

where $\varepsilon = \frac{1}{2} J_2$. Taking the gradient of V_1 we find for the components of the perturbation vector

$$F_x = \varepsilon \frac{3\mu}{r^5} x(-1 + 5\frac{z^2}{r^2}),$$
$$F_y = \varepsilon \frac{3\mu}{r^5} y(-1 + 5\frac{z^2}{r^2}),$$
$$F_z = \varepsilon \frac{3\mu}{r^5} z(-3 + 5\frac{z^2}{r^2}).$$

The equations of motion (D.3.3–D.3.5) become in this case

$$\ddot{r} - \frac{h^2}{r^3} = -\frac{\mu}{r^2} - \varepsilon \frac{3\mu}{r^4}(1 - 3\cos^2(\theta)), \tag{D.4.3a}$$
$$\frac{d}{dt}(r^2\dot{\theta}) - r^2(\dot{\phi})^2 \sin(\theta)\cos(\theta) = \varepsilon \frac{6\mu}{r^3} \sin(\theta)\cos(\theta), \tag{D.4.3b}$$
$$\frac{d}{dt}(r^2\dot{\phi}\sin^2(\theta)) = 0. \tag{D.4.3c}$$

The last equation can be integrated and then expresses that the z-component of angular momentum is conserved. This could be expected from the assumption of axial symmetry. Note that the energy is conserved, so we have two integrals of motion of system (D.4.3). For the system to be integrable we need another independent integral; we have no results available on the existence of such a third integral.

D.5 Harmonic Oscillator Formulation for Motion Around an 'Oblate Planet'

We shall transform equations (D.4.3) in the following way. The dependent variables r and θ are replaced by

$$u = \frac{1}{r} \text{ and } v = \cos(\theta). \tag{D.5.1}$$

The independent variable t is replaced by a time-like variable τ given by

$$\dot{\tau} = \frac{h}{r^2} = hu^2 \quad , \ \tau(0) = 0; \tag{D.5.2}$$

here h is again the length of the angular momentum vector. Note that τ is monotonically increasing, as a time-like variable should, except in the case of radial (vertical) motion. We cannot expect that for all types of perturbations τ runs ad infinitum like t; in other words, equation (D.5.2) may define a mapping from $[0, \infty)$ into $[0, C]$ with C a positive constant. If no perturbations are present, τ represents an angular variable, see Section D.2. In the case of equations (D.4.3) we find, using the transformations (D.5.1-D.5.2),

$$\frac{d^2u}{d\tau^2} + u = \frac{\mu}{h^2} + \varepsilon\frac{6\mu}{h^2}uv\frac{du}{d\tau}\frac{dv}{d\tau} + \varepsilon 3\mu\frac{u^2}{h^2}(1 - 3v^2), \tag{D.5.3a}$$

$$\frac{d^2v}{d\tau^2} + v = \varepsilon\frac{6\mu}{h^2}uv(\frac{dv}{d\tau})^2 - \varepsilon 6\frac{\mu}{h^2}uv(1 - v^2), \tag{D.5.3b}$$

$$\frac{dh}{d\tau} = -\varepsilon\frac{6\mu}{h}uv\frac{dv}{d\tau}. \tag{D.5.3c}$$

Instead of the variable h it makes sense to use as variable h^2 or μ/h^2; here we shall use h. In a slightly different form these equations have been presented by Kyner [164]; note however that the discussion in that paper on the time scale of validity and the order of approximation is respectively wrong and unnecessarily complicated. System (D.5.3) still admits the energy integral but no other integrals are available

Exercise D.5.1. In what sense are our transformations canonical?

Having solved system (D.5.3) we can transform back to time t by solving equation (D.5.2).

D.6 First Order Averaging for Motion Around an 'Oblate Planet'

We have to put equations (D.5.3) in the standard form for averaging using the familiar Lagrange method of variation of parameters. As we have seen in Chapter 1 the choice of the perturbation formulation affects the computational work, not the final result. We find it convenient to choose in this problem the transformation $u, du/d\tau \mapsto a_1, b_1$ and $v, dv/d\tau \mapsto a_2, b_2$ defined by

$$u = \frac{\mu}{h^2} + a_1\cos(\tau + b_1) \ , \ \frac{du}{d\tau} = -a_1\sin(\tau + b_1), \tag{D.6.1a}$$

$$v = a_2\cos(\tau + b_2) \quad , \ \frac{dv}{d\tau} = -a_2\sin(\tau + b_2). \tag{D.6.1b}$$

The inclination i of the orbital plane, which is a constant of motion if no perturbations are present, is connected with the new variable a_2 by $a_2 = \sin(i)$. Abbreviating equations (D.5.3) by

$$\frac{d^2u}{d\tau^2} + u = \frac{\mu}{h^2} + \varepsilon G_1,$$
$$\frac{d^2v}{d\tau^2} + v = \varepsilon G_2,$$

we find the system

$$\frac{da_1}{d\tau} = \frac{2\mu}{h^3}\frac{dh}{d\tau}\cos(\tau + b_1) - \varepsilon G_1 \sin(\tau + b_1), \tag{D.6.2a}$$
$$\frac{db_1}{d\tau} = -\frac{2\mu}{h^3}\frac{dh}{d\tau}\frac{\sin(\tau + b_1)}{a_1} - \varepsilon G_1 \frac{\cos(\tau + b_1)}{a_1}, \tag{D.6.2b}$$
$$\frac{da_2}{d\tau} = -\varepsilon G_2 \sin(\tau + b_2), \tag{D.6.2c}$$
$$\frac{db_2}{d\tau} = -\varepsilon G_2 \frac{\cos(\tau + b_2)}{a_2}. \tag{D.6.2d}$$

We have to add equation (D.5.3c) and in G_1, G_2 we have to substitute variables according to equations (D.6.1). The resulting system is 2π-periodic in τ and we apply first order averaging. The approximations of a, b will be indicated by α, β. For the right hand side of equation (D.5.3c) we find average zero, so that $\mathbf{h}(\tau) = \mathbf{h}(0) + \mathcal{O}(\varepsilon)$ on the time scale $1/\varepsilon$, i.e. for $0 \leq \varepsilon\tau \leq L$ with L a constant independent of ε. Averaging of equations (D.6.2) produces

$$\frac{d\alpha_1}{d\tau} = 0, \tag{D.6.3a}$$
$$\frac{d\alpha_2}{d\tau} = 0, \tag{D.6.3b}$$
$$\frac{d\beta_1}{d\tau} = -\varepsilon \frac{3\mu^2}{\mathrm{h}^4(0)}(1 - \frac{3}{2}\alpha_2^2), \tag{D.6.3c}$$
$$\frac{d\beta_2}{d\tau} = \varepsilon \frac{3\mu^2}{\mathrm{h}^4(0)}(1 - \alpha_2^2). \tag{D.6.3d}$$

So in this approximation the system is integrable. Putting $p = -3\mu^2(1 - \frac{3}{2}\alpha_2^2(0))/\mathrm{h}^4(0)$ and $q = \mu^2(1-\alpha_2^2(0))/\mathrm{h}^4(0)$ we conclude that we have obtained the following first-order approximations on the time scale $1/\varepsilon$

$$\mathrm{u}(\tau) = \frac{\mu}{\mathrm{h}^2(0)} + \mathrm{a}_1(0)\cos(\tau + \varepsilon p\tau + \mathrm{b}_1(0)) + \mathcal{O}(\varepsilon), \tag{D.6.4a}$$
$$\mathrm{v}(\tau) = \mathrm{a}_2(0)\cos(\tau + \varepsilon q\tau + \mathrm{b}_2(0)) + \mathcal{O}(\varepsilon). \tag{D.6.4b}$$

It is remarkable that the original, rather complex, perturbation problem admits such simple approximations. Note however that in higher approximation or on longer time scales qualitatively new phenomena may occur. To illustrate this we shall discuss some special solutions.

Equatorial Orbits

The choice of potential (D.4.1) has as a consequence that we can restrict the motion to the equator plane $z = 0$, or $\theta = \frac{1}{2}\pi$ for all time. In equation (D.5.3b) these solutions correspond to $v = dv/d\tau = 0$, $\tau \geq 0$. Equation (D.5.3a) reduces to

$$\frac{d^2u}{d\tau^2} + u = \frac{\mu}{h^2} + \varepsilon\frac{3\mu}{h^2}u^2. \tag{D.6.5}$$

The time-like variable τ can be identified with the *azimuthal angle* ϕ. Equation (D.5.3c) produces that h is a constant of motion in this case. It is not difficult to show that the solutions of the equations of motion restricted to the equatorial plane in a neighborhood of the oblate planet are periodic. We find

$$u(\tau) = \frac{\mu}{h^2} + \mathrm{a}_1(0)\cos(\tau - \varepsilon\frac{3\mu^2}{h^4}\tau + \mathrm{b}_1(0)) + \mathcal{O}(\varepsilon)$$

on the time scale $1/\varepsilon$. Using the theory of Chapters 2 and 3 it is very easy to obtain higher-order approximations but no new qualitative phenomena can be expected at higher order, as for equatorial orbits and because of symmetry, the higher order terms depend on u (or r) only.

Polar Orbits

The axisymmetry of the potential (D.4.1) triggers off the existence of orbits in meridional planes: taking $\dot{\phi} = 0$ for all time solves equation (D.4.3c) (and more in general (D.3.5) with the assumption of axisymmetry). In this case the time-like variable τ can be identified with θ; equation (D.5.3b) is solved by $v(\tau) = \cos(\tau)$. System (D.6.4) produces the approximation ($s = \frac{3}{2}\mu^2/\mathrm{h}^4(0)$)

$$u(\tau) = \frac{\mu}{h^2(0)} + \mathrm{a}_1(0)\cos(\tau + \varepsilon s\tau + \mathrm{b}_1(0)) + \mathcal{O}(\varepsilon)$$

on the time scale $1/\varepsilon$. Again one can obtain higher-order approximations for the third order system (D.5.3).

The Critical Inclination Problem

Analyzing the averaged system (D.6.3) one expects a resonance domain near the zeros of $d\beta_1/d\tau - d\beta_2/d\tau$. This expectation is founded on our analysis of averaging over spatial variables (Chapter 7) and the theory of higher order resonance in two degrees of freedom Hamiltonian systems (Section 10.6.4). In particular, this is a secondary resonance as discussed in Section 7.6. The resonance domain follows from

$$\frac{d\beta_1}{d\tau} - \frac{d\beta_2}{d\tau} = \varepsilon \frac{3\mu^2}{\mathrm{h}^4(0)}(\frac{5}{2}\alpha_2^2 - 2) = 0$$

or $\alpha_2^2 = \frac{4}{5}$; in terms of the inclination i

$$\sin^2(i) = \frac{4}{5}.$$

This i is called the *critical inclination.* To analyze the flow in the resonance domain we have to use higher-order approximations (such as secondary first-order averaging, Section 7.6). Using different transformations, but related techniques the higher order problem has been discussed by various authors; we mention [83], [72] and [67].

D.7 A Dissipative Force: Atmospheric Drag

In this section we shall study the influence of a dissipative force by introducing atmospheric drag. In the subsequent sections we introduce other dissipative forces producing evolution of two body systems. In the preceding sections we studied Hamiltonian perturbations of an integrable Hamiltonian system. The introduction of dissipative forces presents qualitatively new phenomena; an interesting aspect is however that we can apply the same perturbation techniques. Suppose that the second body is moving through an atmosphere surrounding the primary body. We ignore the lift acceleration or assume that this effect is averaged out by tumbling effects. For the drag acceleration vector we assume that it takes the form

$$-\varepsilon B(r)|\dot{\boldsymbol{r}}|^m \dot{\boldsymbol{r}}, \quad m \text{ a constant.}$$

$B(r)$ is a positive function, determined by the density of the atmosphere; in more realistic models B also depends on the angles and the time. Often one chooses $m = 1$, corresponding to a velocity-squared aerodynamic force law; $m = 0$, B constant, corresponds to linear friction. (Some aspects of aerodynamic acceleration, including lift effects, and the corresponding harmonic oscillator formulation have been discussed in [273]; for perturbation effects in atmospheres see also [106]). Assuming that a purely gravitational perturbation force $\boldsymbol{F}_g$ is present, we have in equation (D.3.1)

$$\boldsymbol{F} = \varepsilon \boldsymbol{F}_g - \varepsilon B(r)|\dot{\boldsymbol{r}}|^m \dot{\boldsymbol{r}}.$$

To illustrate the treatment we restrict ourselves to equatorial orbits. This only means a restriction on $\boldsymbol{F}_g$ to admit the existence of such orbits (as in the case of motion around an oblate planet). Putting $z = 0$ ($\theta = \frac{1}{2}\pi$) we find with $h = r^2\dot{\phi}$, $|\dot{\boldsymbol{r}}|^2 = \dot{r}^2 + h^2 r^{-2}$ from equation (D.3.3) and (D.3.5)

$$\ddot{r} - \frac{h^2}{r^3} = -\frac{\mu}{r^2} + \varepsilon f_1(r,\phi) - \varepsilon B(r)(\dot{r}^2 + h^2 r^{-2})^{\frac{m}{2}} \dot{r},$$
$$\frac{dh}{dt} = \varepsilon f_2(r\phi) - \varepsilon B(r)(\dot{r}^2 + h^2 r^{-2})^{\frac{m}{2}} h,$$

in which f_1 and f_2 are gravitational perturbations to be computed from

$$f_1(r,\phi) = F_{gx}\cos(\phi) + F_{gy}\sin(\phi),$$
$$f_2(r,\phi) = -r(F_{gx}\sin(\phi) - f_{gy}\cos(\phi)).$$

The requirement of continuity yields that f_1 and f_2 are 2π-periodic in ϕ. For example in the case of motion around an oblate planet we have, comparing with equations (D.4.3a) and (D.4.3b), $f_1(r,\phi) = -\frac{3\mu}{r^4}$, $f_2(r,\phi) = 0$. The time-like variable τ, introduced by equation (D.5.2), can be identified with ϕ; putting $u = 1/r$ we find

$$\frac{d^2u}{d\tau^2} + u = \frac{\mu}{h^2} - \varepsilon\frac{f_1(\frac{1}{u},\tau)}{h^2u^2} - \varepsilon\frac{f_2(\frac{1}{u},\tau)}{h^2u^2}\frac{du}{d\tau}, \tag{D.7.1a}$$
$$\frac{dh}{d\tau} = -\varepsilon\frac{f_2(\frac{1}{u},\tau)}{hu^2} - \varepsilon\frac{B(\frac{1}{u})}{u^2}[u^2 + (\frac{du}{d\tau})^2]^{\frac{m}{2}} h^m. \tag{D.7.1b}$$

The perturbation problem (D.7.1) can be treated by first-order averaging as in Appendix B. After specifying f_1, f_2, B and m we transform by (D.6.1a)

$$u = \frac{\mu}{h^2} + a_1\cos(\tau + b_1)\ ,\ \frac{du}{d\tau} = -a_1\sin(\tau + b_1).$$

To be more explicit we discuss the case of equatorial motion around an oblate planet; equations (D.7.1) become

$$\frac{d^2u}{d\tau^2} + u = \frac{\mu}{h^2} + 3\varepsilon\mu\frac{u^2}{h^2}, \tag{D.7.2a}$$
$$\frac{dh}{d\tau} = -\varepsilon\frac{B(\frac{1}{u})}{u^2}[u^2 + (\frac{du}{d\tau})^2]^{\frac{m}{2}} h^m. \tag{D.7.2b}$$

Since the density function B is positive, it is clear that the length of the angular momentum vector h is monotonically decreasing. Transforming we have (cf. equation (D.6.2a))

$$\frac{da_1}{d\tau} = -\varepsilon\frac{2\mu}{u^2}B(\frac{1}{u})h^{m-3}[\frac{\mu^2}{h^4} + a_1^2 + 2\frac{\mu}{h^2}a_1\cos(\tau + b_1)]^{\frac{m}{2}}\cos(\tau + b_1)$$
$$-3\varepsilon\mu\frac{u^2}{h^2}\sin(\tau + b_1), \tag{D.7.3a}$$
$$\frac{db_1}{d\tau} = +\varepsilon\frac{2\mu}{u^2}B(\frac{1}{u})h^{m-3}[\frac{\mu^2}{h^4} + a_1^2 + 2\frac{\mu}{h^2}a_1\cos(\tau + b_1)]^{\frac{m}{2}}\ \frac{\sin(\tau + b_1)}{a_1}$$
$$-3\varepsilon\mu\frac{u^2}{h^2}\frac{\cos(\tau + b_1)}{a_1} \tag{D.7.3b}$$

to which we add equation (D.7.2b); at some places we still have to write down the expression for u. The right hand side of equations (D.7.2b–D.7.3) is 2π-periodic in τ; averaging produces that the second term on the right hand side of (D.7.3a) and the first term on the right hand side of (D.7.3b) vanishes. For the density function B one usually chooses a function exponentially decreasing with r or a combination of powers of r (hyperbolic density law). A simple case arises if we choose

$$B(r) = \frac{B_0}{r^2}.$$

The averaged equations take the form

$$\begin{aligned}
\frac{d\alpha_1}{d\tau} &= -2\varepsilon\mu B_0 h^{m-3} \frac{1}{2\pi} \int_0^{2\pi} [\frac{\mu^2}{h^4} + \alpha_1^2 + 2\frac{\mu}{h^2}\alpha_1 \cos(\tau + \beta_1)]^{\frac{m}{2}} \cos(\tau + \beta_1)\, d\tau, \\
\frac{d\beta_1}{d\tau} &= -3\varepsilon \frac{\mu^2}{h^4}, \\
\frac{dh}{d\tau} &= -\varepsilon B_0 h^m \frac{1}{2\pi} \int_0^{2\pi} [\frac{\mu^2}{h^4} + \alpha_1^2 + 2\frac{\mu}{h^2}\alpha_1 \cos(\tau + \beta_1)]^{\frac{m}{2}}\, d\tau,
\end{aligned}$$

in which α_1, β_1, h are $\mathcal{O}(\varepsilon)$-approximations of a_1, b_1 and h on the time scale $1/\varepsilon$, assuming that we impose the initial conditions $\alpha_1(0) = \mathrm{a}_1(0)$ etc. In the case of a velocity-squared aerodynamic force law ($m = 1$), we have still to evaluate two definite integrals. These integrals are elliptic and they can be analyzed by series expansion (note that $0 < 2\mu\alpha_1/h^2 < \mu^2/h^4 + \alpha_1^2$ so that we can use binomial expansion). Linear force laws are of less practical interest but it can still be instructive to carry out the calculations. If $m = 0$ we find

$$\alpha_1(\tau) = \mathrm{a}_1(0)\ ,\ \beta_1(\tau) = \mathrm{b}_1(0) + \frac{3}{4}\frac{\mu^2}{B_0}(e^{-4\varepsilon B_0 \tau} - 1),\ \mathrm{h}(\tau) = \mathrm{h}(0)e^{-\varepsilon B_0 \tau}$$

which in the original variables corresponds to a spiraling down of the body moving through the atmosphere.

D.8 Systems with Mass Loss or Variable G

We consider now a class of problems in which mass is ejected isotropically from the two-body system and is lost to the system or in which the gravitational 'constant' G decreases with time. The treatment is taken from [274]. It can be shown that the relative motion of the two bodies takes place in a plane; introducing polar coordinates in this plane we have the equations of motion

$$\ddot{r} = -\frac{\mu(t)}{r^2} + \frac{h^2}{r^3}, \tag{D.8.1a}$$

$$r^2\dot{\phi} = h. \tag{D.8.1b}$$

The length of the angular momentum vector is conserved, $\mu(t)$ is a monotonically decreasing function with time. Introducing again $u = 1/r$ and the time-like variable τ by $\dot{\tau} = hu^2$ we find

$$\frac{d^2u}{d\tau^2} + u = \frac{\mu(t(\tau))}{h^2}. \tag{D.8.2}$$

We assume that μ varies slowly with time; in particular we shall assume that

$$\dot{\mu} = -\varepsilon\mu^3 \quad , \; \mu(0) = \mu_0. \tag{D.8.3}$$

In the reference given above one can find a treatment of a more general class of functions μ, and also the case of fast changes in μ has been discussed there. We can write the problem (D.8.2–D.8.3) as

$$\frac{d^2u}{d\tau^2} + u = w, \tag{D.8.4a}$$

$$\frac{dw}{d\tau} = -\varepsilon h^3 \frac{w^3}{u^2}, \tag{D.8.4b}$$

where we put $\mu(t(\tau))/h^2 = w(\tau)$. To obtain the standard form for averaging it is convenient to transform $(u, du/d\tau) \mapsto (a, b)$ by

$$u = w + a\cos(\tau) + b\sin(\tau) \; , \; \frac{du}{d\tau} = -a\sin(\tau) + b\cos(\tau). \tag{D.8.5}$$

We find

$$\frac{da}{d\tau} = \varepsilon \frac{h^3 w^3 \cos(\tau)}{(w + a\cos(\tau) + b\sin(\tau))^2} \; , \; \mathrm{a}(0) \text{ given}, \tag{D.8.6a}$$

$$\frac{db}{d\tau} = \varepsilon \frac{h^3 w^3 \sin(\tau)}{(w + a\cos(\tau) + b\sin(\tau))^2} \; , \; \mathrm{b}(0) \text{ given}, \tag{D.8.6b}$$

$$\frac{dw}{d\tau} = -\varepsilon \frac{h^3 w^3}{(w + a\cos(\tau) + b\sin(\tau))^2} \; , \; \mathrm{w}(0) = \frac{\mu(0)}{h^2}. \tag{D.8.6c}$$

The right hand side of system (D.8.6) is 2π-periodic in τ and averaging produces

$$\begin{aligned}
\frac{d\alpha}{d\tau} &= -\varepsilon h^3 \frac{\alpha W^3}{(W^2 - \alpha^2 - \beta^2)^{\frac{3}{2}}}, \; \alpha(0) = \mathrm{a}(0), \\
\frac{d\beta}{d\tau} &= -\varepsilon h^3 \frac{\beta W^3}{(W^2 - \alpha^2 - \beta^2)^{\frac{3}{2}}} \; , \; \beta(0) = \mathrm{b}(0), \\
\frac{dW}{d\tau} &= -\varepsilon h^3 \frac{W^4}{(W^2 - \alpha^2 - \beta^2)^{\frac{3}{2}}} \; , \; \mathrm{W}(0) = \mathrm{w}(0),
\end{aligned} \tag{D.8.7}$$

where $a(\tau) - \alpha(\tau)$, $b(\tau) - \beta(\tau)$, $w(\tau) - W(\tau) = \mathcal{O}(\varepsilon)$ on the time scale $1/\varepsilon$. It is easy to see that

$$\frac{\alpha(\tau)}{\alpha(0)} = \frac{\beta(\tau)}{\beta(0)} = \frac{W(\tau)}{\mathrm{w}(0)}, \quad (\mathrm{a}(0), \mathrm{b}(0) \neq 0) \tag{D.8.8}$$

and we find

$$\alpha(\tau) = \mathrm{a}(0)e^{-\varepsilon\lambda\tau},$$

with $\lambda = h^3w^3(0)a^2(0)/(\mathrm{w}^2(0) - \mathrm{a}^2(0) - \mathrm{b}^2(0))^{\frac{3}{2}}$. Using again (D.8.8) we can construct an $\mathcal{O}(\varepsilon)$-approximation for $\mathrm{u}(\tau)$ on the time scale $1/\varepsilon$. Another possibility is to realize that $\mathrm{W}(\tau) = \mu(t(\tau))/h^2 + \mathcal{O}(\varepsilon)$ so that with equation (D.8.3)

$$W = \frac{1}{h^2}\left(\frac{1}{\mu(0)^2} + 2\varepsilon t\right)^{-\frac{1}{2}} + \mathcal{O}(\varepsilon) \tag{D.8.9}$$

and corresponding expressions for α and β. We have performed our calculations without bothering about the conditions of the averaging theorem, apart from periodicity. It follows from the averaged equation (D.8.7) that the quantity $(W^2 - \alpha^2 - \beta^2)$ should not be small with respect to ε. This condition is not a priori clear from the original equation (D.8.6). Writing down the expression for the instantaneous energy of the two-body system we have

$$E(t) = \frac{1}{2}\dot{r}^2 + \frac{1}{2}\frac{h^2}{r^2} - \frac{\mu(t)}{r},$$

or, with transformation (D.8.5),

$$E(t(\tau)) = \frac{1}{2}h^2\left(\left(\frac{du}{d\tau}\right)^2 + u^2\right) - \mu u = \frac{1}{2}h^2(a^2 + b^2 - w^2). \tag{D.8.10}$$

Negative values of the energy correspond to bound orbits, zero energy with a parabolic orbit, positive energy means escape. The condition that $W^2 - \alpha^2 - \beta^2$ is not small with respect to ε implies that we have to exclude nearly-parabolic orbits among the initial conditions which we study. This condition is reflected in the conditions of the averaging theorem, see for instance Theorem 2.8.1. Note also that, starting in a nearly-parabolic orbit the approximate solutions never yield a positive value of the energy. The conclusion however, that the process described here cannot produce escape orbits is not justified as the averaging method does not apply to the nearly-parabolic transition between elliptic and hyperbolic orbits. To analyze the situation of nearly-parabolic orbits it is convenient to introduce another coordinate system involving the orbital elements e (eccentricity) and E (eccentric anomaly) or f (true anomaly). We discussed such a system briefly in Section 7.10.1. A rather intricate asymptotic analysis of the nearly-parabolic case shows that nearly all solutions starting there become hyperbolic on a time scale of order 1 with respect to ε; see [266].

D.9 Two-body System with Increasing Mass

Dynamically this is a process which is very different from the process of decrease of mass. In the case of decrease of mass each layer of ejected material takes with it an amount of instantaneous momentum. In the case of increase of mass, we have to make an assumption about the momentum of the material falling in. With certain assumptions, see [274], the orbits are found again in a plane. Introducing the total mass of the system $m = m(t)$ and the gravitational constant G we then have the equations of motion

$$\ddot{r} = -\frac{Gm}{r^2} + \frac{h^2}{m^2 r^3} - \frac{\dot{m}}{m}\dot{r}, \tag{D.9.1a}$$

$$mr^2\dot{\phi} = h. \tag{D.9.1b}$$

Note that (D.9.1b) represents an integral of motion with a time-varying factor. Introduction of $u = 1/r$ and the time-like variable τ by

$$\dot{\tau} = \frac{hu^2}{m}, \ \tau(0) = 0$$

we find that the equation

$$\frac{d^2u}{d\tau^2} + u = w \tag{D.9.2}$$

in which the equation for w has to be derived from $w = Gm^3/h^2$. Assuming slow increase of mass according to the relation $\dot{m} = \varepsilon m^n$ (n a constant) we find

$$\frac{dw}{d\tau} = \varepsilon \frac{3h^{\frac{2n}{3}-1}}{G^{\frac{n}{3}}} \frac{w^{1+\frac{n}{3}}}{u^2}. \tag{D.9.3}$$

It is clear that we can approximate the solutions of system (D.9.2–D.9.3) with the same technique as in the preceding section. The approximations can be computed as functions of τ and t and they represent $\mathcal{O}(\varepsilon)$-approximations on the time scale $1/\varepsilon$.

E

On Averaging Methods for Partial Differential Equations

E.1 Introduction

This appendix is an adaptation and extension of the paper [280].

The qualitative and quantitative analysis of weakly nonlinear partial differential equations is an exciting field of investigation. However, the results are still fragmented and it is too early to present a coherent picture of the theory. Instead we will survey the literature, while adding technical details in a number of interesting cases.

Formal approximation methods, as for example multiple timing, have been successful, for equations on both bounded and unbounded domains. Another formal method that has attracted a lot of interest is Whitham's approach to combine averaging and variational principles [288]; see for these formal methods [281]. At an early stage, a number of formal methods for nonlinear hyperbolic equations were analyzed, with respect to the question of asymptotic validity, in [265].

An adaptation of the Poincaré–Lindstedt method for periodic solutions of weakly nonlinear hyperbolic equations was given in [119]; note that this is a rigorous method, based on the implicit function theorem. An early version of the Galerkin averaging method can be found in [222], where vibrations of bars are studied.

The analysis of asymptotic approximations with proofs of validity rests firmly on the qualitative theory of weakly nonlinear partial differential equations. Existence and uniqueness results are available that typically involve contraction, or other fixed-point methods, and maximum principles; we will also use projection methods in Hilbert spaces (Galerkin averaging).

Some of our examples will concern conservative systems. In the theory of finite-dimensional Hamiltonian systems we have for nearly-integrable systems the celebrated KAM theorem, which, under certain nondegeneracy conditions, guarantees the persistence of many tori in the nonintegrable system. For infinite-dimensional conservative systems we now have the KKAM theorems developed by Kuksin [159, 160]. Finite-dimensional invariant manifolds obtained in this way are densely filled with quasiperiodic orbits; these are the kind of solutions we often obtain by our approximation methods. It is stressed, however, that identification of approximate solutions with solutions covering invariant manifolds makes sense only if the validity of the approximation has been demonstrated.

Various forms of averaging techniques are being used in the literature. They are sometimes indicated by terms like "homogenization" or "regularization" methods, and their main purpose is to stabilize numerical integration schemes for partial differential equations. However, apart from numerical improvements we are also interested in asymptotic estimates of validity and in qualitative aspects of the solutions.

E.2 Averaging of Operators

A typical problem formulation would be to consider the Cauchy problem (or later an initial-boundary value problem) for equations like

$$u_t + Lu = \varepsilon f(u), \quad t > 0, u(0) = u_0. \tag{E.2.1}$$

Here L is a linear partial differential operator, and $f(u)$ represents the perturbation terms, possibly nonlinear.

To obtain a standard form $u_t = \varepsilon F(t, u)$, suitable for averaging in the case of a partial differential equation, can already pose a formidable technical problem, even in the case of simple geometries. However, it is reasonable to suppose that one can solve the "unperturbed" ($\varepsilon = 0$) problem in some explicit form before proceeding to the perturbation problem.

A number of authors, in particular in the former Soviet Union, have addressed problem (E.2.1). For a survey of such results see [189]; see also [247].

There still does not exist a unified mathematical theory with a satisfactory approach to higher-order approximations (normalization to arbitrary order) and enough convincing examples. In what follows we shall discuss some results that are relevant for parabolic equations. For the functional-analytic terminology see [128].

E.2.1 Averaging in a Banach Space

In [122], (E.2.1) is considered in the "high-frequency" form

$$u_t + Lu = F(t/\varepsilon, u, \varepsilon), \quad t > 0, \tag{E.2.2}$$

in which L is the generator of a C^0-semigroup $T_L(t)$ on a Banach space X, $F(s, u, \varepsilon)$ is continuous in s, u, ε, continuous differentiable in u, and almost-periodic in t, uniformly for u in compact subsets of X. The operator L has to be time-independent.

Initial data have to be added, and the authors consider the problem formulations of delay equations and parabolic PDEs. In both cases the operator $T_L(t)$ is used to obtain the variation of constants formula followed by averaging:

$$F_0(v) = \lim_{T\to\infty} \frac{1}{T} \int_0^T F(s,v)ds.$$

The averaged equation is

$$v_t + Lv = F_0(v).$$

Note that the transformation $t \mapsto \varepsilon t$ produces an equation in a more usual shape. Equation (E.2.2) has the equivalent form

$$u_t + \varepsilon Lu = \varepsilon F(t,u,\varepsilon), \quad t > 0. \tag{E.2.3}$$

An interesting aspect is that the classical theorems of averaging find an analogue here. For instance, a hyperbolic equilibrium of the averaged equation corresponds to an almost-periodic solution (or, if F is periodic, a periodic solution) of the original equation. Similar theorems hold for the existence and approximation of tori.

E.2.2 Averaging a Time-Dependent Operator

We shall follow the theory developed by Krol in [156], which has some interesting applications. Consider the problem (E.2.1) with two spatial variables x, y and time t; $f(u)$ is linear. Assume that after solving the unperturbed problem, by a variation of constants procedure we can write the problem in the form

$$\frac{\partial F}{\partial t} = \varepsilon L(t)F, \quad F(x,y,0) = \gamma(x,y). \tag{E.2.4}$$

We have

$$L(t) = L_2(t) + L_1(t), \tag{E.2.5}$$

where

$$L_2(t) = b_1(x,y,t)\frac{\partial^2}{\partial x^2} + b_2(x,y,t)\frac{\partial^2}{\partial x \partial y} + b_3(x,y,t)\frac{\partial^2}{\partial y^2},$$
$$L_1(t) = a_1(x,y,t)\frac{\partial}{\partial x} + a_2(x,y,t)\frac{\partial}{\partial y},$$

in which $L_2(t)$ is a uniformly elliptic operator on the domain, and L_1, L_2 and hence L are T-periodic in t; the coefficients a_i, b_i, and γ are C^∞ and bounded with bounded derivatives.

We average the operator L by averaging the coefficients a_i, b_i over t:

$$\bar{a}_i(x,y) = \frac{1}{T}\int_0^T a_i(x,y,s)\,\mathrm{ds}, \quad \bar{b}_i(x,y) = \frac{1}{T}\int_0^T b_i(x,y,s)\,\mathrm{ds}, \tag{E.2.6}$$

producing the averaged operator $\bar{L}$. As an approximating problem for (E.2.4) we now take

$$\frac{\partial \bar{F}}{\partial t} = \varepsilon \bar{L}\bar{F}, \quad \bar{F}(x,y,0) = \gamma(x,y). \tag{E.2.7}$$

A rather straightforward analysis shows existence and uniqueness of the solutions of problems (E.2.4) and (E.2.7) on the time scale $1/\varepsilon$.

Theorem E.2.1 (Krol, [156]). *Let F be the solution of initial value problem (E.2.4) and $\bar{F}$ the solution of initial value problem (E.2.7). Then we have the estimate $\|F - \bar{F}\| = \mathcal{O}(\varepsilon)$ on the time scale $1/\varepsilon$. The norm $\|.\|$ is the sup norm on the spatial domain and on the time scale $1/\varepsilon$.*

The classical approach to prove such a theorem would be to transform (E.2.4) by a near-identity transformation to an averaged equation that satisfies (E.2.4) to a certain order in ε. In this approach we meet in our estimates fourth-order derivatives of F; this puts serious restrictions on the method. Instead, Ben Lemlih and Ellison [171] and, independently, Krol [156] apply a near-identity transformation to $\bar{F}$ that is autonomous and on which we have explicit information.

Proof [Of Theorem E.2.1] Existence and uniqueness on the time scale $1/\varepsilon$ of the initial value problem (E.2.4) follows in a straightforward way from [105].

We introduce $\tilde{F}$ by the near-identity transformation

$$\tilde{F}(x,y,t) = \overline{F}(x,y,t) + \varepsilon \int_0^t (L(s) - \overline{L})\, \mathrm{d}s \bar{F}(x,y,t). \tag{E.2.8}$$

To estimate $\tilde{F} - \bar{F}$, we use that the integrand in (E.2.8) is periodic with zero average and that the derivatives of $\bar{F}, L(t)$, and $\bar{L}$ are bounded. If t is a number between nT and $(n+1)T$ we have

$$\begin{aligned}
\|\tilde{F} - \bar{F}\|_\infty &= \varepsilon \left\| \int_{nT}^t (L(s) - \bar{L})\, \mathrm{d}s \bar{F}(x,y,t) \right\|_\infty \\
&\le 2\varepsilon T(\|a_1\|_\infty \|\bar{F}_x\|_\infty + \|a_2\|_\infty \|\bar{F}_y\|_\infty + \|b_1\|_\infty \|\bar{F}_{xx}\|_\infty \\
&\quad + \|b_2\|_\infty \|\bar{F}_{xy}\|_\infty + \|b_3\|_\infty \|\bar{F}_{yy}\|_\infty) \\
&= \mathcal{O}(\varepsilon)
\end{aligned}$$

on the time scale $1/\varepsilon$. Differentiation of the near-identity transformation (E.2.8) and using (E.2.7), (E.2.8) repeatedly, produces an equation for $\tilde{F}$:

$$\begin{aligned}
\frac{\partial \tilde{F}}{\partial t} &= \frac{\partial \bar{F}}{\partial t} + \varepsilon (L(t) - \bar{L})\bar{F} + \varepsilon \int_0^t (L(s) - \bar{L})\, \mathrm{d}s \frac{\partial \bar{F}}{\partial t} \\
&= \varepsilon L(t)\tilde{F} + \varepsilon^2 \int_0^t ((L(s) - \bar{L})\bar{L} - L(t)(L(s) - \bar{L}))\, \mathrm{d}s \bar{F} \\
&= \varepsilon L(t)\tilde{F} + \varepsilon^2 \mathcal{M}(t)\bar{F},
\end{aligned}$$

with initial value $\tilde{F}(x,y,0) = \gamma(x,y)$. Here $\mathcal{M}(t)$ is a T-periodic fourth-order partial differential operator with bounded coefficients. The implication is that $\tilde{F}$ satisfies (E.2.4) to order ε^2. Putting

$$\frac{\partial}{\partial t} - \varepsilon L(t) = \mathcal{L},$$

we have

$$\mathcal{L}(\tilde{F} - \bar{F}) = \varepsilon^2 \mathcal{M}(t)\bar{F} = \mathcal{O}(\varepsilon)$$

on the time scale $1/\varepsilon$. Moreover, $(\tilde{F} - \bar{F})(x,y,0) = 0$.
To complete the proof we will use barrier functions and the (real) Phragmén–Lindelöf principle (see for instance [221]). Putting $c = \|\mathcal{M}(t)\bar{F}\|_\infty$ we introduce the barrier function

$$B(x,y,t) = \varepsilon^2 ct$$

and the functions (we omit the arguments)

$$Z_1 = \tilde{F} - F - B, Z_2 = \tilde{F} - F + B.$$

We have

$$\mathcal{L}Z_1 = \varepsilon^2 \mathcal{M}(t)\bar{F} - \varepsilon^2 c \le 0,\ Z_1(x,y,0) = 0,$$
$$\mathcal{L}Z_2 = \varepsilon^2 \mathcal{M}(t)\bar{F} + \varepsilon^2 c \ge 0,\ Z_2(x,y,0) = 0.$$

Since Z_1 and Z_2 are bounded, we can apply the Phragmén–Lindelöf principle, resulting in $Z_1 \le 0$ and $Z_2 \ge 0$. It follows that

$$-\varepsilon^2 ct \le \tilde{F} - F \le \varepsilon^2 ct,$$

so that we can estimate

$$\|\tilde{F} - F\|_\infty \le \|B\|_\infty = \mathcal{O}(\varepsilon)$$

on the time scale $1/\varepsilon$. Since we found already $\|\tilde{F} - \bar{F}\|_\infty = \mathcal{O}(\varepsilon)$ on the time scale $1/\varepsilon$, we can apply the triangle inequality to produce

$$\|F - \bar{F}\|_\infty = \mathcal{O}(\varepsilon)$$

on the time scale $1/\varepsilon$. □

E.2.3 Application to a Time-Periodic Advection-Diffusion Problem

As an application one considers in [156] the transport of material (chemicals or sediment) by advection and diffusion in a tidal basin. In this case the advective flow is nearly periodic, and diffusive effects are small. The problem can be formulated as

$$\frac{\partial C}{\partial t} + \nabla \cdot (\mathbf{u}C) - \varepsilon \Delta C = 0, \quad C(x,y,0) = \gamma(x,y), \tag{E.2.9}$$

where $C(x, y, t)$ is the concentration of the transported material, the flow $\mathbf{u} = \mathbf{u}^0(x, y, t) + \varepsilon\mathbf{u}^1(x, y)$ is given; $\mathbf{u}^0$ is T-periodic in time and represents the tidal flow, $\varepsilon\mathbf{u}^1$ is a small residual current arising from wind fields and fresh water input from rivers. Since the diffusion process is slow, we are interested in a long-time scale approximation.

If the flow is divergence-free the unperturbed ($\varepsilon = 0$) problem is given by

$$\frac{\partial C_0}{\partial t} + \mathbf{u}^0 \cdot \nabla C_0 = 0, \quad C_0(x, y, 0) = \gamma(x, y), \tag{E.2.10}$$

a first-order equation that can be integrated along the characteristics with solution $C_0 = \gamma(Q(t)(x, y))$. In the spirit of variation of constants we introduce the change of variables

$$C(x, y, t) = F(Q(t)(x, y), t). \tag{E.2.11}$$

We expect F to be slowly time-dependent when introducing (E.2.11) into the original equation (E.2.9). Using again the technical assumption that the flow $\mathbf{u}^0 + \varepsilon\mathbf{u}^1$ is divergence-free, we obtain a slowly varying equation of the form (E.2.4). Note that the assumption of divergence-free flow is not essential; it only facilitates the calculations.

Krol [156] presents some extensions of the theory and explicit examples in which the slowly varying equation is averaged to obtain a time-independent parabolic problem. Quite often the latter problem still has to be solved numerically, and one may wonder what, then, the use is of this technique. The answer is that one needs solutions on a long time scale and that numerical integration of an equation in which the fast periodic oscillations have been eliminated is a much safer procedure.

In the analysis presented thus far we have considered unbounded domains. To study the equation on spatially bounded domains, adding boundary conditions does not present serious obstacles to the techniques and the proofs. An example is given below.

E.2.4 Nonlinearities, Boundary Conditions and Sources

An extension of the advection-diffusion problem has been obtained in [127]. Consider the problem with initial and boundary values on the two-dimensional domain $\Omega, 0 \leq t < \infty$,

$$\begin{aligned} \frac{\partial C}{\partial t} + \nabla \cdot (\mathbf{u}C) - \varepsilon \Delta C + \varepsilon f(C) &= \varepsilon B(x, y, t), \\ C(x, y, 0) &= \gamma(x, y), (x, y) \in \Omega \\ C(x, y, t) &= 0, (x, y) \in \partial\Omega \times [0, \infty). \end{aligned}$$

The flow $\mathbf{u}$ is expressed as above, the term $f(C)$ is a small reaction term representing, for instance, the reactions of a material with itself or the settling down of sediment; $B(x, y, t)$ is a T-periodic source term, for instance representing dumping of material.

Note that we have chosen the Dirichlet problem; the Neumann problem would be more realistic but it presents some problems, boundary layer corrections and complications in the proof of asymptotic validity, which we avoid here.

The next step is to obtain a standard form, similar to (E.2.4), by the variation of constants procedure (E.2.11), which yields

$$U_t = \varepsilon L(t)U - \varepsilon f(U) + \varepsilon D(x, y, t), \tag{E.2.12}$$

where $L(t)$ is a uniform elliptic T-periodic operator generated by the (unperturbed) time t flow operator as before, $D(x, y, t)$ is produced by the inhomogeneous term B. Averaging over time t produces the averaged equation

$$\bar{U}_t = \varepsilon \bar{L}\bar{U} - \varepsilon \bar{f}(x, y, \bar{U}) + \varepsilon \bar{D}(x, y) \tag{E.2.13}$$

with appropriate initial-boundary values.

Theorem E.2.1 produces $\mathcal{O}(\varepsilon)$-approximations on the time scale $1/\varepsilon$. It is interesting that we can obtain a stronger result in this case. Using sub- and supersolutions in the spirit of maximum principles ([221]), it is shown in [127] that the $\mathcal{O}(\varepsilon)$ estimate is *valid for all time.* The technique is very different from the validity for all time results in Chapter 5.

Another interesting aspect is that the presence of the source term triggers off the existence of a unique periodic solution which is attracting the flow. In the theory of averaging in the case of ordinary differential equations the existence of a periodic solution is derived from the implicit function theorem. In the case of averaging of this parabolic initial-boundary value problem one has to use a topological fixed-point theorem.

The paper [127] contains an explicit example for a circular domain with reaction term $f(C) = aC^2$, and for the source term B, Dirac delta functions.

E.3 Hyperbolic Operators with a Discrete Spectrum

In this section we shall be concerned with weakly nonlinear hyperbolic equations of the form

$$u_{tt} + Au = \varepsilon g(u, u_t, t, \varepsilon), \tag{E.3.1}$$

where A is a positive, self-adjoint linear differential operator on a separable real Hilbert space. Equation (E.3.1) can be studied in various ways. First we shall discuss theorems in [49], where more general semilinear wave equations with a discrete spectrum were considered to prove asymptotic estimates on the $1/\varepsilon$ time scale.

The procedure involves solving an equation corresponding to an infinite number of ordinary differential equations. In many cases, resonance will make this virtually impossible, the averaged (normalized) system is too large, and we have to take recourse to truncation techniques; we discuss results in [155] on the asymptotic validity of truncation methods that at the same time yield information on the time scale of interaction of modes.

Another fruitful approach for weakly nonlinear wave equations, as for example (E.3.1), is using multiple time scales. In the discussion and the examples we shall compare some of the methods.

E.3.1 Averaging Results by Buitelaar

Consider the semilinear initial value problem

$$\frac{dw}{dt} + \mathcal{A}w = \varepsilon f(w, t, \varepsilon), \quad w(0) = w_0, \tag{E.3.2}$$

where $-\mathcal{A}$ generates a uniformly bounded C^0-group $H(t), -\infty < t < +\infty$, on the separable Hilbert space X (in fact, the original formulation is on a Banach space but here we focus on Hilbert spaces), and f satisfies certain regularity conditions and can be expanded with respect to ε in a Taylor series, at least to some order. A generalized solution is defined as a solution of the integral equation

$$w(t) = H(t)w_0 + \varepsilon \int_0^t H(t-s) f(w(s), s, \varepsilon)\, ds. \tag{E.3.3}$$

Using the variation of constants transformation $w(t) = H(t)z(t)$ we obtain the integral equation corresponding to the standard form

$$z(t) = w_0 + \varepsilon \int_0^t F(z(s), s, \varepsilon)\, ds, \quad F(z, s, \varepsilon) = H(-s) f(H(s)z, s, \varepsilon). \tag{E.3.4}$$

Introduce the average $\overline{F}$ of F by

$$\overline{F}(z) = \lim_{T \to \infty} \frac{1}{T} \int_0^T F(z, s, 0)\, ds \tag{E.3.5}$$

and the averaging approximation $\overline{z}(t)$ of $z(t)$ by

$$\overline{z}(t) = w_0 + \varepsilon \int_0^t \overline{F}(\overline{z}(s))\, ds. \tag{E.3.6}$$

We mention that:

- f has to be Lipschitz continuous and uniformly bounded on $\bar{D} \times [0, \infty) \times [0, \varepsilon_0]$, where D is an open, bounded set in the Hilbert space X.
- F is Lipschitz continuous in D, uniformly in t and ε.

Under these rather general conditions Buitelaar [49] proves that $z(t) - \overline{z}(t) = o(1)$ on the time scale $1/\varepsilon$.

In the case that $F(z, t, \varepsilon)$ is T-periodic in t we have the estimate $z(t) - \overline{z}(t) = \mathcal{O}(\varepsilon)$ on the time scale $1/\varepsilon$.

Remark E.3.1. For the proof we need the concepts of almost-periodic function and averaging in Banach spaces. The theory of complex-valued almost-periodic functions was created by Harald Bohr (see Section 4.6); later the theory was extended to functions with values in Banach spaces by Bochner. Bochner's definition is based on the spectral decomposition of almost-periodic functions. The classical definition by Bohr can be reformulated analogously.

Definition E.3.2 (Bochner's criterion). *Let X be a Banach space. Then $h : \mathbb{R} \to X$ is* **almost-periodic** *if and only if h belongs to the closure, with respect to the uniform convergence on $\mathbb{R}$, of the set of trigonometric polynomials*

$$\left\{ P_n : \mathbb{R} \to X : t \mapsto \sum_{k=1}^{n} a_k e^{i\lambda_k t} | n \in \mathbb{N}, \lambda_k \in \mathbb{R}, a_k \in X \right\}.$$

The following lemma is useful.

Lemma E.3.3 (Duistermaat). *Let K be a compact metric space, X a Banach space, and h a continuous function: $K \times \mathbb{R} \to X$. Suppose that for every $z \in K, t \mapsto h(z,t)$ is almost-periodic, and assume that the family $z \mapsto h(z,t) : K \to X, t \in \mathbb{R}$ is equicontinuous. Then the average*

$$\overline{h}(z) = \lim_{T \to \infty} \frac{1}{T} \int_0^T h(z,s)\,\mathrm{d}s$$

is well defined and the limit exists uniformly for $z \in K$. Moreover, if $\phi : \mathbb{R} \to K$ is almost-periodic, then $t \mapsto h(\phi(t), t)$ is almost-periodic.

Proof See [281, Section 15.9]. □

Another basic result that we need is formulated as follows:

Theorem E.3.4 (Buitelaar, [49]). *Consider (E.3.2) with the conditions given above; assume that X is an associated separable Hilbert space and that $-i\mathcal{A}$ is self-adjoint and generates a denumerable, complete orthonormal set of eigenfunctions. If $f(z,t,0)$ is almost-periodic, $F(z,t,0) = T(-t)f(T(t)z,t,0)$ is almost-periodic and the average $\overline{F}(z)$ exists uniformly for z in compact subsets of D. Moreover, a solution starting in a compact subset of D will remain in the interior of D on the time scale $1/\varepsilon$.*

Proof For $z \in X$, we have $z = \sum_k z_k e_k$, and it is well known that the series $T(t)z = \sum_k e^{-i\lambda_k t} z_k e_k$ (λ_k the eigenvalues) converges uniformly and is in general almost-periodic. From Duistermaat's Lemma E.3.3 it follows that $t \mapsto F(z,t,0)$ is almost-periodic with average $\overline{F}(z)$. The existence of the solution in a compact subset of D on the time scale $1/\varepsilon$ follows from the usual contraction argument. □

Remark E.3.5. That the average $\overline{F}(z)$ exists uniformly is very important in the cases in which the spectrum $\{\lambda_k\}$ accumulates near a point that leads to "small denominators." Because of this uniform existence, such an accumulation does not destroy the approximation. ♡

It turns out that in this framework we can use again the methods of proof as they were developed for averaging in ordinary differential equations. One possibility is to choose a near-identity transformation as used before in Section E.2 on averaging of operators. Another possibility is to use the concept of local averaging.

An example in which we can apply periodic averaging is the wave equation

$$u_{tt} - u_{xx} = \varepsilon f(u, u_x, u_t, t, x, \varepsilon), \quad t \geq 0, \quad 0 < x < 1, \tag{E.3.7}$$

where

$$u(0,t) = u(1,t) = 0, \quad u(x,0) = \phi(x), \quad u_t(x,0) = \psi(x), \quad 0 \leq x \leq 1.$$

A difficulty is often that the averaged system is still infinite-dimensional without the possibility of reduction to a subsystem of finite dimension. A typical example is the case $f = u^3$; see [281] and the discussion in Section E.3.4.

An example that is easier to handle is the Klein–Gordon equation

$$u_{tt} - u_{xx} + a^2 u = \varepsilon u^3, \quad t \geq 0, \quad 0 < x < \pi, \quad a > 0.$$

We can apply almost-periodic averaging, and the averaged system splits into finite-dimensional parts; see Section E.3.3.

A similar phenomenon arises in applications to rod and beam equations. A rod problem with extension and torsion produces two linear and nonlinearly coupled Klein–Gordon equations, which is a system with various resonances. A number of cases were explored in [50].

E.3.2 Galerkin Averaging Results

General averaging and periodic averaging of infinite-dimensional systems is important, but in many interesting cases the resulting averaged system is still difficult to analyze and we need additional theorems. One of the most important techniques involves projection methods, resulting in truncation of the system. This was studied by various authors, in particular in [155].
Consider again the initial-boundary value problem for the nonlinear wave equation (E.3.7). The normalized eigenfunctions of the unperturbed ($\varepsilon = 0$) problem are $v_n(x) = \sqrt{2}\sin(n\pi x), n = 1, 2, \ldots$, and we propose to expand the solution of the initial-boundary value problem for equation (E.3.7) in a Fourier series with respect to these eigenfunctions of the form

$$u(t,x) = \sum_{n=1}^{\infty} u_n(t) v_n(x). \tag{E.3.8}$$

By taking inner products this yields an infinite system of ordinary differential equations that is equivalent to the original problem. The next step is then to truncate this infinite dimensional system and apply averaging to the truncated system. The truncation is known as Galerkin's method, and one has to estimate the combined error of truncation and averaging.
The first step is that (E.3.7) with its initial-boundary values has exactly one solution in a suitably chosen Hilbert space $\mathcal{H}_k = H_0^k \times H_0^{k-1}$, where H_0^k are the well-known Sobolev spaces consisting of functions u with derivatives $U^{(k)} \in L^2[0,1]$ and $u^{(2l)}$ zero on the boundary whenever $2l < k$. It is rather standard to establish existence and uniqueness of solutions *on the time scale* $1/\varepsilon$ under certain mild conditions on f; examples are right-hand sides f such as $u^3, uu_t^2, \sin u, \sinh u_t$. Moreover, we note that:

1. If $k \geq 3, u$ is a classical solution of equation (E.3.7).
2. If $f = f(u)$ is an odd function of u, one can find an even energy integral. If such an integral represents a positive definite energy integral, it is now standard that we are able to prove existence and uniqueness for all time.

In Galerkin's truncation method one considers only the first N modes of the expansion (E.3.8) which we shall call the projection u_N of the solution u on a N-dimensional space. To find u_N, we have to solve a $2N$-dimensional system of ordinary differential equations for the expansion coefficients $u_n(t)$ with appropriate (projected) initial values. The estimates for the error $\|u - u_N\|$ depend strongly on the smoothness of the right hand side f of equation (E.3.7) and the initial values $\phi(x), \psi(x)$ but, remarkably enough, not on ε. Krol [155] finds sup norm estimates on the time scale $1/\varepsilon$ and as $N \to \infty$ of the form

$$\|u - u_N\|_\infty = \mathcal{O}(N^{\frac{1}{2}-k}),$$
$$\|u_t - u_{Nt}\|_\infty = \mathcal{O}(N^{\frac{3}{2}-k}).$$

We shall return later to estimates in the analytic case.
As mentioned before, the truncated system is in general difficult to solve. Periodic averaging of the truncated system produces an approximation $\bar{u}_N$ of u_N and finally the following result result.

Theorem E.3.6 (Galerkin averaging). *Consider the initial-boundary value problem*

$$u_{tt} - u_{xx} = \varepsilon f(u, u_x, u_t, t, x, \varepsilon), \quad t \geq 0, \quad 0 < x < 1,$$

where

$$u(0,t) = u(1,t) = 0, \quad u(x,0) = \phi(x), \quad u_t(x,0) = \psi(x), \quad 0 \leq x \leq 1.$$

Suppose that f is k-times continuously differentiable and satisfies the existence and uniqueness conditions on the time scale $1/\varepsilon, (\phi, \psi) \in \mathcal{H}_k$; if the solution of the initial-boundary problem is (u, u_t) and the approximation obtained by the Galerkin averaging procedure $(\bar{u}_N, \bar{u}_{Nt})$, we have on the time scale $1/\varepsilon$,

$$\|u - \bar{u}_N\|_\infty = \mathcal{O}(N^{\frac{1}{2}-k}) + \mathcal{O}(\varepsilon), \quad N \to \infty, \quad \varepsilon \to 0,$$
$$\|u_t - \bar{u}_{Nt}\|_\infty = \mathcal{O}(N^{\frac{3}{2}-k}) + \mathcal{O}(\varepsilon), \quad N \to \infty, \quad \varepsilon \to 0.$$

Proof See [155]. □

There are a number of remarks:

- Taking $N = \mathcal{O}(\varepsilon^{-\frac{2}{2k-1}})$ we obtain an $\mathcal{O}(\varepsilon)$-approximation on the time scale $1/\varepsilon$. So, the required number of modes decreases when the regularity of the data and the order up to which they satisfy the boundary conditions increases.
- However, this decrease of the number of required modes is not uniform in k. So it is not obvious for which choice of k the estimates are optimal at a given value of ε.
- An interesting case arises if the nonlinearity f satisfies the regularity conditions for all k. This happens for instance if f *is an odd polynomial in* u and with analytic initial values. In such cases the results can be improved by introducing Hilbert spaces of analytic functions (so-called Gevrey classes). The estimates in [155] for the approximations on the time scale $1/\varepsilon$ obtained by the Galerkin averaging procedure become in this case

$$\|u - \bar{u}_N\|_\infty = \mathcal{O}(N^{-1}a^{-N}) + \mathcal{O}(\varepsilon), \quad N \to \infty, \quad \varepsilon \to 0,$$
$$\|u_t - \bar{u}_{Nt}\|_\infty = \mathcal{O}(a^{-N}) + \mathcal{O}(\varepsilon), \quad N \to \infty, \quad \varepsilon \to 0,$$

  where the constant a arises from the bound one has to impose on the size of the strip around the real axis on which analytic continuation is permitted in the initial-boundary value problem.
  The important implication is that because of the a^{-N}-term we need only $N = \mathcal{O}(|\log \varepsilon|)$ terms to obtain an $\mathcal{O}(\varepsilon)$-approximation on the time scale $1/\varepsilon$.
- It is not difficult to improve the result in the case of finite-modes initial values, i.e., the initial values can be expressed in a finite number of eigenfunctions $v_n(x)$. In this case the error becomes $\mathcal{O}(\varepsilon)$ on the time scale $1/\varepsilon$ if N is taken large enough.
- Here and in the sequel we have chosen Dirichlet boundary conditions. It is stressed that this is by way of example and not a restriction. We can also use the method for Neumann conditions, periodic boundary conditions, etc.
- It is possible to generalize these results to higher-dimensional (spatial) problems; see [155] for remarks and [216] for an analysis of a two-dimensional nonlinear Klein–Gordon equation with Dirichlet boundary conditions on a rectangle. In the case of more than one spatial dimension, many more resonances may be present.
- Related proofs for Galerkin averaging were given in [98] and [99]. These papers also contain extensions to difference and delay equations.

To illustrate the general results, we will study now approximations of solutions of explicit problems. These problems are typical for the difficulties one may encounter.

E.3.3 Example: the Cubic Klein–Gordon Equation

As a prototype of a nonlinear wave equation with dispersion consider the nonlinear Klein–Gordon equation

$$u_{tt} - u_{xx} + u = \varepsilon u^3, \quad t \geq 0, \quad 0 < x < \pi, \tag{E.3.9}$$

with boundary conditions $u(0,t) = u(\pi,t) = 0$ and initial values $u(x,0) = \phi(x), u_t(x,0) = \psi(x)$ which are supposed to be sufficiently smooth.

The problem has been studied by many authors, often by formal approximation procedures, see [148].

What do we know qualitatively? It follows from the analysis in [155] that we have existence and uniqueness of solutions on the time scale $1/\varepsilon$ and for all time if we add a minus sign on the right-hand side. In [159] and [32] one considers Klein–Gordon equations as a perturbation of the (integrable) sine–Gordon equation and to prove, in an infinite-dimensional version of KAM theory, the persistence of most *finite-dimensional* invariant manifolds in system (E.3.9). See also the subsequent discussion of results in [40] and [19]. We start with the eigenfunction expansion (E.3.8), where we have

$$v_n(x) = \sin(nx), \quad \lambda_n^2 = n^2 + 1, \quad n = 1, 2, \ldots,$$

for the eigenfunctions and eigenvalues. Substituting this expansion in the equation (E.3.9) and taking the L^2 inner product with $v_n(x)$ for $n = 1, 2, \ldots$ produces an infinite number of coupled ordinary differential equations of the form

$$\ddot{u}_n + (n^2 + 1)u_n = \varepsilon f_n(u), \quad n = 1, 2, \ldots, \infty$$

with

$$f_n(u) = \sum_{n_1, n_2, n_3 = 1}^{\infty} c_{n_1 n_2 n_3} u_{n_1} u_{n_2} u_{n_3}.$$

Since the spectrum is nonresonant (see [252]), we can easily average the complete system or, alternatively, to any truncation number N. The result is that the actions are constant to this order of approximation, the angles are varying slowly as a function of the energy level of the modes.

Considering the theory summarized before, we can make the following observations with regard to the asymptotic character of the estimates:

- In [252] it was proved that, depending on the smoothness of the initial values (ϕ, ψ), we need $N = \mathcal{O}(\varepsilon^{-\beta})$ modes (β a positive constant) to obtain an $\mathcal{O}(\varepsilon^{\alpha})$-approximation $(0 < \alpha \leq 1)$ on the time scale $1/\varepsilon$.

- Note that according to [49], discussed in Section E.3.1, we have the case of averaging of an almost-periodic infinite-dimensional vector field that yields an $o(1)$-approximation on the time scale $1/\varepsilon$ in the case of general smooth initial values.
- If the initial values can be expressed in a finite number of eigenfunctions $v_n(x)$, it follows from Section E.3.2 that the error is $\mathcal{O}(\varepsilon)$ on the time scale $1/\varepsilon$.
- Using the method of two time scales, in [272] an asymptotic approximation of the infinite system is constructed (of exactly the same form as above) with estimate $\mathcal{O}(\varepsilon)$ on the time scale $1/\sqrt{\varepsilon}$. In [271] a method is developed to prove an $\mathcal{O}(\varepsilon)$ approximation on the time scale $1/\varepsilon$, which is applied to the nonlinear Klein–Gordon equation with a quadratic nonlinearity $(-\varepsilon u^2)$.
- In [252] also a second-order approximation is constructed. It turns out that there exists a small interaction between modes with number n and number $3n$, which probably involves much longer time scales than $1/\varepsilon$. This is still an open problem.
- In [40] one considers the nonlinear Klein–Gordon equation (E.3.9) in the rather general form

$$u_{tt} - u_{xx} + V(x)u = \varepsilon f(u), \quad t \geq 0, \quad 0 < x < \pi, \tag{E.3.10}$$

  with V an even periodic function and $f(u)$ an odd polynomial in u. Assuming rapid decrease of the amplitudes in the eigenfunction expansion (E.3.8) and Diophantine (nonresonance) conditions on the spectrum, it is proved that *infinite*-dimensional invariant tori persist in the nonlinear wave equation (E.3.10) corresponding to almost-periodic solutions. The proof involves a perturbation expansion that is valid on a long time scale.
- In [19] one considers the nonlinear Klein–Gordon equation (E.3.9) in the more general form

$$u_{tt} - u_{xx} + mu = \varepsilon\phi(x, u), \quad t \geq 0, \quad 0 < x < \pi, \tag{E.3.11}$$

  and the same boundary conditions. The function $\phi(x, u)$ is polynomial in u, entire analytic and periodic in x, and odd in the sense that $\phi(x, u) = -\phi(-x, -u)$.
  Under a certain nonresonance condition on the spectrum, it is shown in [19] that the solutions remain close to finite-dimensional invariant tori, corresponding to quasiperiodic motion on time scales longer than $1/\varepsilon$.

The results of [40] and [19] add to the understanding and interpretation of the averaging results, and since we are describing manifolds of which the existence has been demonstrated, it raises the question of how to obtain longer time scale approximations.

E.3.4 Example: a Nonlinear Wave Equation with Infinitely Many Resonances

In [148] and [252] an exciting and difficult problem is briefly discussed: the initial-boundary value problem

$$u_{tt} - u_{xx} = \varepsilon u^3, \quad t \geq 0, \quad 0 < x < \pi, \tag{E.3.12}$$

with boundary conditions $u(0,t) = u(\pi,t) = 0$ and initial values $u(x,0) = \phi(x), u_t(x,0) = \psi(x)$ that are supposed to be sufficiently smooth.

Starting with an eigenfunction expansion (E.3.8) we have

$$v_n(x) = \sin(nx), \quad \lambda_n^2 = n^2, \quad n = 1, 2, \ldots,$$

for the eigenfunctions and eigenvalues. The infinite-dimensional system becomes

$$\ddot{u}_n + n^2 u_n = \varepsilon f_n(u), \quad n = 1, 2, \ldots, \infty,$$

with $f_n(u)$ representing the homogeneous cubic right-hand side. The authors note that since there is an infinite number of resonances, after applying the two-time scales method or averaging, we still have to solve an infinite system of coupled ordinary differential equations. The problem is even more complicated than the famous Fermi–Pasta–Ulam problem since the interactions are global instead of nearest-neighbor.

Apart from numerical approximation, Galerkin averaging seems to be a possible approach, and we state here the application in [155] to this problem with the cubic term. Suppose that for the initial values ϕ, ψ we have a finite-mode expansion of M modes only; of course, we take $N \geq M$ in the eigenfunction expansion. Now the initial values ϕ, ψ are analytic and in [155] one optimizes the way in which the analytic continuation of the initial values takes place. The analysis leads to the following estimate for the approximation $\bar{u}_N$ obtained by Galerkin averaging:

$$\|u - \bar{u}_N\|_\infty = \mathcal{O}(\varepsilon^{\frac{N+1-M}{N+1+2M}}), \quad 0 \leq \varepsilon^{\frac{N+1}{N+1+2M}} t \leq 1. \tag{E.3.13}$$

It is clear that if $N \gg M$ the error estimate tends to $\mathcal{O}(\varepsilon)$ and the time scale to $1/\varepsilon$. The result can be interpreted as an upper bound for the speed of energy transfer from the first M modes to higher-order modes.

The Analysis by Van der Aa and Krol

Consider the coupled system of ordinary differential equations corresponding to problem (E.3.12) for arbitrary N; this system is generated by the Hamiltonian $\mathbf{H}^N$. Note that although (E.3.12) corresponds to an infinite-dimensional Hamiltonian system, this property does not necessarily carry over to projections.

Important progress has been achieved by Van der Aa and Krol in [261], who apply Birkhoff normalization to the Hamiltonian system $\mathbf{H}^N$; the normalized Hamiltonian is indicated by $\overline{\mathbf{H}}^N$. This procedure is asymptotically equivalent to averaging. Remarkably enough the flow generated by $\overline{\mathbf{H}}^N$ for arbitrary N contains an infinite number of invariant manifolds.

Consider the "odd" manifold M_1 that is characterized by the fact that only odd-numbered modes are involved in M_1. Inspection of $\overline{\mathbf{H}}^N$ reveals that M_1 is an invariant manifold.

In the same way, the "even" manifold M_2 is characterized by the fact that only even-numbered modes are involved; this is again an invariant manifold of $\overline{\mathbf{H}}^N$.

In [252] this was noted for $N = 3$, which is rather restricted; the result can be extended to manifolds M_m with $m = 2^k q, q$ an odd natural number, k a natural number. It turns out that projections to two modes yield little interaction, so this motivates us to look at projections with at least $N = 6$ involving the odd modes $1, 3, 5$ on M_1 and $2, 4, 6$ on M_2.

In [261] $\overline{\mathbf{H}}^6$ is analyzed, in particular the periodic solutions on M_1. For each value of the energy this Hamiltonian produces three normal mode (periodic) solutions which are stable on M_1. Analyzing the stability in the full system generated by $\overline{\mathbf{H}}^6$ we find again stability.

An open question is whether there exist periodic solutions in the flow generated by $\overline{\mathbf{H}}^6$ that are not contained in either M_1 or M_2.

What is the relation between the periodic solutions found by averaging and periodic solutions of the original nonlinear wave problem (E.3.12)? Van der Aa and Krol [261] compare with results obtained in [101] where the Poincaré–Lindstedt continuation method is used to prove existence and to approximate periodic solutions. Related results employing elliptic functions have been derived in [175]. It turns out that there is very good agreement but the calculation by the Galerkin averaging method is technically simpler.

E.3.5 Example: the Keller–Kogelman Problem

An interesting example of a nonlinear equation with dispersion and dissipation, generated by a Rayleigh term, was presented in [145]. Consider the equation

$$u_{tt} - u_{xx} + u = \varepsilon \left(u_t - \frac{1}{3} u_t^3 \right), \quad t \geq 0, \quad 0 < x < \pi, \tag{E.3.14}$$

with boundary conditions $u(0, t) = u(\pi, t) = 0$ and initial values $u(x, 0) = \phi(x), u_t(x, 0) = \psi(x)$ that are supposed to be sufficiently smooth. As before, putting $\varepsilon = 0$, we have for the eigenfunctions and eigenvalues

$$v_n(x) = \sin(nx), \quad \lambda_n = \omega_n^2 = n^2 + 1, \quad n = 1, 2, \ldots,$$

and again we propose to expand the solution of the initial boundary value problem for (E.3.14) in a Fourier series with respect to these eigenfunctions of the form (E.3.8). Substituting the expansion into the differential equation we have

$$\sum_{n=1}^{\infty} \ddot{u}_n \sin nx + \sum_{n=1}^{\infty} (n^2+1) u_n \sin nx = \varepsilon \sum_{n=1}^{\infty} \dot{u}_n \sin nx - \frac{\varepsilon}{3} \left(\sum_{n=1}^{\infty} \dot{u}_n \sin nx \right)^3 .$$

When taking inner products we have to Fourier analyze the cubic term. This produces many terms, and it is clear that we will not have exact normal mode solutions, since for instance mode m will excite mode $3m$.
At this point we can start averaging, and it becomes important that the spectrum not be resonant. In particular, we have in the averaged equation for u_n only terms arising from $\dot{u}_n^3$ and $\sum_{i\neq n}^{\infty} \dot{u}_i^2 \dot{u}_n$. The other cubic terms do not survive the averaging process; the part of the equation for $n = 1, 2, \ldots$ that produces nontrivial terms is

$$\ddot{u}_n + \omega_n^2 u_n = \varepsilon \left(\dot{u}_n - \frac{1}{4} \dot{u}_n^3 - \frac{1}{2} \sum_{i\neq n}^{\infty} \dot{u}_i^2 \dot{u}_n \right) + \cdots ,$$

where the dots stand for nonresonant terms. This is an infinite system of ordinary differential equations that is still fully equivalent to the original problem.

We can now perform the actual averaging in a notation that contains only minor differences from that of [145]. Transforming in the usual way $u_n(t) = a_n(t) \cos \omega_n t + b_n(t) \sin \omega_n t$, $\dot{u}_n(t) = -\omega_n a_n(t) \sin \omega_n t + \omega_n b_n(t) \cos \omega_n t$, to obtain the standard form, we obtain after averaging the approximations given by (a bar denotes approximation)

$$2\dot{\overline{a}}_n = \varepsilon \overline{a}_n \left(1 + \frac{n^2+1}{16} (\overline{a}_n^2 + \overline{b}_n^2) - \frac{1}{4} \sum_{k=1}^{\infty} (k^2+1)(\overline{a}_k^2 + \overline{b}_k^2) \right) ,$$

$$2\dot{\overline{b}}_n = \varepsilon \overline{b}_n \left(1 + \frac{n^2+1}{16} (\overline{a}_n^2 + \overline{b}_n^2) - \frac{1}{4} \sum_{k=1}^{\infty} (k^2+1)(\overline{a}_k^2 + \overline{b}_k^2) \right) .$$

This system shows fairly strong (although not complete) decoupling because of the nonresonant character of the spectrum. Because of the self-excitation, we have no conservation of energy. Putting $\overline{a}_n^2 + \overline{b}_n^2 = E_n, \quad n = 1, 2, \ldots$, multiplying the first equation by $\overline{a}_n$ and the second equation by $\overline{b}_n$, and adding the equations, we have

$$\dot{E}_n = \varepsilon E_n \left(1 + \frac{n^2+1}{16} E_n - \frac{1}{4} \sum_{k=1}^{\infty} (k^2+1) E_k \right) .$$

We have immediately a nontrivial result: starting in a mode with zero energy, this mode will not be excited on a time scale $1/\varepsilon$. Another observation is that if we have initially only one nonzero mode, say for $n = m$, the equation for E_m becomes

$$\dot{E}_m = \varepsilon E_m \left(1 - \frac{3}{16}(m^2+1)E_m\right).$$

We conclude that we have stable equilibrium at the value

$$E_m = \frac{16}{3(m^2+1)}.$$

More generally, Theorem E.3.4, yields that the approximate solutions have precision $o(\varepsilon)$ on the time scale $1/\varepsilon$; if we start with initial conditions in a finite number of modes the error is $\mathcal{O}(\varepsilon)$, see Section E.3.2. For related qualitative results see [162].

E.4 Discussion

As noted in the introduction, the theory of averaging for PDEs is far from complete. This holds in particular for equations to be studied on unbounded spatial domains. For a survey of methods and references see [281, Chapter 14]. We mention briefly some other results that are relevant for this survey of PDE averaging.

In Section E.2 we mentioned the approach of Ben Lemlih and Ellison [171] to perform averaging in a suitable Hilbert space. They apply this to approximate the long-time evolution of the quantum anharmonic oscillator. Sáenz extends this approach in [230] and [232].

In [185] Matthies considers fast periodic forcing of a parabolic PDE to obtain by a near-identity transformation and averaging an approximate equation plus exponentially small part; as an application certain dynamical systems aspects are explored. Related results are obtained for infinite-dimensional Hamiltonian equations in [186].

An interesting problem arises in studying wave equations on domains that are three-dimensional and thin in the z-direction. In [76] and [75] the thinness is used as a small parameter to derive an approximate set of two-dimensional equations, approximating the original system. The time scale estimates are inspired by Hamiltonian mechanics.

Finally, a remark on slow manifold theory, which has been very influential in asymptotic approximation theory for ODEs recently. There are now extensions for PDEs that look very promising. The reader is referred to [22] and [23].

References

[1] R. Abraham and J.E. Marsden. *Foundations of Mechanics.* The Benjamin/Cummings Publ. Co., Mass. Reading, 1978.

[2] R.H. Abraham and C.D. Shaw. *Dynamics—The Geometry of Behavior I, II.* Inc. Aerial Press, California Santa Cruz, 1983.

[3] V.M. Alekseev. Quasirandom oscillations and qualitative questions in celestial mechanics. *Amer. Math. Soc. Transl.*, 116(2):97–169, 1981.

[4] V. I. Arnol′d. A spectral sequence for the reduction of functions to normal form. *Funkcional. Anal. i Priložen.*, 9(3):81–82, 1975.

[5] V. I. Arnol′d. Spectral sequences for the reduction of functions to normal forms. In *Problems in mechanics and mathematical physics (Russian)*, pages 7–20, 297. Izdat. "Nauka", Moscow, 1976.

[6] V.I. Arnol′d. Instability of dynamical systems with several degrees of freedom. *Dokl. Akad. Nauk. SSSR*, 156:581–585, 1964.

[7] V.I. Arnol′d. Conditions for the applicability, and estimate of the error, of an averaging method for systems which pass through states of resonance during the course of their evolution. *Soviet Math.*, 6:331–334, 1965.

[8] V.I. Arnol′d. *Mathematical Methods of Classical Mechanics*, volume 60. MIR, Springer Graduate Texts in Mathematics, Springer-Verlag, Moscow, New York, 1978.

[9] V.I. Arnol′d. *Geometrical Methods in the Theory of Ordinary Differential Equations.* Springer-Verlag, New York, 1983.

[10] V.I. Arnol′d, V.V. Kozlov, and A.I. Neishstadt. *Mathematical Aspects of Classical and Celestial Mechanics, in Dynamical Systems III (V.I. Arnol′d, ed.).* Springer-Verlag, Berlin etc., 1988.

[11] Zvi Artstein. Averaging of time–varying differential equations revisited. Technical report, 2006.

[12] Alberto Baider. Unique normal forms for vector fields and Hamiltonians. *Journal of Differential Equations*, 78:33–52, 1989.

[13] Alberto Baider. Unique normal forms for vector fields and Hamiltonians. *J. Differential Equations*, 78(1):33–52, 1989.

[14] Alberto Baider and Richard Churchill. The Campbell-Hausdorff group and a polar decomposition of graded algebra automorphisms. *Pacific Journal of Mathematics*, 131:219–235, 1988.

[15] Alberto Baider and Richard Churchill. Unique normal forms for planar vector fields. *Math. Z.*, 199(3):303–310, 1988.

[16] Alberto Baider and Jan A. Sanders. Further reduction of the Takens-Bogdanov normal form. *Journal of Differential Equations*, 99:205–244, 1992.

[17] T. Bakri, R. Nabergoj, A. Tondl, and Ferdinand Verhulst. Parametric excitation in nonlinear dynamics. *Int. J. Nonlinear Mech.*, 39:311–329, 2004.

[18] M. Balachandra and P.R.Sethna. A generalization of the method of averaging for systems with two time-scales. *Archive for Rational Mechanics and Analysis*, 58:261–283, 1975.

[19] D. Bambusi. On long time stability in Hamiltonian perturbations of nonresonant linear pde's. *Nonlinearity*, 12:823–850, 1999.

[20] C. Banfi. Sull'approssimazione di processi non stazionari in meccanica non lineare. *Bolletino dell Unione Matematica Italiana*, 22:442–450, 1967.

[21] C. Banfi and D.Graffi. Sur les méthodes approchées de la mécanique non linéaire. *Actes du Coll. Equ. Diff. Non Lin.*, pages 33–41, 1969.

[22] P.W. Bates, Kening Lu, and C. Zeng. Existence and persistence of invariant manifolds for semiflows in Banach space. *Memoirs AMS*, 135:1–129, 1998.

[23] P.W. Bates, Kening Lu, and C. Zeng. Persistence of overflowing manifolds for semiflow. *Comm. Pure Appl. Math.*, 52:983–1046, 1999.

[24] G. Belitskii. Normal forms in relation to the filtering action of a group. *Trudy Moskov. Mat. Obshch.*, 40:3–46, 1979.

[25] G. R. Belitskii. Invariant normal forms of formal series. *Functional Analysis and Applications*, 13:59–60, 1979.

[26] R. Bellman. *Methods of Nonlinear Analysis I.* Academic Press, New York, 1970.

[27] A. Ben Lemlih. *An extension of the method of averaging to partial differential equations.* PhD thesis, University of New Mexico, 1986.

[28] A. Ben Lemlih and J. A. Ellison. The method of averaging and the quantum anharmonic oscillator. *Phys. Rev. Lett*, 55:1950–1953, 1986.

[29] Martin Bendersky and Richard C. Churchill. A spectral sequence approach to normal forms. In *Recent developments in algebraic topology*, volume 407 of *Contemp. Math.*, pages 27–81. Amer. Math. Soc., Providence, RI, 2006.

[30] M.V. Berry. Regular and irregular motion. In Jorna [142], pages 16–120.

[31] J.G. Besjes. On the asymptotic methods for non-linear differential equations. *Journal de Mécanique*, 8:357–373, 1969.

[32] A.I. Bobenko and S. Kuksin. The nonlinear Klein-Gordon equation on an interval as a perturbed Sine-Gordon equation. *Comment. Meth. Helvetici*, 70:63–112, 1995.

[33] N.N. Bogoliubov, Yu.A. Mitropolskii, and A.M.Samoilenko. *Methods of Accelerated Convergence in Nonlinear Mechanics*. Hindustan Publ. Co. and Springer Verlag, Delhi and Berlin, 1976.

[34] N.N. Bogoliubov and Yu.A. Mitropolsky. The method of integral manifolds in nonlinear mechanics. *Contributions to Differential Equations*, 2:123–196, 1963. Predecessor of Journal of Differential Equations.

[35] N.N. Bogoliubov and Yu.A.Mitropolskii. *Asymptotic methods in the theory of nonlinear oscillations*. Gordon and Breach, New York, 1961.

[36] H. Bohr. *Fastperiodische Funktionen*. Springer Verlag, Berlin, 1932.

[37] M. Born. *The mechanics of the atom*. G. Bell and Sons, London, 1927.

[38] Raoul Bott and Loring W. Tu. *Differential forms in algebraic topology*. Springer-Verlag, New York, 1982.

[39] T. Bountis, H. Segur, and F. Vivaldi. Integrable Hamiltonian systems and the Painlevé property. *Phys. Review A*, 25(3):1257–1264, 1989.

[40] J. Bourgain. Construction of approximative and almost periodic solutions of perturbed linear Schrödinger and wave equations. *GAFA*, 6:201–230, 1996.

[41] A.D. Brjuno. Instability in a Hamiltonian system and the distribution of asteroids. *Math. USSR Sbornik (Mat.Sbornik)*, 1283:271–312, 1970.

[42] Bram Broer. On the generating functions associated to a system of binary forms. *Indag. Math. (N.S.)*, 1(1):15–25, 1990.

[43] Henk Broer, Igor Hoveijn, Gerton Lunter, and Gert Vegter. *Bifurcations in Hamiltonian systems*, volume 1806 of *Lecture Notes in Mathematics*. Springer-Verlag, Berlin, 2003.

[44] H.W. Broer, S-N. Chow, Y. Kim, and G. Vegter. A normally elliptic Hamiltonian bifurcation. *Z. angew. Math. Phys.*, 44:389–432, 1993.

[45] H.W. Broer, I. Hoveijn, G.A. Lunter, and G. Vegter. Resonances in a spring-pendulum: algorithms for equivariant singularity theory. *Nonlinearity*, 11:1269–1605, 1998.

[46] H.W. Broer, G.B. Huitema, and M.B. Sevryuk. *Quasi-periodic motions in families of dynamical systems: order amidst chaos*, volume 1645 of *Lecture Notes Mathematics*. Springer-Verlag, Berlin, Heidelberg, New York, 1996.

[47] H.W. Broer, G.A. Lunter, and G. Vegter. Equivariant singularity theory with distinguished parameters: Two case studies of resonant hamiltonian systems. *Physica D*, 112:64–80, 1998.

[48] H.W. Broer, H.M. Osinga, and G. Vegter. Algorithms for computing normally hyperbolic invariant manifolds. *Z. angew. Math. Phys.*, 48:480–524, 1997.

[49] R.P. Buitelaar. *The method of averaging in Banach spaces*. PhD thesis, University of Utrecht, 1993.

[50] R.P. Buitelaar. On the averaging method for rod equations with quadratic nonlinearity. *Math. Methods Appl. Sc.*, 17:209–228, 1994.

[51] P.F. Byrd and M.B. Friedman. *Handbook of Elliptic Integrals for Engineers and Scientists.* Springer Verlag, 1971.

[52] J. Calmet, W.M. Seiler, and R.W. Tucker, editors. *Global Integrability of Field Theories.* Universitätsverlag Karlsruhe, 2006.

[53] F.F. Cap. Averaging method for the solution of non-linear differential equations with periodic non-harmonic solutions. *International Journal Non-linear Mechanics*, 9:441–450, 1973.

[54] Carmen Chicone. *Ordinary differential equations with applications*, volume 34 of *Texts in Applied Mathematics.* Springer, New York, second edition, 2006.

[55] S.-N. Chow and J.K. Hale. *Methods of bifurcation theory*, volume 251 of *Grundlehren der mathematischen Wissenschaften.* Springer-Verlag, Berlin, Heidelberg, New York, 1982.

[56] S.-N. Chow and J. Mallet-Paret. Integral averaging and bifurcation. *Journal of Differential Equations*, 26:112–159, 1977.

[57] Shui-Nee Chow and Jack K. Hale, editors. *Dynamics of infinite-dimensional systems*, volume 37 of *NATO Advanced Science Institutes Series F: Computer and Systems Sciences*, Berlin, 1987. Springer-Verlag.

[58] A. Clairaut. Mémoire sur l'orbite apparent du soleil autour de la Terre an ayant égard aux perturbations produites par les actions de la Lune et des Planetes principales. *Mém. de l'Acad. des Sci. (Paris)*, pages 521–564, 1754.

[59] A. Coddington and N. Levinson. *Theory of Ordinary Differential Equations.* McGraw-Hill, New-York, 1955.

[60] E.G.D. Cohen, editor. *Fundamental Problems in Statistical Mechanics III.* Elsevier North-Holland, 1975. Proceedings of the 3rd International Summer School, Wageningen, The Netherlands, 29 July-15 August 1974.

[61] E.G.D. Cohen, editor. *Fundamental Problems in Statistical Mechanics.* North Holland Publ., Amsterdam and New York, 1980.

[62] Jane Cronin and Jr. Robert E. O'Malley, editors. *Analyzing multiscale phenomena using singular perturbation methods*, volume 56 of *Proc. Symposia Appl. Math.*, Providence, RI, 1999. AMS.

[63] R. Cushman. Reduction of the 1:1 nonsemisimple resonance. *Hadronic Journal*, 5:2109–2124, 1982.

[64] R. Cushman and J. A. Sanders. A survey of invariant theory applied to normal forms of vectorfields with nilpotent linear part. In Stanton [249], pages 82–106.

[65] R. Cushman, Jan A. Sanders, and N. White. Normal form for the $(2;n)$-nilpotent vector field, using invariant theory. *Phys. D*, 30(3):399–412, 1988.

[66] R.H. Cushman. 1:2:2 resonance. Manuscript, 1985.

[67] Richard Cushman. *Global Aspects of Classical Integrable Systems.* Birkhäuser, Basel, 1997.

[68] Richard Cushman and Jan A. Sanders. Nilpotent normal forms and representation theory of $\mathrm{sl}(2,\mathbb{R})$. In Golubitsky and Guckenheimer [110], pages 31–51.

[69] Pierre-Simon de Laplace. *Traité de Mécanique Céleste*, volume 1–5. Duprat, Courcier-Bachelier, Paris, 1979.

[70] T. De Zeeuw and M. Franx. Structure and dynamics of elliptical galaxies. In *Annual review of Astronomy and Astrophysics*, volume 29, pages 239–274. Annual Rev Inc., 1991.

[71] A. Degasperis and G. Gaeta, editors. *SPT98-Symmetry and Perturbation Theory II*, Singapore, 1999. World Scientific.

[72] A. Deprit. The elimination of the parallax in satellite theory. *Celestial Mechanics*, 24:111–153, 1981.

[73] R.L. Devaney. Homoclinic orbits in hamiltonian systems. *J. Differential Equations*, 21:431–438, 1976.

[74] R.L. Devaney and Z.H. Nitecki, editors. *Classical Mechanics and Dynamical Systems.* Inc. Marcel Dekker, New York, 1981.

[75] R.E. Lee DeVille. Reduced equations for models of laminated materials in thin domains. II. *Asymptotic Analysis*, 42:311–346, 2005.

[76] R.E. Lee DeVille and C. Eugene Wayne. Reduced equations for models of laminated materials in thin domains. I. *Asymptotic Analysis*, 42:263–309, 2005.

[77] J.J. Duistermaat. Bifurcations of periodic solutions near equilibrium points of Hamiltonian systems. In Salvadori [233]. CIME course *Bifurcation Theory and Applications.*

[78] J.J. Duistermaat. Erratum to: [79]. preprint, University of Utrecht, 1984.

[79] J.J. Duistermaat. Non-integrability of the 1:1:2-resonance. *Ergodic Theory and Dynamical Systems*, 4:553–568, 1984.

[80] H. Dullin, A. Giacobbe, and R. Cushman. Monodromy in the resonant swing spring. *Physica D*, 190:15–37, 2004.

[81] W. Eckhaus. New approach to the asymptotic theory of nonlinear oscillations and wave-propagation. *J. Math. An. Appl.*, 49:575–611, 1975.

[82] W. Eckhaus. *Asymptotic Analysis of Singular Perturbations.* North-Holland Publ. Co., Amsterdam, 1979.

[83] M. Eckstein, Y.Y. Shi, and J. Kevorkian. Satellite motion for arbitrary eccentricity and inclination around the smaller primary in the restricted three-body problem. *The Astronomical Journal*, 71:248–263, 1966.

[84] James A. Ellison, Albert W. Sáenz, and H. Scott Dumas. Improved Nth order averaging theory for periodic systems. *Journal of Differential Equations*, 84:383–403, 1990.

[85] C. Elphick et al. A simple global characterization for normal forms of singular vector fields. *Physica D*, 29:95–127, 1987.

[86] L. Euler. De seribus divergentibus. *Novi commentarii ac. sci. Petropolitanae*, 5:205–237, 1754.

[87] L. Euler. *Opera Omnia, ser. I, 14*. Birkhäuser, 1924.

[88] R.M. Evan-Iwanowski. *Resonance Oscillations in Mechanical Systems.* Elsevier Publ. Co., Amsterdam, 1976.

[89] P. Fatou. Sur le mouvement d'un système soumis á des forces á courte période. *Bull. Soc. Math.*, 56:98–139, 1928.

[90] Alex Fekken. On the resonant normal form of a fully resonant hamiltonian function. Technical report, Vrije Universiteit, Amsterdam, 1986. Rapport 317.

[91] N. Fenichel. Asymptotic stability with rate conditions, II. *Ind. Univ. Math. J.*, 26:81–93, 1971.

[92] N. Fenichel. Persistence and smoothness of invariant manifolds for flows. *Ind. Univ. Math. J.*, 21:193–225, 1971.

[93] N. Fenichel. Asymptotic stability with rate conditions. *Ind. Univ. Math. J.*, 23:1109–1137, 1974.

[94] N. Fenichel. Geometric singular perturbations theory for ordinary differential equations. *J. Diff. Eq.*, 31:53–98, 1979.

[95] S. Ferrer, H. Hanssmann, J. Palacián, and P. Yanguas. On perturbed oscillators in 1 : 1 : 1 resonance: the case of axially symmetric cubic potentials. *J. Geom. Phys.*, 40:320–369, 2002.

[96] S. Ferrer, M. Lara, J. Palacián, J.F. San Juan, A. Viartola, and P. Yanguas. The Hénon and Heiles problem in three dimensions. I: Periodic orbits near the origin. *Int. J. Bif. Chaos*, 8:1199–1213, 1998.

[97] S. Ferrer, M. Lara, J. Palacián, J.F. San Juan, A. Viartola, and P. Yanguas. The Hénon and Heiles problem in three dimensions. II: Relative equilibria and bifurcations in the reduced system. *Int. J. Bif. Chaos*, 8:1215–1229, 1998.

[98] M. Fečkan. A Galerkin-averaging method for weakly nonlinear equations. *Nonlinear Anal.*, 41:345–369, 2000.

[99] M. Fečkan. *Galerkin-averaging method in infinite-dimensional spaces for weakly nonlinear problems*, pages 269–279. Volume 43 of Grosinho et al. [115], 2001.

[100] A.M. Fink. *Almost Periodic Differential Equations.* Springer Verlag Lecture Notes in Mathematics 377, Berlin, 1974.

[101] J.P. Fink, W.S. Hall, and A.R. Hausrath. A convergent two-time method for periodic differential equations. *J. Diff. Eqs.*, 15:459–498, 1974.

[102] J. Ford. Ergodicity for nearly linear oscillator systems. In Cohen [60], page 215. Proceedings of the 3rd International Summer School, Wageningen, The Netherlands, 29 July-15 August 1974.

[103] L.E. Fraenkel. On the method of matched asymptotic expansions I, II, III. *Proc. Phil. Soc.*, 65:209–284, 1969.

[104] F. Franklin. On the calculation of the generating functions and tables of groundforms for binary quantics. *American Journal of Mathematics*, 3:128–153, 1880.

[105] A. Friedman. *Partial Differential Equations of Parabolic Type.* Prentice-Hall, Englewood Cliffs, NJ, 1964.

[106] F.T. Geyling and H.R. Westerman. *Introduction to Orbital Mechanics.* Addison-Wesley Publ. Co., Mass. Reading, 1971.

[107] Roger Godement. *Topologie algébrique et théorie des faisceaux.* Hermann, Paris, 1958.

[108] E.G. Goloskokow and A.P. Filippow. *Instationäre Schwingungen Mechanischer Systeme.* Akademie Verlag, Berlin, 1971.

[109] M. Golubitsky, I. Stewart, and D.G. Schaeffer. *Singularities and Groups in Bifurcation Theory II.* Applied Mathematical Sciences 69 Springer-Verlag, New York, 1988.

[110] Martin Golubitsky and John M. Guckenheimer, editors. *Multiparameter bifurcation theory,* volume 56 of *Contemporary Mathematics,* Providence, RI, 1986. American Mathematical Society.

[111] J.H. Grace and M.A. Young. *The Algebra of Invariants.* Cambridge University Press, 1903.

[112] J. Grasman. *Asymptotic methods for relaxation oscillations and applications,* volume 63 of *Appl. Math. Sciences.* Springer-Verlag, Berlin, Heidelberg, New York, 1987.

[113] W. M. Greenlee and R. E. Snow. Two-timing on the half line for damped oscillation equations. *J. Math. Anal. Appl.,* 51(2):394–428, 1975.

[114] B. Greenspan and P.J. Holmes. Repeated resonance and homoclinic bifurcation in a periodically forced family of oscillators. *SIAM J. Math. Anal.,* 15:69–97, 1984.

[115] H.R. Grosinho, M. Ramos, C. Rebelo, and L. Sanches, editors. *Nonlinear Analysis and Differential Equations,* volume 43 of *Progress in Nonlinear Differential Equations and Their Applications.* Birkhäuser Verlag, Basel, 2001.

[116] J. Guckenheimer and P.J. Holmes. *Nonlinear Oscillations, Dynamical Systems and Bifurcations of Vector Fields,* volume 42. Applied Mathematical Sciences Springer Verlag, New York, 1983.

[117] J.D. Hadjidemetriou. Two-body problem with variable mass: a new approach. *Icarus,* 2:440, 1963.

[118] Y. Hagihara. *Celestial Mechanics.* The MIT Press, Mass. Cambridge, 1970-76.

[119] J. Hale. Periodic solutions of a class of hyperbolic equations containing a small parameter. *Arch. Rat. Mech. Anal.,* 23:380–398, 1967.

[120] J.K. Hale. *Oscillations in nonlinear systems.* MacGraw-Hill, repr. Dover Publ., New York (1992), New York, 1963.

[121] J.K. Hale. *Ordinary Differential Equations.* Wiley-Interscience, New York, 1969.

[122] J.K. Hale and S.M. Verduyn Lunel. Averaging in infinite dimensions. *J. Integral Equations and Applications,* 2:463–491, 1990.

[123] Brian C. Hall. *Lie groups, Lie algebras, and representations*, volume 222 of *Graduate Texts in Mathematics*. Springer-Verlag, New York, 2003. An elementary introduction.

[124] Marshall Hall, Jr. *The theory of groups*. The Macmillan Co., New York, N.Y., 1959.

[125] G. Haller. *Chaos Near Resonance*. Springer, New York, 1999.

[126] G. Haller and S. Wiggins. Geometry and chaos near resonant equilibria of 3-DOF Hamiltonian systems. *Physica D*, 90:319–365, 1996.

[127] J.J. Heijnekamp, M.S. Krol, and Ferdinand Verhulst. Averaging in non-linear transport problems. *Math. Methods Appl. Sciences*, 18:437–448, 1995.

[128] D. Henry. *Geometric theory of semilinear parabolic equations*, volume 840 of *Lecture Notes in mathematics*. Springer, Heidelberg, 1981.

[129] David Hilbert. *Theory of algebraic invariants*. Cambridge University Press, Cambridge, 1993. Translated from the German and with a preface by Reinhard C. Laubenbacher, Edited and with an introduction by Bernd Sturmfels.

[130] M. Hirsch, C. Pugh, and M. Shub. *Invariant Manifolds*, volume 583 of *Lecture Notes Mathematics*. Springer-Verlag, Berlin, Heidelberg, New York, 1977.

[131] Chao-Pao Ho. A shadowing approximation of a system with finitely many saddle points. *Tunghai Journal*, 34:713–728, 1993.

[132] G-J. Hori. Theory of general perturbations with unspecified canonical variables. *Publ. Astron. Soc. Japan*, 18:287–296, 1966.

[133] I. Hoveijn. *Aspects of resonance in dynamical systems*. PhD thesis, Utrecht University, Utrecht, Netherlands, 1992.

[134] Igor Hoveijn and Ferdinand Verhulst. Chaos in the 1 : 2 : 3 Hamiltonian normal form. *Phys. D*, 44(3):397–406, 1990.

[135] Robert A. Howland. A note on the application of the Von Zeipel method to degenerate Hamiltonians. *Celestial Mechanics*, 19:139–145, 1979.

[136] J. Humphreys. *Introduction to Lie Algebras and Representation Theory*. Springer-Verlag, New York, 1972.

[137] C.G.J. Jacobi. *C. G. J. Jacobi's Vorlesungen über Dynamik. Gehalten an der Universität zu Königsberg im Wintersemester 1842-1843 und nach einem von C. W. Borchart ausgearbeiteten hefte. hrsg. von A. Clebsch*. Druck & Verlag von George Reimer, Berlin, 1842.

[138] J.H. Jeans. *Astronomy and Cosmogony*. At The University Press, Cambridge, 1928.

[139] R. Johnson, editor. *Dynamical Systems, Montecatini Terme 1994*, volume 1609 of *Lecture Notes in Mathematics*, Berlin, Heidelberg, New York, 1994. Springer-Verlag.

[140] C.K.R.T. Jones. Geometric singular perturbation theory. In Johnson [139], pages 44–118.

[141] C.K.R.T. Jones and A.I. Khibnik, editors. *Multiple-Time-Scale Dynamical Systems*, volume 122 of *IMA volumes in mathematics and its applications.* Springer-Verlag, New York, 2001.

[142] S. Jorna, editor. *Regular and irregular motion*, volume 46. Am. Inst. Phys. Conf. Proc., 1978.

[143] T.J. Kaper. An introduction to geometric methods and dynamical systems theory for singular perturbation problems. In *Analyzing multiscale phenomena using singular perturbation methods*, pages 85–131, 1999.

[144] T.J. Kaper and C.K.R.T. Jones. *A primer on the exchange lemma for fast-slow systems*, pages 85–131. Volume 122 of Jones and Khibnik [141], 2001.

[145] J.B. Keller and S. Kogelman. Aymptotic solutions of initial value problems for nonlinear partial differential equations. *SIAM J. Appl. Math.*, 18:748–758, 1970.

[146] J. Kevorkian. On a model for reentry roll resonance. *SIAM J. Appl. Math.*, 35:638–669, 1974.

[147] J. Kevorkian. Perturbation techniques for oscillatory systems with slowly varying coefficients. *SIAM Review*, 29:391–461, 1987.

[148] J. Kevorkian and J.D. Cole. *Perturbation Methods in Applied Mathematics*, volume 34 of *Applied Math. Sciences.* Springer-Verlag, Berlin, Heidelberg, New York, 1981.

[149] U. Kirchgraber and E.Stiefel. *Methoden der analytischen Störungsrechnung und ihre Anwendungen.* B.G.Teubner, Stuttgart, 1978.

[150] U. Kirchgraber and H. O. Walther, editors. *Dynamics Reported, Volume I.* Wiley, New york, 1988.

[151] Anthony W. Knapp. *Lie groups beyond an introduction*, volume 140 of *Progress in Mathematics.* Birkhäuser Boston Inc., Boston, MA, second edition, 2002.

[152] V.V. Kozlov. *Symmetries, Topology and Resonances in Hamiltonian Mechanics.* Springer-Verlag, Berlin etc., 1996.

[153] B. Krauskopf, H. M. Osinga, E.J. Doedel, M.E. Henderson, J. Guckenheimer, A. Vladimirsky, M. Dellnitz, and O. Junge. A survey of methods for computing (un)stable manifolds of vector fields. *Intern. J. Bif. Chaos*, 15:763–791, 2005.

[154] Bernd Krauskopf and Hinke M. Osinga. Computing geodesic level sets on global (un)stable manifolds of vector fields. *SIAM J. Appl. Dyn. Systems*, 2:546–569, 2003.

[155] M.S. Krol. On a Galerkin-averaging method for weakly non-linear wave equations. *Math. Methods Appl. Sciences*, 11:649–664, 1989.

[156] M.S. Krol. On the averaging method in nearly time-periodic advection-diffusion problems. *SIAM J. Appl. Math.*, 51:1622–1637, 1989.

[157] M.S. Krol. *The method of averaging in partial differential equations.* PhD thesis, University of Utrecht, 1990.

[158] N.M. Krylov and N.N. Bogoliubov. *Introduction to Nonlinear Mechanics (in Russian).* Izd. AN UkSSR, Kiev, 1937. Vvedenie v Nelineinikhu Mekhaniku.

[159] S. Kuksin. *Nearly Integrable Infinite-Dimensional Hamiltonian Systems*, volume 1556 of *Lecture Notes Mathematics.* Springer-Verlag, Berlin, Heidelberg, New York, 1991.

[160] S. Kuksin. Lectures on Hamiltonian Methods in Nonlinear PDEs. In Sergei Kuksin Giancarlo Benettin, Jacques Henrard, editor, *Hamiltonian Dynamics. Theory and Applications: Lectures given at C.I.M.E.-E.M.S. Summer School held in Cetraro, Italy, July 1-10, 1999*, volume 1861 of *Lecture Notes in mathematics*, pages 143–164. Springer, Heidelberg, 2005.

[161] M. Kummer. An interaction of three resonant modes in a nonlinear lattice. *Journal of Mathematical Analysis and Applications*, 52:64–104, 1975.

[162] J. Kurzweil. Van der Pol perturbation of the equation for a vibrating string. *Czech. Math. J.*, 17:558–608, 1967.

[163] Yu. A. Kuznetsov. *Elements of applied bifurcation theory, 3^d ed.*, volume 42 of *Appl. Math. Sciences.* Springer-Verlag, Berlin, Heidelberg, New York, 2004.

[164] W.T. Kyner. A mathematical theory of the orbits about an oblate planet. *SIAM J.*, 13(1):136–171, 1965.

[165] J.-L. Lagrange. *Mécanique Analytique (2 vols.).* edition Albert Blanchard, Paris, 1788.

[166] V.F. Lazutkin. *KAM Theory and Semiclassical Approximations to Eigenfunctions.* Ergebnisse der Mathematik und ihrer Grenzgebiete 24. Springer-Verlag, Berlin etc., 1993.

[167] Brad Lehman. The influence of delays when averaging slow and fast oscillating systems: overview. *IMA J. Math. Control Inform.*, 19(1-2):201–215, 2002. Special issue on analysis and design of delay and propagation systems.

[168] Brad Lehman and Vadim Strygin. Partial and generalized averaging of functional differential equations. *Funct. Differ. Equ.*, 9(1-2):165–200, 2002.

[169] Brad Lehman and Steven P. Weibel. Averaging theory for delay difference equations with time-varying delays. *SIAM J. Appl. Math.*, 59(4):1487–1506 (electronic), 1999.

[170] Brad Lehman and Steven P. Weibel. Fundamental theorems of averaging for functional-differential equations. *J. Differential Equations*, 152(1):160–190, 1999.

[171] A. Ben Lemlih and J.A. Ellison. Method of averaging and the quantum anharmonic oscillator. *Phys. Reviews Letters*, 55:1950–1953, 1985.

[172] A. H. M. Levelt. The semi-simple part of a matrix. In A. H. M. Levelt, editor, *Algoritmen In De Algebra: A Seminar on Algebraic Algorithms.*

Department of Mathematics, University of Nijmegen, Nijmegen, The Netherlands, 1993.

[173] B. M. Levitan and V. V. Zhikov. *Almost periodic functions and differential equations.* Cambridge University Press, Cambridge, 1982. Translated from the Russian by L. W. Longdon.

[174] A.J. Lichtenberg and M.A. Lieberman. *Regular and Stochastic Motion*, volume 38. Applied Mathematical Sciences Springer Verlag, New York, 1983.

[175] B.V. Lidskii and E.I. Shulman. Periodic solutions of the equation $u_{tt} - u_{xx} = u^3$. *Functional Anal. Appl.*, 22:332–333, 1967.

[176] P. Lochak and C. Meunier. *Multiphase Averaging for Classical Systems.* Springer, New York, 1980.

[177] P. Lochak and C. Meunier. *Multiphase averaging for classical systems*, volume 72 of *Applied Mathematical Sciences.* Springer-Verlag, New York, 1988. With applications to adiabatic theorems, Translated from the French by H. S. Dumas.

[178] J.-L. Loday. *Cyclic Homology*, volume 301 of *Grundlehren der mathematischen Wissenschaften.* Springer–Verlag, Berlin, 1991.

[179] A.M. Lyapunov. Problème general de la stabilité du mouvement. *Ann. of Math. Studies*, 17, 1947.

[180] P. Lynch. Resonant motions of the three-dimensional elastic pendulum. *Int. J. Nonlin. Mech.*, 37:345–367, 2001.

[181] R. S. MacKay and J. D. Meiss. *Hamiltonian Dynamical Systems.* Adam Hilger, Bristol, 1987. A collection of reprinted articles by many authors, including the main authors listed above, compiled and introduced by these authors.

[182] David Mumo Malonza. Normal forms for coupled Takens-Bogdanov systems. *J. Nonlinear Math. Phys.*, 11(3):376–398, 2004.

[183] L.I. Mandelstam and N.D. Papalexi. Über die Begründung einer Methode für die Näherungslösung von Differentialgleichungen. *J. f. exp. und theor. Physik*, 4:117, 1934.

[184] L. Martinet, P. Magnenat, and Ferdinand Verhulst. On the number of isolating integrals in resonant systems with 3 degrees of freedom. *Celestial Mech.*, 25(1):93–99, 1981.

[185] K. Matthies. Time-averaging under fast periodic forcing of parabolic partial differential equations: exponential estimates. *J. Diff. Eqs*, 174:133–180, 2001.

[186] K. Matthies and A. Scheel. Exponential averaging for Hamiltonian evolution equations. *Trans. AMS*, 355:747–773, 2002.

[187] Sebastian Mayer, Jürgen Scheurle, and Sebastian Walcher. Practical normal form computations for vector fields. *ZAMM Z. Angew. Math. Mech.*, 84(7):472–482, 2004.

[188] William Mersman. A new algorithm for the Lie transformation. *Celestial Mechanics*, 3:81–89, 1970.

[189] Y.A. Mitropolsky, G. Khoma, and M. Gromyak. *Asymptotic Methods for investigating Quasiwave Equations of Hyperbolic Type.* Kluwer Ac. Publ., Dordrecht, 1997.

[190] Ya.A. Mitropolsky. *Problems of the Asymptotic Theory of Nonstationary Vibrations.* Israel Progr. Sc. Transl., Jerusalem, 1965.

[191] Ya.A. Mitropolsky. *Certains aspects des progrès de la méthode de centrage*, volume 4.4. Edizione Cremonese CIME, Roma, 1973.

[192] Th. Molien. Über die Invarianten der linearen Substitutionsgruppen. *Sitz.-Ber. d. Preub. Akad. d. Wiss., Berlin*, 52, 1897.

[193] J. Moser. Regularization of Kepler's problem and the averaging method on a manifold. *Comm. Pure Appl. Math.*, 23:609–636, 1970.

[194] J. Moser. Stable and random motions in dynamical systems on celestial mechanics with special emphasis. *Ann. Math. Studies*, 77, 1973.

[195] J. Moser and C.L. Siegel. *Lectures on Celestial Mechanics.* Springer-Verlag, 1971.

[196] James Murdock. Nearly Hamiltonian systems in nonlinear mechanics: averaging and energy methods. *Indiana University Mathematics Journal*, 25:499–523, 1976.

[197] James Murdock. Some asymptotic estimates for higher order averaging and a comparison with iterated averaging. *SIAM J. Math. Anal.*, 14:421–424, 1983.

[198] James Murdock. Qualitative theory of nonlinear resonance by averaging and dynamical systems methods. In Kirchgraber and Walther [150], pages 91–172.

[199] James Murdock. Shadowing multiple elbow orbits: an application of dynamcial systems theory to perturbation theory. *Journal of Differential Equations*, 119:224–247, 1995.

[200] James Murdock. Shadowing in perturbation theory. *Applicable Analysis*, 62:161–179, 1996.

[201] James Murdock. *Perturbations: Theory and Methods.* SIAM, Philadelphia, 1999.

[202] James Murdock. On the structure of nilpotent normal form modules. *J. Differential Equations*, 180(1):198–237, 2002.

[203] James Murdock. *Normal forms and unfoldings for local dynamical systems.* Springer Monographs in Mathematics. Springer-Verlag, New York, 2003.

[204] James Murdock. Hypernormal form theory: foundations and algorithms. *J. Differential Equations*, 205(2):424–465, 2004.

[205] James Murdock and Chao-Pao Ho. On shadowing with saddle connections. Unpublished, 1999.

[206] James Murdock and Clark Robinson. Qualitative dynamics from asymptotic expansions: local theory. *Journal of Differential Equations*, 36:425–441, 1980.

[207] James Murdock and Jan A. Sanders. A new transvectant algorithm for nilpotent normal forms. Technical report, Iowa State University, 2006. submitted for publication.

[208] James Murdock and Lih-Chyun Wang. Validity of the multiple scale method for very long intervals. *Z. Angew. Math. Phys.*, 47(5):760–789, 1996.

[209] A.H. Nayfeh. *Perturbation Methods.* Wiley-Interscience, New York, 1973.

[210] A.H. Nayfeh and D.T. Mook. *Nonlinear Oscillations.* John Wiley, New York, 1979.

[211] N.N. Nekhoroshev. An exponential estimate of the time of stability of nearly-integrable Hamiltonian systems. *Russ. Math. Surv. Usp. Mat. Nauk*, 32(6):1–655–66, 1977.

[212] Morris Newman. *Integral matrices.* Academic Press, New York, 1972. Pure and Applied Mathematics, Vol. 45.

[213] E. Noether. Invariante variationsprobleme. *Nachr. v.d. Ges. d. Wiss. zu Göttingen, Math.–phys. Kl.*, 2:235–257, 1918.

[214] Peter J. Olver. *Classical invariant theory.* Cambridge University Press, Cambridge, 1999.

[215] Jacob Palis and Welington de Melo. *Geometric Theory of Dynamical Systems.* Springer, New York, 1982.

[216] H. Pals. The Galerkin-averaging method for the Klein-Gordon equation in two space dimensions. *Nonlinear Analysis*, 27:841–856, 1996.

[217] Lawrence M. Perko. Higher order averaging and related methods for perturbed periodic and quasi-periodic systems. *SIAM Journal of Applied Mathematics*, 17:698–724, 1968.

[218] H. Poincaré. *Les Méthodes Nouvelles de la Mécanique Céleste*, volume I. Gauthiers-Villars, Paris, 1892.

[219] H. Poincaré. *Les Méthodes Nouvelles de la Mécanique Céleste*, volume II. Gauthiers-Villars, Paris, 1893.

[220] Claudio Procesi. *Lie Groups – An Approach through Invariants and Representations.* Universitext. Springer, 2006.

[221] M.H. Protter and H.F. Weinberger. *Maximum Principles in Differential Equations.* Prentice-Hall, Englewood Cliffs, NJ, 1967.

[222] G.G. Rafel. *Applications of a combined Galerkin-averaging method*, pages 349–369. Volume 985 of Verhulst [278], 1983. Surveys and new trends.

[223] B. Rink. Symmetry and resonance in periodic FPU chains. *Comm. Math. Phys.*, 218:665–685, 2001.

[224] C. Robinson. *Stability of periodic solutions from asymptotic expansions*, pages 173–185. In Devaney and Nitecki [74], 1981.

[225] C. Robinson. Sustained resonance for a nonlinear system with slowly varying coefficients. *SIAM J. Math. Anal.*, 14(5):847–860, 1983.

[226] Clark Robinson. Structural stability on manifolds with boundary. *Journal of Differential Equations*, 37:1–11, 1980.

[227] Clark Robinson and James Murdock. Some mathematical aspects of spin-orbit resonance. II. *Celestial Mech.*, 24(1):83–107, 1981.

[228] M. Roseau. *Vibrations nonlinéaires et théorie de la stabilité.* Springer-Verlag, 1966.

[229] M. Roseau. *Solutions périodiques ou presque périodiques des systèmes differentiels de la mécanique non-linéaire.* Springer Verlag, Wien-New York, 1970.

[230] A.W. Sáenz. Long-time approximation to the evolution of resonant and nonresonant anharmonic oscillators in quantum mechanics (erratum in [231]). *J. Math. Phys.*, 37:2182–2205, 1996.

[231] A.W. Sáenz. Erratum to [230]. *J. Math. Phys.*, 37:4398, 1997.

[232] A.W. Sáenz. Lie-series approach to the evolution of resonant and nonresonant anharmonic oscillators in quantum mechanics. *J. Math. Phys.*, 39:1887–1909, 1998.

[233] L. Salvadori, editor. *Bifurcation Theory and Applications*, volume 1057. Springer-Verlag Lecture Notes in Mathematics, 1984. CIME course *Bifurcation Theory and Applications.*

[234] E. Sanchez-Palencia. Méthode de centrage et comportement des trajectoires dans l'espace des phases. *Ser. A Compt. Rend. Acad. Sci.*, 280:105–107, 1975.

[235] E. Sanchez-Palencia. Méthode de centrage - estimation de l'erreur et comportement des trajectoires dans l'espace des phases. *Int. J. Non-Linear Mechanics*, 11(176):251–263, 1976.

[236] Jan A. Sanders. Are higher order resonances really interesting? *Celestial Mech.*, 16(4):421–440, 1977.

[237] Jan A. Sanders. On the Fermi-Pasta-Ulam chain. preprint 74, University of Utrecht, 1978.

[238] Jan A. Sanders. On the passage through resonance. *SIAM J. Math. Anal.*, 10(6):1220–1243, 1979.

[239] Jan A. Sanders. Asymptotic approximations and extension of time-scales. *SIAM J. Math. Anal.*, 11(4):758–770, 1980.

[240] Jan A. Sanders. Melnikov's method and averaging. *Celestial Mech.*, 28(1-2):171–181, 1982.

[241] Jan A. Sanders. Normal forms of 3 degree of freedom hamiltonian systems at equilibrium in the semisimple resonant case. In Calmet et al. [52].

[242] Jan A. Sanders and Richard Cushman. Limit cycles in the Josephson equation. *SIAM J. Math. Anal.*, 17(3):495–511, 1986.

[243] Jan A. Sanders and J.-C. van der Meer. Unique normal form of the Hamiltonian 1 : 2-resonance. In *Geometry and analysis in nonlinear dynamics (Groningen, 1989)*, pages 56–69. Longman Sci. Tech., Harlow, 1992.

[244] Jan A. Sanders and Ferdinand Verhulst. Approximations of higher order resonances with an application to Contopoulos' model problem. In *Asymptotic analysis*, pages 209–228. Springer, Berlin, 1979.

[245] Jan A. Sanders and Ferdinand Verhulst. *Averaging methods in nonlinear dynamical systems.* Springer-Verlag, New York, 1985.

[246] P.R. Sethna and S.M. Schapiro. Nonlinear behaviour of flutter unstable dynamical systems with gyroscopic and circulatory forces. *J. Applied Mechanics*, 44:755–762, 1977.

[247] A.L. Shtaras. The averaging method for weakly nonlinear operator equations. *Math. USSSR Sbornik*, 62:223–242, 1989.

[248] Michael Shub. *Global Stability of Dynamical Systems.* Springer-Verlag, Berlin, Heidelberg, New York, 1987.

[249] Dennis Stanton, editor. *Invariant theory and tableaux*, volume 19 of *The IMA Volumes in Mathematics and its Applications*, New York, 1990. Springer-Verlag.

[250] Shlomo Sternberg. *Celestial Mechanics - Parts I & II. 1st ed.* W A Benjamin, NY, 1969.

[251] A. Stieltjes. Recherches sue quelques séries semi-convergentes. *Ann. de l'Ec. Normale Sup.*, 3:201–258, 1886.

[252] A.C.J. Stroucken and Ferdinand Verhulst. The Galerkin-averaging method for nonlinear, undamped continous systems. *Math. Methods Appl. Sci.*, 9:520–549, 1987.

[253] K. Stumpff. *Himmelsmechanik*, volume 3. VEB Deutscher Verlag der Wissenschaften, Berlin, 1959-74.

[254] J.J. Sylvester. Tables of the generating functions and groundforms for the binary quantics of the first ten orders. *American Journal of Mathematics*, II:223–251, 1879.

[255] V.G. Szebehely and B.D. Tapley, editors. *Long Time Predictions in Dynamics*, Dordrecht, 1976. Reidel. Proc. of ASI, Cortina d'Ampezzo (Italy) , 1975.

[256] E. Tournier, editor. *Computer algebra and differential equations*, volume 193 of *London Mathematical Society Lecture Note Series*, Cambridge, 1994. Cambridge University Press. Papers from the conference (CADE-92) held in June 1992.

[257] J.M. Tuwankotta and F. Verhulst. Symmetry and resonance in Hamiltonian systems. *SIAM J. Appl. Math.*, 61:1369–1385, 2000.

[258] J.M. Tuwankotta and F. Verhulst. Hamiltonian systems with widely separated frequencies. *Nonlinearity*, 16:689–706, 2003.

[259] E. van der Aa. First order resonances in three-degrees-of-freedom systems. *Celestial Mechanics*, 31:163–191, 1983.

[260] E. Van der Aa and M. De Winkel. Hamiltonian systems in $1:2:\omega$–resonance ($\omega = 5$ or 6). *Int. J. Nonlin. Mech.*, 29:261–270, 1994.

[261] E. van der Aa and M.S. Krol. *Weakly nonlinear wave equation with many resonances*, pages 27–42. In [157], 1983.

[262] Els van der Aa and Jan A. Sanders. The $1:2:1$-resonance, its periodic orbits and integrals. In Verhulst [276], pages 187–208. From theory to application.

[263] Els van der Aa and Ferdinand Verhulst. Asymptotic integrability and periodic solutions of a Hamiltonian system in 1 : 2 : 2-resonance. *SIAM J. Math. Anal.*, 15(5):890–911, 1984.

[264] A.H.P. van der Burgh. *Studies in the Asymptotic Theory of Nonlinear Resonance.* PhD thesis, Technical Univ. Delft, Delft, 1974.

[265] A.H.P. van der Burgh. On the asymptotic validity of perturbation methods for hyperbolic differential equations. In Verhulst [276], pages 229–240. From theory to application.

[266] L. van der Laan and Ferdinand Verhulst. The transition from elliptic to hyperbolic orbits in the two-body problem by slow loss of mass. *Celestial Mechanics*, 6:343–351, 1972.

[267] J.-C. van der Meer. Nonsemisimple 1:1 resonance at an equilibrium. *Celestial Mechanics*, 27:131–149, 1982.

[268] B. van der Pol. A theory of the amplitude of free and forced triode vibrations. *The Radio Review*, 1:701–710, 1920.

[269] B. van der Pol. On Relaxation-Oscillations. *The London, Edinburgh and Dublin Philosophical Magazine and Journal of Science*, 2:978–992, 1926.

[270] A. van der Sluis. Domains of uncertainty for perturbed operator equations. *Computing*, 5:312–323, 1970.

[271] W.T. van Horssen. Asymptotics for a class of semilinear hyperbolic equations with an application to a problem with a quadratic nonlinearity. *Nonlinear Analysis TMA*, 19:501–530, 1992.

[272] W.T. van Horssen and A.H.P. van der Burgh. On initial boundary value problems for weakly nonlinear telegraph equations. asymptotic theory and application. *SIAM J. Appl. Math.*, 48:719–736, 1988.

[273] P.Th.L.M. van Woerkom. *A multiple variable approach to perturbed aerospace vehicle motion.* PhD thesis, Princeton, 1972.

[274] Ferdinand Verhulst. Asymptotic expansions in the perturbed two-body problem with application to systems with variable mass. *Celestial Mech.*, 11:95–129, 1975.

[275] Ferdinand Verhulst. On the theory of averaging. In Szebehely and Tapley [255], pages 119–140. Proc. of ASI, Cortina d'Ampezzo (Italy) , 1975.

[276] Ferdinand Verhulst, editor. *Asymptotic analysis*, volume 711 of *Lecture Notes in Mathematics.* Springer, Berlin, 1979. From theory to application.

[277] Ferdinand Verhulst. Discrete symmetric dynamical systems at the main resonances with applications to axi-symmetric galaxies. *Philosophical Transactions of the Royal Society of London*, 290:435–465, 1979.

[278] Ferdinand Verhulst, editor. *Asymptotic analysis. II*, volume 985 of *Lecture Notes in Mathematics.* Springer-Verlag, Berlin, 1983. Surveys and new trends.

[279] Ferdinand Verhulst. Asymptotic analysis of Hamiltonian systems. In *Asymptotic analysis, II* [278], pages 137–183. Surveys and new trends.

[280] Ferdinand Verhulst. On averaging methods for partial differential equations. In Degasperis and Gaeta [71], pages 79–95.

[281] Ferdinand Verhulst. *Methods and Applications of Singular Perturbations, boundary layers and multiple timescale dynamics*, volume 50 of *Texts in Applied Mathematics*. Springer–Verlag, Berlin, Heidelberg, New York, 2005.

[282] Ferdinand Verhulst and I. Hoveijn. Integrability and chaos in Hamiltonian normal forms. In *Geometry and analysis in nonlinear dynamics (Groningen, 1989)*, pages 114–134. Longman Sci. Tech., Harlow, 1992.

[283] V.M. Volosov. Averaging in systems of ordinary differential equations. *Russ. Math. Surveys*, 17:1–126, 1963.

[284] U. Kirchgraber H. O. Walther, editor. *Volume 2 Dynamics Reported.* Chichester B.G. Teubner, Stuttgart and John Wiley & Sons, 1989.

[285] L. Wang, D.L. Bosley, and J. Kevorkian. Asymptotic analysis of a class of three-degree-of-freedom Hamiltonian systems near stable equilibrium. *Physica D*, 88:87–115, 1995.

[286] Alan Weinstein. Normal modes for nonlinear Hamiltonian systems. *Inv. Math.*, 20:47–57, 1973.

[287] Alan Weinstein. Simple periodic orbits. In *Topics in nonlinear dynamics (Proc. Workshop, La Jolla Inst., La Jolla, Calif., 1977)*, volume 46 of *AIP Conf. Proc.*, pages 260–263. Amer. Inst. Physics, New York, 1978.

[288] G. B. Whitham. Two-timing, variational principles and waves. *J. Fluid Mech.*, 44:373–395, 1970.

[289] C.A. Wilson. Perturbations and solar tables from Lacaille to Delambre: the rapprochement of observation and theory. *Archive for History of Exact Sciences*, 22:53–188 (part I) and 189–304 (part II), 1980.

Index of Definitions & Descriptions

General Index

– **K** –

– **L** –

– **M** –

– **N** –

– **O** –

– **P** –

– **t** –

– **u** –

Applied Mathematical Sciences

(continued from page ii)

Applied Mathematical Sciences

(continued from previous page)

116. *Taylor:* Partial Differential Equations II, Qualitative Studies of Linear Equations.
117. *Taylor:* Partial Differential Equations III, Nonlinear Equations.
118. *Godlewski/Raviart:* Numerical Approximation of Hyperbolic Systems of Conservation Laws.
119. *Wu:* Theory and Applications of Partial Functional Differential Equations.
120. *Kirsch:* An Introduction to the Mathematical Theory of Inverse Problems.
121. *Brokate/Sprekels:* Hysteresis and Phase Transitions.
122. *Gliklikh:* Global Analysis in Mathematical Physics: Geometric and Stochastic Methods.
123. *Khoi/Le/Schmitt:* Global Bifurcation in Variational Inequalities: Applications to Obstacle and Unilateral Problems.
124. *Polak:* Optimization: Algorithms and Consistent Approximations.
125. *Arnold/Khesin:* Topological Methods in Hydrodynamics.
126. *Hoppensteadt/Izhikevich:* Weakly Connected Neural Networks.
127. *Isakov:* Inverse Problems for Partial Differential Equations, 2nd ed.
128. *Li/Wiggins:* Invariant Manifolds and Fibrations for Perturbed Nonlinear Schrodinger Equations.
129. *Müller:* Analysis of Spherical Symmetries in Euclidean Spaces.
130. *Feintuch:* Robust Control Theory in Hilbert Space.
131. *Ericksen:* Introduction to the Thermodynamics of Solids, Revised Edition.
132. *Ihlenburg:* Finite Element Analysis of Acoustic Scattering.
133. *Vorovich:* Nonlinear Theory of Shallow Shells.
134. *Vein/Dale:* Determinants and Their Applications in Mathematical Physics.
135. *Drew/Passman:* Theory of Multicomponent Fluids.
136. *Cioranescu/Saint Jean Paulin:* Homogenization of Reticulated Structures.
137. *Gurtin:* Configurational Forces as Basic Concepts of Continuum Physics.
138. *Haller:* Chaos Near Resonance.
139. *Sulem/Sulem:* The Nonlinear Schrödinger Equation: Self-Focusing and Wave Collapse.
140. *Cherkaev:* Variational Methods for Structural Optimization.
141. *Naber:* Topology, Geometry, and Gauge Fields: Interactions.
142. *Schmid/Henningson:* Stability and Transition in Shear Flows.
143. *Sell/You:* Dynamics of Evolutionary Equations.
144. *Nédélec:* Acoustic and Electromagnetic Equations: Integral Representations for Harmonic Problems.
145. *Newton:* The N-Vortex Problem: Analytical Techniques.
146. *Allaire:* Shape Optimization by the Homogenization Method.
147. *Aubert/Kornprobst:* Mathematical Problems in Image Processing: Partial Differential Equations and the Calculus of Variations, 2nd ed.
148. *Peyret:* Spectral Methods for Incompressible Viscous Flow.
149. *Ikeda/Murota:* Imperfect Bifurcation in Structures and Materials.
150. *Skorokhod/Hoppensteadt/Salehi:* Random Perturbation Methods with Applications in Science and Engineering.
151. *Bensoussan/Frehse:* Regularity Results for Nonlinear Elliptic Systems and Applications.
152. *Holden/Risebro:* Front Tracking for Hyperbolic Conservation Laws.
153. *Osher/Fedkiw:* Level Set Methods and Dynamic Implicit Surfaces.
154. *Bluman/Anco:* Symmetries and Integration Methods for Differential Equations.
155. *Chalmond:* Modeling and Inverse Problems in Image Analysis.
156. *Kielhöfer:* Bifurcation Theory: An Introduction with Applications to PDEs.
157. *Kaczynski/Mischaikow/Mrozek:* Computational Homology.
158. *Oertel:* Prandtl–Essentials of Fluid Mechanics, 10th Revised Edition.
159. *Ern/Guermond:* Theory and Practice of Finite Elements.
160. *Kaipio/Somersalo:* Statistical and Computational Inverse Problems.